Energy Planning

Energy Planning

Editor

Dolf Gielen

MDPI • Basel • Beijing • Wuhan • Barcelona • Belgrade • Manchester • Tokyo • Cluj • Tianjin

Editor
Dolf Gielen
International Renewable
Energy Agency
Germany

Editorial Office
MDPI
St. Alban-Anlage 66
4052 Basel, Switzerland

This is a reprint of articles from the Special Issue published online in the open access journal *Energies* (ISSN 1996-1073) (available at: https://www.mdpi.com/journal/energies/special_issues/ energy_planning).

For citation purposes, cite each article independently as indicated on the article page online and as indicated below:

LastName, A.A.; LastName, B.B.; LastName, C.C. Article Title. *Journal Name* **Year**, *Volume Number*, Page Range.

ISBN 978-3-0365-3987-4 (Hbk)
ISBN 978-3-0365-3988-1 (PDF)

Contents

About the Editor

Dolf Gielen has served as the director of the IRENA Innovation and Technology Centre in Bonn since 2011. He studied Chemical Engineering at Technical University Eindhoven and Environmental Sciences at Utrecht University. He received a Ph.D. in Engineering from the Delft University of Technology in 1999. He was a visiting fellow at the National Institute of Sciences in Tsukuba, Japan, from 2000 to 2002, and he has been a non-resident fellow at the Colorado School of Mines Payne Institute since 2018. Dolf Gielen worked for the International Energy Agency from 2002 to 2009 and was chief of the Energy Efficiency and Energy Policy at the United Nations Industrial Development Organization from 2009 to 2011. His main areas of interest are energy policy and planning, energy economics, greenhouse gas emission mitigation, innovative technologies, and sustainability. He is the author of more than 200 peer-reviewed scientific papers and reports.

Editorial

Energy Planning

Dolf Gielen

Innovation and Technology Centre (IITC), International Renewable Energy Agency, Robert-Schuman-Platz 3, 53175 Bonn, Germany; info@irena.org

This Special Issue focuses on progress in energy transition planning. Many national governments as well as sub-national governments have announced goals to achieve net zero by mid-century. In parallel, the UNFCCC has established a mechanism for its parties to formulate and communicate long-term low greenhouse gas emission development strategies (LT-LEDS, or LTS for short) to operationalize the carbon-neutral vision stipulated by the Paris Agreement. As of March 2022, 51 countries have submitted their LTS to the UNFCCC. There is a need to understand better all aspects of how energy supply and energy demand can be fully decarbonized. Given the energy system complexity, there is an important role for models and energy planning in such assessment. Long-term energy scenarios (LTES) are effective tools for policymakers to agree on how to achieve ambitious goals. While LTES have been used for decades to guide energy policy, the energy and socio-economic transformation that is caused by net zero emissions pushes the boundaries of LTES further.

This Special Issue contains 13 papers that contribute to the science of energy planning. The papers cover new country analyses from around the world (Brazil, Colombia, Ecuador, Ghana, Italy, Mexico, Sweden, Thailand, Ukraine, United Kingdom). A number of analyses use well-established energy system analysis methodologies but apply them in combination with new energy technology data in the context of specific countries. The findings suggest there is no one-fits-all solution, and that national circumstances must be considered in the design of energy transition strategies. The papers also contribute to the energy planning methodology:

- Energy and climate policy planning processes: a number of papers discuss how to combine different models to yield policy-relevant results.
- New methodologies are proposed for post-processing of modeling results: One paper discusses the breakdown of global modeling results for countries and sectors. Integrated assessment model (IAM) results are translated into lifestyle impacts, and decomposition analysis is used to identify the role of contributing factors for carbon neutrality.
- One paper assesses the impact of climate change on future solar PV generation.
- Carbon budgets at the sectoral, sub-national, national, and global levels: One paper focuses on the national and sectoral carbon budgets that follow from global models. Two papers focus on sector coupling for electromobility and decarbonization of the chemical and petrochemical industries, respectively. One paper discusses the role of liquid biofuels for transportation.
- One article focuses on the better use of modeling results for policy making, and another focuses on the role of regulation.
- Renewable energy and its technical model representations (geo-spatial and temporal resolutions): one paper discusses methodologies to compare various system aspects in the optimization of high VRE and gas-based power generation. Another paper assesses flexibility enhancement strategies.
- Definition of carbon neutrality and the role of carbon sinks, carbon removal, and carbon leakage for net zero target: one paper discusses the importance of a phase-out

Citation: Gielen, D. Energy Planning. *Energies* **2022**, *15*, 2621. https://doi.org/10.3390/en15072621

Received: 25 March 2022
Accepted: 30 March 2022
Published: 3 April 2022

Publisher's Note: MDPI stays neutral with regard to jurisdictional claims in published maps and institutional affiliations.

of net emissions from agriculture, forestry, and other land uses, reductions in non-carbon greenhouse gases, and land restoration to scale up atmospheric CO_2 removal.

This brief overview shows that energy planning continues to evolve rapidly. IRENA, through its LTES Network and Clean Energy Ministerial LTES initiative, is facilitating dialogue among energy planners and modelers in the government sector who develop and use. These discussions have highlighted that, for the energy and climate scenarios to be truly effective in informing decarbonization pathways and energy planning, some misalignments between the national LTES and global climate scenarios need to be addressed.

A closer look at the findings of the individual papers reveals important new insights:

Hanmer et al. [1] discuss that countries' emission reduction commitments under the Paris Agreement have significant implications for lifestyles. A novel methodology is presented for translating global scenarios into lifestyle implications at the national and household levels, which can be generalized to any service or country and versatile to work with any model or scenario. The 5Ds method post-processes integrated assessment model projections of the sectoral energy demand for the global region to derive energy-service-specific lifestyle change at the household level. The methodology is applied for two energy services (mobility, heating) in two countries (the UK, Sweden), showing how effort to reach zero carbon targets varies between countries and households. Our method creates an analytical bridge between global model output and information that can be used at national and local levels, making the lifestyle implications of climate targets clear.

Saisirirat et al. [2] discuss a detailed Ghana vehicle ownership model with necessary transport parameters to construct an energy demand model to provide insight for reducing GHG emission contributions from road transport through biofuel (both bioethanol and biodiesel) using the Low Emission Analysis Platform (LEAP) modeling framework. The model setup builds on an earlier study for Thailand. Scenarios include alternative (ALT), with up to E20/B20, and extreme (EXT), with up to E85/B50, for new vehicles. Energy demand and GHG emissions were analyzed from Ghana's transport sector data to show potential benefits that accrue from biofuel usage. The findings show that 8.4% and 11.1% of GHG emission reduction in 2030 can be achieved with a 0.13% and 0.27% additional arable land requirement from the ALT and EXT scenarios.

Silva et al. [3] assess climate-related risks and countermeasures in solar power plants in Thailand using thematic analysis with self-administered observations and structured interviews. The findings can inform long-term energy planning to ensure climate adaptation capacity. The analysis points out that floods and storms were perceived as major climate events affecting solar power plants in Thailand, followed by lightning and fires. Several countermeasures are proposed, some of which require extensive investment. The findings show that enabling regulations or financial incentives are needed for the implementation of climate-proofing countermeasures. Public and private sectors need to secure a sufficient budget for fast recovery after severe climate incidents. Measures must be taken to facilitate the selection of climate-resilient sites by improving conditions of power purchase agreements or assisting winning bidders in enhancing the climate adaptability of their sites. These issues should be considered during Thailand's long-term energy planning.

Chen et al. [4] use decomposition analysis to investigate the key contributions to changes in greenhouse gas emissions in different scenarios. Decomposition formulas are derived for the three highest-emitting sectors: power generation, industry, and transportation (both passenger and freight). These formulas were applied to recently developed 1.5 °C emission scenarios by the Integrated Model to Assess the Global Environment (IMAGE), emphasizing the role of renewables and lifestyle changes. The decomposition analysis shows that carbon capture and storage (CCS), both from fossil fuel and bioenergy combustion, renewables, and reducing the carbon intensity provide the largest contributions to emission reduction in the scenarios. Efficiency improvement is also critical, but part of the potential is already achieved in the baseline scenario. The relative importance of different emission reduction drivers is similar for OECD and non-OECD regions, but there are some noteworthy differences. In the non-OECD region, improving efficiency in industry and transport and increasing the share of renewables in power generation are more important in reducing emissions than in the OECD region, while CCS in power generation and electrification of passenger transport are more important drivers in the OECD region.

Gaeta et al. [5] focus on the challenges and opportunities of reaching net zero emissions by 2050 in Italy. To support Italian energy planning, the authors developed energy roadmaps towards national climate neutrality, consistent with the Paris Agreement objectives and the IPCC goal of limiting the increase in the global surface temperature to 1.5 °C. These scenarios identify the correlations among the main pillars for the change of the energy paradigm towards net emissions by 2050. The energy scenarios were developed using TIMES-RSE, a partial equilibrium and technology-rich optimization model of the entire Italian energy system. Subsequently, an in-depth analysis was developed with the sMTISIM, a long-term simulator of power system and electricity markets. The results show that, to achieve climate neutrality by 2050, the Italian energy system will have to experience profound transformations in multiple and strongly related dimensions. A predominantly renewable-based energy mix (at least 80–90% by 2050) is essential to decarbonize most of the final energy consumption. However, the strong increase in non-programmable renewable sources requires particular attention to new flexibility resources needed for the power system, such as Power-to-X. The green fuels produced from renewables via Power-to-X will be a vital energy source for those sectors where electrification faces technical and economic barriers. The findings also confirm that the European "energy efficiency first" principle represents the very first step on the road to climate neutrality.

Petrović et al. [6] analyze the Ukrainian energy system in the context of the Paris Agreement and the 1.5 °C objective. A TIMES-Ukraine model of the whole Ukrainian energy system is deployed to analyze how the energy system may develop until 2050, taking current and future policies into account. The results show the development of the Ukrainian energy system based on energy efficiency improvements, electrification, and renewable energy. The share of renewables in electricity production is predicted to reach between 45% and 57% in 2050 in the main scenarios with moderate emission reduction ambitions and ~80% in the ambitious alternative scenarios. The cost-optimal solution includes a reduction in the space heating demand in buildings by 20% in the frozen policy and 70% in other scenarios, while electrification of industries leads to reductions in energy intensity of 26–36% in all scenarios except for the frozen policy. Energy efficiency improvements and emission reductions in the transport sector are achieved through increased use of electricity from 2020 in all scenarios except for the frozen policy, reaching 40–51% in 2050. The stated policies present a cost-efficient alternative for keeping Ukraine's greenhouse gas emissions at today's level.

Werlang et al. [7] point out the need for energy planning to quickly adapt to provide useful inputs to the regulation activity so that a cost-effective electricity market emerges to facilitate the integration of renewables. This paper analyzes the role of system planning and regulations in two specific elements in energy market design: the concept of firm capacity, and the presence of distributed energy resources, both of which can be influenced by regulation. The analysis quantifies the role of the current regulation in the total cost of the Brazilian and Mexican electricity systems when compared to a reference "efficient" energy planning scenario that adopts standard cost minimization principles and that is well suited to most of the relevant features of the new energy transformation scenario. The findings show two very common features of regulatory designs that can lead to distortions: (i) renewables commonly having a lower "perceived cost" under the current regulations, either due to direct incentives such as tax breaks or due to indirect access to more attractive contracts or financing conditions; and (ii) requirements for reliability that are often defined overly conservatively, overstating the hardships imposed by renewable generation on the existing system and underestimating the potential of technology portfolios.

Nazaré et al. [8] discuss the increase in the need for operating reserves that follows from the penetration of variable renewable energy (VRE) in thermal-dominated systems. In the case of hydro-dominated systems, the cost-effective flexibility provided by hydro-plants facilitates the penetration of VRE, but the compounded production variability of these resources challenges the integration of baseload gas-fired plants. The Brazilian power system illustrates this situation. Given the current competitiveness of VRE, a natural question is the economic value and tradeoffs for expanding the system, opting between baseload gas-fired generation and VRE in an already flexible hydropower system. This paper presents a methodology based on a multi-stage and stochastic capacity expansion model to estimate the optimal mix of baseload thermal power plants and VRE additions. The assessment method considers their contributions for the security of supply, which includes peak, energy, and operating reserves, which are endogenously defined and sized in time-varying and dynamic ways, as well as adequacy constraints. The presented model calculates the optimal decision plan, allowing for the estimation of the economical tradeoffs between baseload gas and VRE supply considering their value for the required services to the system. This allows for a comparison between the integration costs of these technologies on the same basis, thus helping policymakers and system planners to better decide on the best way to integrate the gas resources in an electricity industry that is increasingly renewable. A case study based on a real industrial application is presented for the Brazilian power system.

Correa-Laguna et al. [9] describe the construction of an integrated bottom-up LEAP model tailored to the Colombian case. An integrated model facilitates capturing synergies and intersectoral interactions within the national GHG emission system. Hence, policies addressing one sector and influencing others are identified and correctly assessed. Thus, 44 mitigation policies and mitigation actions were included in the model, in this way identifying the sectors being directly and indirectly affected by them. The mitigation scenario developed in this paper reaches a reduction of 28% in GHG emissions compared with the reference scenario. The importance of including non-energy sectors is evident in the Colombian case, as GHG emission reductions are mainly driven by AFOLU. The model allows for the correct estimate of the scope and potential of mitigation actions by considering indirect, unintended emission reductions in all IPCC categories, as well as synergies with all mitigation actions included in the mitigation scenario. Moreover, the structure of the model is suitable for testing potential emission trajectories, facilitating its adoption by official entities and its application in climate policy making.

Godoy et al. [10] analyze the pathway to develop a clean and diversified electricity mix for Ecuador, covering the demand of three specific development levels of electric transportation. The linear optimization model (urbs) and the Ecuador Land Use and Energy Network Analysis (ELENA) are used to optimize the expansion of the power system in the period from 2020 to 2050. The results show that reaching an electricity mix 100% based on renewable energies is possible and this supply can support a highly electrified transport sector that includes 47.8% of road passenger transportation and 5.9% of road freight transportation. Therefore, the electrification of this sector is a viable alternative for the country to rely on its own energy resources while reinforcing its future climate change mitigation commitments.

Saygin and Gielen [11] assess the techno-economic potential of 20 decarbonization options in the chemical and petrochemical sectors. While previous analyses focused on the production processes, this analysis covers the full product life cycle CO_2 emissions. The analysis elaborates the carbon accounting complexity that results from the non-energy use of fossil fuels, and highlights the importance of strategies that consider the carbon stored in synthetic organic products—an aspect that warrants more attention in long-term energy scenarios and strategies. Average mitigation costs in the sector would amount to USD 64 per ton of CO_2 for full decarbonization in 2050. The rapidly declining renewables cost is one main causes for this low-cost estimate. Renewable energy supply solutions, in combination with electrification, account for 40% of total emission reductions. Annual biomass use grows to 1.3 gigatons, green hydrogen electrolyzer capacity grows to 2435 gigawatts, and recycling rates increase six-fold, while product demand is reduced by a third, compared to the reference case. CO_2 capture, storage, and use equals 30% of the total decarbonization effort (1.49 gigatons per year), where about one third of the captured CO_2 is of biogenic origin. Circular economy concepts, including recycling, account for 16%, while energy efficiency accounts for 12% of the decarbonization needed. Achieving full decarbonization in this sector will increase energy and feedstock costs by more than 35%. The analysis shows the importance of renewable-based solutions, accounting for more than half of the total emission reduction potential, higher than previous estimates.

Teske et al. [12] present two global non-overshoot pathways (+2.0 °C and +1.5 °C) with regional decarbonization targets for the four primary energy sectors—power, heating, transportation, and industry—in 5-year steps up to 2050. The normative scenarios illustrate the effects of efficiency measures and renewable energy use, describe the roles of increased electrification of the final energy demand and synthetic fuels, and quantify the resulting electricity load increases for 72 sub-regions. Non-energy scenarios include a phase-out of net emissions from agriculture, forestry, and other land uses, reductions in non-carbon greenhouse gases, and land restoration to scale up atmospheric CO_2 removal, estimated at -377 Gt CO_2 in 2100. An estimate of the COVID-19 effects on the global energy demand is included, and a sensitivity analysis describes the impacts if implementation is delayed by 5, 7, or 10 years, which would significantly reduce the likelihood of achieving the 1.5 °C goal. The analysis applies a model network consisting of energy system, power system, transport, land use, and climate models.

Carvajal et al. [13] focus on the use of long-term energy scenarios (LTES) in the government sector, and specifically how the new challenges and opportunities brought by the clean energy transition change the way in which governments use LTES. The information tends to remain tacit, and a gap exists in understanding the way to enhance LTES use and development at the government level. The experience from national institutions that are leading the improvement in official energy scenario planning is used as a basis to derive a set of overarching best practices in three ares (i) strengthen LTES development, (ii) effectively use LTES for strategic energy planning, and (iii) enhance institutional capacity for LTES-based energy planning. The best practice experience was collected through the International Renewable Agency's LTES Network activities. The LTES-based energy planning methodologies need to adapt, reflecting the changing landscapes, and that more effective and extensive use of LTES in government needs to be further encouraged.

The energy planning field must continue to evolve rapidly in the coming years as the energy transition is globally high on the political agenda and carries the information that is needed for decision making changes.

In the context of the LTES initiative and network, four areas will be developed further in the coming years [14]:

- Exchange of net zero scenarios between LT-LEDS climate planners and energy planners.
- Collecting best engagement practices with stakeholders in the scenario process. This includes the preparation of model and scenario input parameters as well as the translation of findings into policy-relevant conclusions and actions.
- Stock take of scenario findings—collection and comparison of scenarios as well as identification of scenario gaps for global government decarbonization action.
- Identification of modeling robustness and weaknesses including better representation of aspects such as infrastructure modeling, behavioral change, and the need for systemic innovation for mission-driven change.

Funding: The LTES initiative and network is supported by the Governments of Denmark and Germany.

Conflicts of Interest: The author declares no conflict of interest.

References

1. Hanmer, C.; Wilson, C.; Edelenbosch, O.Y.; Vuuren, D.P. Translating Global Integrated Assessment Model Output into Lifestyle Change Pathways at the Country and Household Level. *Energies* **2022**, *15*, 1650. [CrossRef]
2. Saisirirat, P.; Rushman, J.; Silva, K.; Chollacoop, N. Contribution of Road Transport to the Attainment of Ghana's Nationally Determined Contribution (NDC) through Biofuel Integration. *Energies* **2022**, *15*, 880. [CrossRef]

3.	Silva, K.; Janta, P.; Chollacoop, N. Points of Consideration on Climate Adaptation of Solar Power Plants in Thailand: How Climate Change Affects Site Selection, Construction and Operation. *Energies* **2022**, *15*, 171. [CrossRef]
4.	Chen, H.; Hof, A.; Daioglou, V.; Boer, H.; Edelenbosch, O.; Berg, M.; Wijst, K.-I.; Vuuren, D. Using Decomposition Analysis to Determine the Main Contributing Factors to Carbon Neutrality across Sectors. *Energies* **2022**, *15*, 132. [CrossRef]
5.	Gaeta, M.; Businge, C.; Gelmini, A. Achieving Net Zero Emissions in Italy by 2050: Challenges and Opportunities. *Energies* **2022**, *15*, 46. [CrossRef]
6.	Petrović, S.; Diachuk, O.; Podolets, R.; Semeniuk, A.; Bühler, F.; Grandal, R.; Boucenna, M.; Balyk, O. Exploring the Long-Term Development of the Ukrainian Energy System. *Energies* **2021**, *14*, 7731. [CrossRef]
7.	Werlang, A.; Cunha, G.; Bastos, J.; Serra, J.; Barbosa, B.; Barroso, L. Reliability Metrics for Generation Planning and the Role of Regulation in the Energy Transition: Case Studies of Brazil and Mexico. *Energies* **2021**, *14*, 7428. [CrossRef]
8.	Nazaré, F.; Barroso, L.; Bezerra, B. A Probabilistic and Value-Based Planning Approach to Assess the Competitiveness between Gas-Fired and Renewables in Hydro-Dominated Systems: A Brazilian Case Study. *Energies* **2021**, *14*, 7281. [CrossRef]
9.	Correa-Laguna, J.; Pelgrims, M.; Espinosa, M.; Morales, R. Colombia's GHG Emissions Reduction Scenario: Complete Representation of the Energy and Non-Energy Sectors in LEAP. *Energies* **2021**, *14*, 7078. [CrossRef]
10.	Godoy, J.; Villamar, D.; Soria, R.; Vaca, C.; Hamacher, T.; Ordóñez, F. Preparing the Ecuador's Power Sector to Enable a Large-Scale Electric Land Transport. *Energies* **2021**, *14*, 5728. [CrossRef]
11.	Saygin, D.; Gielen, D. Zero-Emission Pathway for the Global Chemical and Petrochemical Sector. *Energies* **2021**, *14*, 3772. [CrossRef]
12.	Teske, S.; Pregger, T.; Simon, S.; Naegler, T.; Pagenkopf, J.; Deniz, Ö.; Adel, B.; Dooley, K.; Meinshausen, M. It Is Still Possible to Achieve the Paris Climate Agreement: Regional, Sectoral, and Land-Use Pathways. *Energies* **2021**, *14*, 2103. [CrossRef]
13.	Carvajal, P.; Miketa, A.; Goussous, N.; Fulcheri, P. Best Practice in Government Use and Development of Long-Term Energy Transition Scenarios. *Energies* **2022**, *15*, 2180. [CrossRef]
14.	IRENA. Long-Term Energy Scenarios (LTES) Network. Available online: https://www.irena.org/energytransition/Energy-Transition-Scenarios-Network (accessed on 29 March 2022).

Review

Best Practice in Government Use and Development of Long-Term Energy Transition Scenarios

Pablo E. Carvajal, Asami Miketa, Nadeem Goussous * and Pauline Fulcheri

International Renewable Energy Agency—Innovation and Technology Centre, 53113 Bonn, Germany; pablocarvajals@gmail.com (P.E.C.); amiketa@irena.org (A.M.); pauline.fulcheri@outlook.com (P.F.)
* Correspondence: ngoussous@irena.org

Abstract: Long-term energy scenarios (LTES) have been serving as an important planning tool by a wide range of institutions. This article focuses on how LTES have been used (and also devised in some cases) in the government sector, and specifically how the new challenges and opportunities brought by the aspiration for the clean energy transition change the way that governments use LTES. The information tends to remain tacit, and a gap exists in understanding the way to enhance LTES use and development at the government level. To address this gap, we draw on the experience from national institutions that are leading the improvement in official energy scenario planning to articulate a set of overarching best practices to (i) strengthen LTES development, (ii) effectively use LTES for strategic energy planning and (iii) enhance institutional capacity for LTES-based energy planning, all in the context of new challenges associated with the clean energy transition. We present implementation experience collected through the International Renewable Agency's LTES Network activities to exemplify these best practices. We highlight that in the context of the broad and complex challenges of a clean energy transition driven by ambitious climate targets, the LTES-based energy planning methodologies need to evolve, reflecting the changing landscapes, and that more effective and extensive use of LTES in government needs to be further encouraged.

Keywords: long-term energy scenarios; energy planning; energy modelling; clean energy transition; climate scenarios

Citation: Carvajal, P.E.; Miketa, A.; Goussous, N.; Fulcheri, P. Best Practice in Government Use and Development of Long-Term Energy Transition Scenarios. *Energies* **2022**, *15*, 2180. https://doi.org/10.3390/en15062180

Academic Editor: Alessia Arteconi

Received: 24 January 2022
Accepted: 10 March 2022
Published: 16 March 2022

Publisher's Note: MDPI stays neutral with regard to jurisdictional claims in published maps and institutional affiliations.

1. Introduction

The clean energy transition poses a unique challenge, particularly to energy planners [1–3], who must deal with envisioning the changes of the energy system in a context of uncertainty and rapid change. The growing deployment of low-cost renewables, the need for a more integrated, innovative and flexible power grid and the impacts of demand and consumer behaviour through end use electrification are some of the key transition features that energy planners must include in any long-term analysis. The energy transition will also be supported by advanced policy frameworks and market mechanisms, which will generate new business models and fundamentally transform the status quo [4–7]. Expanding pressures to align the economy to low emission carbon pledges and the climate objectives of the Paris Agreement necessitate a more aggressive strategy than previous approaches that sought to stabilise or halve emissions [8–11]. Policy and decision making must have a strategic, forward-looking approach that continually embeds new evolutions and uncertainties in policy, markets and technology.

Long-term energy scenarios (LTES) have been traditionally the building block of national energy planning, supporting the development of national energy plans, national energy outlooks, electricity generation and transmission capacity expansion plans and energy demand analysis [12–27]. LTES can help government planners to prepare for the long-term policy interventions, identify the short-term challenges and opportunities and inform recommendations on where to direct domestic and foreign investment [28,29]. Most recently, LTES use has broadened to the climate community, and they are being used for designing nationally determined contributions (NDC) [30] and, most importantly, long-term low emission development strategies (LT-LEDS) [31], which should be submitted by all signatory countries of the Paris Agreement. While the global and regional LTES also inform national policy debate, the focus of this article is on national LTES, built by or built for governments for their planning purposes, unless otherwise specified.

LTES are mostly developed with energy modelling tools [32–34], which help to develop a mathematical representation of a part or the totality of the energy system. Models allow representing the complex interdependencies within an energy system and its linkages to broader societal and environmental factors and assess the short and long-term impacts of choices of technological pathways and policy choices. However, given the substantial future uncertainties caused by an accelerated energy transition [35–38], using deterministic quantitative models can often produce misleading conclusions. For example, retrospective analysis of the projected solar photovoltaic and wind energy installed capacity has shown a consistent underestimation when compared to current trends [39–41]. It also reflects the fact that as scenario analysis becomes more influential, society may dynamically respond to messages portrayed by such analysis. Even with better modelling approaches, enhanced computational power and refinement of input data, it is impossible to validate long-term scenario results [42–44]. In this sense, model-based scenario analysis benefits have focused on assessing a wide range of pathways and gaining insights from them, rather than aiming to narrow the ranges and to produce "accurate" predictions. The notion of accurate prediction could be misleading given the inherent uncertainties of technology progress in the long run and the dynamic nature of policy interventions.

The energy scenario modelling community, academia and research communities have demonstrated various improvements of modelling approaches, and these are well documented (e.g., [45–49]). To our knowledge, however, the government's application of national-level LTES and the best practices and experience in using them to guide the clean energy transition remain as gaps. Although we recognise that government practices are highly context specific, the objective rules on how governments develop and use LTES can be drawn from learning from others. We therefore see that it is critical to synthesise the tacit knowledge underpinning effective LTES analysis in the government. This paper aims at formalising best practices in using and developing LTES in the government in the context of the clean energy transition. It seeks to complement the recent literature that is studying energy transition scenarios in the context of sustainability [50], geopolitics [51], societal processes [52], modelling methods [53] and economic impacts [54] by engaging with those who rely on scenario-based results to help navigate the energy transition—i.e., government energy planners. We address three currently unmet objectives in the literature: (i) to showcase examples of successful application of LTES in the government, (ii) to establish scenario best practices to address the energy transition through community-wide efforts and (iii) to inform energy planners on effective use scenario-based analysis. We draw on the collective experience of national energy institutions in different countries worldwide that are members of the International Renewable Agency's (IRENA) Long-Term Energy Scenarios Network (LTES Network) (IRENA LTES Network webpage: https://irena.org/energytransition/Energy-Transition-Scenarios-Network (accessed on 23 January 2022)) [55–57], thus providing a global and comprehensive view on how governments are adapting their scenario practices to the requirements of the energy transition.

We categorise a set of best practices into three critical pillars for national LTES, namely (i) strengthening scenario development, (ii) improving scenario use and (iii) identifying capacity-building approaches [53]. While we focus on LTES in the government, the recommendations can be applied in other sectors using scenarios for decision-making.

2. Mental Model of IRENA's LTES Network

The objective of IRENA's LTES network is to advocate the effective use of LTES as a tool for planning to accelerate the energy transition [57]. The process for developing LTES and using them is encompassed within the chain of the energy policy making process (Figure 1), which is also a part of the overarching energy planning practice in the government. The LTES typically (though not always) uses results from energy modelling as inputs, involving energy modellers and analysts to quantify policy implications, draw outlooks and identify uncertainties. The policy making process involves decision-makers who rely on scenario insights to design national planning documents, long-term energy policy and, more recently, climate targets. In contrast, qualitative scenario characteristics—such as future storylines and narratives—are shaped via stakeholder and expert elicitation at different stages. We note that the LTES use and development process, as seen in Figure 1, may also involve feedback loops among the stages.

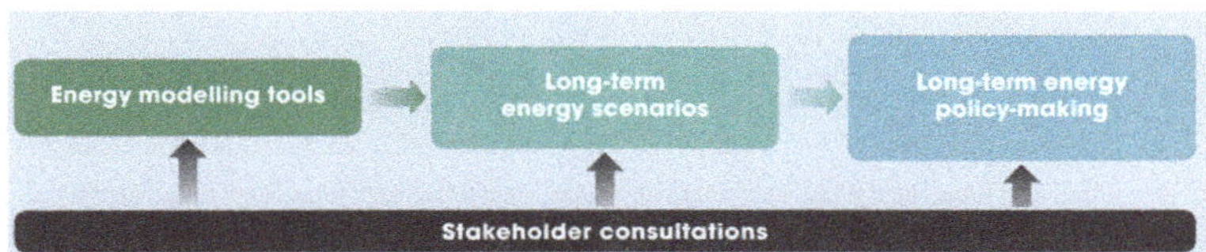

Figure 1. The mental model of IRENA's Long-Term Energy Scenarios network [57].

Three focus areas with respective focus questions were defined to systematically organise the information stemming from the LTES network's activities (Table 1). The collection of national experience in LTES and energy planning processes worldwide is country-specific and has touched upon a broad range of topics. However, we have found common features of good practice that we present in this paper. In strengthening scenario development, the focus has been on the definition of modelling scopes that capture the features of the energy transition. On improving the use of scenarios, the focus has been on strategic use of long-term energy scenarios for long-term policy making. On identifying institutional capacity, the focus has been on distilling best practices to source adequate scenario development abilities and human resources within government institutions.

Table 1. Focus areas of IRENA's LTES network and focus questions.

	Focus Area	Focus Question
1.	Strengthening scenario development	How to develop scenarios to better capture potentially transformational changes?
2.	Improving scenario use	How to use scenarios for better strategic decision-making by governments and investors?
3.	Identifying institutional capacity	How to better enhance institutional capacity for scenario planning?

3. Best Practices for LTES in the Government

A set of best practices should drive LTES development and use to guide the energy transition. The best practices presented here are inspired by the discussions held with scenario experts who participated in IRENA's LTES network activities.

I.   Robust development of LTES

    a.   Establish a strong governance structure: broad participation of stakeholders and stronger coordination across different government institutions are needed.

    b.   Expand the boundaries of scenarios: emerging technologies, business models and disruptive innovations need to be better accounted for in LTES.

II.   Effective use of LTES

    a.   Define the purpose of LTES: clarifying the purpose of LTES is needed as they can be used in different purposes in different contexts, leading to misinterpretation of the results.

    b.   Communicate transparently and effectively: Innovative communication methods can be deployed to transparently share assumptions and results of LTES with stakeholders.

III.   Institutional ownership of LTES capacity

    a.   Develop the appropriate scenario planning capacity: different national circumstances lead to a unique institutional ownership model of LTES capacity, and the right balance of in-house government skills and support from third-party organisations can be identified.

In the following sections, we elaborate more on each best practice and provide examples of how countries implement these best practices. Where appropriate, we layout critical challenges that must be met.

4. Robust Development of LTES

4.1. Establishing a Strong Governance Structure

The process of developing LTES differs throughout regions and contexts. Some governments have established advanced legal frameworks to outline LTES steps, stakeholders and the frequency of scenarios exercises [17,58,59], while others have less stringent guidelines or none at all, implying a more ad hoc approach. The clean energy transition necessitates better coordination and expansive governance of LTES development than before. For example, with distributed energy resources and smart grid technologies, the traditionally passive electricity consumers will be more active players of the energy system, i.e., prosumers, which will potentially influence and be influenced by LTES [60,61]. The massive electrification of end-use sectors with green electricity and the unique spatial and temporal characteristics of variable renewables require better coordination among institutions to operate the power system and to develop scenarios. Cities and regions are becoming part of the scenario process, whereas in the past scenario, planning was a more centralized top-down matter [62,63]. In addition, the link of the energy transition to climate policy requires better coordination amongst different institutional jurisdictions, e.g., energy scenarios developed by ministries of energy versus climate scenarios developed by ministries of environment catering to international climate pledges [64].

This best practice delves into two critical aspects to improve scenario development governance structures—(i) participatory processes; and (ii) coordination between entities.

4.1.1. Participatory Processes

Participatory processes help to increase the legitimacy, acceptance and utility of LTES. Inviting various stakeholders to brainstorm on a possible range of scenarios is central to mapping expectations and creating a mutual vision of the future, which is crucial to discover perspectives that do not include the inherent governmental bias. An additional benefit from participatory processes is that it facilitates guaranteeing a just and inclusive energy transition. Impressive experience of successful participatory processes reaching out to hundreds of stakeholders to develop LTES has been found in Colombia to inform the National Energy Plan 2020–2050 [65] (Figure 2), Denmark to inform the Energy and Climate Outlook 2030 [24], Brazil to develop the National Energy Plan 2050 [12,13], the United Kingdom to inform National Grid's Future Energy Scenarios 2050 study [66], Chile to support its long-term energy and transmission planning governance [17] and in South Africa to update the assumptions of the Integrated Resource Plan 2019 [20].

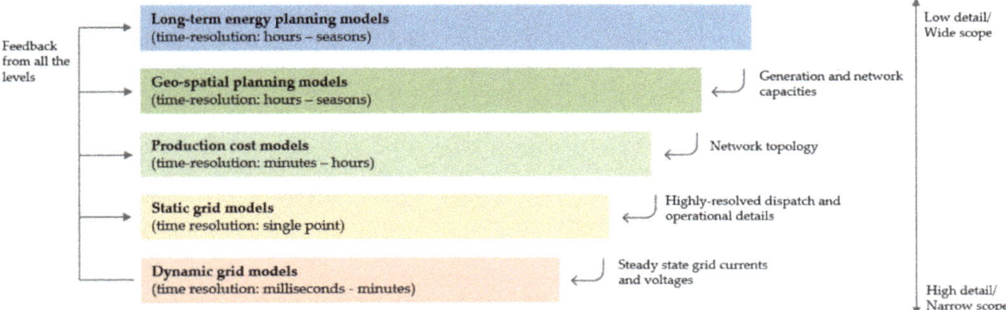

Figure 2. Colombia's scenario development participatory workshops for the national energy plan 2020–2050 carried out by Energy and Mining Planning Unit [67].

4.1.2. Coordination among LTES Entities

Stages of scenario development often involve a range of different institutions, and these may have specific methodologies according to their specific objectives and conditions. Developing national LTES in isolation (i.e., only by one institution) risks misinterpretation and misalignment amongst entities. Therefore, coordination among institutions can help to derive comparable and meaningful conclusions by, for example, approaching the same LTES question from different institutional perspectives. We identify three levels of coordination to improve LTES development particularly relevant to the energy transition. The first level of coordination is required between the institutions developing top-down official LTES (central government and ministries) and institutions conducting bottom-up technical studies for different power system segments. For example, the clean energy transition may demand a greater share of variable renewable energy (VRE) in a power system, which necessitates a range of energy planning models to be deployed in a coordinated manner (Figure 3).

Figure 3. Tools with different scopes and their feedback loops for energy system scenario planning [34].

The second level of coordination is inter-institutional, between different sectors developing scenarios. In the context of the energy transition, this is namely between the climate community in charge of establishing climate targets and the energy planners who traditionally possess more knowledge of energy models and scenarios. National LTES must be aligned with climate target frameworks, such as Nationally Determined Contributions (NDC) [68]. In some instances, NDCs have shown to be more ambitious than LTES developed independently by energy ministries of agencies, or vice versa, which can be contradictory [69].

The third level of LTES coordination is between central and federal governments and between regional and national governing bodies. Here, the availability of local energy resources comes into play and complicates scenario governance. For example, an ambitious regional energy transition scenario may collide with a more conservative national-level scenario. LTES could be employed to study regional diversities and provide granularity to national level exercises, and beyond that to cater to regional planning needs and resource governance [70–73].

Good experience in coordination of LTES planning can be found in the United Arab Emirates National Energy Strategy 2050 development process, including all communities and local governments [74]; Canada's coordination among federal government organisations involved in data and modelling of LTES through the Federal Energy Information Framework, which aids the production of Canada's Energy Futures report (Figure 4) [75]; Costa Rica's 2050 Decarbonisation Plan, including energy and climate LTES [58]; and in the coordination between the European Network of Transmission System Operators for electricity (ENTSO-E) and gas (ENTSOG) to develop a common scenarios report by 2040 [76].

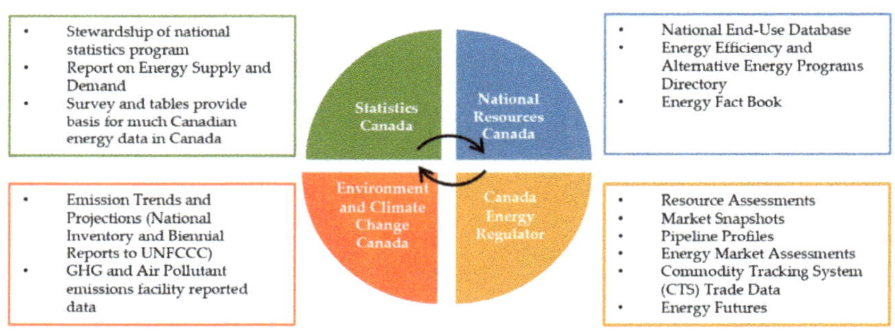

Figure 4. Canada's Federal Energy Information Framework [77].

4.2. Expanding the Boundaries of Scenarios

Quantitative LTES, formulated by modelling tools, reflect the underlying model structure and its scope. To sufficiently reflect the complexity of the energy transition, it is necessary for emerging technologies, business models and disruptive innovations to be better addressed [78,79]. As end-users and end-use technologies increasingly change their roles in the energy system from passive to active, the distinction between supply and demand in the traditional sense is harder to draw [80–82]. However, characterising such disruptive innovations and radical societal changes in scenarios continues to pose a challenge and is at the forefront of energy modelling and consumer behaviour research [83–85]. How and when these innovations will be wholly developed and actively utilised can hardly be determined at the present. Thus, LTES are a valuable tool to explore the consequences of disruptive technologies and ambitious policy choices. Expanding the boundaries of scenarios delves into two essential aspects for national-level LTES—(i) developing LTES showcasing a just energy transition; and (ii) considering innovation in the energy sector.

4.2.1. Scenarios for a Just Energy Transition

The profound socioeconomic transformation that accompanies the energy transition and a low carbon economy raises questions from policymakers concerning the impacts on economic growth, employment, welfare and living conditions [50,51,86,87]. The clean energy transition will unveil "winners" that grasp the opportunities and "losers" that reap the risks; thus, policymakers are interested in pinpointing these groups to ensure adherence to the principle of equity [88]. The COVID-19 pandemic generated renewed attention to sustainable development pathways that can enable a green recovery. Integration of social considerations into scenario analysis is needed to enable policymakers to assess the impacts and timelines of a just transition [89,90]. We also recognise that the transition can be ill used for purposes of green washing [91] and to promote political agendas [92], and thus scenarios can be used to fact check pledges and to establish credible transition plans and targets that will reduce the misuse of transition narratives.

In Germany, a government-appointed commission advised a complete and gradual elimination of coal by 2038 [93,94]. In January 2019, following an extensive consultation procedure, a phase-out plan was presented to offer a €40 billion economic package to affected coal regions, including alternative industry investment projects and state aid for coal workers [68]. In Finland, the VTT Technical Research Centre of Finland Ltd. (VTT) developed a modelling framework to analyse the impacts of the 2030 policies in the country's national economy, energy economy, natural resources, emissions and health (see Figure 5). The European Commission's scenario study, A Clean Planet for all, showcased scenarios with a time horizon of 2050 that considered the interplay between energy, the economy, land use and agriculture and non-CO_2 GHG and air pollution [95] (see Figure 6).

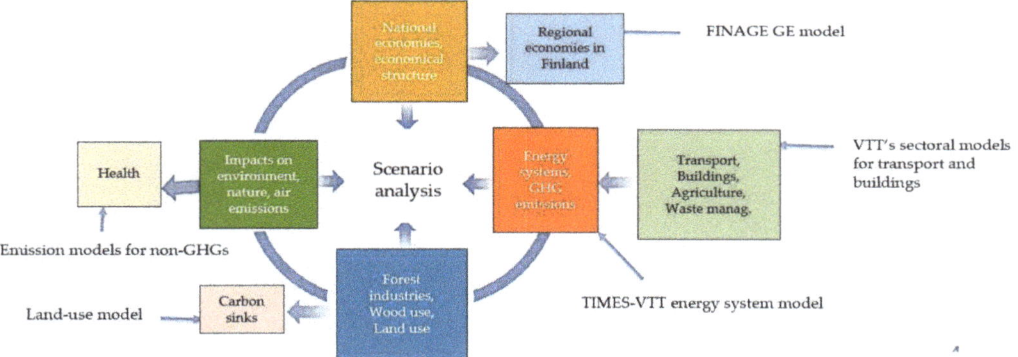

Figure 5. Finland's scenario modelling framework to study the impact of policies.

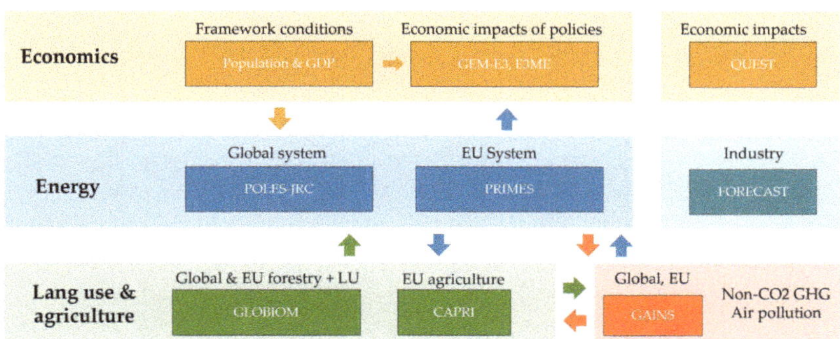

Figure 6. The European Commission modelling suite for integrated modelling of the economy, energy, land use and agriculture and air pollution [96].

4.2.2. Accounting for Innovation in the Energy Sector

Innovation in decentralisation, digitalisation and electrification are crucial components of the energy transition and need to be better accounted for in LTES. For example, amplifying auto-consumption from rooftop solar PV systems through the use of residential battery storage and electric vehicles (EVs) was not prominently considered by model designers 20–30 years ago [97–99]. Hydrogen, a key energy carrier to decarbonise energy-intensive industries, and how it may co-evolve with renewable electricity infrastructure, continues to be highly unaccounted-for in current techno-economic modelling [100–102]. Current scenarios also probably underestimate the growth sector coupling (VRE in transport, buildings and industry) [103–106]. IRENA has identified a set of 31 such innovations within four main categories relevant to the upscaling of variable renewable energy and which are relevant to consider expanding the boundaries of LTES for the transition (see Table 2).

Table 2. The landscape of innovations to integrate variable renewable energy.

Enabling Technologies	Business Models	Market Design	System Operation
1. Utility-scale batteries 2. Behind-the-meter batteries 3. EV smart charging 4. Renewable power-to-heat 5. Renewable power-to-hydrogen 6. IoT 7. AI and big data 8. Blockchain 9. Renewable mini-grids 10. Supergrids 11. Flexibility in conventional power plants	12. Aggregators 13. P2P electricity trading 14. Energy-as-a-service 15. Community-ownership models 16. Pay-as-you-go models	17. Increasing time granularity in electricity markets 18. Increasing space granularity in electricity markets 19. Innovative ancillary services 20. Re-designing capacity markets 21. Regional markets 22. Time-of-use tariffs 23. Market integration of distributed energy resources 24. Net billing schemes	25. Future role of distribution system operators 26. Co-operation between transmission and distribution system operators 27. Advanced forecasting of variable renewable power generation 28. Innovative operation of pumped hydropower storage 29. Virtual power lines 30. Dynamic line rating

The Japanese 5th Strategic Energy Plan 2050 [107] recognises the uncertainty of technological innovation and the ambiguity with regards to changes in conditions. To tackle the issue, it develops multiple-track scenarios that pursue all options, including renewable energy, hydrogen, carbon capture and storage and nuclear power. The Italian Integrated Energy and Climate Plan [108] considers a dimension of research and innovation action through the framework of scenario-building. It focuses on two modelling pillars: the first is concerned with power grids, integration of renewables, auto-production, storage, community energy and aggregators; and the second pillar focuses on facilitating EV adoption. In the United States of America, the National Renewable Energy Lab (NREL) explored targets, factors and innovation that affect electricity sector pathway decisions by 2050 [109].

5. Effective Use of LTES

5.1. Clarifying the Purpose of Scenario-Building

LTES are contextually-dependent, employed for various objectives, and can be used differently depending on the necessary targets. It is crucial to clarify such distinctions to avoid misinterpreting scenario insights. While the inherent objective of scenario development is to offer a snapshot of the energy system of the future, the way that scenarios can be used and applied can vary. For example, scenarios are often developed as a part of the governments' infrastructure planning, such as transmission and capacity expansion investment planning [66,110]. Scenarios are also developed to explore radical transformations and ambitious climate targets [111,112], often as a part of a scientific exercise. Private companies also use scenarios more in the context of market forecasting. Clarifying the purpose of these scenarios can allow policymakers to use their insights correctly and compare them appropriately [113]. We have identified several (contrasting) use-cases of energy scenarios in the context of national energy planning. Figure 7 showcases the three polar distinctions made. However, it is essential to know that those are not considered binary choices but spectra in which the uses of national energy transition scenarios can be defined. The following subsections will elaborate on these distinctions.

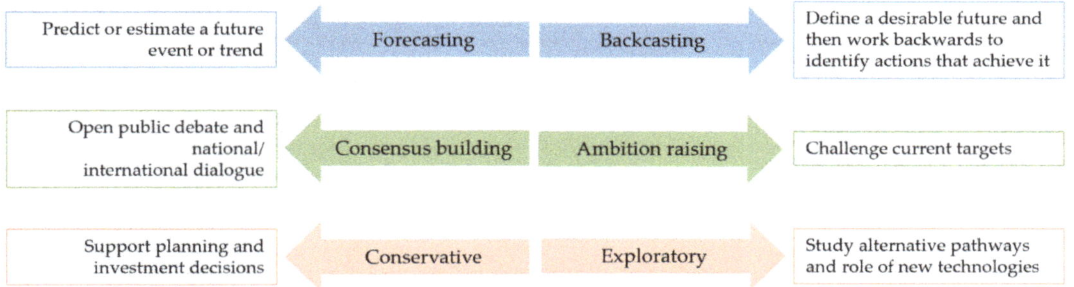

Figure 7. Categorisations of use-cases of LTES [57].

5.1.1. Forecasting and Backcasting

A forecasting-based scenario aims to predict future trends or events, inquiring, "what will happen, given certain decisions and policies?" Strict applications of forecasts are seldom in long-term energy planning (20–30+ years), as the distant horizon make predictions harder to make.

A backcasting-based scenario aims to provide backward pathways from a particular objective or target, in the process determining the policies needed to support this pathway, essentially asking the question, "how can this certain future be achieved?" These scenarios are best suited to study implications of decisions and cost-effective methods to reach national and global targets.

Government scenario developers have been shown to use a combination of backcasting and forecasting methods to produce scenarios. In China, the China National Renewable Energy Centre (CNREC) utilises backcasting scenario analyses to assess policy measures to reach the country's ambitious 2-degree targets by mid-century, as depicted in the 13th five-year plan [114], which is presented in the Annual Renewable Energy Outlook [115,116]. The backcasting is complemented by a forecasting method that reflects current stated short-term policies to ascertain further policy requirements to realize the clean energy transition.

5.1.2. Building Consensus and Raising Ambition

Long-term energy scenarios also act as a tool to initiate discourse and develop consensus on different visions of the future. For example, the Netherlands Environmental Assessment Agency (PBL) develops scenarios to support consensus-building among a wide set of stakeholders (at the national, provincial and municipal levels of governance) to implement climate legislation. PBL developed a model to represent decision-makers with different cost considerations and time horizons to simplify the exploration of options to achieve the clean energy transition [117]. However, a pitfall of developing scenarios with the aim of building consensus is that compromise amongst a diverse range of stakeholders could lead to half-baked or unambitious targets. On the other hand, normative scenarios are used to raise ambition to challenge current targets and stated policies and provide more ambitious pathways to inform national energy planning in line with global climate targets. Based on its REmap analysis and in collaboration with the European Commission, IRENA performed scenario analyses that found that the EU could supply half of its electricity from renewable energy by 2030. This research contributed to the European Council's decision to adopt a more ambitious target of 32% renewable-based energy by 2030. On the national level, other examples of ambition-raising scenario use have been found. In Ireland, projects developed by the International Energy Agency (IEA) Energy Technology Systems Assistance Programme (ETSAP) in co-operation with the University College Cork (UCC) led to developing more ambitious targets being legislated in the 2015 Climate Action and Low Carbon Development Act.

5.1.3. Conservative and Exploratory Scenarios

Conservative scenarios are considered "plausible", which generally contain less ambitious targets, less drastic measures and lower-cost investment options. In contrast, exploratory scenarios push the boundary of opportunities for new and potential technologies, in effect preventing persistent business-as-usual conclusions, showcasing opportunities and potential disruptions, as well as identifying risks for the energy transition. We observed that most government institutions and power system operators are naturally conservative in developing scenarios. For instance, the National Grid—the UK's power system operator—publishes the Future Energy Scenarios report, which develops and identifies a range of conservative scenarios to inform policy and investment decisions [66]. It also acts as a reference point for other scenarios and academic studies. Another example is Ecuador's 10-year electricity master plan, which provides normative scenarios for generation and transmission capacity expansion [118]. Academia, research centres, and non-governmental organisations tend to take a position on the exploratory end of this spectrum [119]. Examples include the "100% renewable energy for Australia" report produced by the Institute of Sustainable Futures (ISF) at the University of Technology Sydney [112], the national deep decarbonization scenarios that the Institute for Sustainable Development and International Relations (IDDRI) carried out for six countries in Latin America [120], the study on climate neutrality for Japan published by Japan's Renewable Energy Institute (REI) and Agora Energiewende [121] and the report from the Indonesian Institute for Essential Services Reform (IESR) [122]. Such studies can help stimulate public debate and challenge government planners to push the assumptions beyond conservative limits.

5.2. Transparent and Effective Communication

Scenarios can most often be used as an effective tool for communication that deciphers the complexities of the energy transition, transforming them into comprehensible and consistent messages. We identified that effectively communicating scenario results ensures the quality and trust of scenarios. Effective communication also includes transparency and accessibility to the underlying data used in models. For the purposes of this paper, communication involves all manners of transmitting information about the scenarios to the public, including publications [24,115,123], news briefs [74,124,125], web-platforms [15,126–129] and events [130–132].

5.2.1. Effective Communication Tools

Communication facilitates the participatory process of developing scenarios, engages a broader set of non-energy stakeholders, and produces straightforward messaging that non-experts can understand. One such method of communicating scenarios is through web-based scenario visualization platforms and calculators. The UK's Department for Business, Energy and Industrial Strategy (BEIS) developed the Mackay Carbon Calculator. This online tool calculates the energy mix and resulting emissions based on various levels of ambition for different sectors with the horizon year of 2100. It provides the public with experience in scenario analysis and pathways and likewise provides BEIS with insights on the public's views [133]. Another such example of an online tool is Exploring Canada's Energy Future, developed by Canada's Energy Regulator (CER), which is a web-based interactive tool based on the CER's Energy Futures report. The platform allows users to navigate by region, sector and type of scenario (demand or power) [75,126]. Innovative methods of communicating directly with policymakers were also showcased among LTES Network members. For example, the Ministry of Energy and Industry of the United Arab Emirates (UAE) created the game "Future Lab" in the context of its Energy Strategy for

2050 [74]. Future Lab allowed senior government officials to test each scenario's systemic opportunities and consequences to learn about the complexities of the future energy system. This was done by providing the officials with experiential insights, with sonic and visual effects simulating future environments, along with other insights including "smelling the burning air of the future" in a fossil-fuel heavy long-term scenario.

5.2.2. Transparent and Publicly Available Information

Transparency of input data, methodology and model assumptions is necessary, as it allows scenarios to be carefully inspected by different stakeholders and allows policymakers to deduce which assumptions and narratives drive certain results. LTES discussions highlighted calls from various stakeholders (government and civil society organisations) for a clear explanation of key model input data, constraints, parameters and scenario outputs in order to avoid potential "black box" scenario approaches. This includes technology cost data, as a majority of scenario developers utilise technology cost projections for the medium and long-term, which can be subject to conservative approaches despite past trends having more drastic rates of change in costs. Availability of such data in a transparent manner is vital to develop trustworthy scenarios that feature input from a diverse set of participants, and to allow feedback and criticism from various experts and stakeholders. Examples of good practices exist across different LTES institutions. In Italy, the National Statistical System, which comprises a broad coalition of private and public bodies, publishes annual reports that contain official statistics used in the country's scenario publications, such as the Integrated National and Climate Plan (INECP) [108]. In Denmark, the Danish Energy Agency issues the Energy Cost and Technology catalogues, which make widely-available yearly updates on costs and technology efficiency, which are used as a reference point for scenario building [134]. In the United States, the National Renewable Energy Laboratory (NREL) publishes the Annual Technology Baseline report, which features current and forecasted cost and data for various technologies for use in the energy sector [135]. In Chile, the Ministry of Energy publishes its five-year Long-term Energy Planning process, including all details about committee formation, methodologies, assumptions and deadlines, as well as annual background reports on data used [17]. These methods, amongst others, create transparency and increase data legitimacy, which builds trust in both the scenario building and policy making processes for the country, and ensures more reliable research on future energy pathways.

6. Institutional Ownership of LTES Capacity

6.1. Building the Correct Type of LTES Capacity in Government

From the government's perspective, LTES development (modelling) capacity can be either insourced or outsourced. Some governments build and maintain in-house modelling capacities within their energy ministries, energy agencies or other government-dependent institutions. Governments can also outsource scenario development to research, technical institutions or consultancies. There is also a middle-of-the-road option where government jointly develops modelling capacity with independent energy agencies or technical institutions. Figure 8 presents a conceptualization of where scenario building capacity can be allocated from the government's perspective. Insourcing or outsourcing are not mutually exclusive, and it is not that one is correct and the other is wrong. Still, we argue that each has distinct advantages and challenges that have been identified in countries that have successfully implemented them (see Table 3).

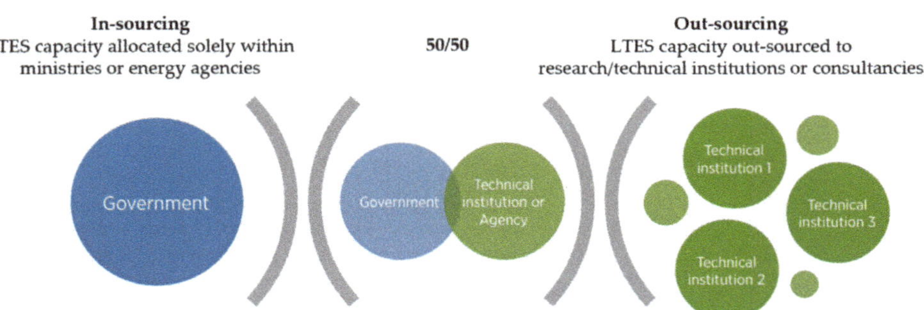

Figure 8. Allocation of LTES development capacity from the government's perspective.

Table 3. Keys for success and advantages and challenges of insourcing and outsourcing LTES development capacity from the government's perspective.

Issue	Insourcing	Outsourcing
Allocation of capacity	Ministries or energy agencies.	Independent technical institutions or consultancies.
Government involvement	Allows for closer and more constant interaction with policymakers.	Can result in intermittent and shorter interactions with policymakers.
Scenario diversity	Tends to have a limited number of scenarios, usually reflecting less exploratory viewpoints.	Tends to cover a wider range of scenarios, reflecting the client's vision and agenda.
Quality of results	Relies on government technical capacity and access to tools and data.	Allows procurement from different high-end commercial tools tailored for purpose.
Response rate	Quick response to pressing government policy needs, subject to the capacity of the team.	May take time to procure scenarios but allows a different execution timing as required by government specifications.
Transparency	Ensures full transparency of inputs and outputs through closer interaction with in-house modelling team.	Tends to be black-box, and proprietary licences may potentially limit full access to the tools.
Cost	Possibly less costly but requires significant efforts to build modelling capacity.	Tends to be expensive to hire commercial consultancy firms.
Keys to success	Quality assurance (e.g., engaging with academia). Team or agency devoted to modelling and scenario development. Establishing an institutional process for systematic updates of LTES.	Absorptive capacity within a government to comprehend modelling outcomes. Full disclosure of scenario data and modelling methodology; Access to enough high-quality research institutions.

6.1.1. Insourcing Scenario Development Capacity

Governments that have succeeded in institutionalising modelling capacity have a dedicated modelling and scenario team, an institutional process for regular updates of LTES, regular engagement with external stakeholders to establish quality assurance, continuous training activities and effective presentation of LTES benefits for decision-making. A key advantage of insourcing is national ownership, which is crucial for developing a solid strategic energy planning process and government buy-in of scenarios. Governments that are now developing internal modelling capacity can begin with using more basic methodologies—for instance, with an accounts-based model rather than a complex energy system optimisation model [136,137]. It is also relevant to discuss the use of proprietary tools (e.g., LEAP [138], TIMES [139]) and non-proprietary tools (e.g., SAM [140], NEMS [141], MESSAGE [142], OSeMOSYS [143]), the usual trade-off being the license costs and user support [144,145]. Insourcing scenario modelling capacity will guarantee that scenario developers experience closer and more frequent communication with high-level governmental energy planners; however, this can likely result in unambitious and conservative scenarios that are heavily influenced by government agendas.

In Mexico, the Secretariat of Energy (SENER) produces a yearly series of LTES [146] and is responsible for the National Energy Strategy [147]. SENER's inhouse energy planning team has benefitted from partnerships with, for example, the Danish Energy Agency for training in the BALMOREL capacity expansion model [148] and with IRENA to produce a roadmap to 2030 [149]. The United Kingdom insources scenario development capacity in the Department of Business, Energy and Industrial Strategy (BEIS) and produces a quality assurance guide for experts performing model analysis in the public sector [150].

Table 4 illustrates the four-step process of quality assured modelling analysis in BEIS. In Brazil, official scenario capacity is allocated in the Energy Research Office (EPE), an independent governmental agency that supports the Ministry of Mines and Energy (MME) in developing scenarios for the National Energy Plan 2050 [12] and the Ten-Year Energy Expansion Plans (PDE) [13]. The Netherlands Environmental Assessment Agency (PBL) is an autonomous government agency that houses its own internal scenario development capacity, resulting in national outlooks and other scenario analyses [151,152].

Table 4. The United Kingdom's four-step quality assurance (QA) framework for modelling in government.

1. Planning	2. Expert Review	3. Analytical Clearance	4. Approval/ Sign-Off
- QA must be factored into project planning. - Outcome: agreed roles, responsibilities, resources and timings; utilisation of appropriate expertise.	- Independent scrutiny of analysis and evidence. - Ongoing revision process. - Drawing on expertise from each relevant discipline. - Peer reviews used to improve work.	- Statement that evidence within the project is adequate for its intended purpose (with any caveats). - Based on peer review opinions and actions taken in response.	- Overall completion of a product. - Factors in clearance statement, in addition to wider factors.

6.1.2. Outsourcing Scenario Development Capacity

Outsourcing scenario development effectively will require strong in-house (government) capacity to comprehend LTES and to ensure good contracting of consultants. Therefore, training scenario users is as important as training scenario developers. Outsourcing allows access to better models and building techniques; the drawback, however, is having black box tools, undermining of internal scenario capacity and creating a lock-in or over-reliance on a few consultancy companies that will fulfil the contractor's desires. Outsourcing also has the advantage of ensuring experts develop the LTES. Therefore, governments who cannot insource may begin with outsourcing and follow that with knowledge transfer activities, which can be supported through collaboration with academia and international institutions. In any case, when outsourcing, full disclosure of scenario data and modelling methodologies is recommended.

In Germany, the Ministry of Economy and Energy (BMWi) has highly-capable internal capacity to both comprehend and build LTES [153]. Yet, the country has the availability of multiple first-class energy research institutions that carry out independent scenario studies which can be compared to gain a wider range of insights [154–157]. The

United Arab Emirates Energy Strategy 2050 was developed using a proprietary modelling tool outsourced to consultancy firms; nevertheless, the energy strategy team in the Ministry of Energy and Industry is now in the process of building capacity to develop scenarios in-house [158]. IRENA's Masterplan Development Support Programme supported the development of the Kingdom of Eswatini Electricity Master Plan [159,160]; the International Energy Agency's Energy Technology Systems Analysis Programme (ETSAP) influenced Portugal's National Action Plan on Climate Change in 2014 and the Republic of Ireland's Climate Action and Low Carbon Development Act in 2015 [161].

7. Conclusions

The application of LTES to draw policy-relevant insight regarding the transition to a clean, sustainable and low-carbon energy and economic system is fraught with challenges. Scenario development models and tools are simplifications of a complex and dynamic real-world energy system, and results must be considered under the condition that transition features, such as higher shares of variable energy, electrification and new markets structures, are being considered in the analysis. However, the development and use of LTES to inform the transition goes beyond the modelling space. It demands robust energy governance, institutionalised energy planning processes and absorptive capacity in government to make use of complex insights. Operating under such technical and governance circumstances requires scenario practitioners to handle results with caution. These challenges notwithstanding, LTES remain a vital tool employed by government agencies as a basis for their decisions, plans and policies, and not only in the energy sector; LTES will surely play a critical supporting role to develop mid-term nationally determined contributions under Article 4.2 and long-term low greenhouse gas emission development strategies under Article 4.19 [162–165] in all participating nations of the Paris Agreement.

Despite the importance of LTES in national energy policymaking, there has been little effort to develop formal guidelines for their application in government. Best LTES practice is typically learned through replicating experience from other countries and apprenticeship with more experienced scenario users from academia, technical institutes, international development bodies and consultancies. By contrast, the literature shows that energy modelling has benefitted from efforts to standardise its approach (e.g., [166,167]), and served as a practical guide for modellers [168–170].

This paper is a first effort to document and formalise best practices regarding the use of LTES in the government in the energy transition context. We view such guidelines as an essential national energy planning resource, which we hope create a set of expectations for LTES-based analysis and the minimum considerations for effective LTES use. Best practice guidelines, however subjective and imperfect, also serve as a benchmark for methodological refinements and future debates.

The best practices listed in this paper draw upon the LTES literature and the first-hand experience from national energy institutions that are transforming their scenario practice as the energy policy landscape is driven by climate action. As the transition continues to unfold, new approaches will likely need to be developed to tackle even more ambitious climate goals and the profound socioeconomic and infrastructure challenges that arise. Best practices for LTES in the government will evolve as the discussion involves more people, tools and models are refined and the climate policy landscape changes.

Author Contributions: Conceptualization, P.E.C.; methodology, A.M.; formal analysis, P.E.C., A.M. and N.G.; writing—original draft preparation, P.E.C. and N.G.; writing—review and editing, A.M.; tables and figures, P.F. All authors have read and agreed to the published version of the manuscript.

Funding: This research received no external funding.

Acknowledgments: We appreciate the kind support of Celine Ashby, who supported early data collection for the county examples.

Conflicts of Interest: The authors declare no conflict of interest.

References

1. Bridge, G.; Bouzarovski, S.; Bradshaw, M.; Eyre, N. Geographies of energy transition: Space, place and the low-carbon economy. *Energy Policy* **2013**, *53*, 331–340. [CrossRef]
2. Solomon, B.D.; Krishna, K. The coming sustainable energy transition: History, strategies, and outlook. *Energy Policy* **2011**, *39*, 7422–7431. [CrossRef]
3. Markard, J. The next phase of the energy transition and its implications for research and policy. *Nat. Energy* **2018**, *3*, 628–633. [CrossRef]
4. Teske, S.; Pregger, T.; Simon, S.; Naegler, T. High renewable energy penetration scenarios and their implications for urban energy and transport systems. *Curr. Opin. Environ. Sustain.* **2018**, *30*, 89–102. [CrossRef]
5. IRENA. *World Energy Transitions Outlook: 1.5 °C Pathway*; International Renewable Energy Agency: Abu Dhabi, United Arab Emirates, 2021. Available online: https://www.irena.org/publications/2021/Jun/World-Energy-Transitions-Outlook (accessed on 23 January 2022).
6. IRENA. *Innovation Landscape Report*; International Renewable Energy Agency: Abu Dhabi, United Arab Emirates, 2019.
7. IRENA. *Global Renewables Outlook: Energy Transformation 2050*; International Renewable Energy Agency: Abu Dhabi, United Arab Emirates, 2020.
8. IPCC. *Global Warming of 1.5 °C. An IPCC Special Report on the Impacts of Global Warming of 1.5 °C above Pre-Industrial Levels and Related Global Greenhouse Gas Emission Pathways, in the Context of Strengthening the Global Response to the Threat of Climate Change, Sustainable Development, and Efforts to Eradicate Poverty*; IPCC: Geneva, Switzerland, 2018. Available online: https://www.ipcc.ch/sr15/ (accessed on 23 January 2022).
9. Mccollum, D.L.; Zhou, W.; Bertram, C.; De Boer, H.-S.; Bosetti, V.; Busch, S.; Després, J.; Drouet, L.; Emmerling, J.; Fay, M.; et al. Energy investment needs for fulfilling the Paris Agreement and achieving the Sustainable Development Goals. *Nat. Energy* **2018**, *3*, 589–599. [CrossRef]
10. Moss, R.H.; Edmonds, J.A.; Hibbard, K.A.; Manning, M.R.; Rose, S.K.; van Vuuren, D.P.; Carter, T.R.; Emori, S.; Kainuma, M.; Kram, T.; et al. The next generation of scenarios for climate change research and assessment. *Nature* **2010**, *463*, 747–756. [CrossRef]
11. UNFCCC. Intended Nationally Determined Contributions (INDCs), INDC Portal. 2015. Available online: http://unfccc.int/focus/indc_portal/items/8766.php (accessed on 2 January 2016).
12. MME; EPE. *Plano Nacional de Energia PNE 2030*; Ministry of Energy and Minerals (MME) and Energy Research Office (EPE): Brasilia, Brazil, 2007. Available online: https://www.epe.gov.br/pt/publicacoes-dados-abertos/publicacoes/Plano-Nacional-de-Energia-PNE-2030 (accessed on 23 January 2022).
13. MME; EPE. *The Ten Year Energy Expansion Plan (PDE) 2029*; Ministério de Minas e Energia, Empresa de Pesquisa Energética: Rio de Janeiro, Brazil, 2020. Available online: https://www.epe.gov.br/sites-pt/publicacoes-dados-abertos/publicacoes/PublicacoesArquivos/publicacao-422/PDE%202029.pdf (accessed on 23 January 2022).
14. Ministerio de Energía de Chile. Ruta Energetica 2018-2022—Liderando la Modernizacion con Sello Ciudadano. 2018. Available online: https://www.energia.gob.cl/sites/default/files/rutaenergetica2018-2022.pdf (accessed on 23 January 2022).
15. Ministerio de Energía de Chile. *Planificación Energética de Largo Plazo—Escenarios Energéticos*; Ministerio de Energía: Santiago, Chile, 2019. Available online: https://www.energia.gob.cl/planificacion-energetica-de-largo-plazo-escenarios-energeticos (accessed on 11 August 2020).
16. Ministerio de Energía de Chile. *Planificación Energética de Largo Plazo 2018–2022*; Gobierno de Chile: Santiago, Chile, 2018. Available online: https://energia.gob.cl/planificacion-energetica-de-largo-plazo-proceso (accessed on 23 January 2022).
17. Ministerio de Energía de Chile. *Long Term Energy Planning—Process*; Ministry of Energy: Santiago, Chile, 2019. Available online: https://www.energia.gob.cl/planificacion-energetica-de-largo-plazo-proceso (accessed on 3 June 2020).
18. Ministerio de Energía de Chile. *Informe de Actualización de Antecedentes 2019*; Ministerio de Energía: Santiago, Chile, 2019. Available online: http://www.energia.gob.cl/sites/default/files/documentos/20191209_actualizacion_pelp_-_iaa_2019.pdf (accessed on 23 January 2022).
19. Republic of South Africa. Request for Comments: Draft Integrated Resource Plan 2018. Available online: http://www.energy.gov.za/IRP/irp-update-draft-report2018/IRP-Update-2018-Draft-for-Comments.pdf (accessed on 23 January 2022).
20. Republic of South Africa. *Integrated Resource Plan (IRP2019)*; Department of Energy: Pretoria, South Africa, 2019. Available online: http://www.energy.gov.za/IRP/2019/IRP-2019.pdf (accessed on 23 January 2022).
21. Ministry of Economic Affairs and Employment for Finland. *Government Report on the National Energy and Climate Strategy for 2030*; Ministry of Economic Affairs and Employment: Helsinki, Finland, 2017. Available online: https://tem.fi/documents/1410877/2769658/Government+report+on+the+National+Energy+and+Climate+Strategy+for+2030/0bb2a7be-d3c2-4149-a4c2-78449ceb1976/Government+report+on+the+National+Energy+and+Climate+Strategy+for+2030.pdf (accessed on 23 January 2022).
22. Ministry of Economic Affairs and Employment for Finland. *Finland's Integrated Energy and Climate Plan*; Ministry of Economic Affairs and Employment: Helsinki, Finland, 2019. Available online: https://ec.europa.eu/energy/sites/ener/files/documents/fi_final_necp_main_en.pdf (accessed on 23 January 2022).

23. Ministry of Economic Affairs and Employment for Finland. *Energy and Climate Roadmap 2050*; Ministry of Economic Affairs and Employment: Helsinki, Finland, 2014. Available online: https://tem.fi/documents/1410877/2769658/Energy+and+Climate+Roadmap+2050.pdf/9fd1b4ca-346d-4d05-914a-2e20e5d33074/Energy+and+Climate+Roadmap+2050.pdf?t=1464241259000 (accessed on 23 January 2022).
24. DEA. *Denmark's Energy and Climate Outlook 2020*; Danish Energy Agency: København, Denmark, 2020. Available online: https://ens.dk/en/our-services/projections-and-models/denmarks-energy-and-climate-outlook (accessed on 23 January 2022).
25. Zhang, Y.; Zhang, X.; Lan, L. Robust optimization-based dynamic power generation mix evolution under the carbon-neutral target. *Resour. Conserv. Recycl.* **2021**, *178*, 106103. [CrossRef]
26. Silvestri, L.; Di Micco, S.; Forcina, A.; Minutillo, M.; Perna, A. Power-to-hydrogen pathway in the transport sector: How to assure the economic sustainability of solar powered refueling stations. *Energy Convers. Manag.* **2021**, *252*, 115067. [CrossRef]
27. Liu, Q.; Sun, Y.; Liu, L.; Wu, M. An uncertainty analysis for offshore wind power investment decisions in the context of the national subsidy retraction in China: A real options approach. *J. Clean. Prod.* **2021**, *329*, 129559. [CrossRef]
28. Lin, B.; Omoju, O.E. Focusing on the right targets: Economic factors driving non-hydro renewable energy transition. *Renew. Energy* **2017**, *113*, 52–63. [CrossRef]
29. Remus, C.; Guran, L.; Platon, D.; Turnock, D. *Foreign Direct Investment and Social Risk in Romania: Progress in Less-Favoured Areas*; Routledge: Abingdon-on-Thames, UK, 2005; pp. 305–348. [CrossRef]
30. UNFCCC. All NDCs. 2021. Available online: https://www4.unfccc.int/sites/NDCStaging/Pages/All.aspx (accessed on 25 June 2021).
31. UNFCCC. All LT-LEDS. 2021. Available online: https://unfccc.int/process/the-paris-agreement/long-term-strategies (accessed on 25 June 2021).
32. Gargiulo, M.; Gallachóir, B. Long-term energy models: Principles, characteristics, focus, and limitations. *WIREs Energy Environ.* **2013**, *2*, 158–177. [CrossRef]
33. Hourcade, J.; Jaccard, M.; Bataille, C.; Ghersi, F. Hybrid modelling: New Answers to Old Challenges—Introduction to the Special Issue of the Energy Journal. International Association for Energy Economics, 2006. Available online: https://hal-enpc.archives-ouvertes.fr/hal-00716778 (accessed on 25 June 2021).
34. IRENA. *Planning for the Renewable Energy Future: Long-Term Modelling and Tools to Expand Variable Renewable Power in Emerging Economies*; International Renewable Energy Agency: Abu Dhabi, United Arab Emirates, 2017. Available online: https://www.irena.org/-/media/Files/IRENA/Agency/Publication/2017/IRENA_Planning_for_the_Renewable_Future_2017.pdf (accessed on 23 January 2022).
35. Bloomfield, H.; Brayshaw, D.; Troccoli, A.; Goodess, C.; De Felice, M.; Dubus, L.; Bett, P.; Saint-Drenan, Y.-M. Quantifying the sensitivity of european power systems to energy scenarios and climate change projections. *Renew. Energy* **2020**, *164*, 1062–1075. [CrossRef]
36. Fortes, P.; Seixas, J.; Simoes, S.; Cleto, J. Long term energy scenarios under uncertainty. In Proceedings of the 2008 5th International Conference on the European Electricity Market, Lisboa, Portugal, 28–30 May 2008; pp. 1–6. [CrossRef]
37. Pye, S.; Li, F.G.; Petersen, A.; Broad, O.; McDowall, W.; Price, J.; Usher, W. Assessing qualitative and quantitative dimensions of uncertainty in energy modelling for policy support in the United Kingdom. *Energy Res. Soc. Sci.* **2018**, *46*, 332–344. [CrossRef]
38. Usher, W.; Strachan, N. An expert elicitation of climate, energy and economic uncertainties. *Energy Policy* **2013**, *61*, 811–821. [CrossRef]
39. Creutzig, F.; Agoston, P.; Goldschmidt, J.C.; Luderer, G.; Nemet, G.; Pietzcker, R.C. The underestimated potential of solar energy to mitigate climate change. *Nat. Energy* **2017**, *2*, 17140. [CrossRef]
40. Xiao, M.; Junne, T.; Haas, J.; Klein, M. Plummeting costs of renewables—Are energy scenarios lagging? *Energy Strat. Rev.* **2021**, *35*, 100636. [CrossRef]
41. Carrington, G.; Stephenson, J. The politics of energy scenarios: Are International Energy Agency and other conservative projections hampering the renewable energy transition? *Energy Res. Soc. Sci.* **2018**, *46*, 103–113. [CrossRef]
42. Betz, G. What's the Worst Case? The Methodology of Possibilistic Prediction. *Anal. Krit.* **2010**, *32*, 87–106. [CrossRef]
43. Craig, P.P.; Gadgil, A.; Koomey, J.G. What Can History Teach Us? A Retrospective Examination of Long-Term Energy Forecasts for the United States. *Annu. Rev. Energy Environ.* **2002**, *27*, 83–118. [CrossRef]
44. DeCarolis, J.; Hunter, K.; Sreepathi, S. The case for repeatable analysis with energy economy optimization models. *Energy Econ.* **2012**, *34*, 1845–1853. [CrossRef]
45. Nielsen, S.K.; Karlsson, K.B. Energy scenarios: A review of methods, uses and suggestions for improvement. *Int. J. Glob. Energy Issues* **2007**, *27*, 302. [CrossRef]
46. Paltsev, S. Energy scenarios: The value and limits of scenario analysis. *WIREs Energy Environ.* **2016**, *6*, e242. [CrossRef]
47. Groves, D.G.; Lempert, R.J. A new analytic method for finding policy-relevant scenarios. *Glob. Environ. Change* **2007**, *17*, 73–85. [CrossRef]
48. Söderholm, P.; Hildingsson, R.; Johansson, B.; Khan, J.; Wilhelmsson, F. Governing the transition to low-carbon futures: A critical survey of energy scenarios for 2050. *Futures* **2011**, *43*, 1105–1116. [CrossRef]
49. Wright, D.; Stahl, B.; Hatzakis, T. Policy scenarios as an instrument for policymakers. *Technol. Forecast. Soc. Chang.* **2020**, *154*, 119972. [CrossRef]

50. Child, M.; Koskinen, O.; Linnanen, L.; Breyer, C. Sustainability guardrails for energy scenarios of the global energy transition. *Renew. Sustain. Energy Rev.* **2018**, *91*, 321–334. [CrossRef]

51. Bazilian, M.; Bradshaw, M.; Gabriel, J.; Goldthau, A.; Westphal, K. Four scenarios of the energy transition: Drivers, consequences, and implications for geopolitics. *WIREs Clim. Change* **2019**, *11*, e625. [CrossRef]

52. Poganietz, W.-R.; Weimer-Jehle, W. Introduction to the special issue 'Integrated scenario building in energy transition research'. *Clim. Change* **2020**, *162*, 1699–1704. [CrossRef]

53. DeCarolis, J.; Daly, H.; Dodds, P.; Keppo, I.; Li, F.; McDowall, W.; Pye, S.; Strachan, N.; Trutnevyte, E.; Usher, W.; et al. Formalizing best practice for energy system optimization modelling. *Appl. Energy* **2017**, *194*, 184–198. [CrossRef]

54. Režný, L.; Bureš, V. Energy Transition Scenarios and Their Economic Impacts in the Extended Neoclassical Model of Economic Growth. *Sustainability* **2019**, *11*, 3644. [CrossRef]

55. IRENA. Exchanging Best Practices to Incorporate Variable Renewable Energy into Long-Term Energy/Power Sector Planning in South America (Technical Workshop Summary Report). 2017. Available online: www.irena.org/-/media/Files/IRENA/Agency/Events/2017/Aug/Summary-Report---IRENA-Regional-Workshop-on-Long-term-Energy-Planning---Buenos-Aires-2017.f?la=en&hash=0FA13A0D0A9F67E354EA27652FD190A24EEED3CB (accessed on 23 January 2022).

56. IRENA. *Power Sector Planning in Arab Countries: Incorporating Variable Renewables*; International Renewable Energy Agency: Abu Dhabi, United Arab Emirates, 2020. Available online: https://www.irena.org/publications/2020/Jan/Arab-VRE-planning (accessed on 23 January 2022).

57. IRENA. *Scenarios for the Energy Transition: Global Experience and Best Practices*; International Renewable Energy Agency: Abu Dhabi, United Arab Emirates, 2020. Available online: https://www.irena.org/publications/2020/Sep/Scenarios-for-the-Energy-Transition-Global-experience-and-best-practices (accessed on 25 June 2021).

58. Costa Rica Gobierno de Bicentenario. *Plan Nacional de Descarbonización 2018–2050*; Gobierno de Costa Rica: San José, Costa Rica, 2019. Available online: https://cambioclimatico.go.cr/wp-content/uploads/2019/02/PLAN.pdf (accessed on 23 January 2022).

59. Ministry of the Environment of Finland. Climate Change Act. 2015. Available online: https://www.finlex.fi/fi/laki/kaannokset/2015/en20150609.pdf (accessed on 23 January 2022).

60. Kaplan, P.O.; Witt, J.W. What is the role of distributed energy resources under scenarios of greenhouse gas reductions? A specific focus on combined heat and power systems in the industrial and commercial sectors. *Appl. Energy* **2018**, *235*, 83–94. [CrossRef] [PubMed]

61. Wu, Y.; Xu, C.; Ke, Y.; Li, X.; Li, L. Portfolio selection of distributed energy generation projects considering uncertainty and project interaction under different enterprise strategic scenarios. *Appl. Energy* **2018**, *236*, 444–464. [CrossRef]

62. Farzaneh, H. Development of a Bottom-up Technology Assessment Model for Assessing the Low Carbon Energy Scenarios in the Urban System. *Energy Procedia* **2017**, *107*, 321–326. [CrossRef]

63. Fichera, A.; Frasca, M.; Palermo, V.; Volpe, R. An optimization tool for the assessment of urban energy scenarios. *Energy* **2018**, *156*, 418–429. [CrossRef]

64. Van Vuuren, D.; Kok, M.T.; Girod, B.; Lucas, P.; de Vries, B. Scenarios in Global Environmental Assessments: Key characteristics and lessons for future use. *Glob. Environ. Change* **2012**, *22*, 884–895. [CrossRef]

65. UPME. National Energy Plan: Energy Principles 2050. 2015. Available online: https://www1.upme.gov.co/Documents/PEN_IdearioEnergetico2050.pdf (accessed on 23 January 2022).

66. National Grid ESO. *Future Energy Scenarios 2020*; National Grid ESO Ltd.: London, UK, 2020. Available online: https://www.nationalgrideso.com/document/173821/download (accessed on 23 January 2022).

67. UPME. Long-term planning process in Colombia. In Proceedings of the Long-Term Energy Scenarios (LTES) for Developing National Clean Energy Transition Plans in Latin America Webinar Series 2021, online, 3 February 2021. Available online: https://www.irena.org/-/media/Files/IRENA/Agency/Events/2021/Feb/Slides_LTES-LATAM/COLOMBIA.pdf?la=en&hash=A78608356DA33512925D452C9DEA99D886B52FC9 (accessed on 23 January 2022).

68. UNEP. Emissions Gap Report 2019. Available online: http://www.unenvironment.org/resources/emissions-gap-report-2019 (accessed on 16 June 2020).

69. IRENA. *NDCs in 2020: Advancing Renewables in the Power Sector and Beyond*; International Renewable Energy Agency: Abu Dhabi, United Arab Emirates, 2019.

70. BID. La Red del Futuro: Desarrollo de una Red Eléctrica Limpia y Sostenible para América Latina. 2017. Available online: https://publications.iadb.org/es/publicacion/14076/la-red-del-futuro-desarrollo-de-una-red-electrica-limpia-y-sostenible-para (accessed on 6 July 2021).

71. Boie, I.; Kost, C.; Bohn, S.; Agsten, M.; Bretschneider, P.; Snigovyi, O.; Pudlik, M.; Ragwitz, M.; Schlegl, T.; Westermann, D. Opportunities and challenges of high renewable energy deployment and electricity exchange for North Africa and Europe—Scenarios for power sector and transmission infrastructure in 2030 and 2050. *Renew. Energy* **2016**, *87*, 130–144. [CrossRef]

72. IRENA. Planning and Prospects for Renewable Power: Eastern and Southern Africa. 2021. Available online: https://www.irena.org/publications/2021/Apr/Planning-and-prospects-for-renewable-power-Eastern-and-Southern-Africa (accessed on 6 July 2021).

73. Johannsen, R.; Østergaard, P.; Maya-Drysdale, D.; Mouritsen, L.K.E. Designing Tools for Energy System Scenario Making in Municipal Energy Planning. *Energies* **2021**, *14*, 1442. [CrossRef]

74. UAE Ministry of Energy and Infrastructure. Vice President Unveils UAE Energy Strategy for Next Three Decades. 2017. Available online: https://bit.ly/3xp0nsk (accessed on 3 June 2020).
75. CER. *Canada's Energy Future 2019*; Canada Energy Regulator: Calgary, AB, Canada, 2019. Available online: https://www.cer-rec.gc.ca/en/data-analysis/canada-energy-future/2019/index.html#:~{}:text=Canada\T1\textquoterights%20Energy%20Future%202019%3A%20Energy,Canadians%20over%20the%20long%20term (accessed on 23 January 2022).
76. ENTSO-E; ENTSOG. *TYNDP 2020 Scenario*; ENTSO-E, ENTSOG: Brussels, Belgium, 2020.
77. NRCan. Long-Term Energy Scenarios in Canada's Clean Energy Transition. In Proceedings of the Long-Term Energy Scenarios (LTES) Campaign Webinar Series 2019, online, 21 March 2019. Available online: https://www.irena.org/renewables/Knowledge-Gateway/webinars/2018/Nov/Webinar-series-on-Long-term-Energy-Scenarios (accessed on 23 January 2022).
78. McDowall, W. Disruptive innovation and energy transitions: Is Christensen's theory helpful? *Energy Res. Soc. Sci.* **2018**, *37*, 243–246. [CrossRef]
79. Tayal, D. Disruptive forces on the electricity industry: A changing landscape for utilities. *Electr. J.* **2016**, *29*, 13–17. [CrossRef]
80. Pettifor, H.; Wilson, C.; Axsen, J.; Abrahamse, W.; Anable, J. Social influence in the global diffusion of alternative fuel vehicles—A meta-analysis. *J. Transp. Geogr.* **2017**, *62*, 247–261. [CrossRef]
81. Edelenbosch, O.Y.; Mccollum, D.L.; Pettifor, H.; Wilson, C.; Van Vuuren, D.P. Interactions between social learning and technological learning in electric vehicle futures. *Environ. Res. Lett.* **2018**, *13*, 124004. [CrossRef]
82. Grubler, A.; Wilson, C.; Bento, N.; Boza-Kiss, B.; Krey, V.; Mccollum, D.L.; Rao, N.D.; Riahi, K.; Rogelj, J.; De Stercke, S.; et al. A low energy demand scenario for meeting the 1.5 °C target and sustainable development goals without negative emission technologies. *Nat. Energy* **2018**, *3*, 515–527. [CrossRef]
83. Ornetzeder, M.; Wächter, P.; Rohracher, H.; Schreuer, A.; Weber, M.; Kubeczko, K.; Paier, M.; Knoflacher, M.; Späth, P. Beyond energy scenarios: Exploring critical socio-economic issues in the transformation of the energy system. In Proceedings of the Sussex Energy Group Conference, Brighton, UK, 25–26 February 2010.
84. Hooper, T. Do Energy Scenarios Pay Sufficient Attention to the Environment? UKERC, February 2018. Available online: https://ukerc.ac.uk/news/energy-scenarios-and-the-environment/ (accessed on 2 July 2021).
85. Weimer-Jehle, W.; Vögele, S.; Hauser, W.; Kosow, H.; Poganietz, W.-R.; Prehofer, S. Socio-technical energy scenarios: State-of-the-art and CIB-based approaches. *Clim. Change* **2020**, *162*, 1723–1741. [CrossRef]
86. Newell, P.; Mulvaney, D. The political economy of the 'just transition'. *Geogr. J.* **2013**, *179*, 132–140. [CrossRef]
87. VanCleef, A. Hydropower Development and Involuntary Displacement: Toward a Global Solution. *Indiana J. Glob. Leg. Stud.* **2016**, *23*, 349. [CrossRef]
88. IRENA. *Global Renewables Outlook*; International Renewable Energy Agency: Abu Dhabi, United Arab Emirates, 2020. Available online: https://www.irena.org/-/media/Files/IRENA/Agency/Publication/2020/Apr/IRENA_Global_Renewables_Outlook_2020.pdf (accessed on 23 January 2022).
89. Jiang, P.; Van Fan, Y.; Klemeš, J.J. Impacts of COVID-19 on energy demand and consumption: Challenges, lessons and emerging opportunities. *Appl. Energy* **2021**, *285*, 116441. [CrossRef]
90. Klemeš, J.J.; Van Fan, Y.; Jiang, P. The energy and environmental footprints of COVID-19 fighting measures—PPE, disinfection, supply chains. *Energy* **2020**, *211*, 118701. [CrossRef]
91. Johnsson, F.; Karlsson, I.; Rootzén, J.; Ahlbäck, A.; Gustavsson, M. The framing of a sustainable development goals assessment in decarbonizing the construction industry—Avoiding "Greenwashing". Renew. Sustain. *Energy Rev.* **2020**, *131*, 110029. [CrossRef]
92. Doiciara, C.; Creţana, R. Pandemic populism: COVID-19 and the rise of the Nationalist AUR party in Romania. *Geogr. Pannonica* **2021**, *25*, 243–259. [CrossRef]
93. BMWi. *Commission on Growth, Structural Change and Employment*; Federal Ministry for Economic Affairs and Energy: Berlin, Germany, 2019. Available online: https://www.bmwi.de/Redaktion/EN/Publikationen/commission-on-growth-structural-change-and-employment.pdf?__blob=publicationFile&v=3 (accessed on 23 January 2022).
94. BMWi. International Dialogue on Global Best Practices for Strategic Long-Term Energy Planning, Presented at the IRENA Pre-Assembly Event. 2020. Available online: https://www.irena.org/energytransition/Energy-Transition-Scenarios-Network/ETS-Net-Events (accessed on 23 January 2022).
95. European Commission. *A Clean Planet for All. A European Strategic Long-Term Vision for a Prosperous, Modern, Competitive and Climate Neutral Economy*; European Commission: Brussels, Belgium, 2018.
96. European Commission. The Use of Modelling Analysis for the EU's Vision for a Long Term Strategy. In Proceedings of the Long-Term Energy Scenarios (LTES) Campaign Webinar Series 2019, online, 21 March 2019. Available online: https://www.irena.org/renewables/Knowledge-Gateway/webinars/2018/Nov/Webinar-series-on-Long-term-Energy-Scenarios (accessed on 23 January 2022).
97. Grimaldo, A.I.; Novak, J. User-Centered Visual Analytics Approach for Interactive and Explainable Energy Demand Analysis in Prosumer Scenarios. In *Computer Vision Systems*; Tzovaras, D., Giakoumis, D., Vincze, M., Argyros, A., Eds.; Springer International Publishing: Cham, Switzerland, 2019; Volume 11754, pp. 700–710. [CrossRef]
98. Riaz, S.; Marzooghi, H.; Verbic, G.; Chapman, A.C.; Hill, D.J. Generic Demand Model Considering the Impact of Prosumers for Future Grid Scenario Analysis. *IEEE Trans. Smart Grid* **2017**, *10*, 819–829. [CrossRef]

99. Giordano, F.; Ciocia, A.; Di Leo, P.; Mazza, A.; Spertino, F.; Tenconi, A.; Vaschetto, S. Vehicle-to-Home Usage Scenarios for Self-Consumption Improvement of a Residential Prosumer with Photovoltaic Roof. *IEEE Trans. Ind. Appl.* **2020**, *56*, 2945–2956. [CrossRef]

100. Ceran, B. Multi-Criteria Comparative Analysis of Clean Hydrogen Production Scenarios. *Energies* **2020**, *13*, 4180. [CrossRef]

101. Kakoulaki, G.; Kougias, I.; Taylor, N.; Dolci, F.; Moya, J.; Jäger-Waldau, A. Green hydrogen in Europe—A regional assessment: Substituting existing production with electrolysis powered by renewables. *Energy Convers. Manag.* **2021**, *228*, 113649. [CrossRef]

102. Quarton, C.J.; Tlili, O.; Welder, L.; Mansilla, C.; Blanco, H.; Heinrichs, H.; Leaver, J.; Samsatli, N.J.; Lucchese, P.; Robinius, M.; et al. The curious case of the conflicting roles of hydrogen in global energy scenarios. *Sustain. Energy Fuels* **2019**, *4*, 80–95. [CrossRef]

103. Bernath, C.; Deac, G.; Sensfuß, F. Impact of sector coupling on the market value of renewable energies—A model-based scenario analysis. *Appl. Energy* **2020**, *281*, 115985. [CrossRef]

104. Gea-Bermúdez, J.; Jensen, I.G.; Münster, M.; Koivisto, M.; Kirkerud, J.G.; Chen, Y.-K.; Ravn, H. The role of sector coupling in the green transition: A least-cost energy system development in Northern-central Europe towards 2050. *Appl. Energy* **2021**, *289*, 116685. [CrossRef]

105. Jadun, P.; McMillan, C.; Steinberg, D.; Muratori, M.; Vimmerstedt, L.; Mai, T. *Electrification Futures Study: End-Use Electric Technology Cost and Performance Projections through 2050*; Technnical Report NREL/TP-6A20-70485; NREL: Golden, CO, USA, 2017. [CrossRef]

106. Ruhnau, O.; Bannik, S.; Otten, S.; Praktiknjo, A.; Robinius, M. Direct or indirect electrification? A review of heat generation and road transport decarbonisation scenarios for Germany 2050. *Energy* **2018**, *166*, 989–999. [CrossRef]

107. METI. *5th Strategic Energy Plan*; METI: Tokyo, Japan, 2018.

108. MISE; MEIT; MIT. Integrated National Energy and Climate Plan (INECP). 2019. Available online: https://ec.europa.eu/energy/sites/ener/files/documents/it_final_necp_main_en.pdf (accessed on 23 January 2022).

109. NREL. *Power Sector Transformation Pathways: Exploring Objectives, Factors, and Technology Innovation to Inform Power Sector Pathway Decisions*; National Renewable Energy Laboratory: Golden, CO, USA, 2020. Available online: https://www.nrel.gov/docs/fy20osti/75138.pdf (accessed on 23 January 2022).

110. Bundesnetzagentur. *Scenario Framework 2021–2035*; Federal Network Agency: Bonn, Germany, 2020. Available online: https://www.bundesnetzagentur.de/SharedDocs/Pressemitteilungen/EN/2020/20200626_NEP.html (accessed on 23 January 2022).

111. IPCC. Fifth Assessment Report. 2014. Available online: https://www.ipcc.ch/assessment-report/ar5/ (accessed on 23 January 2022).

112. UTS. *100% Renewable Energy for Australia: Decarbonising Australia's Energy Sector Within One Generation*; University of Technology Sydney—Institute for Sustainable Futures (UTS-ISF): Sydney, Australia, 2016. Available online: https://www.uts.edu.au/sites/default/files/article/downloads/ISF_100%25_Australian_Renewable_Energy_Report.pdf (accessed on 23 January 2022).

113. EC JRC. *Towards Net-Zero Emissions in the EU Energy System—Insights from Scenarios in Line with the 2030 and 2050 Ambitions of the European Green Deal*; EUR 29981; European Union: Luxembourg, 2020. Available online: https://publications.jrc.ec.europa.eu/repository/handle/JRC118592 (accessed on 23 January 2022).

114. Central Committee of the Communist Party of China. PRC 13th Five-Year Plan (FYP, 2016-2020) on National Economic and Social Development. 2016. Available online: https://www.uschina.org/policy/official-13th-five-year-plan-outline-released (accessed on 23 January 2022).

115. NDRC; CNREC. *China Renewable Energy Outlook 2019*; Energy Research Institute of Academy of Macroeconomic Research/NDRC: Beijing, China, 2019. Available online: https://www.dena.de/fileadmin/dena/Publikationen/PDFs/2019/CREO2019_-_Executive_Summary_2019.pdf (accessed on 23 January 2022).

116. NDRC; CNREC. *China Renewable Energy Outlook 2020*; Energy Research Institute of Academy of Macroeconomic Research/NDRC: Beijing, China, 2020.

117. PBL. Practical Examples and the Impact of the Application of Long-Term Energy Scenarios, and Their Better Use/Development. In Proceedings of the G-STIC, Brussels, Belgium, 29 November 2018. Available online: https://www.irena.org/energytransition/Energy-Transition-Scenarios-Network/ETS-Net-Events (accessed on 23 January 2022).

118. Ministry of Energy and Non-Renewable Natural Resources, Ecuador. Plan Maestro de Electricidad. 2020. Available online: https://www.recursosyenergia.gob.ec/plan-maestro-de-electricidad/ (accessed on 23 January 2022).

119. Teske, S. *Achieving the Paris Climate Agreement Goals: Global and Regional 100% Renewable Energy Scenarios with Non-energy GHG Pathways for +1.5 °C and +2 °C.*; Springer International Publishing: Cham, Switzerland, 2019. [CrossRef]

120. IDDRI. Deep Decarbonization Pathways in Latin America. 2018. Available online: https://www.iddri.org/en/project/deep-decarbonization-pathways-latin-america (accessed on 23 January 2022).

121. Renewable Energy Institute; Agora Energiewende; LUT University. Renewable Pathways to Climate-Neutral Japan. Study on behalf of Renewable Energy Institute and Agora Energiewende. 2021. Available online: https://static.agora-energiewende.de/fileadmin/Projekte/2021/2021_03_JP_2050_study/2021_LUT-Agora-REI_Renewable_pathways_Study.pdf (accessed on 23 January 2022).

122. IESR. *Deep Decarbonization of Indonesia's Energy System: A Pathway to Zero Emissions by 2050*; IESR: Jakarta, Indonesia, 2021.

123. Ministry of Energy and Mineral Resources, Indonesia. Indonesia Energy Outlook. 2020. Available online: https://www.esdm.go.id/en/publication/indonesia-energy-outlook (accessed on 11 August 2020).

124. MME. Plano Nacional de Energia 2050 é Lançado. 2020. Available online: https://www.gov.br/pt-br/noticias/energia-mineraise-combustiveis/2020/12/plano-nacional-de-energia-2050-e-lancado (accessed on 23 January 2022).
125. DEA. Denmark's Baseline Projection 2019. Available online: https://ens.dk/en/press/denmarks-baseline-projection-2019-now-english (accessed on 23 January 2022).
126. CER. *Exploring Canada's Energy Future*; Canada Energy Regulator: Calgary, AB, Canada, 2019. Available online: https://apps2.cer-rec.gc.ca/dvs/?page=landingPage&language=en (accessed on 18 June 2020).
127. DECC. DECC 2050 Calculator. 2013. Available online: http://2050-calculator-tool.decc.gov.uk/#/calculator (accessed on 18 June 2020).
128. DTU. Klimaaftalen. 2020. Available online: https://klimaaftalen.tokni.com (accessed on 18 June 2020).
129. KAPSARC. KAPSARC Transport Analysis Framework (KTAF). 2020. Available online: https://ktaf.kapsarc.org/?locale=en (accessed on 25 June 2020).
130. CSIS Energy and National Security Program. Presentation of the National Energy Board's Canada's Energy Future 2017: Energy Supply and Demand Projections to 2040. 2018. Available online: https://www.youtube.com/watch?v=JGGMozEdQpI (accessed on 25 June 2020).
131. Ministerio de Energía de Chile. Energía Alterna' Interview—Ministerio de Energía: Planificación y Política Energética de Largo Plazo. 2018. Available online: https://www.energia.gob.cl/noticias/nacional/1er-conferencia-internacional-de-calor-y-frio-da-cuenta-de-la-estrategia-nacional-de-chile-en-esta-materia (accessed on 23 January 2022).
132. Government of the Kingdom of Eswatini. IRENA-Eswatini Energy Planning Capacity-Building Programme: Launch Event for the National Eswatini Energy Masterplan 2034. 2018. Available online: https://www.irena.org/events/2018/Oct/IRENA-Eswatini-Energy-Planning-Capacity-Building-Programme-Masterplan-2034-Launch (accessed on 23 January 2022).
133. BEIS. Mackay Carbon Calculator. Available online: https://mackaycarboncalculator.beis.gov.uk/overview/emissions-and-primary-energy-consumption/ (accessed on 23 January 2022).
134. DEA. Technology Data, Energistyrelsen. 2020. Available online: https://ens.dk/en/our-services/projections-and-models/technology-data (accessed on 24 June 2020).
135. NREL. Annual Technology Baseline. 2020. Available online: https://www.nrel.gov/analysis/data-tech-baseline.html (accessed on 9 August 2020).
136. Subhes, B.; Govinda, R.T.; Timilsina, R. A review of energy system models. *Int. J. Energy Sect. Manag.* **2010**, *4*, 494–518. [CrossRef]
137. Ringkjøb, H.-K.; Haugan, P.M.; Solbrekke, I.M.; Ringkjøb, H.-K.; Haugan, P.M.; Solbrekke, I.M. A review of modelling tools for energy and electricity systems with large shares of variable renewables. *Renew. Sustain. Energy Rev.* **2018**, *96*, 440–459. [CrossRef]
138. Stockholm Environment Institute. LEAP (Low Emissions Analysis Platform). 2021. Available online: https://leap.sei.org/ (accessed on 5 July 2021).
139. IEA-ETSAP. TIMES (The Integrated MARKAL-EFOM System). 2010. Available online: https://iea-etsap.org/index.php/etsap-tools/model-generators/times (accessed on 5 July 2021).
140. NREL. SAM (System Advisor Model). 2021. Available online: https://sam.nrel.gov/ (accessed on 5 July 2021).
141. EIA. NEMS (National Energy Modeling System). 2021. Available online: https://www.eia.gov/outlooks/aeo/info_nems_archive.php (accessed on 5 July 2021).
142. International Atomic Energy Agency. MESSAGE (Model for Energy Supply Strategy Alternatives and their General Environmental Impact). 2010. Available online: https://www-pub.iaea.org/MTCD/Publications/PDF/Pub1718_web.pdf (accessed on 5 July 2021).
143. KTH Royal Institute of Technology. OSeMOSYS (Open Source Energy Modelling System). 2008. Available online: http://www.osemosys.org/ (accessed on 5 July 2021).
144. Hilpert, S.; Kaldemeyer, C.; Krien, U.; Günther, S.; Wingenbach, C.; Plessmann, G. The Open Energy Modelling Framework (oemof)—A new approach to facilitate open science in energy system modelling. *Energy Strat. Rev.* **2018**, *22*, 16–25. [CrossRef]
145. Pfenninger, S.; Hirth, L.; Schlecht, I.; Schmid, E.; Wiese, F.; Brown, T.; Davis, C.; Gidden, M.; Heinrichs, H.; Heuberger, C.; et al. Opening the black box of energy modelling: Strategies and lessons learned. *Energy Strat. Rev.* **2018**, *19*, 63–71. [CrossRef]
146. SENER. Prospectivas del Sector Energético, Gobierno de México. 2018. Available online: http://www.gob.mx/sener/documentos/prospectivas-del-sector-energetico (accessed on 26 June 2020).
147. SENER. Estrategia Nacional de Energía, Gobierno de México. 2014. Available online: http://www.gob.mx/sener/documentos/estrategia-nacional-de-energia (accessed on 25 June 2020).
148. DEA Long Term Planning for a Greener Future. In Proceedings of the Long-Term Energy Scenarios (LTES) Campaign Webinar Series 2019, online, 21 March 2019. Available online: https://www.irena.org/renewables/Knowledge-Gateway/webinars/2018/Nov/Webinar-series-on-Long-term-Energy-Scenarios (accessed on 23 January 2022).
149. IRENA.; SENER. Renewable Energy Prospects: Mexico, REmap 2030 Analysis. 2015. Available online: https://www.irena.org/publications/2015/May/Renewable-Energy-Prospects-Mexico (accessed on 23 January 2022).
150. BEIS. *Quality Assurance: Guidance for Models*; The Department of Business, Energy and Industrial Strategy: London, UK, 2018. Available online: https://assets.publishing.service.gov.uk/government/uploads/system/uploads/attachment_data/file/737293/BEIS_QA_Guidance_for_Models.pdf (accessed on 23 January 2022).
151. PBL. Welcome to IMAGE 3.0 Documentation. 2014. Available online: https://models.pbl.nl/image/index.php/Welcome_to_IMAGE_3.0_Documentation (accessed on 26 June 2020).

152. PBL. *Climate and Energy Outlook 2019*; PBL Netherlands Environmental Assessment Agency: The Hague, The Netherlands, 2019. Available online: https://www.pbl.nl/en/publications/climate-and-energy-outlook-2019 (accessed on 23 January 2022).
153. BMU. Expansion Strategy for Renewable Energy, German Federal Ministry for the Environment, Nature Conservation and Nuclear Safety. 2007. Available online: https://elib.dlr.de/56730/1/Nitsch_Leitstudie_2007.pdf (accessed on 23 January 2022).
154. BDI. Klimapfade für Deutschland. 2018. Available online: https://bdi.eu/publikation/news/klimapfade-fuer-deutschland/ (accessed on 19 June 2020).
155. Dena. Integrierte Energiewende. 2020. Available online: https://www.dena.de/de/integrierte-energiewende/ (accessed on 19 June 2020).
156. Leopoldina, Acatech; UNION; BDI; Dena. *Expertise Bündeln, Politik Gestalten—Energiewende Jetzt!* Deutsche Energie-Agentur GmbH (dena): Berlin, Germany, 2019. Available online: https://www.dena.de/fileadmin/dena/Dokumente/Themen_und_Projekte/Energiesysteme/dena-Leitstudie/Expertise_buendeln_Studienvergleich.pdf (accessed on 23 January 2022).
157. Leopoldina; Acatech; UNION. "Sektorkopplung", Energiesysteme der Zukunft. 2020. Available online: https://energiesysteme-zukunft.de/themen/sektorkopplung/ (accessed on 19 June 2020).
158. UAE Ministry of Energy and Infrastructure. UAE National Energy Strategy 2050. In Proceedings of the IRENA LTES Webinar Series, Abu Dhabi, United Arab Emirates, 26 June 2019. Available online: https://www.irena.org/renewables/Knowledge-Gateway/webinars/2018/Nov/Webinar-series-on-Long-term-Energy-Scenarios (accessed on 23 January 2022).
159. IRENA. National Energy Masterplan Development. 2018. Available online: https://www.irena.org/energytransition/Energy-Planning-Support/National-Energy-Master-Plan-Development (accessed on 29 July 2020).
160. IRENA. *Energy Planning and Modelling Support in Africa*; International Renewable Energy Agency: Abu Dhabi, United Arab Emirates, 2020.
161. IEA-ETSAP. Energy Technology Systems Analysis Programme. In Proceedings of the Long-Term Energy Scenarios (LTES) Campaign Webinar Series 2019, online, 21 March 2019. Available online: https://www.irena.org/renewables/Knowledge-Gateway/webinars/2018/Nov/Webinar-series-on-Long-term-Energy-Scenarios (accessed on 23 January 2022).
162. UNFCCC. Paris Agreement, 2015. Available online: https://unfccc.int/files/meetings/paris_nov_2015/application/pdf/paris_agreement_english_.pdf (accessed on 23 January 2022).
163. Rocha, M.; Falduto, C. *Key Questions Guiding the Process of Setting up Long-Term Low-Emissions Development Strategies*; OECD Publishing: Paris, France, 2019. [CrossRef]
164. Roelfsema, M.; Van Soest, H.L.; Harmsen, M.; Van Vuuren, D.P.; Bertram, C.; Elzen, M.D.; Höhne, N.; Iacobuta, G.; Krey, V.; Kriegler, E.; et al. Taking stock of national climate policies to evaluate implementation of the Paris Agreement. *Nat. Commun.* **2020**, *11*, 2096. [CrossRef] [PubMed]
165. Van Vuuren, D.P.; Isaac, M.; Kundzewicz, Z.W.; Arnell, N.; Barker, T.; Criqui, P.; Berkhout, F.; Hilderink, H.; Hinkel, J.; Hof, A.; et al. The use of scenarios as the basis for combined assessment of climate change mitigation and adaptation. *Glob. Environ. Change* **2011**, *21*, 575–591. [CrossRef]
166. Bistline, J.; Budolfson, M.; Francis, B. Deepening transparency about value-laden assumptions in energy and environmental modelling: Improving best practices for both modellers and non-modellers. *Clim. Policy* **2021**, *21*, 1–15. [CrossRef]
167. Howells, M.; Quiros-Tortos, J.; Morrison, R.; Rogner, H.; Niet, T.; Petrarulo, L.; Usher, W.; Blyth, W.; Godínez, G.; Victor, L.F.; et al. Energy System Analytics and Good Governance—U4RIA Goals of Energy Modelling for Policy Support. *in review, preprint* **2021**. [CrossRef]
168. Debnath, K.B.; Mourshed, M. Challenges and gaps for energy planning models in the developing-world context. *Nat. Energy* **2018**, *3*, 172–184. [CrossRef]
169. Lopion, P.; Markewitz, P.; Robinius, M.; Stolten, D. A review of current challenges and trends in energy systems modeling. *Renew. Sustain. Energy Rev.* **2018**, *96*, 156–166. [CrossRef]
170. Prina, M.G.; Manzolini, G.; Moser, D.; Nastasi, B.; Sparber, W. Classification and challenges of bottom-up energy system models—A review. *Renew. Sustain. Energy Rev.* **2020**, *129*, 109917. [CrossRef]

MDPI

Article

It Is Still Possible to Achieve the Paris Climate Agreement: Regional, Sectoral, and Land-Use Pathways

Sven Teske [1,*], Thomas Pregger [2], Sonja Simon [2], Tobias Naegler [2], Johannes Pagenkopf [3], Özcan Deniz [3], Bent van den Adel [3], Kate Dooley [4] and Malte Meinshausen [4]

[1] Institute for Sustainable Futures, University of Technology Sydney (UTS), 235 Jones Street, Sydney, NSW 2007, Australia

[2] Department of Energy Systems Analysis, Institute of Engineering Thermodynamics, German Aerospace Center (DLR), Pfaffenwaldring 38–40, 70569 Stuttgart, Germany; Thomas.Pregger@dlr.de (T.P.); Sonja.Simon@dlr.de (S.S.); Tobias.Naegler@dlr.de (T.N.)

[3] Institute of Vehicle Concepts, German Aerospace Center (DLR), Pfaffenwaldring 38–40, 70569 Stuttgart, Germany; Johannes.Pagenkopf@dlr.de (J.P.); deniz.oezcan@dlr.de (Ö.D.); Bent.vandenAdel@dlr.de (B.v.d.A.)

[4] Australian–German Climate and Energy College, Level 1, 187 Grattan Street, University of Melbourne, Parkville, VIC 3010, Australia; kate.dooley@climate-energy-college.org (K.D.); malte.meinshausen@unimelb.edu.au (M.M.)

[*] Correspondence: sven.teske@uts.edu.au

Citation: Teske, S.; Pregger, T.; Simon, S.; Naegler, T.; Pagenkopf, J.; Deniz, Ö.; van den Adel, B.; Dooley, K.; Meinshausen, M. It Is Still Possible to Achieve the Paris Climate Agreement: Regional, Sectoral, and Land-Use Pathways. *Energies* **2021**, *14*, 2103. https://doi.org/10.3390/en14082103

Academic Editor: Dolf Gielen

Received: 25 February 2021
Accepted: 6 April 2021
Published: 9 April 2021

Publisher's Note: MDPI stays neutral with regard to jurisdictional claims in published maps and institutional affiliations.

Abstract: It is still possible to comply with the Paris Climate Agreement to maintain a global temperature 'well below +2.0 °C' above pre-industrial levels. We present two global non-overshoot pathways (+2.0 °C and +1.5 °C) with regional decarbonization targets for the four primary energy sectors—power, heating, transportation, and industry—in 5-year steps to 2050. We use normative scenarios to illustrate the effects of efficiency measures and renewable energy use, describe the roles of increased electrification of the final energy demand and synthetic fuels, and quantify the resulting electricity load increases for 72 sub-regions. Non-energy scenarios include a phase-out of net emissions from agriculture, forestry, and other land uses, reductions in non-carbon greenhouse gases, and land restoration to scale up atmospheric CO_2 removal, estimated at -377 Gt CO_2 to 2100. An estimate of the COVID-19 effects on the global energy demand is included and a sensitivity analysis describes the impacts if implementation is delayed by 5, 7, or 10 years, which would significantly reduce the likelihood of achieving the 1.5 °C goal. The analysis applies a model network consisting of energy system, power system, transport, land-use, and climate models.

Keywords: climate change; Paris Agreement; 100% renewable energy; 1.5 °C mitigation pathway; energy transition; energy scenario; GHG mitigation; CO_2 emission; non-energy emission; open access book

1. Introduction

Given the challenge that climate change poses for the global community, our research is dedicated to solutions for a low-emission society. To reach a net zero emission society in 2050, we develop normative emission pathways for a temperature rise well below 2 °C. Across various research disciplines, this scenario analysis combines climate, energy, transport, and land use models for a comprehensive picture of the tasks at hand, linking fossil energy emissions to non-energy-related GHG sources and sinks. We depict transition strategies for 100% renewable energy system in all 10 world regions, providing information of the necessary infrastructure, new capacity and investment, which would enable efforts by governments and society to keep climate change well below 2 °C and therefore in line with the Paris Climate Agreement. Earlier results of this research have been presented at Long-term-Scenarios for the Energy Transition (LTES) events [1].

29

Many scenarios have already been constructed and analysed to guide both policy and investment in limiting climate change 'by keeping global temperature rise this century well below 2 degrees Celsius above pre-industrial levels', according to the 2015 Paris Climate Agreement [2]. These published long-term scenarios [3–6] agree that the rapid decarbonization of energy production is required, together with significant negative emissions, throughout the 21st century. Many scenarios rely heavily upon nuclear power and natural-gas- or coal-fired power with carbon capture and storage (CCS) to decarbonize energy production, and negative emissions achieved with bio-energy with carbon capture and storage (BECCS). The 5th Assessment Report (5 AR) and the Special Report on Global warming of 1.5 °C (SR 1.5) of the Intergovernmental Panel on Climate Change (IPCC) also include a large number of representative concentration pathways (RCPs) illustrating a wide range of mitigation strategies [7,8].

Quantitative scenarios are usually constructed with modelling approaches, but always follow explicit or implicit 'if–then' narratives, which should never be understood as future predictions. In the global energy and emission pathways developed and discussed here, we use a storyline-and-modelling approach to make consistent assumptions about the implementation of technologies and to accommodate the multidimensional and multi-perspective character of the decision-making processes. Our intention is to increase the plausibility of the scenarios, rather than to identify supposedly cost-optimal solutions based on uncertain cost assumptions. With this approach, we develop narratives that target a society with net-zero CO_2 emissions by 2050 and construct exemplary normative scenarios that focus on the mitigation of CO_2 in the energy, agriculture, and land-use sectors.

These narratives represent a complementary basis for the difficult political and social decision-making processes required for the comprehensive decarbonization of energy systems. In contrast to previous studies, we have identified the technology paths that are suitable and necessary to achieve the decarbonization of the global energy system, with improvements in efficiency and 100% renewable energies only, by 2050. Limiting possible technologies and avoiding technical carbon dioxide removal (CDR) techniques are justified by the high potential utility of renewable energies and their low specific costs compared with those of nuclear and fossil power plants coupled to CCS and BECCS [9]. Moreover, the environmental effects and social acceptance of the latter options are highly contentious [10]: specifically, the unresolved disposal of radioactive waste in the case of nuclear power [11], and the unresolved doubts about the long-term effectiveness of underground storage of CO_2 in the case of CCS [12].

Therefore, the 2 °C and 1.5 °C scenarios presented here are 'non-overshoot' scenarios that use only widespread and publicly accepted technologies to generate renewable energies or produce green synthetic fuels. The scenarios also fulfil society's obligation to reduce its current energy-related emissions and limit the future energy demand. They meet the overall energy-related CO_2 emission budget of only 590 and 450 gigatonnes of CO_2 (Gt CO_2) respectively between 2015 and 2050, and also consider the non-CO_2 emissions and natural carbon sinks when estimating the overall greenhouse gas (GHG) emissions and related temperature increases. Most of the published 1.5 °C (low) overshoot pathways [13] include negative emission technologies, which buffer the heavy burden of energy transition in some way.

The pathways presented in this paper build upon a recently published scenario study [14]. The detailed assumptions, including technology and cost data, and the results tables can be found in the Supplementary Materials. Both pathways are considered to achieve targets 'well below 2.0 °C', with one representing the upper limit (*2.0 °C Scenario*) and one the lower limit (*1.5 °C Scenario*). The 'reference' (REF) scenario (*5.0 °C Scenario*) is based on the *Current Policies Scenario* published by the International Energy Agency (IEA) [15]. Using a comprehensive emissions accounting system, the pathways for the four major energy sectors—power, heat, transport, and industry—are based on GHG-mitigation strategies for 10 world regions, and focus on the distribution and consumption of energy and related emissions are the main driver of climate change [16]. The increased electrification of transport and heating systems, in order to replace fossil fuels with renewable electricity, plays an important role in those scenarios. The underlying solar and wind potentials were derived from a Geographic Information System (GIS)-based analysis of the required land area to avoid conflicts with other land uses, such as for natural carbon sinks (forests). To further investigate the use of installed capacities in the power systems, the system was modelled at high regional and temporal resolutions. Therefore, the 10 world regions were subdivided into 72 sub-regions to analyse their load developments and storage demands. Atmospheric GHG concentrations and radiative forcing, and their implications for global mean temperature and sea-level rises, were also analysed. To define a sustainable pathway for land-use change and the agricultural sector, we combined the investigation of future energy systems with measures for negative emissions provided by well-established natural land restoration methods. We used reduced-complexity carbon cycle and climate modelling to assess the climatic effects of the calculated emissions pathways. The analysis thus applies an integrated model network consisting of energy systems, power systems, transport, land use, and climate models.

2. Results and Discussion

2.1. Development of Energy Demand Intensities

Starting from the REF scenario, narratives for the demand side of the normative scenarios were developed. The main drivers of the final energy demand in the scenarios are population growth and economic development. The world's population is expected to grow from 7.4 billion in 2015 to 9.8 billion by 2050 [17]. It is assumed that the world's gross domestic product (GDP) will increase, on average, by 3.2% per year in the next three decades [15]. Therefore, our scenarios are based on improvements in efficiency and resulting reductions in demand (Table 1). The implementation of technical efficiency measures plays a significant role in the 1.5 °C Scenario, particularly before 2030. However, both the 1.5 °C and 2.0 °C Scenarios differ only slightly in their final annual energy demands in 2050. In both cases, efficiency measures are required to decouple economic growth and final energy consumption. Conversion losses are reduced, particularly by replacing thermal power generation with renewable technologies. This further reduces the primary energy intensity. The REF Scenario provides the lower benchmarks for efficiency potentials derived from the Current Policies Scenario of the IEA World Energy Outlook [15]. The upper benchmarks for the efficiency potentials for each world region are taken from the literature [18,19], including the low-energy-demand (LED) scenario [20,21]. In the transport sector, a combination of technical measures and modal shifts reduce annual passenger kilometres for private vehicles by 25% in OECD countries under the 1.5 °C scenario. The shift towards electric mobility might be driven by vehicle emission standards and economic incentives to phase out internal combustion engines. It is expected that the acquisition costs for electric cars will be similar to those for cars with combustion engines during the next decade and that maintenance costs will become increasingly competitive.

Table 1. Main strategies and narratives for all regions in each sector of the energy system for the 2.0 °C and 1.5 °C Scenarios compared with the Reference Scenario.

Sector	Main Strategies and Narratives with Different Regional Emphases and Characteristics	Global Average sectoral Demand Intensity	
Industry electricity	Implementation of more-efficient appliances, especially electric drives for compressed air, pumps, fans, and other cross-sectional technologies.	kWh/USD1000 GDP 2015: 55 2050 REF: 36 2050 2.0 °C: 24 2050 1.5 °C: 23	
Industry heating	Electrification of industrial heat will increase from 6% to 34% in 2050 in the 2.0 °C Scenario and to 37% in the 1.5 °C Scenario. Technological improvements, process substitutions, and innovations will be encouraged by favourable conditions and regulative frameworks, allowing rapid technological changes. Integration of waste heat into processes will reduce losses.	MJ/USD1000 GDP 2015: 690 2050 REF: 366 2050 2.0 °C: 185 2050 1.5 °C: 172	
Other sectors (*) electricity	Electricity demand intensities in households, for commercial purposes, and in the service and trade sectors, fisheries, and agriculture will be reduced by the use of most-efficient technologies for lighting, information, communication, cooking, cooling, and hot water. Compared with the REF case, a reduction in specific consumption (depending on region, a slower increase resp.) is assumed for the 1.5 °C pathway over the medium term, as long as fossil power generation dominates.	kWh/USD1000 GDP 2015: 78 2050 REF: 60 2050 2.0 °C: 38 2050 1.5 °C: 37	kWh/capita 2015: 1350 2050 REF: 2370 2050 2.0 °C: 1500 2050 1.5 °C: 1460
Other sectors (*) heating	Share of electric heating will rise from 5% in 2015 to 30% in 2050 in the 2.0 °C Scenario and to 37% in the 1.5 °C Scenario. Final energy use for heating will be reduced and switching heating to low-temperature technologies, such as heat pumps and floor heating. These measures are supplemented with responsible consumption behavior by the consumer, especially in the 1.5 °C Scenario.	MJ/USD1000 GDP 2015: 700 2050 REF: 300 2050 2.0 °C: 180 2050 1.5 °C: 170	MJ/head 2015: 12,600 2050 REF: 11,700 2050 2.0 °C: 7300 2050 1.5 °C: 6700
Transport	Main strategies include electrification and synthetic fuels (hydrogen and synthetic liquid hydrocarbons), depending on the transportation mode. Mode shifts from road and air to more-efficient rail and bus will reduce the share of energy-intensive motorized private transport. Efficiency gains for engines and a moderate use of biofuels will also help to achieve rapid and strong emissions reductions.	MJ/USD1000 GDP 2015: 760 2050 REF: 380 2050 2.0 °C: 130 2050 1.5 °C: 100	MJ/head 2015: 13,000 2050 REF: 15,000 2050 2.0 °C: 5100 2050 1.5 °C: 3900

Narratives are similar in all regions, whereas the demand intensities differ significantly—for both the base year 2015 and the end of the modelling period in 2050. (*) Other Sectors include buildings (residential, commercial, and public services) and the agricultural, forestry, and fishing sectors.

2.2. Demand and Supply Pathways towards +2 °C and +1.5 °C Targets

The role of energy efficiency in decarbonization scenarios is widely documented. The IPCC [8] concluded that '*at the global level, scenarios reaching about 450 ppm CO_2eq are (also) characterized by more rapid improvements in energy efficiency*' and Lovins [22] identified energy efficiency among the most cost-effective ways to reduce carbon emissions. As well as reducing the energy demand by improving energy efficiency, the two energy decarbonization pathways are based on expanding renewable energy supply technologies [21]. Figure 1 shows the resulting final energy demands by sector and scenario, and the primary energy by energy carrier. The measures documented in Table 1 will reduce the total final energy demand to below 280 EJ in 2050 compared with around 540 EJ in the 5 °C (REF) case. Accordingly, annual global primary energy use decreases from 556 EJ in 2015 to about 440 EJ by 2050 under the 2.0 °C Scenario and to 412 EJ under the 1.5 °C Scenario. Both scenarios are non-overshoot scenarios with no CDR technologies, so a rapid reduction in fossil fuels and a significant deployment of renewable energies would already be necessary by 2025. Solar and wind power are the backbones of such an energy system, with complementary contributions from hydro, biomass, and geothermal energy (Figure 1). Compared with today, the installation of renewables-based power and heat generation technologies accelerates significantly.

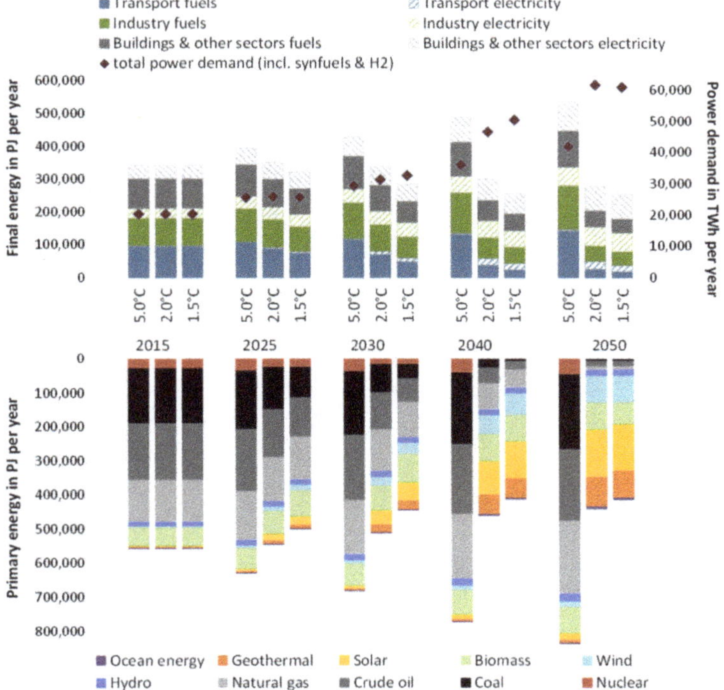

Figure 1. Final energy demand per sector, gross power demand (upper panel), and primary energy supply, including non-energy use (bottom panel), in the scenarios.

Transformation of the Transport Sector

An increase in the efficiency of vehicles with internal combustion engines and a direct electrification rate of 50% by 2050 on world average (Table 2) would be necessary to decrease the final energy consumption of the transport sector by more than 60%, as required for the 2 °C Scenario.

Table 2. Proportions of the final sectoral energy demands met by electricity under the 1.5 °C Scenario. Colors indicate the different electrification shares where red is lower and green is higher.

		2015	2020	2025	2030	2035	2040	2045	2050
DEMAND Transport Electricity Share [%]	OECD North America	0%	0%	5%	18%	35%	45%	51%	54%
	Latin America	0%	0%	2%	9%	36%	48%	53%	53%
	OECD Europe	0%	0%	6%	35%	57%	67%	65%	65%
	Africa	0%	0%	1%	3%	12%	19%	25%	31%
	Middle East	0%	0%	1%	4%	21%	32%	38%	41%
	Eurasia	0%	1%	3%	13%	39%	44%	47%	46%
	Non-OECD Asia	0%	0%	3%	10%	29%	35%	37%	37%
	India	0%	1%	5%	22%	47%	58%	57%	57%
	China	4%	4%	10%	29%	52%	62%	60%	60%
	OECD Pacific	0%	0%	8%	33%	58%	61%	62%	63%
	Global average	1%	1%	5%	17%	38%	46%	49%	50%
DEMAND Industry Electricity Share [%]	OECD North America	29%	28%	28%	32%	40%	49%	51%	55%
	Latin America	23%	23%	25%	29%	33%	40%	47%	56%
	OECD Europe	35%	35%	36%	39%	43%	46%	48%	51%
	Africa	26%	25%	26%	30%	36%	42%	47%	52%
	Middle East	9%	10%	12%	16%	22%	30%	36%	44%
	Eurasia	23%	24%	23%	30%	35%	40%	43%	45%
	Non-OECD Asia	24%	25%	25%	31%	37%	40%	45%	49%
	India	18%	20%	21%	29%	38%	48%	54%	56%
	China	26%	29%	32%	38%	47%	55%	58%	61%
	OECD Pacific	35%	36%	36%	41%	46%	50%	53%	56%
	Global average	26%	27%	28%	33%	40%	47%	51%	54%
DEMAND Buildings Electricity Share [%]	OECD North America	50%	49%	50%	55%	57%	58%	60%	61%
	Latin America	38%	40%	44%	48%	55%	62%	69%	76%
	OECD Europe	31%	33%	34%	38%	47%	49%	51%	53%
	Africa	8%	8%	11%	15%	22%	32%	40%	49%
	Middle East	43%	46%	48%	51%	54%	59%	68%	75%
	Eurasia	18%	19%	20%	23%	26%	28%	31%	35%
	Non-OECD Asia	22%	24%	25%	31%	40%	48%	56%	61%
	India	17%	19%	26%	34%	43%	55%	61%	69%
	China	24%	27%	32%	41%	53%	58%	63%	66%
	OECD Pacific	52%	53%	54%	55%	56%	62%	63%	64%
	Global average	30%	31%	33%	38%	45%	50%	55%	60%

Table 2. *Cont.*

		2015	2020	2025	2030	2035	2040	2045	2050
	OECD North America	10%	12%	35%	65%	86%	96%	99%	100%
	Latin America	33%	39%	53%	67%	84%	96%	100%	100%
	OECD Europe	17%	20%	34%	56%	71%	87%	95%	100%
	Africa	58%	57%	62%	68%	79%	90%	99%	100%
GENERATION	Middle East	1%	4%	13%	27%	54%	79%	99%	100%
Renewable Electricity Share	Eurasia	6%	9%	21%	41%	62%	79%	91%	100%
[%]	Non-OECD Asia	32%	31%	43%	62%	77%	86%	95%	100%
	India	36%	33%	46%	65%	82%	91%	97%	100%
	China	12%	16%	29%	49%	70%	83%	93%	100%
	OECD Pacific	6%	11%	25%	49%	71%	85%	94%	100%
	Global average	18%	21%	35%	56%	74%	88%	96%	100%

An even steeper reduction in the transport demand and more-drastic efficiency-improvement measures (electrification, modal shifts) are required under the 1.5 °C Scenario. In this scenario, the energy demand of the global transport sector is 74% lower in 2050 than under the 5 °C Scenario. Electrification of the vehicle stock and altering transport modes to reduce energy-intensive transport activities will have a high decarbonization effect. However, measures such as the expansion of public transport and the vehicle-sharing infrastructure are equally important. The direct electrification of air and ship transport is particularly limited, so under the 2.0 °C and 1.5 °C Scenarios, electricity-based synthetic fuels increasingly replace fossil fuels in these sectors.

2.3. Regional Differences

How much each renewable energy source supplies to the total demand depends on regional opportunities for and constraints upon deploying renewable energies [21,23]. In the 1.5 °C Scenario, solar heat and power technologies provide over 50% of the total primary energy demand in the global sun belt, because solar is readily available and comparatively inexpensive (see Figure 2). The Middle East exemplifies the solar region model. With low biomass and hydro resources, the Middle East will rely on the development of dispatchable technologies, such as hydrogen production or concentrated solar power, which will help to store the abundance of solar energy. In contrast, in Europe and Eurasia, with their long cold winters, solar contribute only ~20%. Latin America represents a 'bioenergy and hydro region', where biomass provides easily accessible heat and hydro provides dispatchable (balancing) power. A previous detailed analysis for Brazil [24] supports this approach.

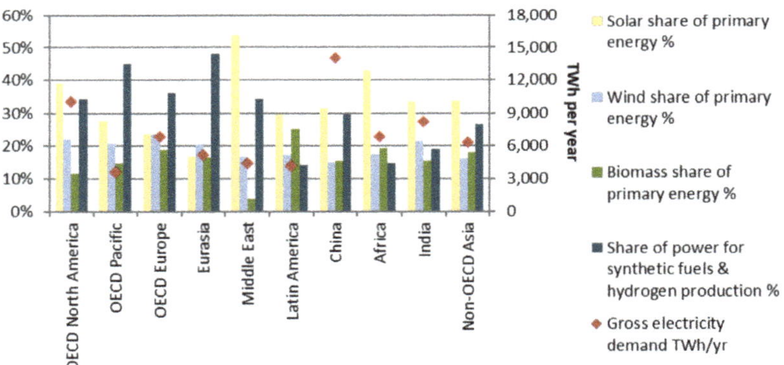

Figure 2. Selected regional characteristics of energy demand and supply in the 1.5 °C Scenario in 2050.

Synthetic fuels, including hydrogen, will become increasingly relevant under the carbon constraint when the share of renewables exceeds 80% (Figure 2). The option to store and distribute energy carriers (hydrogen or synthetic fuels) will be important in this case, especially in regions in which both solar and biomass resources are limited. Eurasia would dedicate almost 50% of its power production for this purpose in the 1.5 °C Scenario. China has the highest power production of all regions in our 1.5 °C Scenario, and would require around 4000 TWh per year in 2050 to generate synthetic gases and fuels to balance the variable renewables, to provide off-season power, and to indirectly electrify the process heat and transport sectors. This large power demand is a special challenge. Recent studies have shown that the regionally integrated deployment of renewable technologies can technically manage this transition, even for the eastern demand centres [25,26]. Because of its industrial lifestyle, North America already has a high proportion of electrification and a large power demand. Therefore, it will also require large amounts of synthetic fuels in our

1.5 °C Scenario. India is another key region for global development and will rely strongly on solar energy for its transformation. In contrast to the various regional requirements for solar, biomass, and synthetic fuels, wind would provide a stable share of 15%–20% in all regions, even under quite different regional assumptions for the synthetic fuel demand.

2.4. Sector Coupling—Electrification Replaces Thermal Processes

Electrification is a key to replacing fossil fuels in thermal processes and combustion engines in all sectors under both high-renewables scenarios. Electrification shares of over 50% of final energy by 2050 (Table 2) will rely on a rapid increase from 21% in 2015 to 38% by 2030, significantly increasing the interactions and interdependencies between power production and power consumption and storage in the transport, industry, and building sectors (Figure 3). As a consequence, the global annual electricity demand increases by 13,600 TWh between 2015 and 2035 and by 7890 TWh between 2035 and 2050 under the 1.5 °C Scenario. Therefore, fossil-fuel-based power generation would be replaced by renewables-based generation, increasing the latter by a factor of 7 between 2015 and 2035. By 2050, ~62,300 TWh per year would be generated from renewables under the 1.5 °C Scenario. This is also required to substantially supply the transport sector. The proportion of electricity in the total final energy used by the transport sector in 2015 was less than 1%, although the proportion was markedly higher in China at 4%, nearly half of which was attributable to the rail system and half to 2- and 3-wheeler vehicles and buses. Under the 1.5 °C Scenario, 38% of the final global transport energy demand needs to be electrified by 2035, although this varies greatly between regions. The industry and building sectors need to double their electrification rates, for both space and process heat, to meet the scenario targets.

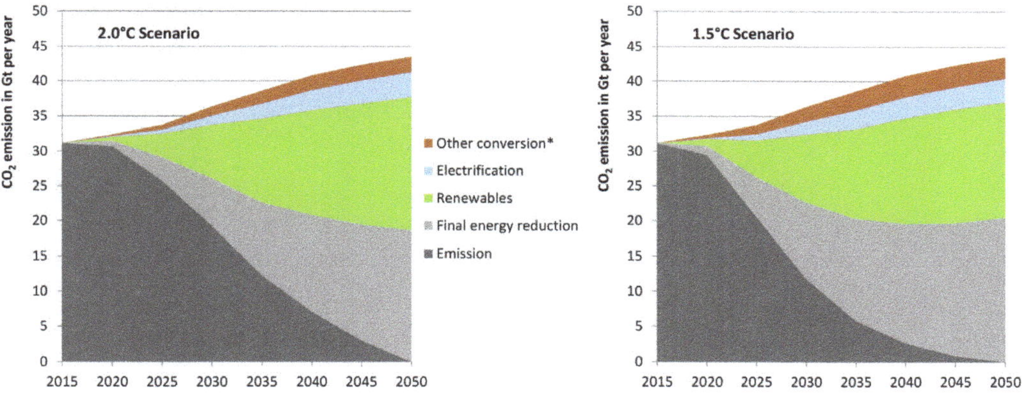

Figure 3. Annual CO_2 emissions reductions in the 2.0 °C (left) and 1.5 °C Scenarios (right) in relation to the 5 °C (REF) Scenario according to the measures implemented. Renewable energies and efficiency measures (including efficiency improvements through electrification) are roughly equally important in all the scenarios. Reduced consumption has an even greater role in the 1.5 °C Scenario. Other conversions (*) include changes in district heating, refineries, coal transformation, and gas transport.

The pace of electrification will differ significantly between the regions. In our scenarios, China, Europe, and OECD Pacific are expected to take the lead in transport-sector electrification because environmental policy incentives are already emerging. In the building sector, electrification will be easiest for regions with low space-heat demands, in warmer climates. In areas with cold winters, such as Eurasia and Europe, investments in heat pumps are required to make electrification possible. Heat grids that integrate biomass, waste heat, and solar collectors are an efficient alternative, but both options require huge improvements in insulation to curb the energy demand. In sub-Saharan Africa, electrification of the building and transport sectors are particular challenging because the rates of electrification and urbanisation are low. However, recent observations indicate that there have been significant improvements in electrification in both rural and urban areas since 2014 [27].

2.5. Power Sector Analysis: Development of Electric Load and Storage Demand

Because electrification is a key transition strategy and wind and photovoltaic sources are highly volatile, we specifically focused on balancing the power systems with historic regional solar and wind data [28,29]. The modelling results show that the average loads increase in all 72 regions under both scenario alternatives. Under the 1.5 °C Scenario, the most significant increases and largest regional differences occur in Africa, where the average load increase between 480% in the northern regions and 750% in the southern regions, reflecting the significant regional differences in access to electricity and the electrification of the transport sector (Figure 4). In OECD Pacific, efficiency measures reduce the average load by 87% in 2030 compared to 2015. By 2050 however, load increases to 116%, as electric mobility and electric process heat in industry are added as new consumers. In most regions, the electrification and thus the load is expected to increase more under the stronger limitations of the 1.5 °C Scenario than in the 2.0 °C Scenario. Only if Middle East, India, and Non-OECD Asia leapfrog on efficiency measures, demand at the end of the modelling period can be leveled out. Flexibility measures, such as fast-reacting dispatch generation capacities and demand-side management, are used in our scenarios to reduce the need for additional transmission and storage capacities, but will not replace them entirely. Under the 2.0 °C Scenario, the global pumped hydro storage capacities increase by 6 GW and battery capacities by 0.8 GW annually between 2015 and 2030, to 244 GW and 12 GW, respectively. By 2050, pumped hydro will increase to 267 GW and batteries to 347 GW of the total installed capacity. By 2050, 197 GW of gas power plants and combined heat and power generation (CHP) capacity will either consume synthetic methane or be retrofitted for hydrogen use. In parallel, the average capacity factor for gas and hydrogen plants will decrease from 29% (around 2600 h/yr) in 2030 to 11% (just under 1000 h/yr) by 2050, providing dispatch power and ancillary services.

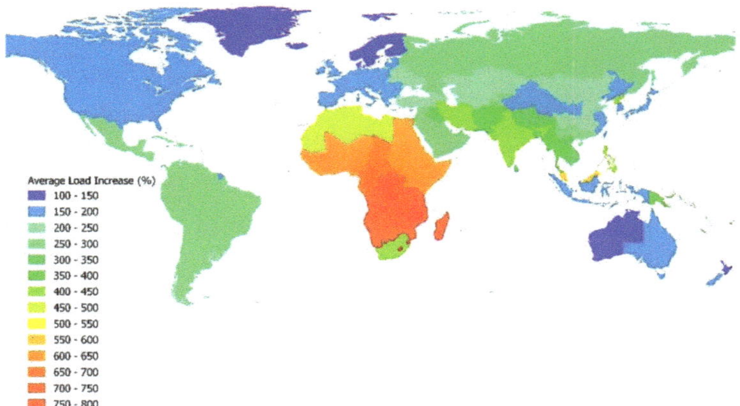

Figure 4. Increases in the average calculated load by 2050 in 72 regions under the 1.5 °C Scenario, in percentages relative to 2020. The average load was calculated across 8760 h per year. The regional ratios between the maximum and minimum loads vary significantly. 'Residual load' in this analysis is the load remaining after the generation of variable renewable power. Negative values indicate that the power generated from solar and wind exceeds the actual load and are exported to other regions, stored, or curtailed. The residual load varies significantly with increased variable generation because maximum load and maximum generation do not occur simultaneously.

2.6. Investment Required and Fuel Cost Savings

By 2050, electricity and synthetic fuels (including hydrogen) will supply 70% of the global final energy required in the 1.5 °C pathway. The overall cumulative investment in power generation required up to 2050 in our scenario is USD 51.1 trillion (USD 1.42 trillion annually on average), which is USD 30.7 trillion more than under the REF scenario, under which an investment of USD 20.4 trillion (USD 0.58 trillion annually) will be required. The overall fuel cost savings in the same scenario will add up to USD 28.8 trillion over the same period, or USD 0.8 trillion per year. Total fuel cost savings in the 1.5 °C pathway alone will cover 90% of the additionally necessary investments in renewable power generation in the 1.5 °C pathway.

The *levelized costs of electricity* (LCOE) of the global power sector under the REF Scenario (without including the costs of CO_2 emissions) are calculated to increase from USD 60 per MWh in 2015 to USD 79 per MWh in 2050. In comparison, the 2.0 °C Scenario will increase the generation costs to USD 77 per MWh by 2030, with a following reduction to USD 70 per MWh by 2050. Under the 1.5 °C Scenario, electricity generation costs will peak at USD 81 per MWh and decrease to USD 70 per MWh—equal to that in the 2.0 °C Scenario.

According to a recent market survey [9], the current LCOE is USD 192 per MWh for nuclear generation, USD 152 per MWh for coal-fired power generation, and USD 68 for gas-fired power (excluding CCS costs). As of 2019, there are two CCS facilities combined with power generation—the 115 MW coal-fired Boundary Dam plant in Canada [30] and the 240 MW gas-fired Petra Nova plant in the USA [31], at which CO_2 capture per tonne costs approximately USD 100 and USD 65, respectively, [32] although only a portion of all fugitive CO_2 emissions is captured. These are significantly higher than the cost of power generation from carbon-neutral renewables—USD 42 per MWh for utility-scale solar photovoltaic and USD 54 per MWh for onshore wind. The cost for sequestration with BECCS is approximately USD 100–200 per tonne CO_2 [33], and it has a limited mitigation potential of 1 Gt CO_2 per year [34]. In comparison, the cost of natural land restoration is <USD 100 per tonne CO_2, with an average potential of 7 Gt CO_2 per year [35,36].

2.7. Distribution of Carbon Emissions

We performed an ex-post analysis of the distribution of CO_2 emissions in the scenarios based on the technical transitions in the energy system. Compared with the IPCC RCPs, the pathways fall within the P1 category (IPCC SR1.5—P1 scenarios are defined as scenarios with lower energy demand up to 2050, due to innovations in social life, business, and technology. At the same time living standards increase and levelize. A leaner energy system facilitates rapid decarbonization of energy supply. Afforestation is the only CDR option considered; neither fossil fuels with CCS nor BECCS are used. OECD Pacific: Japan, South Korea, Australia, and New Zealand. The calculation of inter-regional exchange capacity requirements in MW is also possible, but beyond the scope of this article). Efficiency and electrification strategies strictly limit CO_2 emissions in both low temperature rise scenarios. The cumulative energy-related CO_2 emissions under the 5.0 °C Scenario between 2015 and 2050 are 1341 Gt CO_2, about three times higher than those under the 1.5 °C Scenario (449 Gt CO_2). The OECD regions, China, and India account for over 60% of all emissions under all scenarios, and the cumulative emissions of the combined OECD countries equal those of China. In the low-emission scenarios, the power sector dominates, accounting for one third of all cumulative energy-related carbon emissions, predominantly arising from the necessary phase-out times required for recently built fossil/coal-fired power plants. The industry and transport sectors follow, accounting for 20–25% each. The building/other sectors contribute 10% of carbon emissions under our scenarios. The carbon intensities for all sectors are shown in Table 3. The proportions of renewable electricity generated increase, leading to significant reductions in carbon intensity on the supply side, a prerequisite for low carbon intensities in all other sectors. Carbon emissions then plateau by 2025 and decrease thereafter in the 1.5 °C pathway.

Table 3. Carbon intensity by sector under the 1.5 °C Scenario. Carbon intensities for the industry, building, and transport sectors exclude the electricity consumed in these sectors. The prerequisites for reduced carbon intensity in the industry sector include infrastructural changes, such as renewables-based process heat generation technologies and co-generation. For the transport sector, infrastructural changes are required, such as charging networks for electric vehicles and the expansion of electricity-based public transport. The colour indicates the different carbon intensities by sector and region where red is high, yellow more average values and green low carbon intensity.

t CO$_2$/PJ		2015	2020	2025	2030	2035	2040	2045	2050
OECD North America	TRANSPORT	30,567	29,239	19,312	9134	2994	712	222	0
Latin America		24,086	22,260	17,317	12,460	4682	319	5	0
OECD Europe		21,353	20,018	15,026	5352	1563	18	0	0
Africa		14,023	13,800	13,219	12,615	8969	5274	401	0
Middle East		24,556	24,670	21,164	17,143	8102	2353	6	0
Eurasia		12,571	12,695	11,301	8300	2861	11	6	0
Other Non-OECD Asia		19,305	19,673	16,560	11,117	4322	1269	0	0
India		11,297	12,300	10,587	5288	1765	513	0	0
China		11,914	13,711	10,307	6964	2508	334	2	0
OECD Pacific		22,475	20,662	13,432	6623	2007	445	0	0
Global average		19,648	19,331	14,482	8878	3641	1057	75	0
OECD North America	INDUSTRY	42,072	43,611	35,094	21,504	11,578	4247	853	0
Latin America		36,945	32,813	22,414	13,092	6328	2177	125	0
OECD Europe		40,393	38,716	32,526	23,713	16,861	10,115	3796	0
Africa		37,228	35,765	28,767	17,582	10,537	3384	191	0
Middle East		55,432	51,509	46,876	38,930	30,223	17,373	907	0
Eurasia		44,482	42,274	29,360	19,198	14,782	9877	5000	0
Non-OECD Asia		51,422	51,069	39,170	24,973	13,618	8979	3935	0
India		52,443	51,840	40,147	24,769	11,766	6214	2782	0
China		83,734	81,910	68,450	51,480	30,655	18,203	7727	0
OECD Pacific		47,587	46,810	37,406	26,349	16,443	8356	3702	0
Global average		58,941	56,882	46,116	32,117	18,837	10,485	3870	0
OECD North America	BUILDINGS	28,813	28,158	19,287	10,353	3829	930	42	0
Latin America		24,869	19,539	14,038	9339	5313	1371	167	0
OECD Europe		32,810	30,840	23,200	13,997	9554	5002	1971	0
Africa		8112	7962	6265	4499	2819	779	77	0
Middle East		32,690	31,171	27,763	23,070	17,403	8489	313	0
Eurasia		13,241	13,020	8139	3469	2812	1891	878	0
Non-OECD Asia		19,648	19,331	14,482	8878	3641	1057	75	0
India		16,926	14,363	9111	5303	1672	1235	640	0
China		35,078	30,321	20,007	6094	4745	1242	269	0
OECD Pacific		29,442	28,004	23,326	17,943	8146	4799	1509	0
Global average		25,907	23,925	16,946	9702	5742	2462	631	0

Table 3. *Cont.*

	t CO$_2$/PJ	2015	2020	2025	2030	2035	2040	2045	2050
OECD North America		139,196	125,786	50,187	14,120	5751	1577	119	0
Latin America		78,282	54,046	26,633	10,838	4470	1563	0	0
OECD Europe		97,367	73,323	40,254	22,912	16,018	7056	2608	0
Africa		204,093	173,995	109,773	52,020	16,706	2978	156	0
Middle East		213,418	203,593	144,467	86,342	22,917	5718	243	0
Eurasia	POWER	208,031	157,983	109,225	56,858	36,448	23,753	11,015	0
Non-OECD Asia		177,243	168,415	92,100	33,049	22,213	14,113	5417	0
India		279,508	234,522	116,945	52,299	21,537	3894	1473	0
China		142,179	115,016	77,626	34,766	8703	4726	2216	0
OECD Pacific		155,566	116,407	69,032	33,201	16,628	11,498	5344	0
Global average		150,579	127,401	74,485	34,763	14,788	6423	2353	0

2.8. Land-Use and Non-CO$_2$ Emission Mitigation Scenarios

2.8.1. Land-Sector Emissions

The land-sector emissions presented here are derived from a new probabilistic scenario based on four different land restoration pathways: reforestation, forest ecosystem restoration, sustainable use of forests, and agroforestry [37]. These pathways are based on the premise that the better management of terrestrial ecosystems, including the restoration of degraded natural ecosystems, will allow previously lost carbon stocks to be restored [38–40]. The global aggregated sequestration potential was calculated from the median values for an ensemble of draws for each sequestration pathway and climatic domain (temperate/boreal or tropical/sub-tropical), resulting in a theoretical potential of 151.9 Gt of carbon (C) by 2150 and a maximum carbon density cap of 377 Gt CO$_2$ to 2100 [41]. The four sequestration pathways were aggregated from country-level data for the five Representative Concentration Pathway (RCP) regions (Table 4), and can be considered to approximate biome-average sequestration rates if they are supported by specific land-use policies.

Table 4. Net carbon mitigation from land-use management pathways for 1.5 °C: 2020–2100.

Region [42]	Gt C/year	2020	2030	2040	2050	2060	2070	2080	2090	2100
Asia		0.30	0.05	−0.32	−0.36	−0.35	−0.30	−0.25	−0.16	−0.10
Eastern Europe and Former Soviet Union (REF)		0.00	−0.13	−0.27	−0.28	−0.27	−0.26	−0.25	−0.22	−0.19
Middle East and Africa (MAF)	LAND-USE	0.33	−0.19	−0.55	−0.57	−0.53	−0.42	−0.29	−0.14	−0.06
OECD 1990 Countries (OECD 90)		0.00	−0.18	−0.34	−0.34	−0.32	−0.28	−0.23	−0.18	−0.14
Latin America and Caribbean (LAM)		0.17	−0.27	−0.62	−0.62	−0.55	−0.42	−0.27	−0.14	−0.06
Annual global total		0.79	−0.81	−2.11	−2.17	−2.01	−1.68	−1.28	−0.84	−0.56
Cumulative global total		0.79	0.63	−15.35	−37.17	−58.20	−76.65	−91.18	−101.4	−108.25

Under the 1.5 °C pathway analysis [37], the effects of these different land-use options will sequester up to 32 Gt C by mid-century. The full extent of the net mitigation shown in Table 4 is required to achieve the 1.5 °C Scenario, whereas for the 2.0 °C Scenario, only a third of the sequestration potential is required. The 1.5 °C pathway is consistent with comparable scenarios in the literature [41], which showed mitigation rates of up to −2 Gt C per year from 2040 to 2050. The land-use-related emission and sequestration rates of the 2.0 °C and 1.5 °C pathways in the present study are within the range of currently published scenario distributions (CMIP6 CEDS and IPCC SR1.5 database [13]).

2.8.2. Non-CO$_2$ Emissions

Non-CO$_2$ emissions were modelled based on the other main GHGs (CH$_4$ and N$_2$O), fluorinated gases, and aerosols. The pathways for CH$_4$ and N$_2$O emissions were derived with a quantile regression method [37], resulting in long-term emission levels that track towards the lower end of the distributions of published scenarios (see above CMIP6 CEDS and IPCC SR1.5 database). They show a decline and plateau in CH$_4$ emissions and a slight increase in N$_2$O emissions over the course of the century, associated with agricultural activities [41]. Our quantile regression method assumes a phase out of halocarbon and fluorinated gases over the next 10–20 years, although it does not include the residual levels of background emissions [41]. In our 1.5 °C scenario, sulphate aerosol emissions are set below the SSP1 1.9 scenario, whereas NO$_x$ emissions are between the levels in the SSP 1 2.6 and SSP 1 1.9 scenarios [41]. Emissions of black and organic carbon are not as low as those in the lower SSP scenarios because these emission sources correlate less strongly with fossil-fuel burning, and a reduction in both black and organic carbon emissions will offset the warming and cooling effects of each [41].

2.9. Sensitivity Analysis: The Risk of Delay and the Possible Impact of COVID-19

At the time of writing (June 2020), the COVID-19 pandemic had reduced the global energy demand in an unprecedented way. Initial projections for 2020 [43] estimate a drop in the global primary energy demand of 5%. The oil and coal demands are projected to decline by 8% each, in response to reduced transport services and industrial activities, respectively. Global gas consumption is anticipated to decrease by 4%. Overall energy-related CO_2 emissions are expected to fall by around 8% in 2020. If these forecasts come true, the global energy sector would be almost exactly on the 1.5 °C pathway in terms of the energy demand and overall fossil fuel consumption. Compared with the 1.5 °C Scenario assumptions for 2020, the actual use of coal in 2020 would be 3% lower and that of oil 2% lower, whereas the gas demand would still be 2% higher. However, the energy-related CO_2 emissions would still be 1.1 Gt above those required in the 1.5 °C pathway for 2020. In terms of the electricity-generating capacities from renewable energies, both solar photovoltaic and wind power are consistent with the 1.5 °C trajectory if the market volume for new installations of both technologies in 2019 (115 GW for photovoltaic and 60 GW for wind [44]) are maintained in 2020. However, these developments are being affected by the COVID 19 pandemic, and a short-term decline in technology expansion might be possible. The extent to which the pandemic and the subsequent efforts to revive the global economy can support long-term changes in policy or a restructuring of the global economy remains to be seen. Various socio-economic storylines assume that a rapid return to "business as usual" will prevail in many areas [45].

The delayed implementation of permanent measures that tend us towards the 1.5 °C pathway will lead to additional energy-related carbon emissions. In this analysis, we assume that the calculated CO_2 reduction pathways (2.0 °C and 1.5 °C) will begin 5, 7, or 10 years later than anticipated, and that emissions during that time will remain at the level of the year 2019. In this section, we quantify the additional cumulative carbon emissions that will result from these delays. The energy sector itself will be unable to compensate for those emissions, but will have to rely on society's willingness to pay for net emission reduction technologies, such as BECCS and DACCS, and their inherent additional energy demands. Figure 5 shows the results (in billion metric tonnes) for the 10 world regions. If China delays the implementation of the 1.5 °C pathway by 5 years, an additional 45 billion tonnes of CO_2 will be released, more than the total annual global CO_2 emissions (33 billion tons) in 2019 [46]. The global CO_2 budget under the 1.5 °C Scenario (66% probability) will be surpassed by 13% if all OECD countries delay their decarbonization pathways by 5 years. The cumulative CO_2 emissions of China will equal those of all OECD countries, whereas those of India will equal those of OECD Pacific (OECD Pacific: Japan, South Korea, Australia, and New Zealand) (2015–2050, 1.5 °C scenario). Figure 5 shows the impact on global CO_2 emissions if a whole sector delays the implementation of the 1.5 °C decarbonization pathway. A 5-year delay by the power sector will result in 50 billion tonnes of additional CO_2 emissions.

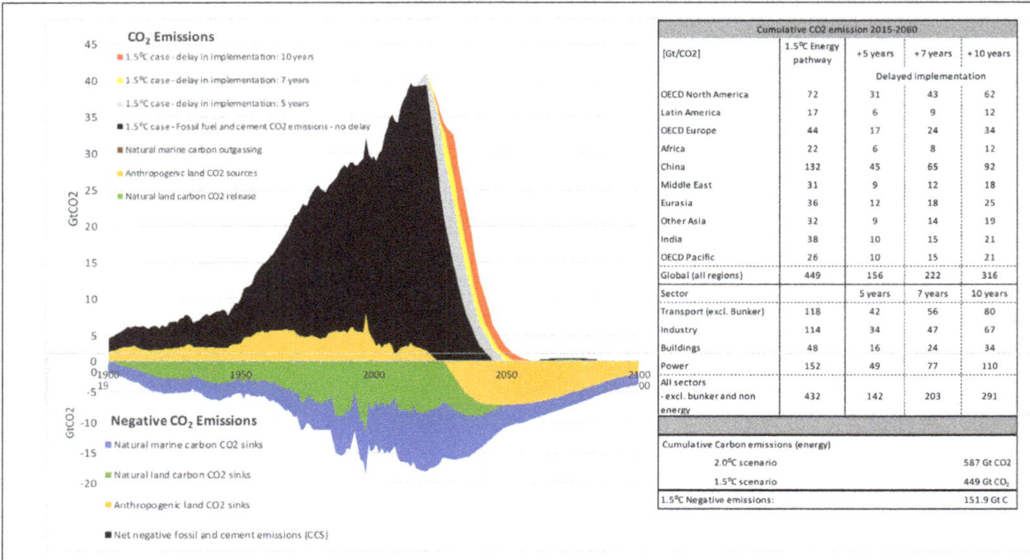

Cumulative CO2 emission 2015-2060				
[Gt/CO2]	1.5°C Energy pathway	+5 years	+7 years	+10 years
		Delayed implementation		
OECD North America	72	31	43	62
Latin America	17	6	9	12
OECD Europe	44	17	24	34
Africa	22	6	8	12
China	132	45	65	92
Middle East	31	9	12	18
Eurasia	36	12	18	25
Other Asia	32	9	14	19
India	38	10	15	21
OECD Pacific	26	10	15	21
Global (all regions)	449	156	222	316
Sector		5 years	7 years	10 years
Transport (excl. Bunker)	118	42	56	80
Industry	114	34	47	67
Buildings	48	16	24	34
Power	152	49	77	110
All sectors - excl. bunker and non energy	432	142	203	291

Cumulative Carbon emissions (energy)	
2.0°C scenario	587 Gt CO2
1.5°C scenario	449 Gt CO$_2$
1.5°C Negative emissions:	151.9 Gt C

Figure 5. Annual and cumulative CO$_2$ emissions under the 1.5 °C Scenario by region and sector; additional cumulative CO$_2$ emissions if implementation is delayed (5, 7, or 10 years); and carbon sinks from land, oceans, and additional land restoration. Additional cumulative CO$_2$ emissions by region and sector were calculated on the assumption that CO$_2$ emissions will remain flat for this region and/or sector for the corresponding time period. Annual emissions are compared between the 1.5 °C pathway and the delayed implementation pathway, and the difference is summed over the entire period (2015–2060).

3. Conclusions

To comply with the Paris Climate Agreement and maintain the global temperature 'well below +2.0 °C', the rapid decarbonization of the energy sector with currently available technologies is necessary, and also possible. The normative scenarios developed here avoid an emissions overshoot by combining the transformation to a fully renewable energy supply with the utilization of the available efficiency potentials in all energy sectors to reduce the total demand. Significant electrification of the transport and heating sectors before 2030 is essential to meet the Paris goals in both scenarios presented here. Increased electrification will require sector coupling, demand-side management, and multiple forms of storage (heat and power), including synthetic fuels. Accelerating the implementation of renewable heat technologies is equally important, because half the global energy supply may still derive from thermal processes by 2050. The fundamental transition of the global energy sector shown in our pathways will only be possible with significant policy changes and energy market reforms. The COVID-19 pandemic is both an opportunity for and a threat to this transition. The International Renewable Energy Agency (IRENA), the International Energy Agency (IEA), and the International Monetary Fund (IMF), as well as various government and non-governmental organisations, are demanding stimulation packages for a sustainable economic recovery in order to create new employment in the renewable energy and energy efficiency industries. Despite the expectation of a rapid economic recovery and existing emergencies, new frameworks for fundamental changes in energy use and supply would be required, so that a quick return to business as usual would be avoided.

However, our scenario analysis demonstrates that maintaining the global temperature 'well below +2.0 °C' cannot be achieved by the decarbonization of the energy sector alone, but will also require significant changes in land use, including the rapid phase-out of deforestation and significant reforestation. These measures are not alternative options to the decarbonization of the energy sector, but shall be implemented in parallel. If governments fail to act and mitigation is delayed, we face a serious risk of exceeding the carbon budget. Under the 1.5 °C Scenario, the additional emissions arising from delayed action (Figure 5) can be compensated if we rely more strongly on atmospheric CO_2 removal via biospheric sequestration—in land and forests.

Without additional delay, only one-third of the total estimated CO_2 removal potential will be required in the 2 °C-compatible pathway—leaving space to increase the amount of removal and still meet the 2 °C objective, albeit with the greater risk that a reliance on biospheric removal entails. However, our 1.5 °C pathway already requires all this biospheric sequestration potential—so a delay in mitigation action will put the 1.5 °C pathway out of reach. The idea of compensating emission overshoots in the long term by additional tree planting is unrealistic because the potential for terrestrial carbon sequestration and storage is limited by the amount of carbon previously lost from the biosphere through land conversion [38,47]. Our 1.5 °C pathway tends towards the upper end of this terrestrial carbon sink capacity. Therefore, significant extension of this already covered land sequestration potential is not possible without options that could be described as "geo-engineering", such as establishing large tree plantations beyond ecosystem boundaries—a solution more vulnerable to the reversal of stored carbon [39]; or geological storage, such as via BECCS—an option likely to transgress planetary boundaries at the gigaton scale required for the 1.5 °C pathways [48]. Delayed mitigation action that is justified by sequestration, and which thus shifts the burden to the land sector, brings a higher risk of mitigation failure and temperature overshoot [49]. In our scenarios, the land-use sequestration pathways complement very ambitious energy-mitigation pathways. Sequestration of CO_2 is therefore regarded as necessary to compensate for past emissions and not for current or future emissions

4. Reflections on Ways of Implementation

Achieving a 2.0 °C or 1.5 °C target requires substantial and long-lasting policy changes in order to unlock the necessary investments in the energy sector. A refocused investment strategy towards emerging and green technologies could also support the recovery of the global economy after the pandemic. Solar photovoltaic and onshore wind energy, in particular, are not only cost competitive with conventionally generated energy, but are increasingly least cost options [9]. The volume of global investment in renewables decreased from USD 328 billion in 2017 to USD 289 billion in 2018 [50] and increased to USD 301.7 billion in 2019 [44], which is still 9% below the 2017 levels, even though total installed capacity increased in the same time [51].

The barriers to the deployment of renewable energies are diverse and country-specific. Therefore, the implementation targets vary significantly across the world regions.

The scenario studies show very clearly that the biggest challenge for North America, Europe and the Pacific region will be to rapidly reduce the high energy intensities, i.e., in particular to significantly reduce energy waste inherent in the industrialized lifestyle. Incentives to avoid rebound effects and to save energy in private consumption are not yet visible anywhere. Europe has above all variable renewable resources and must optimize their integration into the energy system through extensive flexibility measures [24,52,53]. However, Europe also has the promising option of sourcing energy imports from resource-rich regions in North Africa and the Middle East, which has long been under discussion [54–56]. In the OECD Pacific region, imports and exports of synthetic fuels (e.g., between Japan and Australia) could be a likely strategy to support 100% renewable energy systems [57]. In Latin America, an important strategy is to improve the sustainability of renewable resource use by redirecting traditional biomass to efficient and low-emission uses. In addition, it may be important to limit the expansion of large hydropower to minimize negative social and environmental impacts. Both narratives are reflected in our alternative goal-oriented paths.

Particularly in the Earth's sunbelt, the development of hydrogen and synthetic fuel technologies could not only cover domestic demand for chemical energy sources at moderate costs, but also open up future export markets [58,59]. In many countries, however, greater political stability would be a prerequisite for large investment in renewable energy and fuel production. The scenario development has clearly demonstrated that developments in China and India also have a major impact on global energy change. While China is already taking a leading position in several transformation processes, but the speed of its actions is not yet sufficient to achieve long-term goals [60].

The results of our scenarios show that the regional targets for the energy and land-use sectors can provide high-level mid- and long-term policy objectives and therefore investment security. In the energy sector, a combination of regional targets for electrification in all demand sectors (see Table 2) and targets for the maximum carbon intensity for each sector (see Table 3) provide a framework for the medium- and long-term measures required to convert the energy supply, including the energy infrastructure. Binding targets for land use will regulate the areas required for the future protection and restoration of carbon sinks and stocks (e.g., forests) and could also define the expansion of areas for renewable energy generation.

5. Similarities to Published Analysis, Research Limitations and Further Research Requirements

Our results in the energy sector are supported by results of other high renewable energy penetration scenarios [61,62]. However, the role of storage technologies. renewable fuels–such as hydrogen and synthetic fuels–and the extend of electrification of industrial process heat varies significantly. Furthermore, the presented 1.5 °C mitigation pathways do not relay on CCS and/or BECCS and used nature-based carbon sinks instead. The global scale of our energy pathways represents a research limitation as regional differences needed to be simplified. Future load curves are speculative as load management as well as utilization of storage technologies requires more research. The industry sector—with a focus on renewable energy supply options for high temperature process heat—requires more research as well. Decarbonisation pathways for specific industry sectors are required. Finally, the integration of non-energy GHG pathways, land-use change emission pathways and energy scenarios of high resolution need to be improved as current models—especially those used in the IPCC assessment reports are still simplistic.

6. Methodology

In our analysis, we considered the complete energy sector in detail, including electricity, heating and cooling, and transport. We also included a perspective on the non-energy use of fuels and the emission reductions arising from land-use changes, and provided a complete picture of all GHG emissions, extending the focus far beyond CO_2 and the energy sector. This was achieved by integrating a set of assessment models for both the energy and non-energy GHG sectors. The results of the various emission modelling tasks are embedded within the reduced-complexity model MAGICC7 (see e.g., [63]), which allowed the derivation of probabilistic temperature projections with which to assess the likelihood of maintaining the global temperature below 2.0 °C or 1.5 °C. The following section summarizes the applied models and their interactions (Figure 6).

6.1. Non-Energy GHG Emissions Scenarios

We complemented the CO_2 emission pathways from the energy system modeling with non-energy-related GHG emissions. To model the non-energy sector, we used different approaches, first to derive the land-use CO_2 emissions and then to derive the emissions of other GHGs and aerosols. In the first approach, we used a (probabilistic) scenario of land-use emissions based on four narrative land-use pathways, and in the second, we used a newly extended statistical regression method. The following two paragraphs describe, in more detail, the methods used in these approaches.

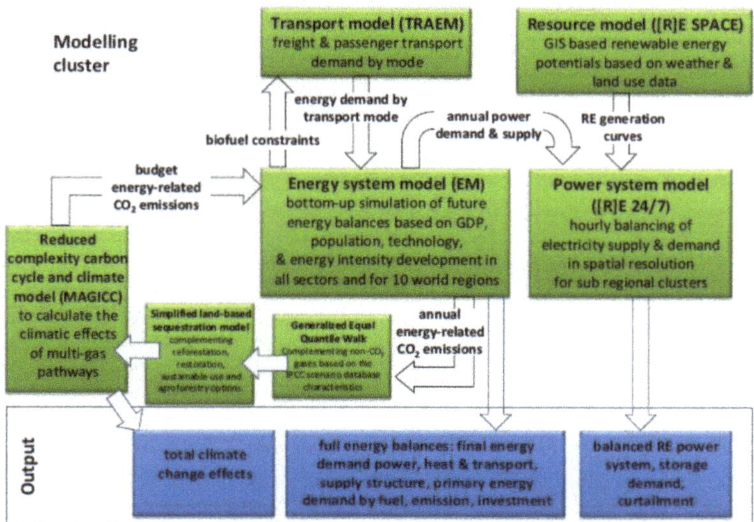

Figure 6. Interaction of the models used in this study.

6.1.1. Generalized Equal Quantile Walk (GQW)

A statistical analysis of 811 multi-gas emission pathways published by the Intergovernmental Panel on Climate Change (IPCC) [7,49] was carried out in order to complete the energy-related CO_2 emission paths with scenarios of other relevant greenhouse gases. The method is an extension of the Equal Quantile Walk method [64] which calculates the median value of greenhouse gases (excluding CO_2) as a function of CO_2 paths in 5-year steps. Further details on this methodology are published in [14].

6.1.2. Land-Based Sequestration Pathways

CO_2 sequestration can be achieved through improved land use such as "restoration of the forest ecosystem", "reforestation", "sustainable forest use" and "agroforestry". Under the assumption that declassified carbon stocks can be restored through sustainable forest use, protected area management and improved land use with the aim of restoring carbon stocks, significant amounts of atmospheric CO_2 can be removed [38,65,66].

Four different sequestration pathways were defined based on literature research and available data from FAO statistics. Assuming that after several years of sustainable land management, a defined amount of carbon is bound annually and thus become carbon sinks. Ultimately, an equilibrium of atmospheric CO_2 is reached. When this equilibrium is reached depends on the type of ecosystem [67]. The phase of transition from a carbon sink to equilibrium is defined as the "phase-out" period.

A maximum of the mean carbon density was assumed based on bio-averaged values for the carbon density of undisturbed forest ecosystems per hectare [68], rather than on average global biome values [69]. The land use sequestration scenarios were calculated up to the year 2300, while the energy scenarios were only calculated until 2050 and the non-energy-related GHGs until 2100. The longer scenario period was necessary to apply the upper limit for the additional carbon density and to quantify the potential for CO_2 sequestration on land. Further details on the methodology are documented in [14].

6.2. Modelling the Energy Sector

To model the energy sector, we combined two complementary approaches: highly spatially and geographically resolved power system modelling and long-term pathway development on an annual basis. The Energy System Model (EM) provides annual energy balances for the complete energy system. The model includes the energy demands for the industry, transport, residential, and others sectors, based on the external inputs of population, GDP, and energy intensity. The industry, residential, and other sectors are each represented by a set of heat, power, and co-generation technologies for all relevant fuel types. The transport sector is supplemented with a detailed transport model (TRAEM), including mobility demand and supply, based on transport technologies (e.g., vehicle types) and mobility services. The IEA World Energy Balances 2017 are the basis for calibrating the 2015 energy demand by region and sector in the model [70]. The conversion-and-power sector in the EM is complemented with a detailed power analysis. EM provides the power demand according to application type as an input parameter for the power system modelling. Based on this input, [R]E 24/7 calculates the necessary infrastructure for the power supply system. This suite of models was used to develop normative, target-oriented long-term scenarios. Starting from the base year and the identified desirable future in 2050 of net zero CO_2 emissions, narratives for suitable transformation pathways were developed. Climate targets in terms of the cumulative CO_2 emissions were set for both the 2.0 °C and a 1.5 °C scenario. To meet these, we constructed bottom-up scenarios covering a switch in the supply technologies. The scenarios are based on detailed input datasets that consider defined CO_2 mitigation and technology expansion targets and limits, potentials and costs for renewables and fossil energy sources, and specific technical parameters for electricity, heat, and fuel generation in the energy systems. We applied a technology transition to all the energy sectors using a gradual approach. We identified the largest remaining emitters based on an ex-post analysis of CO_2 emissions and the gap to reaching the overall CO_2 budget. We then applied additional measures for an accelerated transition towards renewable energy technology. This iterative process was repeated until the carbon budget limitation was achieved.

6.2.1. Transport Model (TRAEM)

The TRAnsport Energy Model (TRAEM) calculates energy demand pathways, broken down into 10 world regions. Based on a passenger–km (pkm) and tonne–km (tkm) activity-based approach, these energy demands were integrated into a global model. The model calculates the final energy demand as the product of specific transport demand of each transport mode with the powertrain-specific energy demand. The model determines the transport energy demand for electricity and various fuels per year in 5-year intervals from 2015 to 2050, with no system or ownership cost-optimization. Total energy demand in the REF Scenario (5.0 °C) follows the IEA World Energy Outlook 2017 Current Policies Scenario [15] up to 2040. Based on the 2035–2040 change rates energy demand was extrapolated linearly to 2050 on reginal level. The was alternative scenarios were adjusted from 2020 to 2050 according to the respective carbon budgets. We attributed biofuels a GHG emission factor of zero because we assume that CO_2 is fixed in the upstream process at the same level as the downstream CO_2 emission. The same applies for CO_2 emissions from synthetic fuel use. The model distinguishes different road passenger transport modes (light-duty vehicles are separated into small, medium, and large cars, 2- and 3-wheelers, and buses), rail passenger transport (urban, regional, and high-speed trains), and aviation (domestic and international passenger flights). Road freight (light-, medium-, and heavy-duty trucks), rail freight, and navigation freight transport were also considered.

Energy intensities per activity varies between the regions, based on the occupancy/load rates of the passenger transport modes or freight vehicles. Total energy demand is then the sum of all demand in all transport modes. The transport data were derived from historical and current transport activity data from statistics, complemented by region specific literature (for example, data on vehicle stock or occupancy rates in selected world regions). The German Aerospace Center (DLR) vehicle databases served as source for for energy intensity per transport. More information on this database and more details and the key assumptions can be found in [71].

6.2.2. Energy System Model (EM)

The scenarios are developed within a mathematical accounting system, specifically developed for the energy sector. It models development ways for energy demand and supply, considering development pathways of potentials, specific fuel consumption, technology and fuel costs, emissions, and limitations by physical flows between a set of technology processes. The data availability and the objectives of the analysis significantly influence the model architecture and approach.

The scenarios are implemented in Mesap/PlaNet, an energy simulation platform, which hosts the global energy system model developed by the DLR [72,73]. The accounting framework calculates detailed and consistent energy system balances, starting from demand and working all the way back to primary energy supply. It consists of two independent modules:

- the flow calculation module with a physical balance of energy supply and demand on annual basis;
- the cost calculation module, for corresponding investment, generation and supply costs.

The model integrates and combines a whole range of different technical options for the transformation of energy systems. The ex-post evaluation of power cost calculation is implemented via the Mesap platform's standard tool and applied to all scenarios. The Model features a database for managing the input parameters and the output for the different scenarios after simulation. The graphical interface serves for structuring the modelled system and defining the quantitative interdependences between individual elements at different structural depths. Details of the structure and relevant model equations are given in the literature [21,74]. The energy flows of the energy system are balanced in the model on an annual basis. These flows connect technologies in each sector to process chains and includes all relevant energy carriers, using linear equations. The model then balances demand and supply by sequentially solving this equation system. The scenario period is disaggregated to 5-year steps until 2050. Further details about the methodology of the Energy System Model (EM) are published in [14]. The main outputs of the model are:

- primary and final energy demands, disaggregated by fuel, technology, and energy sector, according to the classification by the International Energy Agency (IEA);
- required energy required, applied technology and the financial investment for electricity, heating, and mobility (transport)t;
- total cost of energy for the power system;
- energy-related CO_2 emissions over the scenario period.

6.3. Modelling the Power Sector

The power system analysis [R]E 24/7 is a mathematical accounting system that assess the requirements for electricity storage (the calculation of inter-regional exchange capacity requirements in MW is also possible, but beyond the scope of this article). It simulates the electricity system on an hourly basis and at geographic resolution. The methodology of the [R]E 24/7 model has been developed by UTS/ISF [75–78]. It specifically implements the hourly distribution (load curves and storage) and the geographic distribution of power demand and supply.

Hourly load curves for the residential, industry, and transport sectors were synthetically produced on the basis of the annual electricity demands for 2020, 2030, 2040, and 2050 (EM results), technology- and sector-specific energy intensity factors, regional GDP [79], and population data. Load curves for households were determined using nine different household categories, with various degrees of electrification and equipment. To calculate the load curves for business and industry, eight statistical industrial-sector categories were used: agriculture (1), manufacturing (2), mining (3), iron and steel production (4), cement industry (5), construction industry (6), chemical industry (7), and service and trade (8). Each sector had a defined energy intensity, expressed in energy per dollar GDP (MJ/USD_{GDP}), which was converted to electrical units (kW/USD$_{GDP}$) based on an estimated fuel efficiency factor, the electricity share, and operational hours per year. The load curve for the transport sector was calculated from the energy intensities for all electricity-consuming transport modes and hydrogen and synthetic fuel production, divided by the average annual utilization according to the technology (in h/yr). All three sectorial load curves were standardized: the load curves for the household and transport sectors in kilowatts per person (kW/capita) and the industry load curves in kilowatts per dollar GDP (kW/USD GDP). These standard curves were multiplied by the GDP data for each regional population. The standardized sectorial load curves for households and transport were multiplied by the population numbers derived with GIS mapping of each cluster. The standardized load curves for each of the eight industry sectors were multiplied by the corresponding shares of the total GDP values accorded these sectors by region. Because some data for each cluster were unavailable, the eight regional industry load curves were distributed per capita. In the last step, all sectorial load curves (households, transport, and industry) were summed. The spatial distribution of the projected GDP by industry sector remained unchanged in the 72 sub-regions over the years modelled (2020–2050).

The calculated load curves were compared with a cascade of power-generation technologies. The dispatch orders of the power-plant technologies can be changed. If demand and generation are congruent, no subsequent power-plant technologies are required, and the production for these hours will be zero. For variable solar and wind power generation, meteorological data with hourly resolution are required for each cluster (see [80,81]). Further details about the methodology are documented in [14].

Supplementary Materials: The following are available online at https://www.mdpi.com/article/10.3390/en14082103/s1.

Author Contributions: Conceptualization, S.T. and T.P.; methodology, T.P., S.S., J.P., K.D., M.M.; software, T.N., S.T.; validation, S.S., J.P., K.D.; formal analysis, T.P.; investigation, T.P., S.S., T.N., J.P., Ö.D., B.v.d.A., K.D. and M.M.; resources, S.T.; data curation, T.P., S.S., J.P., M.M.; writing—original draft preparation, S.T. and T.P.; writing—review and editing, S.T. and T.P.; visualization, S.T. and T.P.; supervision, S.T.; project administration, S.T.; funding acquisition, S.T. All authors have read and agreed to the published version of the manuscript.

Funding: This research was funded by the Leonardo DiCaprio Foundation, Sustainable Markets Foundation, 45 West 36th Street, 6th Floor, New York, N.Y.

Institutional Review Board Statement: Not applicable.

Informed Consent Statement: Not applicable.

Data Availability Statement: Data available from the Supplementary Materials.

Conflicts of Interest: The authors declare no conflict of interest.

References

1. IRENA. *Scenarios for the Energy Transition: Global Experiences and Best Practices*; International Renewable Energy Agency: Abu Dhabi, United Arab Emirates, 2020; ISBN 978-92-9260-267-3.
2. United Nations Framework Convention on Climate Change (UNFCCC). The Paris Agreement. Available online: http://unfccc.int/paris_agreement/items/9485.php (accessed on 12 March 2018).
3. Rogelj, J.; Popp, A.; Calvin, K.V.; Luderer, G.; Emmerling, J.; Gernaat, D.; Fujimori, S.; Strefler, J.; Hasegawa, T.; Marangoni, G.; et al. Scenarios towards limiting global mean temperature increase below 1.5 °C. *Nat. Clim. Chang.* **2018**, *8*, 325–332. [CrossRef]
4. Kriegler, E.; Riahi, K.; Bauer, N.; Schwanitz, V.J.; Petermann, N.; Bosetti, V.; Marcucci, A.; Otto, S.; Paroussos, L.; Rao, S.; et al. Making or breaking climate targets: The AMPERE study on staged accession scenarios for climate policy. *Technol. Forecast. Soc. Chang.* **2015**, *90*, 24–44. [CrossRef]
5. International Energy Agency (IEA). *World Energy Outlook 2018*; OECD Publishing: Paris, France, 2018. [CrossRef]
6. Rogelj, J.; Elzen, M.D.; Höhne, N.; Fransen, T.; Fekete, H.; Winkler, H.; Schaeffer, R.; Sha, F.; Riahi, K.; Meinshausen, M. Paris Agreement climate proposals need a boost to keep warming well below 2°C. *Nature* **2016**, *534*, 631–639. [CrossRef] [PubMed]
7. Intergovernmental Panel on Climate Change (IPCC). *AR5 Synthesis Report: Climate Change 2014*; IPCC: Geneva, Switzerland, 2014. Available online: https://www.ipcc.ch/report/ar5/syr/ (accessed on 24 March 2021).

8. Intergovernmental Panel on Climate Change (IPCC). *Global Warming of 1.5 °C*; IPCC: Geneva, Switzerland, 2014. Available online: https://www.ipcc.ch/site/assets/uploads/sites/2/2019/06/SR15_Full_Report_High_Res.pdf (accessed on 24 March 2021).

9. Lazard. Lazard's Levelized Costs of Energy Analysis—Version 13.0. Available online: https://www.lazard.com/media/451086/lazards-levelized-cost-of-energy-version-130-vf.pdf (accessed on 24 November 2019).

10. Fridahlab, M.; Lehtveerac, M. Bioenergy with carbon capture and storage (BECCS): Global potential, investment preferences, and deployment barriers. *Energy Res. Soc. Sci.* **2018**, *42*, 155–165. [CrossRef]

11. Ramana, M.V. Technical and social problems of nuclear waste. Wiley Interdiscip. *Rev. Energy Environ.* **2018**, *7*, e298. [CrossRef]

12. Shaffer, G. Long-term effectiveness and consequences of carbon dioxide sequestration. *Nat. Geosci.* **2010**, *3*, 464–467. [CrossRef]

13. Intergovernmental Panel on Climate Change (IPCC). *Global Warming of 1.5 °C*; (See above), SR1.5 p.132, Table 2.6.; IPCC: Geneva, Switzerland, 2019. Available online: https://www.ipcc.ch/site/assets/uploads/sites/2/2019/06/SR15_Full_Report_High_Res.pdf (accessed on 24 March 2021).

14. Teske, S. (Ed.) *Achieving the Paris Climate Agreement Goals—Global and Regional 100% Renewable Energy Scenarios with Non-Energy GHG Pathways for +1.5 °C and +2 °C*; Springer eBook: New York, NY, USA, 2019. [CrossRef]

15. International Energy Agency (IEA). *World Energy Outlook 2017*; OECD Publishing: Paris, France, 2017. [CrossRef]

16. Intergovernmental Panel on Climate Change (IPCC). *AR5 Climate Change 2014: Mitigation of Climate Change*; IPCC: Geneva, Switzerland, 2014. Available online: https://www.ipcc.ch/report/ar5/wg3/ (accessed on 24 March 2021).

17. United Nations, Department of Economic and Social Affairs, Population Division. *World Population Prospects: The 2017 Revision*; United Nations: New York, NY, USA, 2017. Available online: https://population.un.org/wpp/ (accessed on 30 April 2018).

18. Graus, W.J.; Blomen, E.; Worrell, E. Global energy efficiency improvement in the long term: A demand- and supply-side perspective. *Energy Effic.* **2011**, *4*, 435–463. [CrossRef]

19. Kermeli, K.; Graus, W.J.; Worrell, E. Energy efficiency improvement potentials and a low energy demand scenario for the global industrial sector. *Energy Effic.* **2014**, *7*, 987–1011. [CrossRef]

20. Grubler, A.; Wilson, C.; Bento, N.; Boza-Kiss, B.; Krey, V.; Mccollum, D.L.; Rao, N.D.; Riahi, K.; Rogelj, J.; De Stercke, S.; et al. A low energy demand scenario for meeting the 1.5 °C target and sustainable development goals without negative emission technologies. *Nat. Energy* **2018**, *3*, 515–527. [CrossRef]

21. Pregger, T.; Simon, S.; Naegler, T.; Teske, S. Main Assumptions for Energy Pathways. In *Achieving the Paris Climate Agreement Goals: Global and Regional 100% Renewable Energy Scenarios with Non-Energy GHG Pathways for +1.5 °C and +2 °C*; Teske, S., Ed.; Springer International Publishing: New York, NY, USA, 2019; pp. 93–130.

22. Lovins, A.B. How big is the energy efficiency resource? *Environ. Res. Lett.* **2018**, *13*, 09041. [CrossRef]

23. Teske, S.; Pregger, T.; Naegler, T.; Simon, S.; Pagenkopf, J.; van den Adel, B.; Deniz, Ö. Energy Scenario Results. In *Achieving the Paris Climate Agreement Goals: Global and Regional 100% Renewable Energy Scenarios with Non-Energy GHG Pathways for +1.5 °C and +2 °C*; Teske, S., Ed.; Springer International Publishing: New York, NY, USA, 2019; pp. 175–401.

24. Gils, H.C.; Simon, S.; Soria, R. 100% renewable energy supply for Brazil—The role of sector coupling and regional development. *Energies* **2017**, *10*, 1859. [CrossRef]

25. Xiao, M.; Simon, S.; Pregger, T. Scenario analysis of energy system transition—A case study of two coastal metropolitan regions, eastern China. *Energy Strategy Rev.* **2019**, *26*, 100423. [CrossRef]

26. Xiao, M.; Wetzel, M.; Pregger, T.; Simon, S.; Scholz, Y. Modeling the supply of renewable electricity to metropolitan regions in China. *Energies* **2020**, *13*, 3042. [CrossRef]

27. Falchetti, G.; Pachauri, S.; Byers, E.; Danylo, O.; Parkinson, S.C. Satellite observations reveal inequalities in the progress and effectiveness of recent electrification in sub-Saharan Africa. *One Earth* **2020**, *2*, 364–379. [CrossRef]

28. Müller, R.; Pfeifroth, U.; Träger-Chatterjee, C.; Trentmann, J.; Cremer, R. Digging the METEOSAT Treasure—3 decades of solar surface radiation. *Remote Sens.* **2015**, *7*, 8067–8101. [CrossRef]

29. The Renewables.ninja. RE-N DB 2018, online database for hourly time series for solar and wind data for a specific geographical position. Available online: https://www.renewables.ninja/ (accessed on 7 July 2018).

30. Sask Power. Boundary Dam Carbon Capture Project. Available online: https://www.saskpower.com/Our-Power-Future/Infrastructure-Projects/Carbon-Capture-and-Storage/Boundary-Dam-Carbon-Capture-Project (accessed on 20 March 2020).

31. Armpriester, A.W.A. *Parish Post Combustion CO2 Capture and Sequestration Project*; Final Public Design Report for the U.S. Department of Energy, Office of Scientific and Technical Information, 17 February 2017. Available online: https://www.osti.gov/biblio/1344080 (accessed on 20 March 2020).

32. Global CCS Institute. *Global Status of CCS 2019—Targeting Climate Change*. (p. 24, Figure 8). Available online: https://www.globalccsinstitute.com/wp-content/uploads/2019/12/GCC_GLOBAL_STATUS_REPORT_2019.pdf (accessed on 20 March 2020).

33. Fuss, S.; Lamb, W.F.; Callaghan, M.W.; Hilaire, J.; Creutzig, F.; Amann, T.; Beringer, T.; Garcia, W.D.O.; Hartmann, J.; Khanna, T.; et al. Negative emissions—Part 2: Costs, potentials and side effects. *Environ. Res. Lett.* **2018**, *13*, 063002. [CrossRef]

34. Turner, P.A.; Mach, K.J.; Lobell, D.B.; Benson, S.M.; Baik, E.; Sanchez, D.L.; Field, C.B. The global overlap of bioenergy and carbon sequestration potential. *Clim. Chang.* **2018**, *148*, 1–10. [CrossRef]

35. Griscom, B.W.; Adams, J.; Ellis, P.W.; Houghton, R.A.; Lomax, G.; Miteva, D.A.; Schlesinger, W.H.; Shoch, D.; Siikamäki, J.V.; Smith, P.; et al. Natural climate solutions. *Proc. Natl. Acad. Sci. USA* **2017**, *114*, 11645–11650. [CrossRef]

36. Roe, S.; Streck, C.; Obersteiner, M.; Frank, S.; Griscom, B.; Drouet, L.; Fricko, O.; Gusti, M.; Harris, N.; Hasegawa, T.; et al. Contribution of the land sector to a 1.5 °C world. *Nat. Clim. Chang.* **2019**, *9*, 817–828. [CrossRef]

37. Teske, S.; Pregger, T.; Simon, S.; Naegler, T.; Pagenkopf, J.; van den Adel, B.; Meinshausen, M.; Dooley, K.; Briggs, C.; Dominish, E.; et al. Methodology. In *Achieving the Paris Climate Agreement Goals: Global and Regional 100% Renewable Energy Scenarios with Non-Energy GHG Pathways for +1.5 °C and +2 °C*; Teske, S., Ed.; Springer International Publishing: New York, NY, USA, 2019; pp. 25–78.
38. Mackey, B.; Prentice, I.C.; Steffen, W.; House, J.I.; Lindenmayer, D.; Keith, H.; Berry, S. Untangling the confusion around land carbon science and climate change mitigation policy. *Nat. Clim. Chang.* **2013**, *3*, 552–557. [CrossRef]
39. Seddon, N.; Turner, B.; Berry, P.; Chausson, A.; Giradin, C. Grounding nature-based climate solutions in sound biodiversity science. *Nat. Clim. Chang.* **2019**, *9*, 84–87. [CrossRef]
40. Lewis, S.L.; Wheeler, C.E.; Mitchard, E.T.A.; Koch, A. Regenerate natural forests to store carbon. *Nature* **2019**, *568*, 25–28. [CrossRef]
41. Meinshausen, M.; Dooley, K. Mitigation Scenarios for Non-energy GHG. In *Achieving the Paris Climate Agreement Goals: Global and Regional 100% Renewable Energy Scenarios with Non-Energy GHG Pathways for +1.5 °C and +2 °C*; Teske, S., Ed.; Springer International Publishing: New York, NY, USA, 2019; pp. 79–91.
42. Global Energy Assessment 2012. *Annex II, Technical Guidelines: Common Terms, Definitions and Units Used in GEA*; Cambridge University Press: Cambridge, UK; New York, NY, USA; International Institute for Applied Systems Analysis: Laxenburg, Austria; p. 1816. Available online: https://www.iiasa.ac.at/web/home/research/Flagship-Projects/Global-Energy-Assessment/GEA_Annex_II.pdf (accessed on 24 March 2021).
43. International Energy Agency (IEA). Sustainable Recovery—World Energy Outlook Special Report. In *Collaboration with the International Monetary Fund*; OECD Publishing: Paris, France, 2020. Available online: https://www.iea.org/reports/sustainable-recovery (accessed on 24 March 2021).
44. Renewable Energy Policy Network for the 21st Century (REN21). *Renewables 2020 Global Status Report*; REN21 Secretariat: Paris, France, 2020. Available online: https://www.ren21.net/gsr-2020/ (accessed on 24 March 2021).
45. Wells, C.R.; Sah, P.; Moghadas, S.M.; Pandey, A.; Shoukat, A.; Wang, Y.; Wang, Z.; Meyers, L.A.; Singer, B.H.; Galvani, A.P. Impact of international travel and border control measures on the global spread of the novel 2019 coronavirus outbreak. *PNAS* **2020**, *117*, 7504–7509. [CrossRef]
46. International Energy Agency (IEA). Global CO2 Emissions in 2019. Available online: https://www.iea.org/articles/global-co2-emissions-in-2019 (accessed on 11 February 2020).
47. Arneth, A.; Sitch, S.; Pongratz, J.; Stocker, B.D.; Ciais, P.; Poulter, B.; Bayer, A.D.; Bondeau, A.; Calle, B.P.L.; Chini, L.P.; et al. Historical carbon dioxide emissions caused by land-use changes are possibly larger than assumed. *Nat. Geosci.* **2017**, *10*, 79–84. [CrossRef]
48. Heck, V.; Gerten, D.; Lucht, W.; Popp, A. Biomass-based negative emissions difficult to reconcile with planetary boundaries. *Nat. Clim. Chang.* **2018**, *8*, 151–155. [CrossRef]
49. Intergovernmental Panel on Climate Change (IPCC). *Climate Change and Land*; IPCC: Geneva, Switzerland, 2019. Available online: https://www.ipcc.ch/srccl/ (accessed on 24 March 2021).
50. Renewable Energy Policy Network for the 21st Century (REN21). *Renewables 2019 Global Status Report*; REN21 Secretariat: Paris, France, 2019. Available online: https://www.ren21.net/gsr-2019/ (accessed on 30 March 2020).
51. International Renewable Energy Agency. Renewable Energy Statistics 2020. Available online: https://www.irena.org/Statistics/ (accessed on 24 March 2021).
52. Brown, T.; Schlachtberger, D.; Kies, A.; Schramm, S.; Greiner, M. Synergies of sector coupling and transmission reinforcement in a cost-optimised, highly renewable European energy system. *Energy* **2018**, *160*, 720–739. [CrossRef]
53. Borowski, P.F. Zonal and Nodal Models of Energy Market in European Union. *Energies* **2020**, *13*, 4182. [CrossRef]
54. Trieb, F.; Schillings, C.; Pregger, T.; O'Sullivan, M. Solar electricity imports from the Middle East and North Africa to Europe. *Energy Policy* **2012**, *42*, 341–353. [CrossRef]
55. Pregger, T.; Lavagno, E.; Labriet, M.; Seljom, P.; Biberacher, M.; Blesl, M.; Trieb, F.; O'Sullivan, M.; Gerboni, R.; Schranz, L.; et al. Resources, capacities and corridors for energy imports to Europe. *Int. J. Energy Sector Manag.* **2011**, *5*, 125–156. [CrossRef]
56. Dii Desert Energy. A North Africa—Europe Hydrogen Manifesto. 2019. Available online: https://dii-desertenergy.org/wp-content/uploads/2019/12/Dii-hydrogen-study-November-2019.pdf (accessed on 1 October 2020).
57. PV Magazine. Australia and Japan Agree to Hydrogen Future. 24 January 2020 Blake Matich. Available online: https://www.pv-magazine-australia.com/2020/01/24/australia-and-japan-agree-to-hydrogen-future/ (accessed on 10 October 2020).
58. European Commission. A Hydrogen Strategy for a Climate-neutral Europe. Brussels, 8.7.2020 COM(2020) 301 final. Available online: https://ec.europa.eu/energy/sites/ener/files/hydrogen_strategy.pdf (accessed on 12 October 2020).
59. Timmerberg, S.; Kaltschmitt, M. Hydrogen from renewables: Supply from North Africa to Central Europe as blend in existing pipelines—Potentials and costs. *Appl. Energy* **2019**, *237*, 795–809. [CrossRef]
60. Xiao, M.; Simon, S.; Pregger, T. Energy System Transitions in the Eastern Coastal Metropolitan Regions of China—The Role of Regional Policy Plans. *Energies* **2019**, *12*, 389. [CrossRef]
61. Jacobson, M.Z.; Delucchi, M.A.; Bauer, Z.A.; Goodman, S.C.; Chapman, W.E.; Cameron, M.A.; Bozonnat, C.; Chobadi, L.; Clonts, H.A.; Enevoldsen, P.; et al. 100% Clean and Renewable Wind, Water, and Sunlight All-Sector Energy Roadmaps for 139 Countries of the World. *Joule* **2017**, *1*, 108–121. [CrossRef]

62. Breyer, C.; Bogdanov, D.; Aghahosseini, A.; Gulagi, A.; Child, M.; Oyewo, A.S.; Farfan, J.; Sadovskaia, K.; Vainikka, P. Solar photovoltaics demand for the global energy transition in the power sector. *Prog. Photovolt.* **2018**, *26*, 505–523. [CrossRef]

63. Meinshausen, M.; Raper, S.C.B.; Wigley, T.M.L. Emulating coupled atmosphere-ocean and carbon cycle models with a simpler model, MAGICC6—Part 1: Model description and calibration. *Atmos. Chem. Phys.* **2011**, *11*, 1417–1456. [CrossRef]

64. Meinshausen, M.; Hare, B.; Wigley, T.M.L.; van Vuuren, D.; den Elzen, M.G.J.; Swart, R. Multi-gas emission pathways to meet climate targets. *Clim. Chang.* **2006**, *75*, 151–194. [CrossRef]

65. DeCicco, J.M.; Schlesinger, W.H. Reconsidering bioenergy given the urgency of climate protection. *Proc. Natl Acad. Sci. USA* **2018**, *115*, 9642–9645. [CrossRef] [PubMed]

66. Luyssaert, S.; Schulze, E.-D.; Börner, A.; Knohl, A.; Hessenmöller, D.; Law, B.E.; Ciais, P.; Grace, J. Old-growth forests as global carbon sinks. *Nature* **2008**, *455*, 213–215. [CrossRef]

67. Houghton, R.A.; Nassikas, A.A. Negative emissions from stopping deforestation and forest degradation, globally. *Glob. Chang. Biol.* **2018**, *24*, 350–359. [CrossRef]

68. Keith, H.; Mackey, B.G.; Lindenmayer, D.B. Re-evaluation of forest biomass carbon stocks and lessons from the world's most carbon-dense forests. *Proc. Natl Acad. Sci. USA* **2009**, *106*, 11635–11640. [CrossRef]

69. Liu, Y.Y.; Van Dijk, A.I.J.M.; De Jeu, R.A.M.; Canadell, J.G.; McCabe, M.F.; Evans, J.P.; Wang, G. Recent reversal in loss of global terrestrial biomass. *Nat. Clim. Chang.* **2015**, *5*, 470–474. [CrossRef]

70. International Energy Agency (IEA). *World Energy Balances 2017*; OECD Publishing: Paris, France, 2017. [CrossRef]

71. Pagenkopf, J.; van den Adel, B.; Deniz, Ö.; Schmid, S. Transport Transition Concepts. In *Achieving the Paris Climate Agreement Goals: Global and Regional 100% Renewable Energy Scenarios with Non-Energy GHG Pathways for +1.5 °C and +2 °C*; Teske, S., Ed.; Springer International Publishing: New York, NY, USA, 2019; pp. 131–159.

72. Schlenzig, C. Energy planning and environmental management with the information and decision support system MESAP. *Int. J. Glob. Energy Issues* **1998**, *12*, 81–91. [CrossRef]

73. Seven2one. *Mesap/PlaNet Software Framework*; Seven2one Modelling, Mesap4, Release 4.14.1.9; Seven2one Informationssysteme GmbH: Karlsruhe, Germany, 2012.

74. Simon, S.; Naegler, T.; Gils, H.C. Transformation towards a renewable energy system in Brazil and Mexico—Technological and structural options for Latin America. *Energies* **2018**, *11*, 907. [CrossRef]

75. Dunstan, C.; Fattal, A.; James, G.; Teske, S. *Towards 100% renewable energy for Kangaroo Island*; UTS-ISF: Sydney, Australia, 2016; Prepared by the Institute for Sustainable Futures, University of Technology Sydney (with assistance from AECOM); Final report prepared for ARENA, Renewables SA, and Kangaroo Island Council.

76. Teske, S. Bridging the Gap between Energy and Grid Models, Developing an Integrated Infrastructural Planning Model for 100% Renewable Energy Systems in order to Optimize the Interaction of Flexible Power Generation, Smart Grids and Storage Technologies. Ph.D. Thesis, University of Flensburg, Frensburg, Germany, 2015.

77. James, G.; Rutovitz, J.; Teske, S. *Storage Requirements for Reliable Electricity in Australia*; University of Technology Sydney, UTS-ISF: Sydney, Australia, 2017; Report prepared by the Institute for Sustainable Futures for the Australian Council of Learned Academies.

78. Teske, S.; Morris, T.; Nagrath, K. *100% Renewable Energy for Tanzania—Access to Renewable Energy for all Within one Generation*; University of Technology Sydney, UTS-ISF: Sydney, Australia, 2017; Report prepared by ISF for Bread for the World.

79. Central Intelligence Agency Library. World Factbook, Online Database. Available online: https://www.cia.gov/library/publications/the-world-factbook/fields/2195.html (accessed on 7 July 2018).

80. Pfenninger, S.; Staffell, I. Long-term patterns of European PV output using 30 years of validated hourly reanalysis and satellite data. *Energy* **2016**, *114*, 1251–1265. [CrossRef]

81. Staffell, I.; Pfenninger, S. Using bias-corrected reanalysis to simulate current and future wind power output. *Energy* **2016**, *114*, 1224–1239. [CrossRef]

energies

MDPI

Article

Achieving Net Zero Emissions in Italy by 2050: Challenges and Opportunities

Maria Gaeta *, Corine Nsangwe Businge and Alberto Gelmini

RSE—Ricerca sul Sistema Energetico, 20134 Milan, Italy; corine.nsangwebusinge@rse-web.it (C.N.B.);
alberto.gelmini@rse-web.it (A.G.)
* Correspondence: maria.gaeta@rse-web.it

Abstract: This paper contributes to the climate policy discussion by focusing on the challenges and opportunities of reaching net zero emissions by 2050 in Italy. To support Italian energy planning, we developed energy roadmaps towards national climate neutrality, consistent with the Paris Agreement objectives and the IPCC goal of limiting the increase in global surface temperature to 1.5 °C. Starting from the Italian framework, these scenarios identify the correlations among the main pillars for the change of the energy paradigm towards net emissions by 2050. The energy scenarios were developed using TIMES-RSE, a partial equilibrium and technology-rich optimization model of the entire Italian energy system. Subsequently, an in-depth analysis was developed with the sMTISIM, a long-term simulator of power system and electricity markets. The results show that, to achieve climate neutrality by 2050, the Italian energy system will have to experience profound transformations on multiple and strongly related dimensions. A predominantly renewable-based energy mix (at least 80–90% by 2050) is essential to decarbonize most of the final energy consumption. However, the strong increase of non-programmable renewable sources requires particular attention to new flexibility resources needed for the power system, such as Power-to-X. The green fuels produced from renewables via Power-to-X will be a vital energy source for those sectors where electrification faces technical and economic barriers. The paper's findings also confirm that the European "energy efficiency first" principle represents the very first step on the road to climate neutrality.

Keywords: climate change; I-LTS; Paris Agreement; energy scenarios; renewable energy; 2050 carbon neutrality; energy transition; GHG mitigation; energy planning; TIMES model

Citation: Gaeta, M.; Nsangwe Businge, C.; Gelmini, A. Achieving Net Zero Emissions in Italy by 2050: Challenges and Opportunities. *Energies* **2022**, *15*, 46. https://doi.org/10.3390/en15010046

Academic Editor: Devinder Mahajan

Received: 17 November 2021
Accepted: 18 December 2021
Published: 22 December 2021

Publisher's Note: MDPI stays neutral with regard to jurisdictional claims in published maps and institutional affiliations.

1. Introduction

At the end of COP21, the Paris Agreement codified the aspiration to keep the increase in global average temperature to well below 2 °C above pre-industrial levels and to continue efforts to limit the temperature rise to 1.5 °C. It was adopted by 196 parties at COP21 in Paris on 12 December 2015, although it went into effect on 4 November 2016 [1].

As stated by the IPCC scientists: "human activities are estimated to have caused approximately 1.0 °C of global warming above pre-industrial levels. Global warming is likely to reach 1.5 °C between 2030 and 2052 if it continues to increase at the current rate" [2]. Around 40 billion tons of CO_2 are emitted worldwide every year (including deforestation). About half of these emissions are accumulated in the atmosphere and contribute to global warming. To stabilize Earth's temperature, climate science climatology indicates that global greenhouse gas emissions must be first reduced and then eliminated completely. In particular, in emission scenarios compatible with the 1.5 °C target, net global CO_2 emissions reach zero around 2050, and a balance must be achieved between the absorptions and emissions of all GHGs in the second half of the century [3–5].

Therefore, the Paris Agreement is the global political response to the evidence illustrated by the IPCC. To better frame the efforts towards this long-term goal, Article 4 of the Paris Agreement invites countries to update their nationally determined contributions (NDCs)

and elaborate long-term low greenhouse gas emission development strategies (LTS) [1]. In November 2018, the European Commission put forward its goal of a climate-neutral Europe by 2050, in line with the Governance Regulation [6]. Both the European Parliament and the European Council approved this proposal, and in March 2020, the European Union communicated its long-term strategy and the EU target of net zero emissions by 2050 to the UNFCCC [7]. At the same time, the European Commission has put forward a legislative proposal that aims to make the climate neutrality goal of 2050 legally binding within the EU ("European climate law" [8]). The proposal was amended in September 2020, in line with the European Green Deal, by adding an interim climate target consisting of a reduction of net emissions of at least 55% from 1990 levels by 2030, including removals from the LULUCF sector [9]. In December 2020, the European Union communicated its updated NDC to the UNFCCC, including the target of −55% on all greenhouse gases by 2030 [10].

In this context, the Energy Union and Climate Action Governance Regulation established a process for the preparation of national decarbonization strategies by 2050 of member states, and an update every 10 years thereafter, consistent with member states' integrated national energy and climate plans for the period 2021–2030 [11].

In January 2021, the Italian government published the Italian *Long-Term Strategy (I-LTS) on Reducing Greenhouse Gas Emissions*. This process was promoted by several Italian Ministries such as the Ministry of the Environment, Land and Sea; the Ministry of Economic Development; the Ministry of Infrastructure and Transport; and the Ministry of Agricultural, Food and Forestry Policies, and involved many actors from research centers and academia with a crucial role [12]. In fact, the I-LTS is based on studies and analyses of a technical working group made up by RSE (Ricerca sul Sistema Energetico), ISPRA (Istituto superiore per la protezione e la ricerca ambientale), GSE (Gestore Servizi Energetici), Politecnico di Milano, ENEA (Agenzia nazionale per le nuove tecnologie, l'energia e lo sviluppo economico sostenibile) and CMCC (Centro euro-Mediterraneo sui Cambiamenti Climatici). Each subject contributed its own tools and skills to the study, analyzing all the correlations, synergies and implications of the strategy.

This paper describes the studies carried out by RSE underpinning the Italian government's LTS. The energy pathways towards 2050 Italian carbon neutrality presented in this paper are based on a sensitivity quantitative analysis performed on the different availabilities of low carbon options in the long-term. In particular, these scenarios try to grasp the impact of decreasing levels of fossil fuel consumption, from a partial to an almost complete phase-out, according to different hypotheses on the future availability of technological options (i.e., CCS, hydrogen in steel industry), renewable resource potential in the national territory (i.e., photovoltaics development) and possible behavioral changes. A predominantly renewable-based energy mix (at least 80–90% by 2050) is essential to decarbonize most of the final energy consumption, while the residual consumption of fossil fuels must be accompanied by CO_2 capture. However, the strong increase of non-programmable renewable sources requires a completely different electricity system from the current one in terms of both generation mix and system management. Particular attention is therefore given to aspects such as sector coupling and the modeling of new forms of flexibility, like Power-to-X. The change in the energy paradigm of a carbon neutral scenario also highlights the spread of new zero emissions synthetic fuels like hydrogen and other renewable fuels of non-biological origin, which will be essential for decarbonizing the hard-to-abate sectors.

Many decarbonization scenarios in line with the Paris Agreement have been analyzed in recent years, providing very important guidelines [13,14].

The scenarios described in this paper identify the correlations among the main pillars required for the shift of the energy paradigm towards net emissions by 2050 in the Italian context and allowed us to draw lessons and principles which are valid and applicable to many other contexts. The main results of this research will also be presented at a Long-term-Scenarios for the Energy Transition (LTES) event by the IRENA [15]. The challenge was to understand how an energy system almost completely devoid of fossil sources could function, what important changes are needed and what kind of technological and renewable source developments can be expected. From a modeling perspective, the main challenge was to understand how to model new forms of flexibility while solving any critical issues deriving from the system and technological innovations introduced.

2. Materials and Methods

2.1. Scenario Analysis

The decarbonization process is characterized by multiple factors and uncertainties that make it difficult to identify a univocal evolution of the system itself, especially with medium–long term time horizons. However, the scenario analysis allows us to explore different possible paths to reach a condition of "climate neutrality" by 2050 by analyzing trajectories based on alternative assumptions. This approach allows to identify main actions, objectives and sectors with more potential for decarbonization interventions and provide indications on infrastructural and technological needs. Scenario analysis enables the decision makers to orient themselves through this extreme complexity and uncertainty thanks to a quantitative assessment of the impacts of energy–environmental objectives and policies, the evidence of any overlaps, indications on the most promising sectors of intervention and infrastructural and technological needs.

Scenarios are not forecasts but alternative pictures of how the future could unfold and can be constrained by specific objectives to be achieved. A scenario is a coherent, internally consistent, and reasonable description of how the future might develop. It is based on a coherent and consistent set of assumptions about the key relationships and driving variables (e.g., fuel prices, technology evolution) [16].

A scenario can represent a vision of the future towards which to strive and which therefore represents an objective. In this case, the scenario is useful for understanding the conditions (and their timing) that allow to realize a new future. This makes it possible to clarify the conditions that must occur in each stage of the time trajectory considered in order to achieve that result.

2.2. Energy and Power System Model Used

Quantitative scenarios require the use of models, i.e., a formal and mathematical representation of an energy system.

To carry out the scenario analyses of this study, we used the TIMES_RSE model, an energy model of the Markal TIMES family [17]. TIMES (The Integrated MARKAL-EFOM System) is a model generator developed within the IEA-ETSAP (Energy Technology Systems Analysis Program), an international research network using energy scenarios modeling to carry out detailed energy and environmental analyses [18]. In this framework, RSE has developed the TIMES_RSE model, which represents the entire Italian energy system.

TIMES_RSE is a technology rich, bottom-up model of intertemporal optimization that minimizes the total cost for the whole energy system over the entire time horizon of satisfying given demands for energy services, subject to environmental and technological or policy constraints. The equilibrium solution is found using linear programming techniques. The objective function is to minimize the global cost (more accurately, the minimum loss of surplus) required to supply a given amount of energy services. In addition to the 5 end-use sectors (agriculture, industry, residential, commercial and transport), TIMES also explicitly considers two intermediate sectors (refinery and power sector). The time horizon covered goes up to 2060. The model is used to explore the alternative evolution paths of the energy system under different technological, economic, environmental or policy assumptions and to evaluate the effectiveness of environmental and energy policies and their impact on the energy system [19].

Using the TIMES_RSE model, we identified the best mix of energy sources and technologies that allows to meet the projected demand for energy services over the entire time horizon, and to achieve the goal of a complete decarbonization by 2050 at a minimum system cost. A predominantly renewable-based energy mix is essential to decarbonize most of the final energy consumption. However, the strong increase of non-programmable renewable sources requires particular attention to power system management and new flexibility resources.

The TIMES_RSE model only considers 12 time slices within a year, a time resolution not sufficient to describe the operability of the power system and the production variability of renewable sources over days and seasons. Therefore, this work followed a two-step approach: first, the scenario analysis for the overall Italian energy system developed with the TIMES_RSE national energy model set out the total electricity demand and indicative generation mix; then, outputs and constraints from the national model were used as inputs for a detailed study of the impact on the Italian power system, and its specific requirements were carried out with a dedicated simulation model, the sMTSIM [20–22].

The sMTSIM (Stochastic Medium Term SIMulator developed by RSE) is a power market simulator detailed on the national market zones (north, center-north, center-south, south, Sardinia, Sicily). It determines the generation set hourly dispatch and the zonal clearing prices of the day ahead market over an annual time horizon. The solution is found by calculating the hourly marginal price for each market zone, the fuel consumption and cost for each thermal power unit, emissions of CO_2 (and other pollutants) and related costs for emission allowances, revenues, variable profits and market shares of the modelled generation companies, as well as the power flows on the interconnections between market zones. The sMTSIM can also provide information about the level of inter-zonal congestion, the overgeneration amount, energy not supplied and the lack of available reserve capacity.

The impact analysis on the power system follows this flow path/loop (see Figure 1):

1. We start from the generation capacity by sources, electricity demand by sectors and Power-to-X capacity from energy scenario by the national TIMES model;
2. We regionalize the national scenario with a multi-regional TIMES model, MONET [23];
3. The sMTSIM runs a power system hourly simulation. In this simulation the system resources (including storage plants and P2X) are dispatched;
4. The emerging criticalities of the simulation results are assessed (congestion, inability to cover load peaks, excess production of variable renewables, lack of reserve margins, etc.);
5. Through appropriate further simulations, the effectiveness of possible interventions capable of mitigating the criticalities detected is assessed;
6. The optimal set of criticality mitigation interventions is selected, and the final electrical scenario is determined, including the estimate of related investments;
7. The main results of the final power scenario (as plant operation hours, new additional flexible capacity, electricity import variation, storage and infrastructure needs) are reported in the TIMES models for the elaboration of the overall final energy scenario.

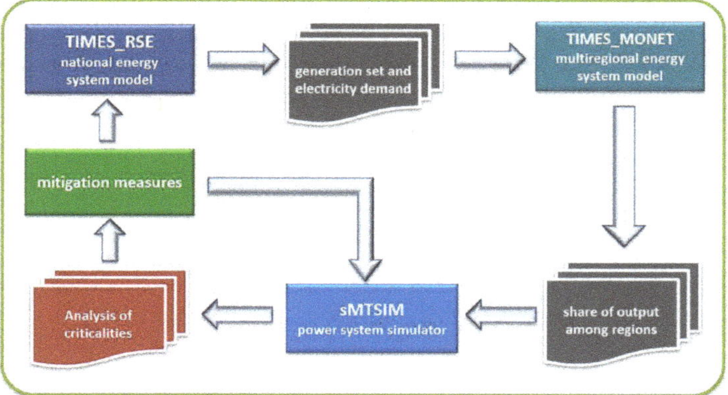

Figure 1. Methodology for analyzing the impact of energy scenarios on power system.

2.3. Key Variables and Main Drivers

Various carbon neutral scenarios have been created by RSE with the following purpose:

- Outlining a strategic path that, with a system vision, takes into consideration aspects of economic and social sustainability, efficiency, and the rational, fair and sustainable use of natural resources;
- Identify the pillars of decarbonization by 2050.

TIMES model, starting from exogenous inputs on the evolution of "main drivers" affecting the evolution of the energy system, is able to determine the optimal combination (i.e., minimum cost) of energy sources and technologies that can satisfy a predetermined demand for energy services (heating/cooling, process heat, motive power, lighting, etc.). The optimization is, of course, bound by the availability of certain resources (technical potentials, capacity of the import infrastructure, natural resources, etc.) whose limits are specified ex ante.

The main assumptions underlying the energy scenario elaborated in this study concern the future trends of some key variables, or drivers, that quantitatively guide the evolution of the system and the energy service demands. The main ones are:

- Demographic dynamics (population and number of families);
- Economic development (evolution of GDP and sectoral added values);
- The cost of energy (international prices of fossil fuels);
- Lifestyle and behaviors.

The development of the system is also influenced by the national and EU political framework, especially in the medium-term, and by the possible technological evolution in terms of efficiency and costs.

To create the scenarios underlying the I-LTS, the drivers in line with the most recent national and international forecasts were used, in particular the drivers of the POTEnCIA Central scenario developed by JRC [24]:

- Demographic growth, the Italian population decreasing in the long-term (Table 1). A gradual population reduction was assumed that would bring the total population below 60 million from 2040. Reducing the members per unit, the number of households would register a slight increase in the projection period 2018–2050, in line with what has been observed in the last few years, reaching an average of 2.2 inhabitants per family in 2050.

Table 1. Evolution of Italian demographic drivers, millions of inhabitants.

	2020	2025	2030	2040	2050
Population	60.5	60.3	60.2	59.8	58.8
N° of households	26.2	26.4	26.5	26.5	26.6

- GDP and added values, consistent with demographic growth with less pronounced growth than in recent years, international fuel prices, CO_2 price of ETS price. GDP and sectoral added values are considered a proxy of production activity and therefore are the variables that drive the energy demand of the industrial, service and freight transport sectors. Overall, at the national aggregate level, GDP grows at an average annual rate of 0.70% in the 2018–2050 projection period (Tables 2 and 3).

Table 2. Evolution of Italian main macroeconomics drivers, average annual rate, %.

	20–25	25–30	30–35	35–40	40–45	45–50
GDP	0.69	0.38	0.29	0.43	0.85	1.20
V.A Agriculture	0.28	0.29	0.25	0.39	0.79	1.11
V.A Service	0.72	0.40	0.30	0.44	0.88	1.24
V.A. Industry	0.64	0.31	0.23	0.34	0.70	0.97

Table 3. Evolution of international fossil fuel prices, € per barrels of oil equivalent, EUR 2016/boe.

	2020	2025	2030	2040	2050
Oil	80.6	91.5	100.8	111.3	116.5
Natural Gas	51.9	56.1	61.1	67.3	69.9
Coal	15.4	18.4	22.0	24.3	25.9

-   Transport activity, the hypotheses on the evolution of the demand for passenger and freight mobility are particularly significant for the definition of a scenario and come from the POTEnCIA Central scenario developed by JRC—Joint Research Center. The projection of the demand for mobility services (passenger mobility and goods handling) is projected over the years according to different rates and depends on variables such as gross domestic product (GDP), population and oil prices. The scenario also contemplates travel containment measures (smart working, teleworking, videoconferencing) and shared transport measures (car sharing/pooling) which reduce the demand for passenger transport (−27% of journeys by car, +60% of journeys by train and increased use of buses in public transport) (Table 4). All these measures, together with population decline projections, lead to a 40% reduction in the number of cars on the roads compared to current levels. Rail transport is also intensifying for goods, which helps to partially contain the increase in demand for goods transported by road and by ship.

Table 4. Evolution of the demand for passenger and freight mobility in 2050, pre- and post-modal shift policies, Gpkm (passenger) and Gtkm (freight).

	Unit	2018	2050 Pre-	2050 Post-
Passenger cars	Gpkm	723	615	525
Motorcycles	Gpkm	32	40	37
Public road transport	Gpkm	103	114	150
Passenger rail	Gpkm	62	82	98
Aviation	Gpkm	74	130	105
Trucks	Gtkm	127	164	134
Freight rail	Gtkm	21	30	37
Inland navigation	Gtkm	65	72	73

2.4. Scenario Definition

The main binding target of the Italian Long Term Strategy, which characterized all the scenarios analyzed, is: "Net zero greenhouse gas emissions by 2050". Three different scenarios (with the TIMES_RSE model) and a sensitivity were built with this target to analyze the impact of the increasingly challenging assumptions in terms of available alternative energies, increasing photovoltaic (PV) potential and the introduction of new synthetic zero-emissions fuels in the industry sector. The energy scenarios analyzed in this paper are:

*   **LTS A**: This first scenario envisages a cap on the PV equal to 200 GW. Fossil fuels continue to be used in power generation and in certain industrial productions thanks to the carbon capture and storage (CCS) technologies (e.g., coal for steel and oil in petrochemicals).
*   **LTS B**: This scenario considers a maximum photovoltaic capacity up to 250 GW and reduces the possibility of using petroleum products and waste for energy generation, thanks to a greater use of circular economy. Industry is also experiencing an important evolution, with the use of natural gas instead of coal in the production of integrated steel and oil products in cement.
*   **LTS C**: This scenario further increases the potential of PV (275 GW), but the main differences are in the industry where the use of petroleum products is severely limited;

in particular, coal is completely eliminated and hydrogen is introduced for steel mills and some other industries (such as glass, chemicals and ceramics).

- **LTS Cs**—sensitivity: The sensitivity of the LTS C scenario to test even more challenging assumptions for the electricity system and evaluate its resilience: 100% generation from RES with higher PV penetration (up to 300 GW), no generation from fossil gas (even if with CCS) and a reduction of imported e-fuels.

The analysis carried out shows how all these scenarios achieve climate neutrality with an energy system in equilibrium, despite the profound transformations required, especially in the electricity system, in line with the in-depth analysis of the European Commission [14]. However, each presents its challenges in terms of costs, construction difficulties, uncertainty about the necessary technologies and the sectors most affected by the transformation.

These scenarios were created trying to satisfy the Italian energy needs without resorting to an increase in the import of energy sources compared to current levels. No analyses have been made on the future prices of commodities in neighboring countries (for example, electricity or hydrogen import from North Africa).

3. Scenario Results

In this paragraph, the results of the TIMES-RSE, TIMES-MONET and the sMTSIM optimization loop are presented. The optimization of the Italian energy system is assessed under the binding constrains presented in Section 2.2, pursuing a minimum system cost objective function. The main outputs of the three models' elaborations are here presented: the carbon emissions trends and energy consumption by fuel and sector; the main technologies deployed; the penetration of electricity and renewables; the evolution of the power sector in terms of technological capacity and production; the development of hydrogen and other green fuels in terms of production and consumption from end-use sectors. Some of the data presented were obtained from further elaborations outside the modeling suite (e.g., energy intensity indexes).

3.1. Carbon Emissions Pathways

By extending the virtuous energy–environmental dynamics envisaged by the National Integrated Energy and Climate Plan (I-NECP) to 2050 [25], Italian GHG emissions can be reduced by about 60% compared to 1990 levels. In 2050, there are around 220 Mton CO_2 eq remaining which, when taking into account the removal of the LULUCF sector, fall just below 200 Mton (net) CO_2 eq. Of these residual emissions in 2050, about 70% comes from "energy uses". Looking at "non-energy uses" covering the remaining 30%, it emerges, as already noted in the I-NECP, that it is substantially difficult to compress emissions from agriculture/livestock and industrial processes [12].

Climate neutrality in 2050 will be a tough challenge for Italy: there will be incompressible residual emissions deriving mostly from industrial processes, the use of solvents and F-gases, waste and the agricultural and livestock sector, the so-called "hard-to-abate sectors". The residual emissions can be offset with CO_2 sequestration and the absorption of "natural sinks" [14]. However, policies to combat fires and sustainable soil management will have to be implemented to maintain and increment the absorption capacity of these sinks [26].

The analysis carried out in support of the I-LTS shows that climate neutrality by 2050 is only possible with a change in the energy paradigm.

From the scenarios carried out, it emerges that each sector will have to contribute to the emission reduction according to its own peculiarities (Figure 2), in particular:

- The energy industry, in particular the power sector, has the potential to eliminate its own emissions; indeed, it can even contribute to climate neutrality with negative emissions, in line with [27];
- The manufacturing industry can contract its emissions. It needs alternative fuels and CCS for emissions deriving from industrial "processes", but it is difficult to achieve net zero emissions, in line with [28];

- Transport and other energy sectors will necessarily have to eliminate their emissions by resorting to all possible options (efficiency, electrification, renewables and green fuels), in line with [29,30];
- Agriculture will be the most difficult sector to decarbonize, and there will be residual emissions to be compensated with sinks [31];
- The waste sector will be able to reduce emissions by resorting to forms of circular economy [32];
- The analysis carried out by the I-LTS working group (in particular ISPRA) considers that, with adequate policies to combat fires and sustainable management of the soil, the emission removal capacity of the LULUCF sector can be brought back to an all-time high of 45 Mton CO_2 eq [12];
- CCS will be a necessary technology to offset the atmospheric emissions from hard-to-abate sectors [28]. The amount of CO_2 captured to CCS will obviously depend on the production methods, the emergence of new technological solutions and changes in the lifestyle of citizens. In the I-LTS, a variable quantity of 20–40 Mton CO_2 eq captured to CCS has been estimated (analysis carried out by ISPRA).

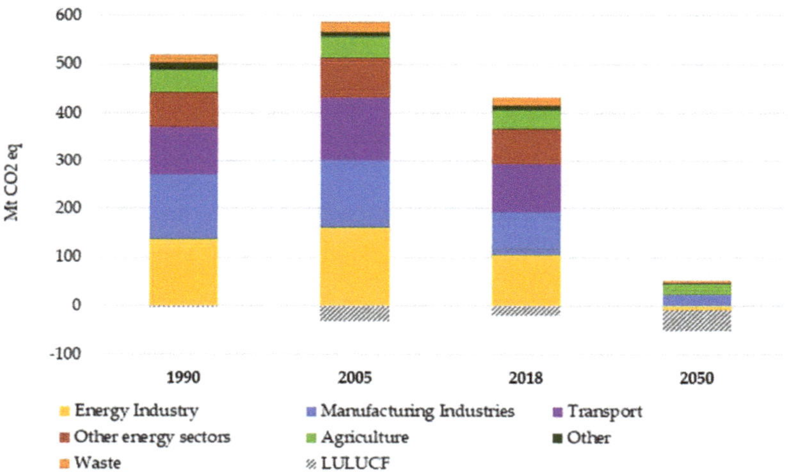

Figure 2. Greenhouse gas emissions and removal in I-LTS; Mt CO_2 eq [25].

In this emissions framework, the role of the energy sector is undoubtedly very important [14]. Starting from the residual emission gap identified within the framework of the extended I-NECP scenario, several simulations were conducted to identify the combinations, synergies and critical issues of the potential levers to achieve climate neutrality by 2050. Reductions in net emissions can be achieved through different portfolios of the mitigation measures, in accordance with the scenarios of the European Long term strategy and illustrated in Figure 3.

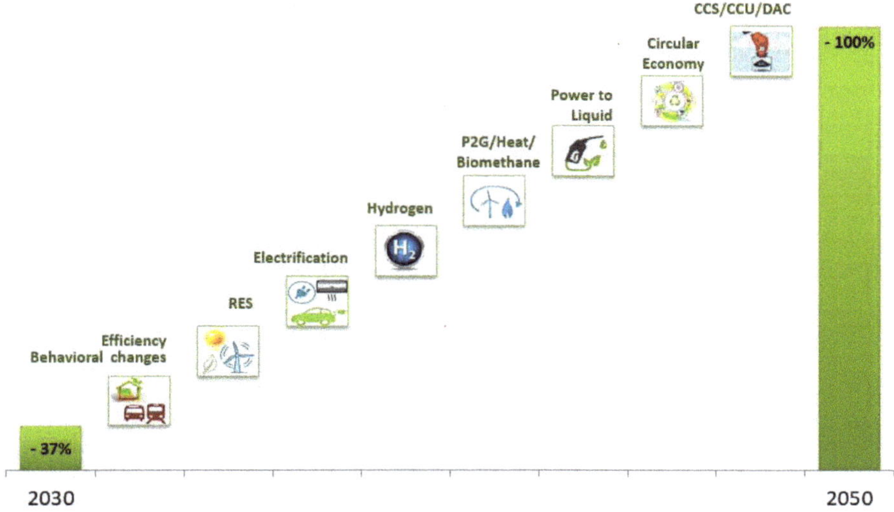

Figure 3. Main decarbonization options.

To achieve climate neutrality by 2050, the system will have to resort to fundamental levers with strong synergies between them:

- Energy efficiency (in full consistency with the European "energy efficiency first" principle) accompanied by behavioral changes (more public passenger mobility and consumption awareness in residential buildings) and a circular economy [7];
- A new energy mix consisting of renewable sources and carbon-free energy carriers such as hydrogen, but also all synthetic fuels derived from hydrogen and electricity (Power-to-X);
- Complete power sector decarbonization with renewable sources, but also the use of carbon capture and storage and use (CCS/CCUS) associated with biomass plants (BECCS) to achieve negative emissions;
- A significant electrification of up to 55% of final consumption: in particular, electrification increases in buildings, especially for heating and cooking, and in the transport sector, driven by the high penetration of electric vehicles for cars and buses.

3.2. The Role of Energy Efficiency in a Carbon Neutral Italy

Rational energy use is a paradigm that must necessarily underpin any initiative towards carbon neutrality, as is widely documented in the literature [33–35]. The European Commission itself has made the energy efficiency first principle a solid foundation on which the 2050 decarbonization scenarios for the European Union were built [14].

In fact, the current energy consumption trends would be unsustainable in the future, as they could deplete the natural, technical and economic potential of energy resources, thus leading to disastrous consequences for the economy. Hence, the reduction of energy demand becomes a tool to protect the national energy system from future uncertainty (e.g., energy price volatility, geopolitical crises), which is a key aspect in the Italian context where energy dependance reaches 80%.

The expected evolution of the Italian primary energy consumption is shown in Figure 4, where the current situation is compared to the optimization results from the TIMES-RSE model. In the 2050 scenarios, primary energy consumption is reduced to 90–101 Mtoe, which represents a cut of over 30% compared to 2019 levels [36]. One could have expected an even steeper reduction of primary energy given the ambitious decarbonization goals. However, with the expansion of the power sector, the unprecedented growth of carbon-free fuels (e.g., hydrogen, biomethane) generates new streams of energy consumption and transformation (see Section 3.4).

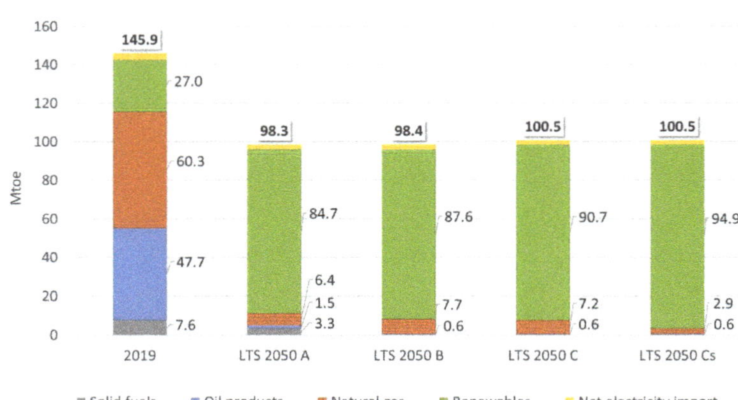

Figure 4. Expected evolution of primary energy consumption (Mtoe): comparison between the current state (2019) and the LTS 2050 decarbonization scenarios.

Compared with today, the 2050 primary energy mix appears completely transformed. The share of fossil fuels is drastically reduced from 79% in 2019 to less than 13% in the LTS scenarios. In such scenarios, the consumption of solid fuels is minimized to 0.6–3 Mtoe, depending on whether the scenarios explore a complete phase-out of coal and its replacement with solid fuels obtained from circular economy measures (the LTS B, LTS C and LTS Cs scenarios) or whether some industrial processes (steel, petrochemical) continue to use fossil fuels combined with CCS technologies (the LTS A scenario). As regards oil products, the LTS B, LTS C and LTS Cs scenarios assume they are eliminated from energy use, while the LTS A scenario considers a marginal use in industry. Natural gas follows a declining path as well, after playing a crucial role in the energy transition towards the Italian 2030 targets [25]. The primary energy consumption of this commodity is reduced by 87–95% compared to 2019, with the highest reduction corresponding to a complete phase-out from the power sector.

Renewables become the backbone of the Italian energy system with a share of 86–94% in the primary energy mix. Such contribution is tripled compared to 2019 and results from the growing penetration of wind and photovoltaic technologies to meet the accelerating electrification of energy demand. These findings are supported in the research by a strong general agreement that renewables will play a key role in decarbonization pathways. According to a study on the future electrification of the Italian energy system, by 2050, more than 80–90% of the electricity will be provided by RES, with the highest shares related to the deployment of emission reduction measures such as a CO_2 price [37]. Another study indicates that in 2050, Italy can potentially achieve approximately 86% penetration of RES in its electricity supply [38]. Our results are also consistent with the general context of EU decarbonization pathways. In the EU Long Term Strategy [14], a new energy system dominated by renewables emerges (51–62% of gross inland consumption), moving away from fossil fuels.

The role of bioenergy (biogas, biomethane and biomass) is also significant. They are used not only for the decarbonization of final thermal energy use, but especially in power generation, where their combination with CCS and CCU technologies generates negative emissions, a crucial element to reach the carbon neutrality target. The importance of carbon capture technologies is also highlighted in the 1.5TECH scenario for EU decarbonization [14], where by 2050 CCS is expected to represent 5% of the total net electricity generation and to be mostly associated with biomass power generation to generate negative emissions. However, as highlighted by Fajardi and Mac Dowell [39], a whole-systems analysis for the value chain of bioenergy with CO_2 capture and storage (BECCS) is needed (from cultivation to transport and energy conversion). In fact, BECCS could lead to both positive and negative carbon emissions depending on the conditions of its deployment [40–42].

The contraction in primary energy consumption directly affects the national energy intensity indicator (Figure 5). Compared to 2019, the LTS scenarios reduce energy intensity by an average of 44% because of both energy efficiency and the effect of the expected economic growth.

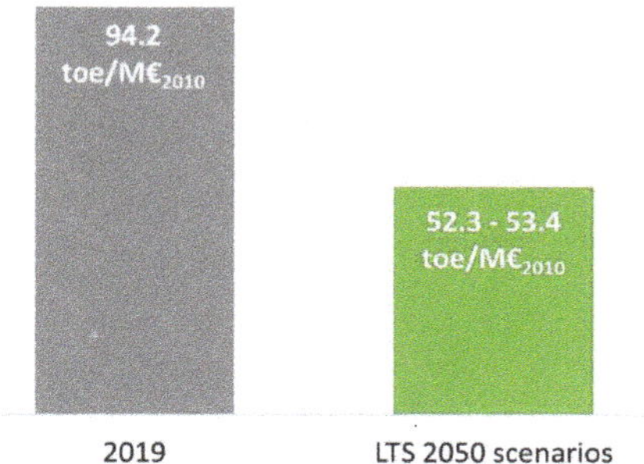

Figure 5. Energy intensity of GDP: comparison between the range of LTS scenarios and 2019.

The gross inland consumption to population ratio also decreases significantly (Figure 6). The reduction of gross inland consumption is over 10 times higher than the expected population decline rate, thus generating a 30% cut in per capita energy demand.

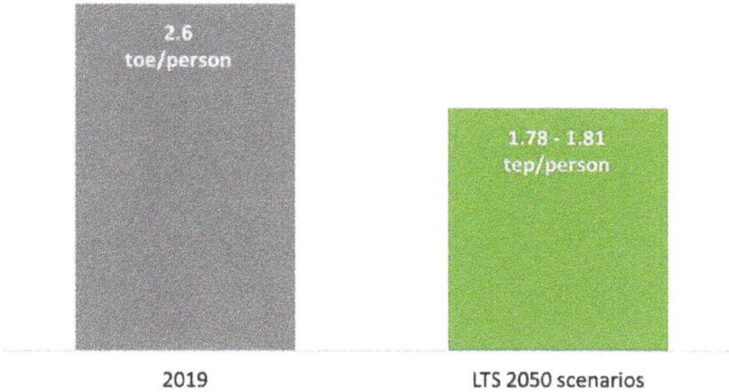

Figure 6. Gross inland energy consumption per capita for LTS 2050 scenarios and current state (2019).

Even greater changes in the structure of Italian energy demand can also be seen in the evolution of final energy consumption (Figure 7).

In the LTS scenarios, the structure and size of final consumption in all these sectors undergo profound transformations to pursue the goal of decarbonizing the entire economy. In these scenarios, final consumption is reduced to 68–71 Mtoe, a contraction of about 40% compared to the situation in 2019. The electricity vector displaces fossil fuels, becoming the main source of final energy consumption by an amount between 37 and 39 Mtoe, more than 50% higher than in 2019. The intense electrification of consumption is accompanied by an extraordinary growth of renewable sources, whose penetration in end-uses reaches 21–24 Mtoe. In detail, the contribution of renewable sources has almost tripled compared to 2019.

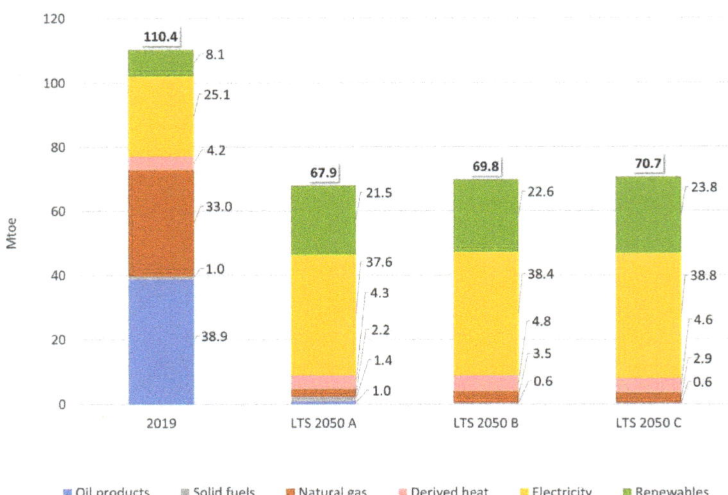

Figure 7. Evolution of final energy consumption by energy carrier: comparison between the current state (2019) and the decarbonization scenarios.

Such a transformation of final energy consumption results from the different decarbonization pathways experienced by transport, industry, residential, tertiary and agricultural sectors (Figure 8).

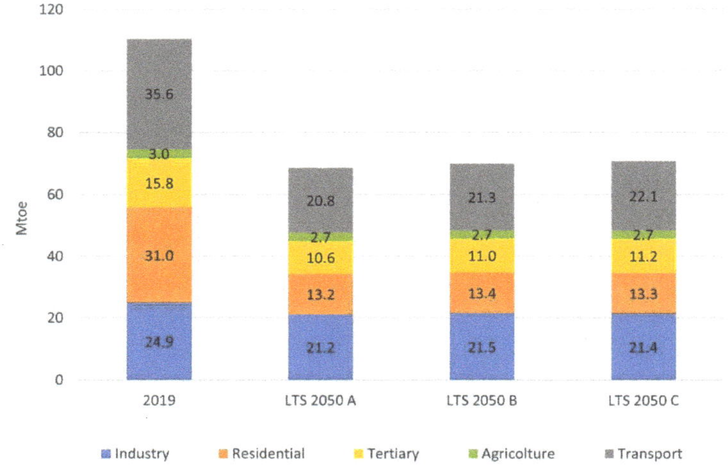

Figure 8. Expected evolution of final energy consumption by sector: comparison between the current state (2019) and the decarbonization scenario.

In the LTS scenarios, the energy consumption of transport amounts to 21–22 Mtoe, almost half compared to 2019. A major role is played by measures aiming at reducing private car use and ownership, such as smart working, car sharing and the modal shift towards public transport and soft mobility (cycling, walking). Addressing the demand of freight transport is also important: measures such as the optimization of logistic chains and the minimization of empty runs are compelling under the 2050 scenarios. However, profound transformations in the transport sector can also be found on the supply side in terms of innovation and technological change. The penetration of electricity contributes to increase the energy efficiency of the vehicle stock. Meanwhile, hydrogen and other electricity-based synthetic fuels (e-fuels) are required to replace fossil fuels in those transport segments where technical and economic barriers of a direct electrification exist, including aviation and marine transport.

The agricultural sector is characterized by a reduction in the consumption of petroleum products, which are replaced by electricity and renewable sources, whose consumption doubles compared to 2019. The combination of these effects leads to a 10% reduction of energy consumption in the sector.

As regards industry, the contribution of fossil fuels is drastically reduced and replaced by a dramatic electrification and thermal energy produced by renewable cogeneration and the direct use of biomass, biomethane and hydrogen. The combination of these factors allows for significant gains in terms of energy efficiency, with an average 15% energy savings in 2050 compared to 2019.

Energy consumption in the residential sector stands at 13 Mtoe, which represents a 58% cut compared to 2019. These enormous energy savings are explained by the high renovation rates of Italian residential buildings, which are currently characterized by mediocre energy performance on average. In fact, 52% of the residential building stock was built before 1970, which exacerbates the energy intensity of the sector, especially with regards to space heating. A 2% yearly renovation rate is required in residential buildings to achieve the 2050 targets, 80% of which is identifiable as deep renovation. This represents a challenging effort if compared to the 0.9–1% rate evaluated in the Italian NECP.

Similar results can be found in the energy scenarios supporting the European Long Term Strategy [14], where the projected yearly renovation rates in the residential sector range between 1.7–1.8%, depending on the decarbonization ambition of the different scenarios. In fact, old buildings are peculiar not only to Italy but to many other countries in the European Union, where on average about 35% of residential and tertiary buildings are over 50 years old and almost 75% were built before energy performance standards were established. Another key element in the decarbonization of the residential sector is electrification, which increases by about 45% compared to 2019 thanks to the adoption of heat pumps in heating and domestic hot water, the two most energy intensive services for dwellings.

Moreover, in the tertiary sector, energy efficiency leads to the phase-out of oil products and natural gas (replaced mainly by renewable sources and secondly by electricity). These effects are visible in the containment of the otherwise rampant energy consumption of services, driven by the economic growth of the sector: the reduction in energy consumption in the sector is on average 30% compared to 2019.

3.3. Electrification as a Pillar of Decarbonization Pathways

To achieve climate neutrality by 2050, besides a radical transformation of energy consumption, a need for the significant electrification of end-uses emerges as well (Figure 9). By 2050, electricity will have a central role in the energy system, growing from a 22% share of final consumption in 2019 to almost 55%. Electricity will play a key role across all sectors, though with different patterns and specific challenges.

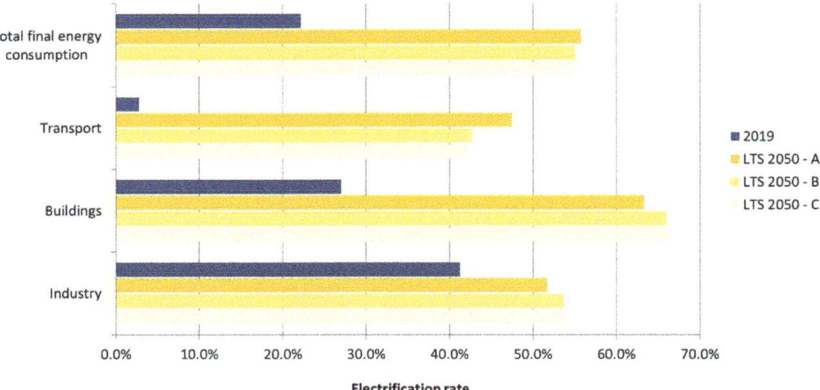

Figure 9. Electrification rates by sector in the 2050 LTS scenarios.

Transport is the sector where electricity is expected to bring the greatest change. Electrification will increase in 30 years from 3% (2019) to over 40% (2050), thus becoming the main energy carrier in the sector. Mostly cars will be affected by this huge transformation, leading to around 20 million battery electric vehicles in 2050 (Table 5).

Table 5. Electric cars in Italy in 2019 and the 2050 LTS scenarios.

	2019	**LTS 2050 Scenarios**
Number of electric cars	~39,000 (BEV [1] + PHEV [2])	~19–20 million 100% BEV

[1] Battery electric vehicle; [2] Plug-in hybrid electric vehicle.

As regards heavy-duty vehicles, electrification is hindered by technical and economic barriers which inevitably prompt the deployment of alternative fuels to reach the decarbonization goals. In fact, traveling long distances while transporting large amounts of goods requires high capacity batteries to satisfy the extensive autonomy and power demand of heavy-duty vehicles, which, however, is detrimental to investments costs.

The buildings sector experiences a significant increase in electricity penetration as well. While energy consumption in the tertiary sector has been historically dominated by electricity, it is in the residential sector that the greatest electrification occurs. The energy service most affected by this transformation is undoubtedly space heating, which nowadays is mainly provided by natural gas boilers. In the decarbonization scenarios, fossil fuel-fired heat generators are replaced by reversible electric heat pumps, which operate throughout the year, providing space heating and cooling as well as domestic hot water. It is estimated that around 70% of residential households will use electric heat pumps in 2050.

Furthermore, in the residential sector, electricity also replaces natural gas in cooking systems, where most meals will be prepared using induction hobs.

Electrification also reaches agriculture, where electric farm machinery is introduced after the phase-out of oil products.

In the industrial sector, electrification increases up to 54% of energy consumption. Electricity penetration is significant not only in less energy intensive sectors, but also in steel production via direct reduced iron. Options such as electric arc furnaces, robotization, digitalization and additive manufacturing contribute to increasing the demand for electricity in the sector.

Alongside the more traditional end-use sectors, the growth of electricity consumption also affects innovative and flexible loads (Figure 10). These mainly consist of Power-to-X plants that produce a variety of new fuels and energy carriers, including biomethane, hydrogen and liquid synthetic fuels, as well as heat. Electricity consumed by Power-to-X amounts to 160–230 TWh, which makes these facilities the largest contributor to the increase in electricity demand in 2050.

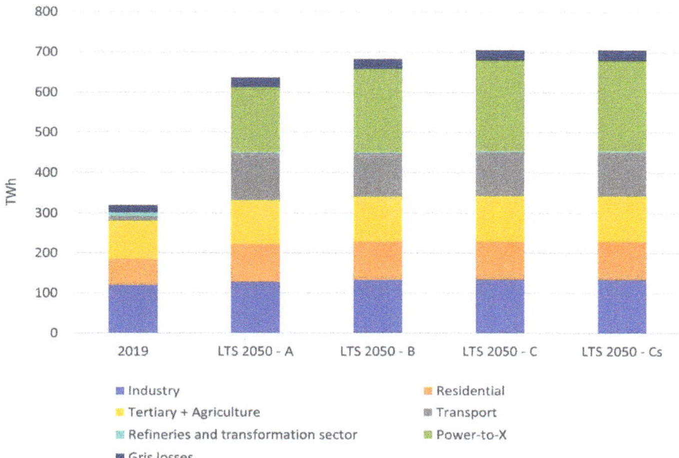

Figure 10. Evolution of electricity consumption in the 2050 LTS scenarios.

3.4. The New Challenges of the Power System

In the decarbonized scenarios, the power system undergoes a deep revolution both in its operating mode and in its role within the energy system. In fact, if in recent years electrification has mainly represented a way to improve the energy efficiency, in the 2050 scenarios it takes a more prominent role in the decarbonization of all the other sectors as well. More precisely, the power sector will have to substantially contribute to all these goals:

- Further electrification of energy uses in order to obtain more efficiency and emissions reduction in all the sectors;
- Synthetic fuel generation without CO_2 emissions (according to the European Hydrogen Strategy [43], EU industry is taking up the challenge and has developed an ambitious plan to reach 2×40 GW of electrolyzers by 2030);
- Direct heat generation without accounting CO_2;
- Opportunity of CO_2 capture from some huge emission source;
- Removal of CO_2 from the atmosphere using the non-dispatchable renewable overgeneration of electricity.

3.4.1. Renewables to Decarbonize the Power System and the Power System Flexibility Issue

In order to reach the decarbonization goal and exploit the decarbonization potential of electrification of final uses, the energy model indicates that the energy system needs, first of all, to have a decarbonized power generation. As our scenarios conclude, by 2050 all the scenarios foresee a renewable generation exceeding 95% of total production, split among the following sources:

- Mainly solar source—between 200 and 300 GW;
- A strong expansion of wind generation—50 GW, of which 16–17 GW is offshore (medium and high deep water);
- New sources like waves and tides—1.5 GW;
- The traditional hydro-, geo-thermal and bio-energies are maintained or developed up to a total of 43 GW.

The flexible fossil fuel production will be limited under 30 TWh (worst case) and always with CCS. In order to reach the decarbonization target, the energy system will need a significant amount of electric renewable generation for both the direct consumption of electricity and indirect use for the production of e-fuels. Generation from solar and wind sources could therefore reach between 440 and 550 TWh (Figure 11).

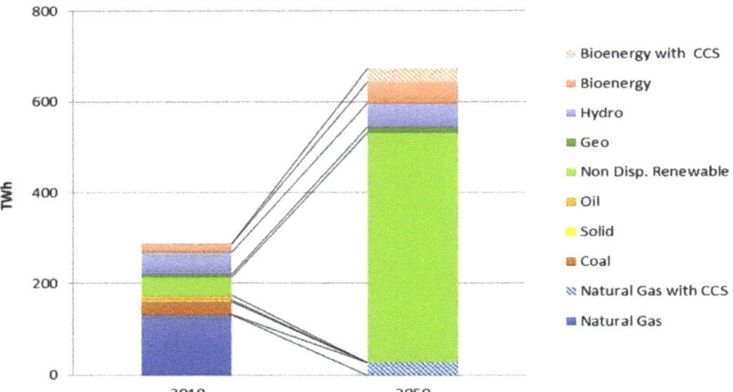

Figure 11. Expected evolution of electric energy production (TWh): comparison between the current state (2018) and the LTS 2050 decarbonization scenarios.

This new configuration of the electrical system leads to the installation of a high storage capacity for the electrical system, with hydro-pumped storage and batteries both on the network and distributed. They have several uses: moving renewable production surpluses into hours with higher demand, or also by P2X plants, and providing huge amounts of power for flexibility and fast backup services.

For pumped storage plants, a potential of approximately 10 GW, inclusive of marine plants, could be developed in addition to the 7 GW existing today. For batteries, the installation varies between 28 GW and 38 GW.

The flexibility of the power system, i.e., the ability to keep production and consumption always safely balanced, even in case of unexpected shocks, is indeed a critical issue to addresses for the evolution of a power system with a high penetration of renewable sources. In a traditional power system setting, flexibility needs were covered, in all their dimensions, by dispatchable thermoelectric and hydroelectric plants. Those plants were perfectly capable of providing all the desired flexibility (classified by various delivery speeds and durations). The need for flexibility is, in fact, an articulated set of specific needs from the intra-hourly time scale (driven by the uncertainty of renewable production forecasts) to the annual scale (driven by the seasonal variation of demand and PV potential generation).

Therefore, a huge flexible consumption capacity and storage capacity will be required. The flexible loads (flexible demand) are mainly from P2X plants transforming electricity into other energy carriers such as green methane, hydrogen and liquid fuels, as well as heat. These are responsible for the electricity consumption, in 2050, of between 160 and 230 TWh (depending on the different scenarios considered), which doubles the total electric demand from current levels (up to 600–700 TWh).

As the installed capacity of non-programmable electric renewables, characterized by intermittent generation, increases, there will be many hours in the year in which electricity production will exceed demand. After granting the daily hourly balance of final electricity consumption (e.g., via batteries and pumped storage), further overgeneration can be transformed into heat or into hydrogen and subsequently into synthetic fuels based on hydrogen and CO_2. The simulations of the power system carried out for the year 2050 based on the decarbonization scenarios show that, in order to give the necessary flexibility to the electricity system, P2X plants cannot operate as basic plants, but only during hours of excess production from intermittent renewables; consequently, they reach a load factor of only approximately 2000 equivalent hours per year.

Another important element of flexibility, on the demand side, will be the recharging of electric vehicles with systems capable of modulating the recharge in the most suitable hours, and even exploiting the car batteries for network services (the so-called "vehicle-to-grid").

Finally, new potential for flexible use of electricity can be realized through seasonal heat storage systems applied to district heating networks. In this way, the heat produced by the surplus of PV production during spring or summer will be accumulated to the supply of heat in the winter season (power-to-heat technology). This system provides additional useful flexibility for the electrical system, and it contributes to the decarbonization of the building sector.

As regards the network infrastructures, there are clear criticalities due to the doubling of electricity consumption compared to today, and even more so due to the quadrupling of the installed, largely non-programmable, power capacity. Therefore, important investments will be necessary, which will have to accompany the development of the system. Even more important will be the planning and location of the P2X plants. The coupling with the gas network will allow to overcome both the transport limits of the national grid and the limited storage capacity of the electricity system, opening up to the much higher potential for energy storage on a daily scale and the possibility of the seasonal accumulation of the gas system.

3.4.2. Sector Coupling and the Role of P2X

The presence of significant quantities of electrolyzers or other P2X systems provides many advantages to the entire energy system, but only if they are correctly located and if their functioning is driven by the needs of the power system. In fact, electrolyzers have a value that goes beyond their pivotal role in the sector coupling (i.e., the joint point for transferring renewable energy from the electricity system to other sectors of consumption). The production of hydrogen obtained solely from the surpluses of renewable electricity production has many advantages: (1) It enables a greater electricity production from renewable sources (supporting the installation of new renewable capacity even in situations otherwise not convenient); (2) It guarantees electric energy at very low marginal cost for the production of H2 (which would be produced with lower variable costs than methane reforming); (3) It reduces the investments required for the power system, including a lower need for batteries.

In addition to these, the flexibility and the speed in the load ramp rate of the electrolyzers (rapid loading ramp and start up speed, unlimited duration), as well as the significance of the power levels involved, makes them particularly interesting for the supply of FRR (frequency restoration reserve) and RR (replacement reserve) services for the grid frequency control. Some initiatives for large scale demonstration have been started in the H2020 context with the H2Future project [44], and in other initiatives (i.e., the HyBalance project, Denmark, [45]). Moreover, ENTSO-E considers this topic in its *Development and Innovation (RDI) Implementation Report 2021–2025* [46] as a guiding instrument for the collaborative research program of transmission system operators (TSOs) in the coming five years. The plants can be made available to switch on to provide a downward reserve or to prepare to increase the upward reserve margins, or they can reduce the load when in operation to participate in the upward reserve. An even more interesting point is that the electrolyzers can provide such regulation services with very competitive variable costs with respect to both more flexible final consumption and non-programmable renewable production, and often even with respect to traditional generation (especially on the upward services). Finally, participation in such services would allow to increase the economic returns of the electrolyzers, boosting the investment case to profitable levels or, in the worst case, allowing a lower need for economic support for this technology.

On the other hand, the production of hydrogen concentrated in few hours and, even worse, in some seasons, causes greater problems in the supply of hydrogen to end users; in fact, expensive hydrogen storage solutions will be needed both in the short term (days) to deal with days of scarce renewable electricity generation, and in the medium term, in seasons with generally poor production (winter and autumn) needing to resort to seasonal storage or other energy carriers. The feasibility (technical and economic) of seasonal storage, such as geological storage, is highly uncertain at the moment. It is necessary to arrange transport to and from the storage site, with energy costs and the need for infrastructures, and it also introduces losses both in energy, by compression and decompression, and in part of the gas itself introduced into the storage, which would no longer be recoverable.

In order to have a first quantification of the H2 seasonal storage capacity the system needs, we compare the H2 production profile with the H2 demand profile during the year. In Figure 12 are shown the weekly average profiles (moving average) of the exceedance in electric renewable generation which is used by the P2X plant and the unused overgeneration. In Figure 13 is shown the weekly average profile of H2 production from electrolyzers and biomass gasification plants, and the area corresponding to the seasonal storage capacity needed to meet the H2 final users' demand is indicated. In the scenarios analyzed in 2050, considering a best case of a very flexible use of H2 from biomass gasification, the minimum hydrogen storage capacity should be slightly more than 1 Mtep (380 kt H2). In the worst case of an inflexible and flat H2 bio production, the storage capacity rises to 1.6 Mtep (Figure 14).

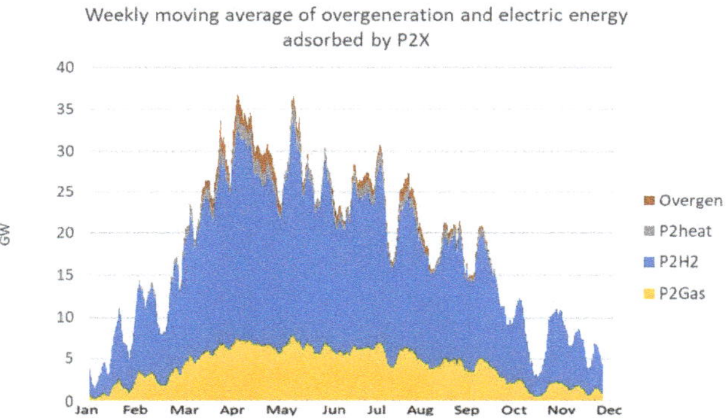

Figure 12. Profiles of renewable electricity surplus consumption in the I-LTS decarbonization scenarios by 2050.

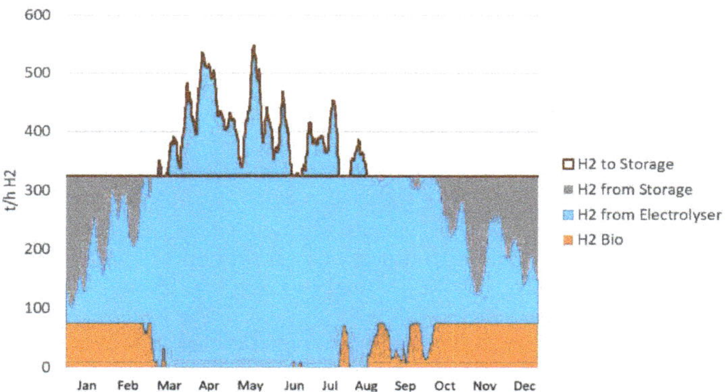

Figure 13. Weekly moving average of H2 production with flexible generation of H2 from biomass gasification.

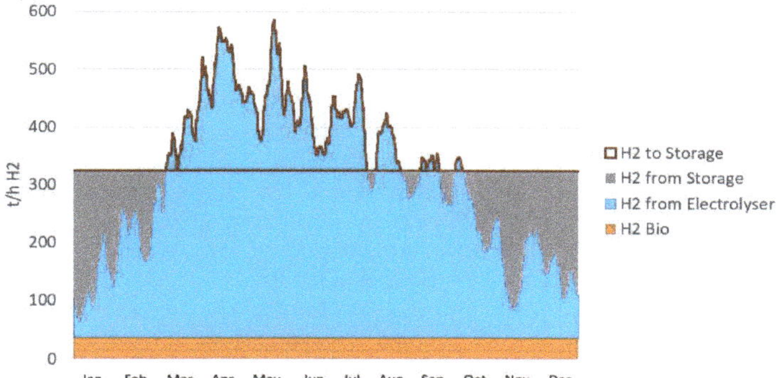

Figure 14. Weekly moving average of H2 production with flat generation of H2 from biomass gasification.

3.5. Hydrogen and Synthetic Fuels: A Disruptive Force

In the LTS scenarios elaborated with the TIMES_RSE model, renewable electricity generation will decarbonize most of the energy consumption by 2050, but not all fossil fuels can be replaced with electricity. Hydrogen has a real potential to fill part of this gap as a vector for the storage and exploitation of renewable electricity, in accordance with [43]. It can be used directly for the decarbonization of non-electrical uses, or it can be transformed into alternative zero-emission fuels in combination with carbon deriving from zero-emission biogenic forms. In our study, we have considered a sensitivity by raising the emission reduction objective. The hydrogen role in the energy sector is strictly linked to the level of the climate-altering emissions reduction objective: an emission target of at least 50% reduction compared to 1990 is needed to promote the diffusion of the hydrogen vector from a few pilot plants and experimental applications at more extensive levels of use (Figure 15).

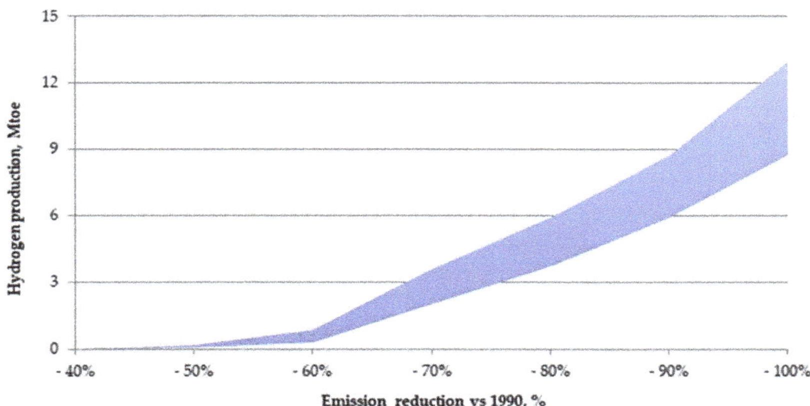

Figure 15. Production of hydrogen linked to the reduction of GHG emissions; Mtoe and %.

The hydrogen development is closely symbiotic and integrated with the growth and spread of non-programmable renewable electricity sources. Deeper decarbonization of the energy system requires an increasingly green power system in order to electrify end-uses with a zero-emission vector [38]. The presence of significant overgeneration resulting from intermittent renewables production, together with the expected cost reduction of electrolyzers, make green hydrogen economically competitive between 2030 and 2040. Hydrogen development is therefore boosted by an increasing share of non-programmable generation sources and their cost reduction, together with decreasing electrolyzer costs and improved supply chain logistics [43]. The large-scale adoption of hydrogen (or hydrogen-derived fuels and products), as in full decarbonization scenarios, can in turn fuel a significant increase in the demand for renewable energy generation [47]. On the other hand, stringent emission constraints make hydrogen convenient if not even necessary for those applications that cannot be electrified (hard-to-abate).

In all the scenarios analyzed, the first commercial application of hydrogen is, together with certain industrial applications, in heavy road transport and trains in the decade 2030–2040, while passenger mobility, mainly cars, decarbonizes through electrification. By 2050, the use of hydrogen in the transport sector becomes significant in our scenarios due to the expected decrease in the costs of the technologies. In the decade 2030–2040, the Power-to-X (P2X) technologies appear for the transformation of electricity into hydrogen and subsequently into synthetic fuels based on hydrogen and CO_2, with an initial greater diffusion of P2L (power-to-liquid) compared to P2G (power-to-gas) (Figure 16).

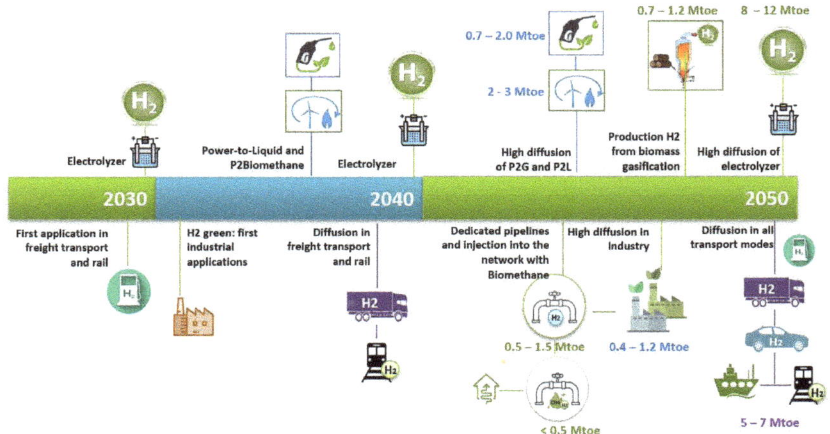

Figure 16. Evolution of hydrogen energy uses in LTS scenarios; Mtoe.

Driving the initial promotion of P2L is the increased demand for decarbonization options for the freight sector, and liquid e-fuels are produced with similar characteristics to gasoline, diesel, naphtha or jet fuel. These types of e-fuels have a simpler storage of hydrogen and an easier integration with the existing logistics infrastructures (such as refueling infrastructures, tanks, etc.) so that the product management phase is more convenient than hydrogen as it is, at least until the necessary infrastructures for hydrogen transport are created. Figure 17 shows the range of use of hydrogen by sector in the various complete decarbonization paths analyzed.

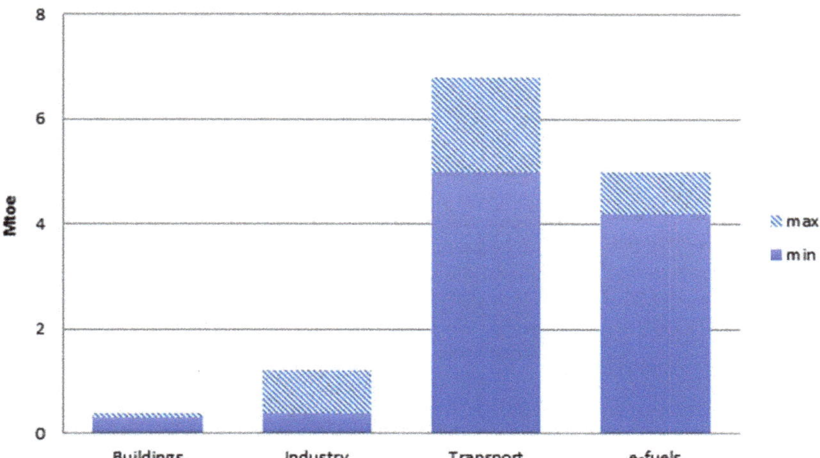

Figure 17. Hydrogen breakdown in the I_LTS scenarios by 2050.

4. Discussion of Decarbonization Challenges

This paper presents the energy scenarios developed by RSE to support the Italian government in outlining a national plan for a complete decarbonization by 2050, in accordance with the Paris Agreement. The TIMES-RSE model was used to represent and optimize the Italian energy system, as well as to investigate the role of different sectors, technological evolutions, innovative energy carriers and the paradigm shifts needed. Moreover, future challenges and technical and social criticalities are highlighted.

The following elements and trends emerge from the four energy scenarios analyzed:

- The power sector has the potential not only to achieve carbon neutrality, but even to generate negative emissions when biomass power plants are associated with CO_2 capture and storage technologies;
- Transport, buildings and agriculture sectors can completely abate their emissions by 2050 as a result of massive energy savings, electrification and the deployment of green fuels such as hydrogen, biomethane and methanol;
- The industrial sector can greatly reduce its combustion emissions thanks to energy efficiency, electrification, alternative fuels and carbon capture technologies, but difficulties remain with the emissions deriving from industrial processes;
- The residual emissions from industrial hard-to-abate sectors, agriculture and livestock can be offset with the absorption of natural wells.

Given these premises, to achieve climate neutrality by 2050 the Italian energy system will have to resort to the following elements that are strongly interlinked:

- The drastic change in the energy mix in favor of carbon-free sources;
- Energy efficiency accompanied by behavioral changes that affect passenger, mobility and energy consumption in the buildings sector;
- A significant electrification of end-uses.

The scenarios are built upon the fundamental energy efficiency first principle, whereby energy savings play a key role in Italian decarbonization. The reduction of energy consumption is expected to reach −30% and −40% for primary and final energy consumption, respectively, with respect to the 2019 situation. However, the expansion of the transportation sector and the energy demand for the production of new alternative carbon-free fuels (hydrogen, e-fuels) reduce the energy saving potentials for primary energy consumption.

Another fundamental aspect to tackle climate neutrality is a significant electrification of end-use sectors. Electricity becomes the main energy carrier in all sectors, especially in transport, where electric vehicles rank first in the share of cars and buses.

Where electrification encounters technical and economic barriers, alternative fuels are needed not only in transport but also in some industrial sectors.

Renewables are crucial for the decarbonization of the energy sector. By 2050, renewable energies are expected to make up to 80–90% of primary energy consumption, thus completely overcoming fossil fuels. The latter will still have a marginal role in hard-to-abate sectors, but their use will necessarily be paired with carbon capture technologies. The RES-e generation by solar and wind sources will reach values between 440 and 550 TWh. This unprecedented renewable generation will need to be assessed from different perspectives, such as the natural and technical potential of wind and solar resources on the national territory, as well as the impact of renewable plants on the environment and landscape, but also the consumption of the raw materials necessary for their construction [48].

In addition to the technological and environmental aspects, it is necessary to solve the problem of power systems balancing, which becomes extraordinarily complex in the scenarios analyzed. In the power system, the electricity production doubles, the renewable capacity installed increases tenfold and the power peak production is four times the historical levels. An extremely complex and smart management of the electricity system will therefore be required, based on the best possible coordination of all flexible resources. It is necessary that all potential flexible resources (both production and consumption) will provide flexibility, contributing according to their own characteristics. Obviously, an important upgrade of transmission and distribution networks and the use of an expensive system of batteries is also indispensable.

A large share of the new generation capacity installed will be directly connected to the distribution grids, which therefore will have to be significantly reinforced [49]; however, more importantly, distribution grids have to evolve in order to manage the greater distributed generation sources and loads compared to today.

Among the various forms of flexibility required by the electricity system, the need for seasonal storage, to transfer summer and spring excess production from photovoltaics to the winter months, is growing enormously. This problem cannot reach a solution within the border of the power system (only the few largest hydroelectric reservoirs could provide some seasonal accumulations of water while pumping, and batteries only work in the short term). The solution suggested here is to use the excesses of renewable electricity production, especially in the seasons with greater producibility, for the production of other energy commodities such as green hydrogen, liquid fuels of renewable origin, synthetic methane and heat for district heating systems with seasonal heat storage.

These uses (generically referred to as P2X) provide several benefits: the power system benefits take advantage of their flexibility, especially seasonal; on the other hand, the receiving energy system obtains a contribution for its decarbonization. It follows that, in order to provide the required flexibility, the P2X plants will have a limited operation, constrained in the hours of excess renewable generation (approximately around 2000 equivalent hours per year). Therefore, the commodity generated must have a transport and storage system adequate to a production concentrated in certain periods of the day/year. At the moment, this is doable for synthetic methane, for instance, or other liquid e-fuels. On the other hand, hydrogen faces greater difficulties for both transport (which has higher costs than other fuels) and storage (which has very high costs for short-term storage facilities and limited availability of seasonal storage, and which is also very concentrated).

Finally, in addition to the criticalities highlighted so far, of an energy system which is so radically different from the current one, the main challenge of the decarbonization process is probably the rapid pace required for the transition. It surely represents a challenge in terms of the industrial investments required, the infrastructure development and the risk of stranded costs, as well as the possible decommissioning or repurposing of some parts of the natural gas distribution network due to the electrification of heating systems, for instance.

However, let us not forget that the challenges become opportunities to reach the net zero goal. Analyzing scenarios of profound decarbonization helps governments and key stakeholders to understand where to turn their efforts, and where to invest in the research and development of technologies. With our analyses, we have seen an important role for hydrogen and its derivatives, and this has allowed us to open a debate in the world of research on the role that this vector could have in Italy. These analyses, and the challenges that are highlighted, allow us to understand which sectors will have a fundamental role, and for the public decision-maker, it is important to understand where to direct the main incentives or reforms for decarbonization.

The high amount of variable renewable sources in the power system allows us to think about how to change or upgrade infrastructure or market rules well in advance, without being caught unprepared by a transition that is becoming ever faster.

The decarbonization scenarios represent a future of the national energy system with a totally different framework from the current one. Such a deep revolution of the energy system raises several questions about its actual implementation feasibility. The challenges are not only technological, but, above all, they are related to the amount of investments required, to the complexity of the system that is going to be built, to the life habit changes that it entails, to the phase-out of some energy commodities or energy infrastructures and the possible consequences that may affect both the energy system itself and other systems.

So, research and innovation (e.g., better and more efficient storage systems and cheaper hydrogen generation) is very important to reduce the costs. The TIMES model can also estimate the system costs and investments necessary for decarbonization. Those costs are an important criterion for choosing between the scenarios and to understand the real impacts of the different decarbonization pathways. In this paper, there are no estimates of the order of magnitude of the additional costs because we plan to investigate all the economic dimensions, from technological investments to macroeconomic impacts, in a future study.

These questions may not find a complete answer today; instead, they must become a starting point, and the reference for current research themes. In fact, the work on long-term scenarios will proceed well beyond this study, as the Italian Long-Term Strategy will have to be officially updated in 2025. However, other energy and environmental policy documents (such as the Italian Green Deal, NECP and Hydrogen Strategy) will also be revised in order to achieve the 2050 objectives. New results of studies on innovative technologies and alternative fuels will be the basis for further and future sensitivities on the complete decarbonization scenarios.

These types of studies and analyses are, in fact, always updated, as the overall objectives and sub-targets established at the international level for the pursuit of decarbonization and the containment of temperature rise continues to evolve.

Author Contributions: Conceptualization, M.G., C.N.B. and A.G.; Data curation, M.G. and C.N.B.; Formal analysis, M.G., C.N.B. and A.G.; Methodology, M.G. and A.G.; Supervision, M.G.; Visualization, C.N.B. and A.G.; Writing—original draft, M.G., C.N.B. and A.G.; Writing—review & editing, M.G., C.N.B. and A.G. All authors have read and agreed to the published version of the manuscript.

Funding: This research has been financed by the Research Fund for the Italian Electrical System under the Contract Agreement between RSE S.p.a. and the Ministry of Economic Development—General Directorate for the Electricity Market, Renewable Energy and Energy Efficiency, Nuclear Energy in compliance with the Decree of 16 April 2018. http://www.ricercadisistema.it/#/.

Institutional Review Board Statement: Not applicable.

Informed Consent Statement: Not applicable.

Data Availability Statement: Not applicable.

Conflicts of Interest: The authors declare no conflict of interest.

References

1. United Nations Framework Convention on Climate Change (UNFCC). The Paris Agreement. Available online: https://unfccc.int/sites/default/files/english_paris_agreement.pdf (accessed on 13 November 2019).
2. Masson-Delmotte, V.; Zhai, P.; Pörtner, H.-O.; Roberts, D.; Skea, J.; Shukla, P.R.; Pirani, A.; Moufouma-Okia, W.; Péan, C.; Pidcock, R. (Eds.) IPCC, 2018: Summary for Policymakers. Global Warming of 1.5 °C. IPCC Special Report. Available online: https://www.ipcc.ch/site/assets/uploads/sites/2/2019/05/SR15_SPM_version_report_LR.pdf (accessed on 4 November 2021).
3. Rogelj, J.; Shindell, D.; Jiang, K.; Fifita, S.; Forster, P.; Ginzburg, V.; Handa, C.; Kheshgi, H.; Kobayashi, S.; Kriegler, E.; et al. Mitigation pathways compatible with 1.5 C in the context of sustainable development. In *Global Warming of 1.5 C. An IPCC Special Report on the Impacts of Global Warming of 1.5 C Above Pre-Industrial Levels and Related Global Greenhouse Gas Emission Pathways, in the Context of Strengthening the Global Response to the Threat of Climate Change, Sustainable Development, and Efforts to Eradicate Poverty*; Masson-Delmotte, V., Zhai, P., Pörtner, H.-O., Roberts, D., Skea, J., Shukla, P., Pirani, A., Moufouma-Okia, W., Péan, C., Pidcock, R., et al., Eds.; IPCC: Geneva, Switzerland, 2018; Chapter 2.
4. Intergovernmental Panel on Climate Change (IPCC). *Climate Change 2013: The Physical Science Basis: Working Group I Contribution to the Fifth Assessment report of the Intergovernmental Panel on Climate Change*; Cambridge University Press: New York, NY, USA, 2014.
5. Intergovernmental Panel on Climate Change (IPCC). *Climate Change 2014: Mitigation of Climate Change: Working Group III Contribution to the Fifth Assessment Report of the Intergovernmental Panel on Climate Change*; Cambridge University Press: New York, NY, USA, 2014.
6. European Commission (EC). A Clean Planet for All. A European Strategic Long-Term Vision for a Prosperous, Modern, Competitive and Climate Neutral Economy. Communication COM (2018) 773final. 2018. Available online: https://eur-lex.europa.eu/legal-content/EN/TXT/PDF/?uri=CELEX:52018DC0773 (accessed on 11 October 2021).
7. European Commission (EC). Long-Term Low Greenhouse Gas Emission Development Strategy of the European Union and Its Member States: Submission by Croatia and the European Commission on Behalf of the European Union and its Member States, Zagreb, 6 March 2020. 2020. Available online: https://unfccc.int/sites/default/files/resource/HR-03-06-2020%20EU%20Submission%20on%20Long%20term%20strategy.pdf (accessed on 9 November 2021).
8. European Commission (EC). Proposal for a Regulation of The European Parliament and of The Council Establishing the Framework for Achieving Climate Neutrality and Amending Regulation (EU) 2018/1999 (European Climate Law). COM(2020) 80 final. Bruxelles, 4 March 2020. Available online: https://ec.europa.eu/info/sites/default/files/commission-proposal-regulation-european-climate-law-march-2020_en.pdf (accessed on 9 November 2021).
9. European Commission (EC). Amended Proposal for a Regulation of the European Parliament and of the Council on Establishing the Framework for Achieving Climate Neutrality and Amending Regulation (EU) 2018/1999 (European Climate Law). COM(2020) 563 final, 17.09.20. Available online: https://opac.oireachtas.ie/AWData/Library3/Documents%20Laid/pdf/CCAEdoclaid161020_161020_103405.pdf (accessed on 9 November 2021).
10. European Commission (EC). The Update of the Nationally Determined Contribution of the European Union and its Member States. Submission by Germany and the European Commission on Behalf of the European Union and its Member States. Available online: https://www4.unfccc.int/sites/ndcstaging/PublishedDocuments/European%20Union%20First/EU_NDC_Submission_December%202020.pdf (accessed on 29 September 2021).
11. European Commission (EC). Regulation (EU) 2018/1999 of the European Parliament and of the Council of 11 December 2018 on the Governance of the Energy Union and Climate Action, amending Regulations (EC) No 663/2009 and (EC) No 715/2009, Directives 94/22/EC, 98/70/EC, 2009/31/EC, 2009/73/EC, 2010/31/EU, 2012/27/EU and 2013/30/EU of the European Parliament and of the Council, Council Directives 2009/119/EC and (EU) 2015/652 and repealing Regulation (EU) No 525/2013 of the European Parliament and of the Council (Text with EEA Relevance.). Available online: https://eur-lex.europa.eu/legal-content/EN/TXT/PDF/?uri=CELEX:32018R1999&from=EN (accessed on 28 September 2021).
12. Ministero dell'Ambiente e della Tutela del Territorio e del Mare, Ministero dello Sviluppo Economico, Ministero delle Infrastrutture e dei Trasporti, Ministero delle Politiche agricole, Alimentari e Forestali. Strategia Italiana di Lungo Termine Sulla Riduzione Delle Emissioni dei Gas a Effetto Serra. 2021. Available online: www.minambiente.it/sites/default/files/lts_gennaio_2021.pdf (accessed on 9 November 2021).
13. International Energy Agency (IEA). Net Zero by 2050 A Roadmap for the Global Energy Sector. 2021. Available online: https://iea.blob.core.windows.net/assets/deebef5d-0c34-4539-9d0c-10b13d840027/NetZeroby2050-ARoadmapfortheGlobalEnergySector_CORR.pdf (accessed on 7 October 2021).
14. European Commission (EC), In-Depth Analysis in Support of the Commission Communication COM (2018) 773. A Clean Planet for all. A European Strategic Long-Term Vision for a Prosperous, Modern, Competitive and Climate Neutral Economy. Available online: https://ec.europa.eu/clima/system/files/2018-11/com_2018_733_analysis_in_support_en.pdf (accessed on 7 October 2021).
15. International Renewable Energy Agency (IRENA). 2nd Webinar Series on National Experience in Long-Term Energy Scenario (LTES) Use and Development. Available online: https://irena.org/events/2021/Nov/2nd-Webinar-Series-on-National-Experience-in-LTES-Use-and-Development (accessed on 21 October 2021).
16. Nakicenovic, N.; Alcamo, J.; Davis, G.; de Vries, B.; Fenhann, J.; Gaffin, S.; Gregory, K.; Grubler, A.; Jung, T.Y.; Kram, T.; et al. Special Report on Emissions Scenarios. A Special Report of IPCC Working Group III. 2000. Available online: https://escholarship.org/uc/item/9sz5p22f (accessed on 5 November 2021).

17. Lanati, F.; Gaeta, M.; Gelmini, A.; Gatti, A.; Mazzocchi, L. "Studi a Supporto Della Governance del Sistema Elettrico ed Energetico Nazionale" RSE, RDS report n° 18007604, 2019, (In Italian). Available online: http://www.rse-web.it/applications/webwork/site_rse/local/doc-rse/evento11-7/VolumeProdottiEmblematiciRicercadiSistema-Luglio2019.pdf (accessed on 9 November 2021).
18. IEA-ETSAP. Available online: https://iea-etsap.org/ (accessed on 21 October 2021).
19. IEA-ETSAP. Optimization Modeling Documentation. Available online: https://iea-etsap.org/index.php/documentation (accessed on 21 October 2021).
20. Siface, D.; Vespucci, M.T.; Gelmini, A. Solution of the mixed integer large scale unit commitment problem by means of a continuous Stochastic linear programming model. *Energy Syst.* **2014**, *5*, 269–284. [CrossRef]
21. Siface, D.; Gelmini, A.; Vespucci, M.T. Development of the stochastic medium-term market simulation model s-MTSIM. In Proceedings of the PMAPS 2012, Istanbul, Turkey, 10–14 June 2012.
22. Siface, D.; Gelmini, A.; Benini, M.; Vespucci, M.T. Assessment of the impact of RES uncertainty on the Italian power system in a 2020 scenario by means of stochastic programming. In Proceedings of the 25th European Conference on Operational Research, Vilnius, Lithuania, 8–11 July 2012.
23. Lanati, F.; Gelmini, A.; Gargiulo, M.; De Miglio, R. Il modello energetico multiregionale MONET. RSE, RDS Report No. 12001033. 2012. Available online: http://www.rse-web.it/documenti/documento/314728 (accessed on 21 October 2021).
24. Mantzos, L.; Wiesenthal, T.; Neuwahl, F.; Ròzsai, M.; Joint Research Center (JRC). The POTEnCIA Central Scenario: An EU Energy Outlook to 2050. 2019. Available online: https://publications.jrc.ec.europa.eu/repository/handle/JRC118353 (accessed on 12 March 2021).
25. Ministero dello Sviluppo Economico, Ministero dell'Ambiente e della Tutela del Territorio e del Mare, Integrated National Energy and Climate Plan. 2019. Available online: https://www.mise.gov.it/images/stories/documenti/it_final_necp_main_en.pdf (accessed on 12 March 2021).
26. Worden, J.R.; Bloom, A.; Pandey, S. Reduced biomass burning emissions reconcile conflicting estimates of the post-2006 atmospheric methane budget. *Nat. Commun.* **2017**, *8*, 1–11. [CrossRef] [PubMed]
27. Arnaut de Pee Dickon Pinner Occo Roelofsen Ken Somers Eveline Speelman Maaike Witteveen. Decarbonization of Industrial Sectors: The Next Frontier. June 2018 McKinsey&Company. Available online: https://www.mckinsey.com/~{}/media/mckinsey/business%20functions/sustainability/our%20insights/how%20industry%20can%20move%20toward%20a%20low%20carbon%20future/decarbonization-of-industrial-sectors-the-next-frontier.pdf (accessed on 9 November 2021).
28. Sergey, P.; Jennifer, M.; Haroon, K.; Howard, H. Hard-to-Abate Sectors: The role of industrial carbon capture and storage (CCS) in emission mitigation. *Appl. Energy* **2021**, *300*, 117322. [CrossRef]
29. EASAC Policy Report 37. *Decarbonisation of Transport: Options and Challenges*; European Academies' Science Advisory Council: Brussels, Belgium, 2019; ISBN 978-3-8047-3977-2.
30. NREL. Transportation Decarbonization Research. Available online: https://www.nrel.gov/transportation/transportation-decarbonization.html (accessed on 30 September 2021).
31. Scherer, L.; Verburg, P.H. Mapping and linking supply- and demand-side measures in climate-smart agriculture. A review. *Agron. Sustain. Dev.* **2017**, *37*, 66. [CrossRef]
32. Gallego-Schmid, A.; Chen, H.M.; Sharmina, M.; Mendoza, J.M.F. Links between circular economy and climate change mitigation in the built environment. *J. Cleaner Prod.* **2020**, *260*, 121115. [CrossRef]
33. Intergovernmental Panel on Climate Change (IPCC). *Global Warming of 1.5 °C*; IPCC: Geneva, Switzerland, 2014; Available online: https://www.ipcc.ch/site/assets/uploads/sites/2/2019/06/SR15_Full_Report_High_Res.pdf (accessed on 30 November 2021).
34. Lovins, A.B. How big is the energy efficiency resource? *Environ. Res. Lett.* **2018**, *13*, 09041. [CrossRef]
35. Kermeli, K.; Graus, W.J.; Worrell, E. Energy efficiency improvement potentials and a low energy demand scenario for the global industrial sector. *Energy Effic.* **2014**, *7*, 987–1011. [CrossRef]
36. Eurostat Databases. Available online: https://appsso.eurostat.ec.europa.eu/nui/show.do?dataset=nrg_bal_c&lang=en (accessed on 15 March 2021).
37. Jafari, M.; Delmastro, C.; Grosso, D.; Bompard, E.; Botterud, A. Electrify Italy: The Role of Renewable Energy. *Gas* **2022**, *2030*, 2050. Available online: https://www.energy-proceedings.org/wp-content/uploads/2020/01/AEAB2019_paper_140.pdf (accessed on 20 October 2021).
38. Bompard, E.; Botterud, A.; Corgnati, S.; Huang, T.; Jafari, M.; Leone, P. An electricity triangle for energy transition: Application to Italy. *Appl. Energy* **2020**, *277*, 115525. [CrossRef]
39. Fajardyab, M.; Dowell, N.M. Can BECCS deliver sustainable and resource efficient negative emissions? *Energy Environ. Sci.* **2017**, *10*, 1389–1426. [CrossRef]
40. McLaren, D. A comparative global assessment of potential negative emissions technologies. *Process. Saf. Environ. Prot.* **2012**, *90*, 489–500. [CrossRef]
41. Pour, N.; Webley, P.A.; Cook, P.J. A Sustainability Framework for Bioenergy with Carbon Capture and Storage (BECCS) Technologies. *Energy Procedia* **2017**, *114*, 6044–6056. [CrossRef]
42. Gough, C.; Garcia-Freites, S.; Jones, C.; Mander, S.; Moore, B.; Pereira, C. Challenges to the use of BECCS as a keystone technology in pursuit of 1.5 °C. *Glob. Sustain.* **2018**, *1*. [CrossRef]

43. European Commission. Communication from the Commission to the European Parliament, the Council, the European Economic and Social Committee and the Committee of the Regions. A Hydrogen Strategy for a Climate-Neutral Europe. COM (2020) 301 Final. Brussels, 8.7.2020. Available online: https://ec.europa.eu/energy/sites/ener/files/energy_system_integration_strategy_.pdf (accessed on 7 May 2021).

44. Andreas Eichhorn (VERBUND Energy4Business), Planning and Scheduling the Electrolyzer in the Austrian Electricity and Balancing Market, Green Hydrogen for Industry—Regulatory Workshop, 11 February 2021 (VERBUND, CEER, ACER). Available online: https://www.ceer.eu/documents/104400/7155071/1_Presentation+1.pdf/23cce532-7f53-e9d9-3538-1eefb1dcead0?version=1.3 (accessed on 12 March 2021).

45. HyBalance Project, "Background". Available online: https://hybalance.eu (accessed on 7 May 2021).

46. ENTSO-E, RDI Implementation Report 2021–2025 Project Concepts for Delivering the ENTSO-E RDI Roadmap 2020–2030. 2021. Available online: https://consultations.entsoe.eu/r-i/entso-e-rdi-implementatation-report-2021-2025/supporting_documents/1_entsoe_rdi_implementation_report_for%20public%20consultation.pdf (accessed on 14 October 2021).

47. Daiyan, R.; MacGill, I.; Amal, R. Opportunities and Challenges fro Renewable Power-to-X. *ACS Energy Lett.* **2020**, *5*, 3843–3847. [CrossRef]

48. Gargiulo, A.; Carvalho, M.L.; Girardi, P. Life Cycle Assessment of Italian Electricity Scenarios. *Energies* **2020**, *13*, 3852. [CrossRef]

49. Rossi, M.; Rossini, M.; Viganò, G.; Migliavacca, G.; Siface, D.; Faifer, I.; Hergun, H.; Bakken Sperstad, I. Planning of distribution networks considering flexibility of local resources: How to deal with transmission system services. In Proceedings of the CIRED 2021 Conference, Geneva, Switzerland, 20–23 September 2021.

Article

Colombia's GHG Emissions Reduction Scenario: Complete Representation of the Energy and Non-Energy Sectors in LEAP

Juan David Correa-Laguna [1,2,*], **Maarten Pelgrims** [1,2], **Monica Espinosa Valderrama** [3] and **Ricardo Morales** [3]

[1] VITO, The Flemish Institute for Technological Research, Boeretang 200, 2400 Mol, Belgium; maarten.pelgrims@vito.be
[2] EnergyVille, Thor Park 8310-8320, 3600 Genk, Belgium
[3] Civil and Environmental Engineering, Universidad de los Andes, Bogotá 111711, Colombia; m.espinosa28@uniandes.edu.co (M.E.V.); r.moralesb@uniandes.edu.co (R.M.)
* Correspondence: juan.correalaguna@vito.be; Tel.: +32-1433-6709

Citation: Correa-Laguna, J.D.; Pelgrims, M.; Espinosa Valderrama, M.; Morales, R. Colombia's GHG Emissions Reduction Scenario: Complete Representation of the Energy and Non-Energy Sectors in LEAP. *Energies* **2021**, *14*, 7078. https://doi.org/10.3390/en14217078

Academic Editors: Dolf Gielen and Rajender Gupta

Received: 18 June 2021
Accepted: 25 October 2021
Published: 29 October 2021

Abstract: The signatory countries of the Paris Agreement must submit their updated Intended National Determined Contributions (INDCs) to the UNFCCC secretariat every five years. In Colombia, this activity was historically carried out with a wide set of diverse non-interconnected sector-specific models. Given the complexity of GHG emissions reporting and the evaluation of mitigation actions on a national scale, the need for a centralized platform was evident. Such approach would allow the integration and analysis of potential interactions among sectors, as well as to guarantee the homogeneity of assumptions and input parameters. In this paper, we describe the construction of an integrated bottom-up LEAP model tailored to the Colombian case, which covers all IPCC sectors. An integrated model facilitates capturing synergies and intersectoral interactions within the national GHG emissions system. Hence, policies addressing one sector and influencing others are identified and correctly assessed. Thus, 44 mitigation policies and mitigation actions were included in the model, in this way, identifying the sectors directly and being indirectly affected by them. The mitigation scenario developed in this paper reaches a reduction of 28% of GHG emissions compared with the reference scenario. The importance of including non-energy sectors is evident in the Colombian case, as GHG emission reductions are mainly driven by AFOLU. The first section describes the GHG emissions context in Colombia. Next, we describe the model structure, main input parameters, assumptions, considerations, and used LEAP functionalities. Results are presented from a GHG emissions accounting and energy demand perspective. The model allows for the correct estimate of the scope and potential of mitigation actions by considering indirect, unintended emissions reductions in all IPCC categories, as well as synergies with all mitigation actions included in the mitigation scenario. Moreover, the structure of the model is suitable for testing potential emission trajectories, facilitating its adoption by official entities and its application in climate policymaking.

Keywords: decarbonization; INDC; LEAP; energy modeling; long-term scenarios; GHG inventory

1. Introduction

Committing to greenhouse gas emissions (GHG) reductions at the national and local level is necessary to minimize the climatic effects of global warming and increase the chances of not exceeding 2 °C in global temperature increase. To reach that goal, the signatory countries of the Paris Agreement are committed to periodically submitting their updated Intended Nationally Determined Contributions (INDC) [1]. Colombia has a GHG emissions profile dominated by the Agriculture, Forestry, and Other Land Use (AFOLU) sector, which in 2014 accounted for 54% of total emissions [2]. Deforestation, through the uncontrolled expansion of the agricultural and livestock frontier towards forested areas, is one of the leading GHG emission sources in the country. In 2014, the transport sector was responsible for 12% of the national GHG emissions, while energy industries accounted for 10% [2]. The Colombian Low-Carbon Development Strategy

Energies **2021**, *14*, 7078. https://doi.org/10.3390/en14217078   https://www.mdpi.com/journal/energies

(CLCDS) provided the framework for the discussion processes and modeling effort leading to the previous INDC formulation in 2015 [3]. In this process, several stakeholders such as the Ministry of Environment and Sustainable Development (MADS), the Ministry of Foreign Affairs (MFA), the National Planning Department (DNP), and academia had an important role.

For Colombia, the process of compiling and communicating GHG emissions accounting and scenarios was typically carried out by a wide set of diverse sector-specific models [3], which were then aggregated to build the INDC. Although in 2014 MADS and the UK government developed the Carbon Calculator 2050 [4], covering the most relevant sectors, the tool was not adopted by each ministry involved in defining future climate and GHG emissions scenarios. The Carbon Calculator had some limitations in capturing annual variations, possible synergies between sectors, and representing all sectors with a high level of detail, except for major cases (e.g., transport sector energy demand and fuel production). For instance, it did not have a dispatch module for the power sector based on a time-slice approach. Moreover, the representation of new technologies was time-consuming and cumbersome. To facilitate the integration and analysis of potential interactions among sectors and to guarantee the homogeneity of the general assumptions, the need for a centralized national system model was evident. Thus, the scenarios definition and development process is strengthened, as has been pointed out in the IRENA's long-term energy scenarios (LTES) [5].

This paper describes the process of developing—in the Long Emissions Analysis Platform software (LEAP)—a Colombia-tailored model (COL-NDC) to formulate the baseline emission trajectory for Colombia's 2020 INDC update and assess future energy needs, as part of a project jointly requested by the Colombian government and the World Bank. Previous LEAP models have been developed for Colombia focused on the energy sector [6–8]. Other studies have used LEAP to analyze the GHG emissions reductions in Colombia and other Latin American countries (i.e., Mexico, Chile, Panama) [9–13]. Conversely to these models, the COL-NDC model includes all energy and non-energy sectors, which provides a holistic approach to GHG emissions accounting and exploration of decarbonization scenarios. The model covers the emissions from all categories defined by the Intergovernmental Group of Experts on Change Climate (IPCC), which are Energy, Industrial Processes (IPPU), AFOLU, and Waste. A unique model is capable of handling interactions among mitigation measures adopted by different sectors (e.g., fugitive emissions reduction due to less extracting activities, which are a result of mitigation actions in-demand sectors). LEAP was chosen as it facilitates the construction of several scenarios using an accounting simulation approach, it can include non-energy sectors, it allows each sector to be modeled with a different approach according to the available data (e.g., top-down, bottom-up), and it does not require a technology-rich database. However, the tool also has some limitations, such as capturing the total system cost, endogenously defining the marginal price of products (e.g., steel price, space heating), and choosing the most cost-optimal scenario based on techno-economic parameters.

2. Methodology and Data

2.1. Data Gathering

Under the Colombian INDC update process, all the relevant Colombian ministries and several governmental organizations were involved in the design of the COL-NDC model structure, data pretreatment, definition of scenarios, and assumptions. Figure 1 shows the interaction and role of the stakeholders during the process, as well as their contributions.

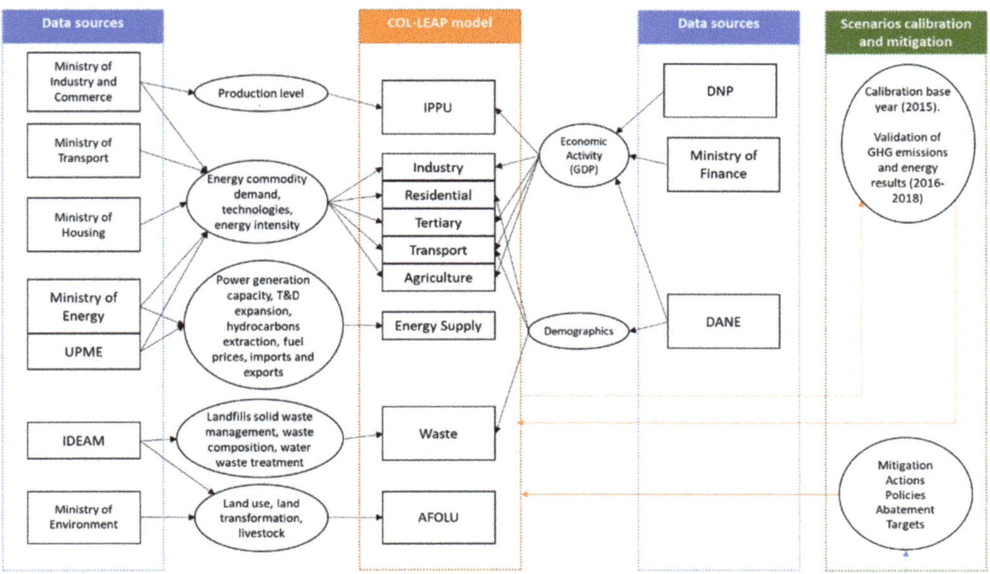

Figure 1. Stakeholders interactions and data exchange.

The COL-NDC model includes general macroeconomic and demographic parameters for the reference scenario as well as for the individual mitigation actions (see Table 1). The population is one of the main drivers of energy and GHG emissions. Therefore, the model includes the distribution of people living in rural and urban areas, as well as the size of households in both areas. National and sectoral GDP are used as main drivers for the industry, agriculture, for the energy demand of tertiary sectors; as well as for the stock of vehicles, industrial waste, and IPPU activity levels. Sectoral GDP projections are established by the DNP through the Colombian Computable General Equilibrium Model for Climate Change (MEG4C) [14].

Table 1. Macroeconomic and demographic assumptions.

Parameter	Units	2015	2020	2030	2050	Source
Population	Million	46.4	50.3	55.7	61.9	[15]
Urban areas	%	75.4%	76.0%	76.8%	76.0%	[15]
Rural areas	%	24.6%	24.0%	23.2%	24.0%	[15]
Urban household size	people	3.4	3.2	2.9	2.4	[15,16]
Rural household size	people	4.0	3.9	3.7	3.2	[15,16]
Annual GDP growth *	%	2.30%	3.40%	3.50%	-	[17,18]

* In 2020: −5.5% due to COVID-19. From 2021 to 2025: on average 5.2%.

2.2. LEAP Tool

The LEAP has been widely used for policy and scenario-based analysis, as well as for energy planning [19]. LEAP is an accounting-type simulation tool, which considers all energy requirements in the supply and transformation sector needed to meet future energy demands and report the associated GHG emissions. Additionally, GHG emissions also account for the non-energy sectors based on activity data and specific emissions factors (e.g., livestock, nitrogen content in fertilizers, biomass from deforestation). While LEAP was initially more energy-system oriented, it has undergone several updates to include additional features such as land use, indirect GHG effects (e.g., health, air quality), and emission cost of non-energy sectors. LEAP offers high flexibility to define the model topology and the possibility to use bottom-up, top-down, and stock-turnover modeling approaches. Nevertheless, it is not possible to optimize the entire system based on a technology-rich approach as is the case of other modeling tools such as TIMES-MARKAL [20]. However, LEAP can quickly reflect the implementation of policies and mitigation actions, which eases the abatement potential and scope assessment of policies and mitigation actions by the comparison of several scenarios.

2.3. Structure of Colombian NDC LEAP Model

The COL-NDC model includes historical data from 2010 to 2014 to compare the trends of the projected period (2015-2050). Since the last official GHG inventory in Colombia dates from 2014[2], this was selected as the base year for energy, activity data, and GHG emissions calibration. The period 2015-2018 is used to compare energy demand and emissions results from the model with official reports. The model uses several modeling approaches (e.g., top-down bottom-up, stock) based on the available data. For example, road transport is modeled considering the existing fleet (stock), vintage and exit curves, and annual sales. Conversely, the waste sector uses a top-down approach based on population and production of waste per capita. For IPPU, activity data is exogenously calculated and fed into the model, where GHG emissions are calculated considering default emissions factors. In the case of AFOLU, an already existing model for AFOLU, which is very detailed and flexible was used for the land use categories. Therefore, land and fertilizer-related emissions are endogenously calculated. On the other hand, emissions linked to livestock farming were completely modeled within LEAP using the number of animals and specific emissions factors by region and type of livestock. Figure 2 presents the general structure of the COL-NDC model and the main links among sectors. Global warming potential (GWP) with a horizon of 100 years is calculated taking into account the fifth assessment report (AR5) of the IPCC [21]. The emission factors (EF) for fuels are taken from a study carried out by the Ministry of Energy in Colombia to characterize the fuels used within the country [22,23]. When data is incomplete, default values from the IPCC guidelines are used. To facilitate the accountability of ministry-specific emissions and the compliance of their targets, GHG emissions are directly allocated in LEAP to the different ministries employing LEAP tags (Tags can be used to organize results that belong to more than one branch in LEAP).

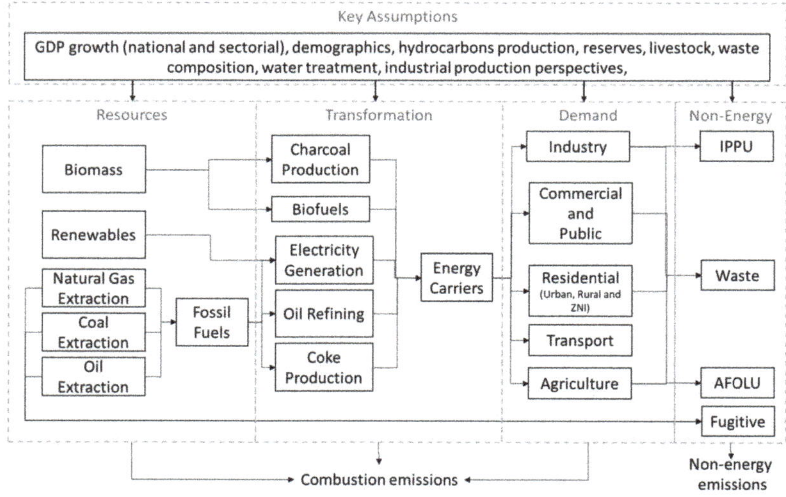

Figure 2. Colombia LEAP model structure.

2.3.1. Energy Demand Sectors

Industry

The structure of the manufacturing industry subsectors is based on the Useful Energy Balance (UEB) (Useful Energy Balance makes reference to the useful energy used by end-use (e.g.: cooking, lighting), taking into account the efficiency of the technology and the final energy. This balance differs to the national energy balance (BECO)) [24], which disaggregates energy demand by seven end-uses, energy vectors, and the efficiency level, as is shown in Table 2. The energy demand of sectors not covered by the UEB follows the energy mix reported in the Colombian Energy Balance (BECO) [25]. The activity levels are linked to sectors In this way, it is possible to calculate the useful energy intensity by end-use in terms of MJ/COP for each industrial sub-sector and specific end-use as the ratio of the useful energy demand of the specific end-use and the sectoral GDP. For the reference scenario, it is assumed that useful energy intensity will remain constant, and no major changes in fuel mix are expected.

Table 2. Industry structure by levels.

Sub-Sector	End Uses	Fuels	Equipment
1A2a—Iron and steel	1. Direct Heating	1. Bagasse	1. Existing efficiency
1A2b—Non-ferrous metals	2. Indirect	2. Coal	2. Best efficiency
1A2c—Chemicals	Heating	3. Natural gas	available in
1A2d—Pulp, paper, and printing	3. Machine Drive	4. Firewood	Colombia
1A2e—Food, beverages, and tobacco	4. Refrigeration	5. Oil	3. Best efficiency
1A2f—Non-metallic minerals	5. Cooling	6. Waste	available
1A2g—Transport equipment	6. Lighting	7. Charcoal	worldwide
1A2h—Machinery	7. Others	8. Coke	
1A2j—Wood and wood products		9. Diesel	
1A2l—Textiles and leather		10. Fuel oil	
1A2m—Industry not specified		11. LPG	
1A2i—Mining and quarrying		12. Gasoline	
1A2k—Construction		13. Kerosene	

Transport

The transport sector is initially split into aviation, road transport, railways, shipping, and others (pipelines and off-road transportation) [26]. Road transport is further divided into additional categories (e.g., public, private, passenger, and freight transport). A top-down approach is selected for aviation, rail, and shipping due to the lack of information to further disaggregate their activity level. For these categories, national GDP drives the increase of energy demand. Conversely, road transport is modeled with a higher disaggregation level, considering the size of the vehicle fleet, fuel efficiency, and average annual activity. LEAP assesses annual GHG emissions based on fleet stock, vehicle activity (km/vehicle-year), and fuel consumption [27]. The total annual fleet in the base year was obtained from the national transport statistics (RUNT) [28]. RUNT data is used to derive vintage and exit curves for light passenger vehicles, motorcycles, light freight vehicles, buses, and trucks. The fleet converted to Compressed Natural Gas (CNG) is obtained from the statistics of the gas union [29] and RUNT. To obtain annual activity by category, the parameters included in two previous national studies are used [30,31]. For the freight sector, annual average activity data for trucks and tractors is taken from the database of the Ministry of Transport [28]. The equivalence between the BECO transport and IPCC categories is used to obtain the total vehicle-kilometers (VKTs) and their distribution by fuel. Average fuel consumption by category is defined according to the European Environmental Agency [32], the Fuel Economy database of the Department of Energy and the United States Environmental Agency [33], previous national studies [34–37] and confidential data provided by the Ministry of Transport (Table 3 summarizes the values used in the model).

Table 3. Fuel economy by fuel and type of vehicle for the base year [27,32–37].

Type	Natural Gas	Diesel	Gasoline
Units	MPG	MPG	MPG
Car	18.2	21.6	19.3
Bus	5.2	6.1	5.1
Medium Truck	5.8	8.8	7.9
Big Truck	3.9	5.9	5.3
Pick up	18.2	21.6	19.3
Micro Bus	11.9	13.9	12.5
Motorcycle	-	75.5	57.3
Taxi	18.2	21.6	19.3
Tractor	3.5	5.3	4.7

In the reference scenario, it is assumed that fuel economy annually improves by 1% between 2015 and 2030 for the new fleet, which is a conservative value considering that for emerging countries, there was an annual improvement of 1.2% between 2005 and 2017 [34]. For trucks and tractors, an annual improvement of 0.5% is considered in line with the improvements reported in similar markets globally [35]. It is assumed that the total vehicle fleet increases according to the GDP and population. Thus, the private transport fleet is modeled in terms of motorization rates using a Gompertz function [36] and two previous studies for Colombia [30,31]. This implies that the speed with which the fleet of light passenger vehicles has been growing, especially motorcycles, decelerates in the following decades. Freight transport fleet grows as a function of total GDP. Conversely, the projection of the public transport fleet responds to the coverage goals of this segment in urban transport, taking into account the participation of public transport according to the case study of the INDC 2015 [31].

Tertiary, Residential, and Agriculture

For the tertiary sector, the useful energy intensity for each end-use is defined from the UEB and sectoral GDP (see Table 4). The tertiary sector accounts for 5% of the total final demand in Colombia [25] and 60% of the national GDP [37]. Due to the variation in consumption patterns and expected GDP growth, this sector is broken down into the commercial and public sectors. It is assumed that useful energy intensity remains constant in the reference scenario. The energy mix is assumed not to undergo significant changes, following the trend of the last years [25].

Table 4. Useful energy intensity in MJ/COP [24,37].

End-Use	Commercial	Public
Water Heating	0.0143	0.0488
Cooking	0.0068	0.0024
Lighting	0.0034	0.0009
Machine Drive	0.0033	0.0019
Air Conditioning	0.0128	0.0046
Refrigerators	0.0010	0.0003
Others	0.0039	0.0054

The residential sector follows a similar approach to the one used for the tertiary sector. However, the demand, in this case, is attributed to households. Since consumption patterns and fuel mix vary between households, this sector is divided into urban and rural households. Data on the size of rural and urban households is obtained from the National Department of Statistics (DANE) [38]. In the case of the residential sector, we define the useful energy intensity in terms of households in urban and rural areas (See Table 5). Due to the lack of information on the future development of useful energy intensity, it is assumed that these values will remain constant. Access to various energy services is an important determinant of energy consumption in the residential sector. According to the National Quality of Life Survey of 2015 (NQLS), 97.2% of urban households and 97.5% of rural households have kitchen facilities in their homes, while the proportion of households with a water heater is 24.5% and 4.1%, respectively [38]. In the case of TV, 92% of urban households have a television at home and it is projected that it will increase to 94% by 2030, reaching 97% coverage in 2050. In 2015, 87% of the urban homes have a refrigerator, 67% a washing machine, and 5% air conditioning (AC), while the respective figures for rural homes were 63.3%, 28.8%, and 1.2% [38]. With the increase of household income over time, access to these goods will increase. In 2030, it is expected that in urban areas 95% of households will have refrigerators, 85% will have washing machines and 10% will acquire AC. It is assumed that the adoption rate in rural areas will evolve similarly.

Table 5. Useful energy intensity in MJ/household [15,24].

End-Use	Urban	Rural
Cooking	2446	2661
Water heating	918	917
Lighting	57	40
TV	92	52
Air Conditioning	3599	3599
Refrigerators	403	403
Wash machine	119	119
Air Fan	49	64.5
Others	294	206

The energy demand of the agriculture sector in the BECO is used to determine the final energy intensity for each fuel used within the sector in terms of kJ/COP. Thermal energy intensity is established at 300 kJ/COP, electrical energy intensity at 43 kJ/COP, and machine drive intensity at 96 kJ/COP. We assume that final energy intensity will remain constant and sectoral GDP will be the main driver.

2.3.2. Supply and Transformation Sectors

The COL-NDC model is designed to represent the official projections of local production, imports, and exports of crude oil, oil derivatives, natural gas, and coal according to the official figures published by the Ministry of Mines and Energy [39–41].

The Power Sector

Power generation capacity is the one established in the Transmission Generation Expansion Plan 2016 (TGEP) [42]. Historical electricity generation and technical parameters of the plants are obtained from public reports by the power market operator (XM) and the Ministry of Mines and Energy [43]. Table 6 shows the efficiency by technology, calculated as the average ratio of the historical fuel consumption and electricity generation.

Table 6. The efficiency of power generation plants by technology [43].

	Diesel	Coal	Fuel Oil	Gas	Jet Gasoline	Fuel Mix
Efficiency	29%	32%	23%	44%	26%	33%

The power system is modeled reflecting the official expansion of the system up to 2030 as defined by the TGEP [42]. Table 7 shows the generation capacity in the reference scenario in 2030, complemented by the optimization feature available in LEAP [44].

Table 7. Reference power capacity mix [MW], [42].

	Hydro	Gas	Coal	Small Hydro	Biomass	Wind	Solar	Geothermal	Other	Total
2030	13,520	4470	1930	1260	0	362	90.5	0	88.3	21,720

The model also includes the energy for self-consumption, as well as the losses due to the transmission and distribution of energy in the national grid (SIN) and non-interconnected zones (ZIN). According to historical data, self-consumption is approximately 3% of the electricity generated and electricity losses are around 11% ($\pm$ 1%) [25]. National energy statistics show that the electricity generated by auto- and cogeneration is consumed mainly in the extraction of oil and natural gas (55%), followed by industry (40%) and injections into the SIN (5%) [25]. The average efficiency in the COL-NDC model for auto- and cogeneration plants is in line with reports of XM [45] and the National Energy Planning Unit (UPME) [46]. As Table 8 shows, in auto- and cogeneration natural gas, bagasse and diesel are the main fuels. The IPCC guidelines indicate that emissions must be accounted for in the sector where electricity from auto- and cogeneration is consumed. Consequently, a specific electricity commodity (Electricity_AUT_COG) is defined in the COL-NDC model to differentiate it from electricity from the national grip (Electricity_SIN). A specific EF is defined for the consumption of Electricity_AUT_COG, reflecting the fuel mix in the auto- and cogeneration module. Installed power generation capcity in ZNI was approximately 242 MW in 2019 [47,48], of which 96% were Diesel power plants and the remaining 4% renewable sources. To consider the trend growth of renewable sources in these areas, a conservative compound annual growth rate (CAGR) of 3% is assumed for the reference scenario.

Table 8. Fuel mix in auto- and cogeneration units (%) [25].

	2010	2011	2012	2013	2014	2015
Bagasse	14.9	13.4	13.9	14.2	16.3	16.2
Coal	8.9	9.1	9.6	7.9	9.2	9.2
Natural Gas	49	47.5	46.7	48.4	47.5	47.8
Hydro	0.6	0.7	0.6	0.6	0.6	0.5
Oil	11.9	15.6	15.6	13.1	11.9	11.8
Diesel	13.4	12.5	12.4	14.4	13.3	13.3
LPG	1.2	1.2	1.2	1.4	1.3	1.2

Fossil Fuel Extraction

The use of diesel and gasoline in coal mining is represented by intensity factors, that is, TJ of fuel used per TJ of coal produced (see Table 9). Exports are included as a restriction to be fulfilled by the model, according to the export levels defined by UPME [41]. Currently, more than 75% of the national oil production is exported. However, oil exports are expected to decrease, driven by higher local demand and current oil reserves levels. Annual oil production capacity is included in the model to ensure that it reflects official production projections according to the Liquid Fuel Supply Plan—2019 (LFSP) [39]. Without new additional reserves, national production is extinguished in the long term, and Colombia becomes a net importer in the reference scenario. The model includes the refining capacity of 400 kbps [39], and no expansion is foreseen for the reference scenario. The extraction of natural gas in Colombia occurs mainly to supply domestic demand. Therefore, existing reserves (14EJ) were included in the base year [25,40], which will be depleted depending on the internal demand and the extraction capacity.

Table 9. Energy intensity in coal mining [25].

	Units	2010–2015
Coal production 2010-2015	[TJ]	14,701,512
Natural gas use in mining	[TJ]	8266
Diesel use in mining	[TJ]	88,303
Gasoline use in mining	[TJ]	584
Natural Gas Intensity	[TJ/TJ $_{Coal}$]	0.000562
Diesel Intensity	[TJ/TJ $_{Coal}$]	0.006006
Gasoline Intensity	[TJ/TJ $_{Coal}$]	0.000040

Other Fuels

Two independent modules are created for bioethanol and biodiesel production, which are limited to the current national capacity. These modules are created to enable Diesel-Biodiesel and Gasoline-Ethanol mixture modeling. The mix at the national level for the period 2010-2018 is obtained from historical data (3–7%) [25]. The model automatically calculates the EF of the mixed fuel discounting the biofuel energy share.

There are two classes of solid fuel production in Colombia: coke and charcoal. These processes are modeled considering the required auxiliary fuels and the EF related to the product based on IPCC values [26].

Fugitive

EF related to fugitive emissions are taken from IPCC default values (Tier 1) [26] and the average EF determined in Colombia for coal mining [49]. Fugitive emissions activity data is associated with the extraction of coal, oil, and natural gas, which are endogenous results in LEAP. Other parameters such as the amount of oil and gas transported and stored, the number of exploring wells and wells in service are obtained from historical values provided by the Ministry of Mines and Energy and included as average factors related to production level.

2.3.3. IPPU

IPPU in LEAP is based solely on the IPCC structure, for which information has been reported in the national GHG emissions inventory [2]. For mineral industries, data is obtained from UPME, the Colombian Mining Information System (SIMCO), the Annual Manufacturing Survey (EAM), and DANE. In the case of the chemical industry, production activity is directly obtained from companies within the sector, the national oil company (ECOPETROL), and the National Association of Businesses of Colombia (ANDI) (Confidential data provided during the World Bank PMR-Colombian NDC update project). Cement and ammonia production are currently operating at their maximum capacity and no expansion is foreseen. Therefore, the production will remain constant. The production of other sectors such as steel, ferroalloys, lubricants, glass, and lime are expected to grow in line with the sectoral GDP. For the case of Ozone Depleting Substances (ODS) substitutes, IDEAM and the Ozone Technical Unit (UTO) provided the emission time series of the respective substances for each subcategory.

2.3.4. AFOLU

In the case of livestock, the COL-NDC model considers 10 regions in Colombia, as the management of herds, feed and manure are different. For each region, specific CH_4 emission factors are used, both for enteric fermentation and manure management. Activity data and EF are provided by the Ministry of Agriculture and the Ministry of Environment.

Indirect N_2O emissions related to land use are calculated directly based on fertilization data, using the IPCC default factors for volatilization and leaching [26]. Projections based on historical values are used for the number of animals and the use of fertilizers as shown in Table 10. During a transition period of 20 years, land units are treated as converted land, and after those 20 years, the converted land units will be reported as land remaining as such. Total emissions from fuelwood extraction from all sectors are based on the energy demand for fuelwood resulting in the Energy sector in LEAP and translated into emissions in AFOLU.

Table 10. Expected growth and projections for AFOLU categories.

	Annual Growth	**Source**
Livestock	4.0% for Birds	FENAVI (National Federation of Poultry Farmers)
	1.5% for Pigs	PorkColombia (National Pig Farming Fund)
	According to historical annual growth	
Land burned	1% in biomass in cropland and grasslands 3% per year for forest land	IDEAM
Deforestation	2.9% for forested lands 1.27% for croplands 1.02% for grasslands	IDEAM-(SMBYC)MEDS (Reference Level of Forest Emissions) [50]
Forest plantations	According to the National Forest Development Plan	IDEAM

2.3.5. Waste

The GHG projection for the waste sector depends largely on population growth, while the industrial waste categories are driven by sector-specific economic growth. The main waste disposal systems currently used in Colombia are sanitary landfilling, open dumping, waste burning through incineration, open burning, and wastewater treatment. The sector follows a bottom-up approach where regional landfills have been individually modeled to reflect available disaggregated data into the model. The solid waste disposal category is based on the First Order Decay (FOD) methodology to estimate solid waste emissions coming from landfills [51]. As waste emissions are impacted by climatic parameters, the base structure of the waste module is divided into four climate zones relevant for the Colombian case (i.e.,: moist & wet tropical climate, wet temperate climate, dry tropical climate, and dry temperate climate) [51]. Waste incineration is linked to the activities of specific sectors or input assumptions within LEAP (e.g., coal extraction, population, sectoral GDP). Eight technologies are modeled for domestic wastewater treatment, differentiating between urban and rural areas. On the other hand, industrial wastewater is divided into seventeen industrial activities (e.g., sugar, pulp, and paper, food). Wastewater treatment is based on the IPCC Tier-1 methodology to estimate all the wastewater-related emissions.

3. Scenarios

3.1. Reference Scenario

The main drivers in the reference scenario are population and GDP, which are common to all scenarios (see Table 1). The relationship between these drivers and the growth of each sector is described in Section 2.3. In this scenario, mitigation policies established or implemented after 2015 are not included. Social phenomena such as migration to urban areas, the reduction of the size of households, and the increase of power purchase are reflected in the number of future urban and rural households, saturation rates of households' appliances and electronic devices, and motorization rates for private passenger vehicles (see Section 2.3.1). Moreover, the population has an impact on waste production and livestock activity, among others. The reference scenario accounts for the economic impact of COVID-19, which has a direct effect on energy consumption—mostly in the industry, agriculture, and tertiary sectors—as well as in process-related emissions (IPPU) and industrial waste. Section 2 describes in more detail the assumptions and considerations for the reference scenario in each sector.

3.2. Mitigation Scenario

The mitigation scenario is the aggregation of individual mitigation actions. Since the potential of some measures is limited, these are grouped by affinity (e.g., energy efficiency measures, waste treatment measures) and pertinence (i.e., NAMA coffee and NAMA Panela energy efficiency). The mitigation scenario covers 44 measures which are listed in Table 11 (for more information see Appendix A, Tables A3–A6), proposed by each responsible ministry in the Colombian government as the result of previous and ongoing projects [52,53]. In LEAP, each mitigation action is individually modeled to assess its actual mitigation potential and limitations, as well as possible intersectoral synergies. When LEAP combines the individual mitigation actions into one aggregated scenario, such scenario inherits the parameters of the mitigation portfolio. Thus, in the case of mutual excluding mitigation actions (e.g., coal replacement with natural gas, and complete electrification of end-use), LEAP uses the expression of the last mitigation action in the inheritance order, therefore, the order must reflect the hierarchy, or priority, of the measures.

Table 11. Mitigation portfolio included in the mitigation scenario.

Sector	Mitigation Measure	Sector	Mitigation Measure
Energy	NAMA Refrigerators	Energy	Metro Bogotá
	Efficient new buildings		Intercity train Metropolitan Area
	Thermal districts		Compressors in pipelines
	Agriculture energy efficiency		Glycol use optimization
	Carbon tax		Recovery in storage tanks
	Demand management	IPPU	ODS substitutes
	Sustainable cement		Chemical industry
	Brick Development	Waste	Coffee and panela wastewater
	Industry Efficiency		Use of biogas in landfills
	Fuel replacement industry		Biogas management water treatment
	Thermal generator efficiency		Biogas burning in landfills
	Diversification Capacity Generation		Recycling of plastic paper and glass
	Mining energy efficiency		Biological mechanical treatment
	Energy Efficiency Refineries	AFOLU	Deforestation reduction
	NAMA TOD		AMTEC rice
	Aviation performance improvements		NAMA Coffee (land use)
	Scrapping and cargo fleet renewal program.		NAMA Panela (land use)
	Urban logistics improvements		Forest plantations
	NAMA TANDEM		Cocoa crops
	Freight transport—River/Road		Ecological restoration
	Freight transport—Train/Road		Efficient wood stoves
	Electric mobility program		NAMA Livestock

4. Results

4.1. Reference Scenario

National GHG emissions in the reference scenario are 346 MtCO$_2$-eq in 2030. AFOLU is responsible for 50% of the emissions, followed by Energy (36%). Table 12 shows the emissions of the reference scenario by IPCC category. Between 2015 and 2030, total GHG emissions grow with a Compound Annual Growth Rate (CAGR) of 2.7%, while the economy grow by approximately 3.5% each year. The growth rate of emissions changes after 2025, mainly driven by carbon sinks on land the remains as such and the reduction of deforestation, which compensates the increasing trend of the industry (CARG: 3.3%), the tertiary (CARG: 3.5%), and the transport sector (CARG: 3.6%). This tendency is due to the expected economic growth, the increase of the purchasing power, as well as the number of households, intensified by the reduction in the average number of people per household. Emissions associated with fuel combustion are mainly due to Diesel and coal, this reflects the increase of energy demand of the transport sector, heat demand in industry, and the use of coal power plants after 2025.

Table 12. GHG emissions results by IPCC category in reference scenario in MtCO$_2$eq.

IPCC Category	2015	2020	2025	2030
1—Energy	87	88	106	125
2—IPPU	9	11	15	18
3—AFOLU	118	170	186	175
4—Waste	19	22	25	28
Total	**233**	**291**	**332**	**346**

In terms of energy, total energy demand rises 724 TJ (+55%) from 2015 to 2046 PJ in 2030. National energy intensity decreases by 0.7% between 2015 and 2030, from 1.642 kJ/COP to 1.630 kJ/COP, being the tertiary sector the one with the highest change (-5%). Conversely, the energy per capita presents an upwards trend, increasing from 28.5 MJ/capita in 205 to 36.7 MJ/capita in 2030. Table 13 shows the increase in demand by energy vector and their participation in 2015 and 2030. The most relevant energy vectors that increase the most are Diesel (+75%), gasoline (+72%), and electricity (+56%). Electricity demand in 2030 is 347 PJ, which is comparable with official results, 323 TJ (PEN-scenario-T1) [54] and 378 TJ (XM-Demand Forecast) [55].

Table 13. Energy demand by energy vector in 2015 and 2030 in the reference scenario.

	2010		2030	
	PJ	Share	PJ	Share
Coal	87	8%	99	5%
Natural Gas	170	15%	296	15%
Wood	154	14%	146	8%
Gasoline	148	13%	340	18%
Diesel	223	20%	468	24%
Coke	16	1%	1	0%
LPG	29	3%	45	2%
Kerosene	36	3%	82	4%
Electricity	190	17%	335	17%
Other fossil	22	2%	12	1%
Other	53	5%	93	5%
Total	**1126**		**1916**	

4.2. Mitigation Scenario

In 2030, total GHG emissions in the mitigation scenario decrease by 96 MtCO$_{2eq}$ to 250 MtCO$_{2eq}$, equivalent to a reduction of 28%. There is a heterogeneous distribution of the

GHG emissions reduction among IPCC sectors, as can be seen in Figure 3. In the mitigation scenario, 79% of the reductions are attributed to AFOLU, mainly by the reduction of deforestation. The remaining reduction is distributed in Energy (18%), IPPU (2%), and Waste (1%). Figure 4 presents the energy demand and GHG emissions in 2030 by sector. In most cases, there is a reduction of energy demand due to energy efficiency measures, technology replacement and the switch to more efficient fuels (e.g., from coal and firewood to natural gas or electricity). However, the increase of natural gas demand leads to an increase of emissions upstream, namely production, pipelines energy consumption, and fugitive emissions. Moreover, GHG emissions decrease in almost all energy demand sectors by an average of 12% in 2030. The transport sector can reduce 6 MtCO$_{2eq}$ by a combination of modal changes, electric vehicles, and improvements in the logistics of freight transport.

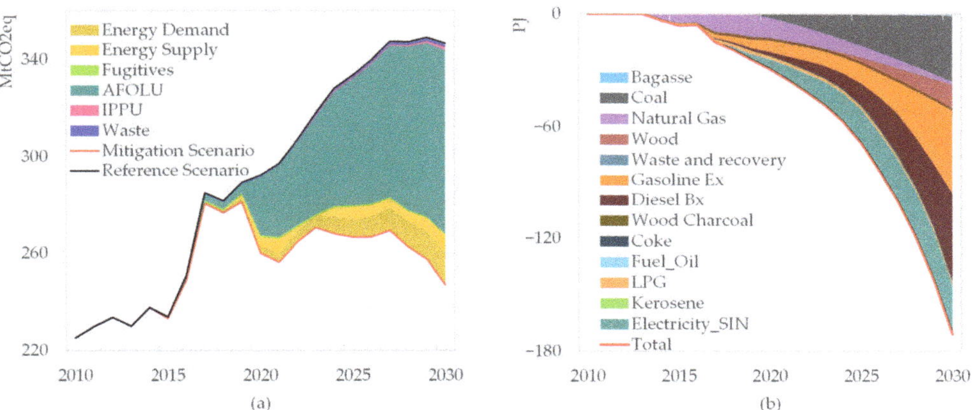

Figure 3. (**a**) GHG emission reductions by IPCC category and (**b**) energy demand changes by energy vector in the mitigation scenario compared with the reference scenario.

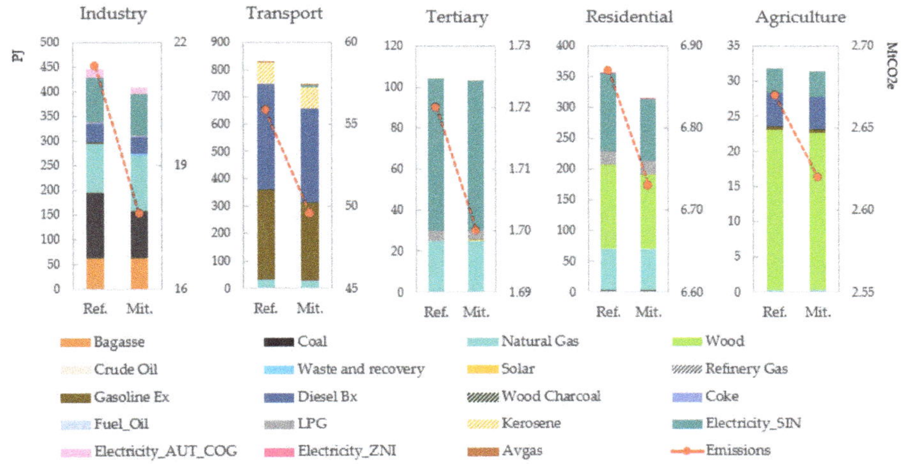

Figure 4. Energy demand and GHG emissions by demand sector in the reference and mitigation scenarios. Right axis in PJ and left axis in MtCO$_{2eq}$.

The potential of individual measures might differ from the actual mitigation in the mitigation scenario due to the implementation of other mitigation alternatives. For example, by reducing the emission factor of the power sector due to the penetration of non-conventional renewable energies (indirect measure), the potential for reducing energy efficiency measures decreases (direct measure). A second case of these synergies occurs when the measures directly affect the same sector or category. For example, in the transport sector, a transport modal change measure such as the construction of a subway in the main cities (initial measure) will result in a reference scenario different than the initial one for the following measures. Hence, an additional measure such as promoting the use of bicycles will have a lower number of possible users than in the case of not having the subway as a means of transport. Table 14 shows the individually estimated emission reduction potential and the actual potential in the mitigation scenario by IPCC category.

Table 14. Mitigation potential of mitigation actions by IPCC category, comparing individual (standalone) and combined potential in mitigation action in $MtCO_{2eq}$.

	Individual Potential	**Mitigation Scenario**	**Variation**
Energy	21.1	17.2	−18.1%
IPPU	1.4	1.6	+12.1%
AFOLU	75.8	75.7	−0.2%
Waste	1.3	1.2	−9.5%
Total	**99.6**	**95.7**	**−3.9%**

5. Conclusions

LEAP demonstrates to be an adequate tool to keep complete historical GHG emissions inventories and build future scenarios. It also allows to correctly assess the actual mitigation potential of some mitigation actions when interacting within a mitigation portfolio in a combined scenario. Although in 2020 total emissions start decreasing, this is mainly due to improvements in land management (deforestation and conservation). Therefore, there is still room for improvement in energy demand sectors, which keep an upward trend. By 2028, modal changes in passenger transport have a relevant effect on Diesel and gasoline demand. This highlights the importance of mass public transportation systems in the main cities from a climate perspective. The industry has the chance to replace the use of coal for thermal uses with natural gas, which might be a solution for the transition towards a low carbon scenario in the long term. Since the use of firewood in the residential sector slightly changes, the promotion of cleaner ways of cooking could have a considerable impact on GHG emissions and population health. Non-energy sectors should be carefully modeled to properly capture real intersectoral synergies since LEAP does not include those interactions by default. However, this approach also presents some limitations. Non-energy sectors must be modeled from a user-defined approach, which increases the computation burden of the model and the risk of mistakes. Moreover, a detailed representation of intersectoral connection in the model according to the user considerations might prevent replicability and lead to neglecting possible interactions. Lastly, scenarios and results are highly susceptible to main assumptions and user expectations as LEAP does not include optimization of demand and non-energy modules-based technology-rich alternatives, which might lead to the definition of less likely scenarios.

Author Contributions: Methodology, J.D.C.-L. and M.E.V.; software, J.D.C.-L. and M.E.V.; validation, J.D.C.-L., M.E.V., M.P. and R.M.; formal analysis, J.D.C.-L. and M.E.V.; data curation, J.D.C.-L. and M.E.V.; writing—original draft preparation, J.D.C.-L. and M.E.V.; writing—review and editing, R.M.; visualization, J.D.C.-L.; supervision, M.P., R.M.; project administration, M.P.; All authors have read and agreed to the published version of the manuscript.

Funding: This research was funded by the World Bank, within the Colombia Partnership for Market Readiness program.

Institutional Review Board Statement: Not applicable.

Informed Consent Statement: Not applicable.

Conflicts of Interest: The authors declare no conflict of interest.

Appendix A Complementary Tables

Table A1. Refineries energy demand [25].

		2010	2011	2012	2013	2014	2015	Average
Refined Oil	[PJ]	666	680	678	631	548	542	624
Energy consumption	[PJ]	51	56	56	55	47	53	53
Energy Intensity	[PJ/PJ]	0.077	0.083	0.082	0.087	0.085	0.098	0.085
Share								
Diesel	*[%]*	*0.02*	*0.04*	*0.06*	*0.01*	*0.02*	*0.02*	*0.03*
Fuel Oil	*[%]*	*5.55*	*5.65*	*0.56*	*0.45*	*0.03*	*0.05*	*2.08*
Refinery Gas	*[%]*	*29.87*	*28.57*	*28.52*	*22.11*	*13.72*	*12.75*	*22.83*
LPG	*[%]*	*2.93*	*2.44*	*3.80*	*5.45*	*7.41*	*16.20*	*6.31*
Natural Gas	*[%]*	*61.61*	*63.29*	*67.05*	*71.98*	*78.82*	*70.97*	*68.74*
Gasoline	*[%]*	*0.00*	*0.00*	*0.00*	*0.00*	*0.00*	*0.01*	*0.00*
Kerosene	*[%]*	*0.00*	*0.00*	*0.00*	*0.00*	*0.00*	*0.00*	*0.00*

Source: Ministry of Mines and Energy.

Table A2. Fugitive emission factors for mining in m3/ton.

Region	Mining		Post-Mining	
	CO_2	CH_4	CO_2	CH_4
Cundinamarca	0.077	13.03	0.018	3.909
Boyacá	0.077	7.17	0.018	2.151
N Santander	0.077	7.17	0.018	2.151
Antioquia	0.077	2.93	0.018	0.879
V Cauca	0.077	2.93	0.018	0.879
Cauca	0.077	2.93	0.018	0.879
Casanare	0.077	1.95	0.018	0.585
Average Underground	**0.077**	**8.926**	**0.018**	**2.678**
Cesar	-	0.89	-	0.267
La Guajira	-	0.89	-	0.267
Santander	-	0.4	-	0.12
Córdoba	-	0.59	-	0.177
Average surface	**-**	**0.888**	**-**	**0.266**

Source: [22,23,26,56].

Table A3. Mitigation actions of the energy sector included in the mitigation scenario.

Scope	Mitigation Measure	Explanation	Target 2030	Assumptions	Implementation in LEAP
Residential sector	NAMA Refrigerators	Change the coolant used in national production and imports of refrigerators, which would reduce the electricity demand of the refrigerators stock.	More efficient refrigerators: 60% of national stock	Replacement will be the result of the natural replacement of obsolete stock.	Change of the share of technology in the residential sector

Table A3. *Cont.*

Scope	Mitigation Measure	Explanation	Target 2030	Assumptions	Implementation in LEAP
Residential and tertiary	Efficient new buildings	Improve the efficiency of new buildings by better materials use and novel design techniques.	20% reduction of energy intensity [MJ/m2] foal all new buildings	The area will grow by 23 million m3. Savings are only in terms of electricity demand.	Reduction of energy intensity in proportion to the expected area of new buildings and efficiency targets and penetration of more efficient technologies.
Tertiary sector	Thermal districts	Avoid the installation of air conditioning systems by the thermal district in public and commercial buildings.	90 Million refrigeration tonnes	Without thermal districts, conventional AC would be used.	Switch demand from the AC category to thermal districts module by reducing energy intensity [MJ/COP]
Agriculture and fishing	Agriculture efficiency (panela and coffee NAMA)	Change of diesel engines for electric engines, and increase the use of biomass for thermal processes in a more efficient manner.	Replace 50% of fossil fuels with biomass in coffee and panela crops, and improve the efficiency of the thermal process by 2%	The measure could be equally implemented in all farms/production sites.	Modification of energy intensity factor in proportion to the share of coffee and panela energy demand in the agriculture-fishing sector.
Transport, industry, and supply	Carbon tax	Impose a tax on fossil fuels in certain sectors such as transport, refineries, and industry. Demand will respond to price increase according to specific elasticities defined for each sector.	US$7/tCO$_2$	Lineal and general demand elasticity to fuel price increase	Modification of energy intensity in relation to sector-specific elasticity and CO$_2$ content of energy vector.
All demand sectors	Demand management	Promote demand response through the introduction of aggregators and incentives.	Reduce by 20% the difference between the peak and valley of the annual electricity demand load curve.	Demand management will be possible with aggregators, smart meters, and incentives to the industry.	Change of the system load curve
Industry	Sustainable cement	Increase the use of biomass and solid waste in the kiln.	15% of kiln energy needs cover with biomass and waste	It is possible to replace coal with biomass and waste without modifying the kiln	Change of fuel mix
	Brick Development	Replacement of coal and liquid fossil fuels with natural gas and biomass.	Fuel mix: 60% natural gas and 40% charcoal and firewood in thermal processes	Current technology can operate with future fuel mix	Change of fuel mix
	Industry Efficiency	Promote energy efficiency programs aiming to improve production practices, and to a lesser extend equipment.	Technologies with better efficiency will reach 30% indirect heat and other end-uses.	It is possible to replace 30% of the technologies (e.g.,: engines, boilers, compressors). Replacement also reflects changes in production behavior.	Share of the best technology in Colombia/international

Table A3. *Cont.*

	Fuel replacement industry	Replacement of coal and liquid fossil fuels with natural gas (when suitable)	Replace 20% of liquid fossil fuels with natural gas	There is a different potential by sector	Change of fuel mix
Electricity supply	Thermal generator efficiency	Preventive and corrective maintenance to augment the efficiency of coal and natural gas power plants.	Increase by 2% the energy efficiency of coal and natural gas power plants	The measure will apply to all coal and natural gas power plants. Maintenance will correct efficiency degradation due to normal operation.	Increase the efficiency of the technology (coal or natural gas) to the desired level.
	Diversification Capacity Generation	Increase the penetration of wind and solar in the generation mix. Additionally, include biogas and geothermal in the generation mix.	Capacity defined in PEN 2015	The power capacity proposed in the Colombian energy plan will match the future electricity demand	Change the exogenous capacity according to the Colombian energy plan (PEN)
Coal extraction	Mining energy efficiency	Improve the efficiency of the mining activities without changes in technologies nor fuel mix.	Reduce energy intensity of electricity and diesel by 1%.	Changes in production techniques/processes reach energy reductions without technology changes	Change of auxiliary fuel intensity
Oil refining	Energy Efficiency Refineries	Improve the efficiency of refining activities without changes in technologies nor fuel mix.	Reduce by 16% energy intensity of refineries (feedstock not included)	Changes in production techniques/processes reach energy reductions without technology changes	Change of auxiliary fuel intensity
Transport	NAMA_TOD	Nationally Appropriate Mitigation Action—Transport Oriented Development (TOD).	The goal is to implement four TOD projects in four cities. The goal is to reduce motorized activity in 2030 with respect to BAU: Passenger light: 0.7%; Taxis: 0.6%; Buses: 0.4%; Medium trucks: 0.01%.	Despite this type of intervention take time to consolidate, it was assumed they will be in place since 2021 and there will be results in emissions since then.	Modal share changes. We create a technology to represent non-motorized modes, with no energy consumption.
	Aviation performance improvements	Performance-Based Navigation (PBN) in domestic aviation.	The mitigation action proposes to cover 60% of the national airports, to improve the fuel efficiency of the commercial flights.	It was assumed some airports won't be able to implement PBN in the next years, so the action affects only a proportion of the domestic operations.	Reduction in fuel intensity factors.

Table A3. *Cont.*

Scrapping and cargo fleet renewal program.	It consists of disintegrating and renovating the oldest vehicles in the cargo fleet. It affects trucks with a gross vehicle weight greater than 10.5 tons and more than 20 years old.	The program seeks to renovate 57,000 trucks between 2015 and 2030.	Older trucks tend to be used much less than new trucks, and that might affect the potential to reduce CO_2eq. It was assumed that the program is accompanied by the optimization of freight operations, so in the end, the net effect is positive.	It is represented using the scrappage and fraction of scrapped replaced functions.
Urban logistics improvements	Urban logistics improvements in the main cities in the country.	These mitigation actions seek to improve the operation of urban logistics in the main cities in the country.	It was assumed that the potential to improve current practices is significant. It is assumed that a national program will be able to cover almost 100% of the operations since the beginning of the action in 2017. This action depends on many external factors, and this is not captured by the assumptions in the model.	Modal share changes. We create a technology to represent avoided activity per year.
NAMA_TANDEM	Nationally Appropriate Mitigation Action—Active transport and travel demand management (TAnDem).	It seeks to promote the use of non-motorized modes in urban passenger transport. The goal is to reduce motorized activity in 2030 with respect to BAU: Passenger light: 0.6%; Taxis: 1.6%; Motorcycles: 0.2%.	It is assumed that the action is generating benefits in GHG emissions since its beginning in 2019. The potential was modeled considering the effects of similar projects in Colombia and Latin America.	Modal share changes. We create a technology to represent non-motorized modes, with no energy consumption.
Multimodal freight transport—River/Road	Increase the participation of waterborne transport in the freight segment.	By modal substitution, the goal is to reduce between 30,000–132,000 t CO_2eq per year in the period 2016–2030.	It is assumed that the main benefits will come from the proportion of freight transport by the river, but there is also an opportunity to improve the road complementary segment.	Modal share changes in road transport.Increase in fuel intensity factors for navigation.
Multimodal freight transport—Train/Road	Increase the participation of rail transport in the freight segment.	By modal substitution, the goal is to reduce between 9000–112,000 t CO_2eq per year in the period 2021–2030.	It is assumed that the main benefits will come from the proportion of freight transport by train, but there is also an opportunity to improve the road complementary segment.	Modal share changes in road transport.Increase in fuel intensity factors for trains.

Table A3. *Cont.*

	Electric mobility program	Increase the participation of electric vehicles.	In terms of activity (VKTs) in 2030 there is this participation of electricity: Passenger light: 22%; Taxis: 5%; Buses: 10%; Medium trucks: 8%.	It is assumed the incentives and other complementary programs will be implemented on time to reach the goal in the electric fleet by 2030.	Sales share changes in road transport.
	Metro Bogotá	The first line of the Bogotá Metro.	By modal substitution, the goal is to reduce 132,000 t CO_2eq per year in the period 2028–2030.	It is assumed that the substitution effects will be gradual and so will be the effects in emissions reduced.	Sales share changes in road transport.Modal share changes.
	Intercity train Metropolitan Area of Bogota	Regional tram to serve the Metropolitan Area of Bogotá.	By modal substitution, the goal is to reduce 32,000 t CO_2eq per year in the period 2024–2030.	It is assumed that the substitution effects will be gradual and so will be the effects in emissions reduced.	Sales share changes in road transport.Modal share changes.
Fugitives	Compressors in natural gas activities	Improve the sealing of compressors in the extraction and transportation of natural gas.	20% less emission in venting	Works on compressors will be lineal from 2018 to reach the target in 2030.	Reduction of emission factor
	Glycol use optimization	Reduce fugitive emissions by optimizing the use of the glycol.	Reduction of emissions by 2%	The reduction of the emission factor reflects the potential assed in some wells in Colombia. This can be extrapolated to the total national production.	Reduction of emission factor
	Recovery in storage tanks	Recovery of fugitive emissions in storage facilities and preventing gas leakages by continuous inspections.	13% less emissions in distribution	All storage facilities might reach the same level of reduction as the pilot projects in some facilities in Colombia have done *.	Reduction of emission factor

Table A4. Mitigation actions of IPPU included in the mitigation scenario.

Scope	Measure	Explanation	Target 2030	Assumptions	Implementation in LEAP
Substances	ODS substitutes	Reduce the use and management of ODS substances	Reduce the use of the most polluting HFCs by 15%	The replacement of substance with other HFCs is possible without affecting the performance of cooling technologies and there will be market acceptance.	Change in production activity
Process emissions	Chemical industry	Reduce process emissions in the industry by improvements in reactions	Reduce 10% process emission in nitric acid production	An emission factor lower than the standard IPCC is possible by improvements in production.	Reduction of emission factor

Table A5. Mitigation actions of AFOLU included in the mitigation scenario.

Scope	Measure	Explanation	Target 2030	Assumptions	Implementation in LEAP
Wastewater	Nama coffee and panela wastewater treatment	Wastewater treatment of coffee and panela farms	5% of water treated with septic tank and burning of 4ktCH4.	All farms are similar and have access to wastewater treatment facilities close to production.	Change in the share of technologies in residential-rural and industry-coffee/sugar wastewater management
Solid Waste	Use of biogas in landfills	Use of landfill gas for the production of electricity	3% of CH4 emissions in major landfills	The production of electricity in landfills covers local electricity demand. Surplus of electricity is neglected.	CH4 recovery variable in function of emission in reference scenario
	Biogas management water treatment	Recovery of CH4 in wastewater treatment plants to destroy CH4 molecules and emit CO_2.	35% wastewater treated with plants with CH4 recovery	Recovered CH4 is used to partially cover sites own energy requirements (electricity and heat)	Change in the share of technologies in residential-urban wastewater management
	Biogas burning in landfills	Recovery of CH4 in landfills to destroy CH4 molecules and emit CO_2.	1.5% of CH4 emissions	Combustion is efficient and most CH4 molecules are destroyed	CH4 use variable in function of emission in reference scenario.
	Recycling of plastic paper and glass	Increase the recycling rate of plastic, paper and glass at national level.	15% in major landfills	Recycling is possible in landfills linked to the five biggest cities. Waste sorting is done outside the landfill facilities.	Change in the amount of solid waste (plastic, glass, and paper disposed) in landfills used in the calculation of emissions.
	Biological mechanical treatment systems	Composting of organic component of municipal solid waste to prevent CH4 emissions.	5% of biological part of waste in major landfills	Organic waste is extracted at the entrance of the landfill	Reduction of the amount of municipal solid waste reaching landfills, and a proportional increase in the solid waste treated by mechanical biological treatment plants.

Table A6. Mitigation actions of Waste sector included in the mitigation scenario.

Scope	Measure	Explanation	Target 2030	Assumptions	Implementation in LEAP
Land use	Deforestation reduction	Deforestation rates are reduced following the ambitions included in the NREF	Reduction of 40kha/yr	Reduction of deforestation linked to illegal activities, intensive agriculture and intensive mining, among others is possible by policies, regulation and surveillance of protected areas.	Results are fed from AFOLU model (exogenous) into subcategories 3B1
	AMTEC rice	Implementation of AMTEC mode for rice production: Volumetric water consumption management; reduction in the use of fertilizers in the productive system; and management of harvest residues.	80% of crops	De adoption of the AMTEC method by rice producer will not present opposition	Results are fed from AFOLU model (exogenous) into subcategories 3C4 y 3C5
	NAMA Coffee (land use)	Implement strategies for the mitigation of GHG generated in the production, harvest and post-harvest stages of Colombian coffee at the farm level.	1.2 kHa/yr. of crops with shade	The benefits of crops with shade are the same in all regions and conditions	Reduction of fertilizer used by coffee crops.
	NAMA Panela (land use)	Encourage the efficient use of synthetic fertilizers and promote the reduction of burns	1500 sugar mills with 800 ha of restoration	Data from the "Andina" region are extrapolated to the national level (14.8 tCO_2/ha/year.)	Results are fed from AFOLU model (exogenous) into subcategories 3B2bi
	Forest plantations	Increased establishment of forest plantations in non-forest areas prior to planting	15 kha/yr	Plantation harvesting is within a cycle equal or less than one year.	Results are fed from AFOLU model (exogenous) into subcategories 3B2a.
	Cocoa crops	Increase in areas dedicated to the cultivation of cocoa under agroforestry systems (SAF), and land rehabilitation.	80k Ha	Given that for the productive sector only 7.6% of the productive units use chemical fertilizers and 6.5% apply organic fertilizers, the use of fertilizers will not be taken into account in the quantification of emission reductions.	Results are fed from AFOLU model (exogenous) into subcategories 3B2a.
	Ecological restoration	Reforestation of already deforested lands	1 million Ha	Land will be restored and protected 20 years. Then, land will pass to the category of land that remains as it is.	Results are fed from AFOLU model (exogenous) into subcategories 3B2a.

Table A6. Cont.

Scope	Measure	Explanation	Target 2030	Assumptions	Implementation in LEAP
Biomass use	Efficient wood stoves	Provide more efficient firewood stoves to households that currently use firewood for cooking	700,000 new stoves	People will continue using firewood for cooking. New stoves will improve efficiency and reduce wood consumption per capita in rural areas.	Increase penetration of effect stoves in demand sector (residential—rural).
Livestock	NAMA Livestock	Reduce GHG emissions generated in livestock production and increase carbon removals from ago-ecosystems dedicated to livestock by intensifying the production of livestock systems and increasing efficiency (less land for animal farming).	38% of livestock farms. Emission factor reduction of 0.55%. And 68kHa of livestock farming to be restored.	1% les fertilizer for 27% of cattle.Almost all land restauration is attributed to 1 of the 10 defined regions,	Reduction of CH4 emissions by enteric fermentation. Reduction of nitrogen fertilizers. And carbon sequestration in soils and biomass from a series of measures (From AFOLU land model)

References

1. Intended Nationally Determined Contributions (INDCs) | UNFCCC. Available online: https://unfccc.int/process-and-meetings/the-paris-agreement/nationally-determined-contributions-ndcs/indcs (accessed on 10 September 2021).
2. IDEAM and Clima Soluciones SAS. Biennial Update Report (BUR). BUR 2. National Inventory Report. April 2019. Available online: https://unfccc.int/sites/default/files/resource/NIR_BUR2_Colombia.pdf (accessed on 6 March 2020).
3. Delgado, R.; Espinosa, M.; Cadena, A.I.; Sandova, J.; Alvarez, C. Formulation of a Nationally Determined Contribution to the mitigation of Climate Change. Colombian case. In Proceedings of the 39th IAEE International Conference, Bergen, Norway, 19–22 June 2016; Available online: https://www.iaee.org/en/publications/proceedingsabstractpdf.aspx?id=13678 (accessed on 23 September 2020).
4. Available Educational Version of the Carbon Calculator | Calculadora Colombia. 2050. Available online: https://calculadora2050.minambiente.gov.co/en/noticias/available-educational-version-carbon-calculator (accessed on 15 September 2021).
5. IRENA. *Scenarios for the Energy Transition: Global Experience and Best Practices*; IRENA: Abu Dhabi, United Arab Emirates, 2020; p. 88.
6. United Nations-ECLAC. Assessment of Development Account Project 06/07 AM Strengthening National Capacities to Design and Implement Sustainable Energy Policies for the Production and Use of Biofuels in Latin America and the Caribbean. 2015. Available online: https://www.cepal.org/sites/default/files/publication/files/40163/S1501327_en.pdf (accessed on 13 June 2021).
7. Nieves, J.A.; Aristizábal, A.J.; Dyner, I.; Báez, O.; Ospina, D.H. Energy demand and greenhouse gas emissions analysis in Colombia: A LEAP model application. *Energy* **2019**, *169*, 380–397. [CrossRef]
8. Nieves, J.A.; Aristizábal, A.J.; Dyner, I.; Báez, O.; Ospina, D. Energy Analysis of the Tertiary Sector of Colombia and Demand Estimation Using LEAP. In Proceedings of the 2018 ICAI Workshops (ICAIW), Bogota, Colombia, 1–3 November 2018; pp. 1–6. [CrossRef]
9. Arango-Aramburo, S.; Veysey, J.; Martínez-Jaramillo, J.E.; Díez-Echavarría, L.; Calderón, S.L.; Loboguerrero, A.M. Assessing the impacts of nationally appropriate mitigation actions through energy system simulation: A Colombian case. *Energy Effic.* **2020**, *13*, 17–32. [CrossRef]
10. Simsek, Y.; Sahin, H.; Lorca, Á.; Santika, W.G.; Urmee, T.; Escobar, R. Comparison of energy scenario alternatives for Chile: Towards low-carbon energy transition by 2030. *Energy* **2020**, *206*, 118021. [CrossRef]
11. Islas-Samperio, J.M.; Manzini, F.; Grande-Acosta, G.K. Toward a Low-Carbon Transport Sector in Mexico. *Energies* **2020**, *13*, 84. [CrossRef]
12. Castrejón, D.; Zavala, A.M.; Flores, J.A.; Flores, M.P.; Barrón, D. Analysis of the contribution of CCS to achieve the objectives of Mexico to reduce GHG emissions. *Int. J. Greenh. Gas Control* **2018**, *71*, 184–193. [CrossRef]
13. McPherson, M.; Karney, B. Long-term scenario alternatives and their implications: LEAP model application of Panama's electricity sector. *Energy Policy* **2014**, *68*, 146–157. [CrossRef]
14. DNP. Manual Modelo de Equilibio General Computable (MEG4C). 2015. Available online: https://colaboracion.dnp.gov.co/CDT/Ambiente/2015%20Manual%20MEG4C.pdf (accessed on 16 June 2021).
15. DANE. Proyecciones y retroproyecciones de población. In *Proyecciones y Retroproyecciones de Población 1950–2070*; DANE: Bogotá, Colombia, 2020.

16. UPME. *Plan Energético Nacional 2020–2050*; UPME: Bogotá, Colombia, 2019.

17. DNP. *Proyección de PIB Sectoriales*; National Planning Department of Colombia (DNP): Bogotá, Colombia, 2020.

18. MinHacienda. *Marco Fiscal de Mediano Plazo 2019*; Ministry of Finance and Public Credit: Bogotá, Colombia, 2019.

19. Heaps, C.G. *LEAP: The Low Emissions Analysis Platform*; Stockholm Environment Institute: Somerville, MA, USA, 2020; Available online: https://leap.sei.org (accessed on 30 April 2021).

20. ETSAP. Overview of TIMES Modelling Tool. 2020. Available online: https://iea-etsap.org/index.php/etsap-tools/model-generators/times (accessed on 9 June 2021).

21. IPCC. *Climate Change 2014: Synthesis Report*; IPCC: Geneva, Switzerland, 2014.

22. IDEAM; PNUD; MADS; DNP; CANCILLERÍA. *Segundo Informe Bienal de Actualización de Colombia ante la Convención Marco de las Naciones Unidas para el Cambio Climático (CMNUCC)*; IDEAM: Bogotá, Colombia, 2018.

23. Pulido, A.D.; Chaparro, N.; Granados, S.; Ortiz, E.; Rojas, A.; Torres, C.F.; Turriago, J.D. *Informe de Inventario Nacional de GEI de Colombia*; Instituto de Hidrología, Meteorología y Estudios Ambientales IDEAM, Programa de las Naciones Unidas para el Desarrollo PNUD: Bogotá, Colombia, 2019.

24. UPME. *Primer Balance de Energía Útil Para Colombia y Cuantificación de las Pérdidas Energéticas Relacionadas y la Brecha de Eficiencia Energética*; UPME: Bogotá, Colombia, 2018.

25. UPME. *Beco: Consulta*; UPME: Bogotá, Colombia, 2018.

26. IPCC. *Directrices del IPCC de 2006 Para los Inventarios Nacionales de Gases de Efecto Invernadero. Volumen 3: Procesos Industriales y uso de Productos*; IPCC: Geneva, Switzerland, 2006.

27. LEAP. *LEAP User Guide. Low Emissions Analysis Platform (LEAP)*; Stockholm Environment Institute: Stockholm, Sweden, 2020.

28. Ministerio de Transporte. *Transporte en Cifras 2013–2018*; Ministerio de Transporte: Bogotá, Colombia, 2019.

29. Naturgas. *Estadísticas Sobre Conversiones a GNV*; Naturgas: Bogotá, Colombia, 2020.

30. Behrentz, E.; Espinosa, M.; Joya, S.; Peña, C.; Prada, A. *Productos Analíticos Para Apoyar la Toma de Decisiones Sobre Acciones de Mitigación a Nivel Sectorial: Curvas de Abatimiento Para Colombia Documento Genera*; Universidad de los Andes: Bogotá, Colombia, 2014.

31. Cadena, A.; Bocarejo, J.; Rodriguez, M.; Rosales, R.; Arguello, R.; Delgado, R.; Florez, E.; Espinosa, M.; Lombo, C.; Lopez, H.; et al. *Upstream Analytical Work to Support Development of Policy Options for Mid- and Long-Term Mitigation Objectives in Colombia—Informe Producto C*; Universidad de los Andes: Bogotá, Colombia, 2016.

32. EEA. EMEP/EEA Air Pollutant Emission Inventory Guidebook. 2019. Available online: http://efdb.apps.eea.europa.eu (accessed on 20 June 2020).

33. USDE-EPA. Fuel Economy. Government Source for Fuel Economy Information. 2020. Available online: www.fueleconomy.gov (accessed on 20 June 2020).

34. IEA. *Fuel Economy in Major Car Markets—Technology and Policy Drivers 2005–2017*; IEA: Paris, France, 2019.

35. ICCT. *Estimating the Fuel Efficiency Technology Potential of Heavy-Duty Trucks in Major Markets Around the World*; ICCT: London, UK, 2016.

36. Dargay, J.; Gately, D.; Sommer, M. Vehicle Ownership and Income Growth, Worldwide: 1960–2030 Author(s): Joyce Dargay, Dermot Gately and Martin Sommer Published by : International Association for Energy Economics Stable. *Energy* **2007**, *28*, 143–170. Available online: http://www.jstor.org/stable/41323125 (accessed on 1 April 2016).

37. DANE. *Cuentas Nacionales: Agregados Macroeconomicos—Base 2015*; DANE: Bogotá, Colombia, 2020.

38. DANE. *Encuesta Nacional de Calidad de Vida—ECV 2015*; DANE: Bogotá, Colombia, 2015.

39. UPME. *Plan Indicativo de Abastecimiento de Combustibles Líquidos*; UPME: Bogotá, Colombia, 2019.

40. UPME. *Estudio Técnico Para el Plan de Abastecimiento de Gas Natural*; UPME: Bogotá, Colombia, 2020; p. 143.

41. UPME; Ministerio de Minas y Energía. Análisis prospectivo del mercado nacional e internacional del carbón térmico, metalúrgico y antracita producido en Colombia. 2020. Available online: https://www1.upme.gov.co/simco/Cifras-Sectoriales/EstudiosPublicaciones/Analisis_prospectivo_mercado_nal_internal_carbon_termico.zip (accessed on 24 September 2021).

42. UPME. Plan de Expansión de Referencia Generación—Transmisión 2016–2030. National Official Plan. 2016. Available online: http://www.upme.gov.co/Fotonoticias/Plan_GT_2016-2030_Preliminar_21-11-2016.pdf (accessed on 19 May 2020).

43. XM. XM Portal BI—Oferta. 2020. Available online: http://portalbissrs.xm.com.co/Paginas/Home.aspx# (accessed on 9 June 2021).

44. Veysey, J. NEMO: The Next Energy Modeling System for Optimization. 2020. Available online: https://www.sei.org/projects-and-tools/tools/nemo-the-next-energy-modeling-system-for-optimization/ (accessed on 30 April 2021).

45. XM. *Informe Seguimiento Cogeneradores Resolución CREG 005 de 2010*; XM: Medellín, Colombia, 2019.

46. UPME. *Capacidad Instalada de Autogeneración y Cogeneración en Sector de Industria, Petróleo, Comercio y Público del País*; UPME: Bogotá, Colombia, 2014.

47. Superintendencia de Servicios Publicos. Sistema Unico de Información—Superservicios. Database. Available online: http://www.sui.gov.co/web/datos-abiertos (accessed on 15 June 2021).

48. IPSE. *Monitoreo y Fortalecimiento Empresarial en la Colombia No Interconectada*; IPSE: Cali, Colombia, 2019.

49. UPME. Factores de Emisión de los Combustibles Colombianos—FECOC. 2016. Available online: http://www.upme.gov.co/calculadora_emisiones/aplicacion/calculadora.html (accessed on 19 May 2021).

50. Minambiente; IDEAM. *Propuesta de Nivel de Referencia de las Emisiones Forestales por Deforestación en Colombia Para Pago por Resultados de REDD+ bajo la CMNUCC*; Minambiente: Bogotá, Colombia; IDEAM: Bogotá, Colombia, 2019.
51. IPCC. 2006 IPCC Guidelines for National Greenhouse Gas Inventories CHAPTER 3—SOLID WASTE DISPOSAL V5. 2006. Available online: https://www.ipcc-nggip.iges.or.jp/public/2006gl/pdf/5_Volume5/V5_3_Ch3_SWDS.pdf (accessed on 25 May 2020).
52. Minambiente, Colombia. Contribución Prevista y Nacionalmente Determinada (indc) de Colombia-Documento de Soporte. 2015. Available online: https://www.minambiente.gov.co/images/cambioclimatico/pdf/documentos_tecnicos_soporte/Contribuci%C3%B3n_Nacionalmente_Determinada_de_Colombia.pdf (accessed on 25 September 2021).
53. Minambiente, Colombia. NDC de COLOMBIA—ACTUALIZACIÓN 2020 Versión Para Consulta Pública. 2020. Available online: http://www.andi.com.co/Uploads/Documento%20NDC%20para%20consulta%20ciudadanos.pdf (accessed on 25 September 2021).
54. UPME. *Plan Energetico Nacional Colombia: Ideario Energético 2050*; UPME: Bogotá, Colombia, 2015.
55. Pronóstico de demanda. Available online: https://www.xm.com.co/Paginas/Consumo/pronostico-de-demanda.aspx (accessed on 29 September 2021).
56. IPCC. 2006 IPCC Guidelines for National Greenhouse Gas Inventories. Chapter 4: Fugitive Emissions, V2.4. 2006. Available online: https://www.ipcc-nggip.iges.or.jp/public/2006gl/pdf/2_Volume2/V2_4_Ch4_Fugitive_Emissions.pdf (accessed on 25 September 2021).

Article

Exploring the Long-Term Development of the Ukrainian Energy System

Stefan N. Petrović [1,*], Oleksandr Diachuk [2], Roman Podolets [2], Andrii Semeniuk [2], Fabian Bühler [1], Rune Grandal [1], Mourad Boucenna [1] and Olexandr Balyk [3,4]

[1]   Centre for Global Cooperation, The Danish Energy Agency, DK-1577 Copenhagen, Denmark; fnbr@ens.dk (F.B.); rngl@ens.dk (R.G.); mdba@ens.dk (M.B.)

[2]   Institute for Economics and Forecasting of the National Academy of Sciences of Ukraine, 01011 Kyiv, Ukraine; diachuk@ief.org.ua (O.D.); podolets@ief.org.ua (R.P.); a_semeniuk@ief.org.ua (A.S.)

[3]   MaREI, The SFI Research Centre for Energy, Climate and Marine, Environmental Research Institute, University College Cork, T23 XE10 Cork, Ireland; olexandr.balyk@ucc.ie

[4]   School of Engineering, University College Cork, T12 HW58 Cork, Ireland

*   Correspondence: snpc@ens.dk

Abstract: This study analyses the Ukrainian energy system in the context of the Paris Agreement and the need for the world to limit global warming to 1.5 °C. Despite ~84% of greenhouse gas emissions in Ukraine being energy- and process-related, there is very limited academic literature analysing long-term development of the Ukrainian energy system. This study utilises the TIMES-Ukraine model of the whole Ukrainian energy system to address this knowledge gap and to analyse how the energy system may develop until 2050, taking into current and future policies. The results show the development of the Ukrainian energy system based on energy efficiency improvements, electrification and renewable energy. The share of renewables in electricity production is predicted to reach between 45% and 57% in 2050 in the main scenarios with moderate emission reduction ambitions and ~80% in the ambitious alternative scenarios. The cost-optimal solution includes reduction of space heating demand in buildings by 20% in frozen policy and 70% in other scenarios, while electrification of industries leads to reductions in energy intensity of 26–36% in all scenarios except frozen policy. Energy efficiency improvements and emission reductions in the transport sector are achieved through increased use of electricity from 2020 in all scenarios except frozen policy, reaching 40–51% in 2050. The stated policies present a cost-efficient alternative for keeping Ukraine's greenhouse gas emissions at today's level.

Keywords: energy systems modelling; scenario analysis; TIMES-Ukraine; decarbonisation; paris agreement; electrification; renewable energy; energy efficiency

Citation: Petrović, S.N.; Diachuk, O.; Podolets, R.; Semeniuk, A.; Bühler, F.; Grandal, R.; Boucenna, M.; Balyk, O. Exploring the Long-Term Development of the Ukrainian Energy System. *Energies* **2021**, *14*, 7731. https://doi.org/10.3390/en14227731

Academic Editor: Dolf Gielen

Received: 31 August 2021
Accepted: 3 November 2021
Published: 18 November 2021

Publisher's Note: MDPI stays neutral with regard to jurisdictional claims in published maps and institutional affiliations.

1. Introduction

The world needs to reach net-zero carbon emissions by 2050 to limit global warming to 1.5 °C [1]. Ukraine is one of the 196 Parties to the Paris Agreement that aims at keeping a global temperature rise in the 21st century to well below 2 °C above pre-industrial levels and to pursue efforts to limit the temperature increase to 1.5 °C [2].

Ukraine's greenhouse gas (GHG) emissions amounted to approximately 332 Mt CO_{2-eq} in 2019 [3]. Out of these, ~66% were energy related, 18% came from industrial process and product use, 13% from agriculture, and the remaining 4% originated in waste management and land use, land-use change and forestry (LULUCF). Ukraine's GHG emissions have remained relatively stable since 2000, constituting approximately 38% of the 1990 level in 2019 [3]. The Ukraine 2050 Low Emission Development Strategy [4], the only long-term strategic document to 2050, projects GHG emissions to constitute 31% of the 1990 level in 2050 in its most ambitious scenario (Table A1). Ukraine has recently updated its nationally determined contribution (NDC) to 65% economy-wide GHG emissions reduction compared to 1990 by 2030 and indicated its consistency with a trajectory to achieve net zero GHG emissions by 2060 [5].

The Energy balance of Ukraine is dominated by fossil fuels. They accounted for about 71% of the total primary energy supply (TPES) in 2019, with the remainder covered by nuclear (~25%) and renewables and waste (~5%) [6]. Ukraine depends on energy imports, which account for 33% of its natural gas, 50% of its coal and 83% of its oil consumption [7]. In the final energy consumption, the largest share is held by natural gas (27%), followed by oil and oil products (22%), electricity (20%), heat (15%), coal (12%) and biofuels and waste (4%) [6]. More than half of the electricity production is from nuclear power [7]. The largest share of final energy consumption is in the residential and industrial sectors, 28% and 33% respectively [6].

Ukraine has a number of energy and climate related targets in its legislation. The adopted legislation targets environmental policy in general [8], energy [9], transport [10], renewable energy [11], energy efficiency [12] and heat supply [13]. Table A2 presents key targets from these documents. None of the targets go beyond 2035.

Long-term energy system modelling and scenarios are commonly used in the literature to analyse the development of national energy systems. The studies often apply energy system models representing the whole energy system of a country [14–16] or a wider geographical region [17–19] to analyse the role of different technologies (e.g., residential heat pumps [20], hydrogen [19], storage [21], district heating [22] or heating technologies [23]), policy measures (e.g., increasing efficiency [24], electrification [25,26] and nuclear reduction [27]) and other developments (e.g., growth in data centres [28]) in reducing GHG emissions or energy transition, often focusing on the period until 2050. Some articles address the uncertainties inherent to long-term energy planning studies such as the present one [29–31]. Some studies combine several models to obtain deeper sectoral insights and analyse the effects on the wider economy [32–34]. Others focus on a single sector independently of whether the model includes only the sector [35–39] or the whole energy system [40–42]. Representing the whole energy system is important when analysing decarbonisation pathways as it allows the representation of trade-offs between sectors under resource constraints [16].

There is limited academic literature analysing long-term Ukraine energy system development published in international journals. Chepeliev et al. [43] soft-linked TIMES-Ukraine and Ukrainian Computable General Equilibrium models to assess low-emission development scenarios for Ukraine. Nevertheless, long-term energy system modelling in Ukraine has been used to support the development of national strategies [4,44,45] and produced technical reports [46].

The Ukraine Energy Outlook 2021 is an independent study of the Ukrainian energy system, utilising the energy system model TIMES-Ukraine [47], which analyses the possible development of the energy system in Ukraine under various scenarios, taking into account enacted policies, technological development and environmental considerations. The Outlook is part of a long standing effort of the Danish Energy Agency to support long-term energy planning processes in partner countries in the framework of government-to-government cooperation through application of modelling tools and practices. It presents a quantitative assessment and comparison of different energy sector development pathways with regards to various indicators (including investment cost, energy intensity and GHG emissions), enabling the benefits and drawbacks of respective scenarios to be evaluated. The Outlook can thus be used as a technical reference when planning new measures in the climate and energy sectors, to assess the impact of policy measures and to provide inputs to national energy strategy documents in order to achieve national energy and climate goals. Previously, Energy Outlooks have been developed for China [48–52], Indonesia [53] and Vietnam [54]. The findings of the Ukraine Energy Outlook 2021 are presented and further developed in this study.

The aim of this paper is to fill the gap in the academic literature on the possible long-term energy system development in Ukraine given current policies, using long-term energy system modelling, as well as to develop an indication of possible outcomes of measures aimed at decarbonisation and energy transition. The remainder of the paper is organised as follows. Sections 2 and 3 describe the TIMES-Ukraine model and the scenarios analysed, respectively. Section 4 presents both sector-specific results and results for the whole energy system across main scenarios, as well as the results of the alternative scenarios. Finally, this is followed by a discussion and conclusions section.

2. Materials and Methods

2.1. TIMES Modelling Framework

The Integrated MARKAL-EFOM System (TIMES) modelling framework [55–57] is developed and maintained by the Energy Technology Systems Analysis Programme (ET-SAP), an International Energy Agency (IEA) Technology Collaboration Programme which was established in 1976. TIMES models follow the bottom-up energy system modelling approach [58,59] and are typically single or multi-regional models of local [60], national [16,25], multinational [61] or global [62] energy systems. They are often characterised by databases allowing ample technology choice and are used for both medium- and long-term energy systems analysis and planning. The TIMES code, implemented in General Algebraic Modelling System (developed by GAMS Software GmbH, Frechen, Germany), is open-source and is distributed under GNU General Public License version 3 [63].

TIMES assumes perfectly competitive markets and full foresight is typically used for the entire modelling horizon. However, it is also possible to perform an analysis in myopic mode or using the rolling horizon optimisation approach. The objective function is to maximise the total surplus by minimising the total system costs discounted to the reference year. It includes the following components: investment costs, fixed and variable operation and maintenance (O&M) costs, import costs and export revenues, taxes and subsidies, as well as the residual value of technologies (salvage value) at the end of the modelling horizon. The type of inputs used to build TIMES models are exogenous energy service demand projections, supply curves (i.e., resource potentials and costs), policies, as well as both technical and economic parameters for various technologies. Outputs of TIMES include optimal investments, operation and import (export) levels that are both time- and region-specific (for multi-regional models). Costs, environmental indicators, marginal prices and commodity flows are also included alongside the optimal solution.

2.2. TIMES-Ukraine Model

The TIMES-Ukraine energy system model is a member of the the MARKAL/TIMES model family. It includes a comprehensive characterisation of the Ukrainian energy system suitable for representing the energy dynamics in the long-term [47]. The model structure follows the methodological approach of the State Statistics Service of Ukraine [64] on energy statistics, which is harmonised with the IEA and Eurostat, and includes a representation of more than 1800 technologies. The TIMES-Ukraine model divides the energy system into seven sectors (Figure 1).

The database of the model contains economic and energy statistics data for 2005–2019. Any of the following years can be used as a starting point for the optimisation (i.e., calibration years): 2005, 2009, 2012 and 2015. Key input data for the model are regularly updated and include annual energy production statistics, international trade, information on power plant and boiler performance, etc.

Based on their energy intensity, industrial users in the model are categorised into two groups. Energy-intensive energy branches are represented through technologies that are product-specific. For the rest, a generic representation is adopted based on the four process types: thermal processes, electric engines, electrochemical processes and other processes.

The transport sector in TIMES-Ukraine includes a representation of the following types of transportation: pipelines, road, railway, navigation and aviation. Energy services related to passenger and freight transport are delivered by road and rail transport technologies.

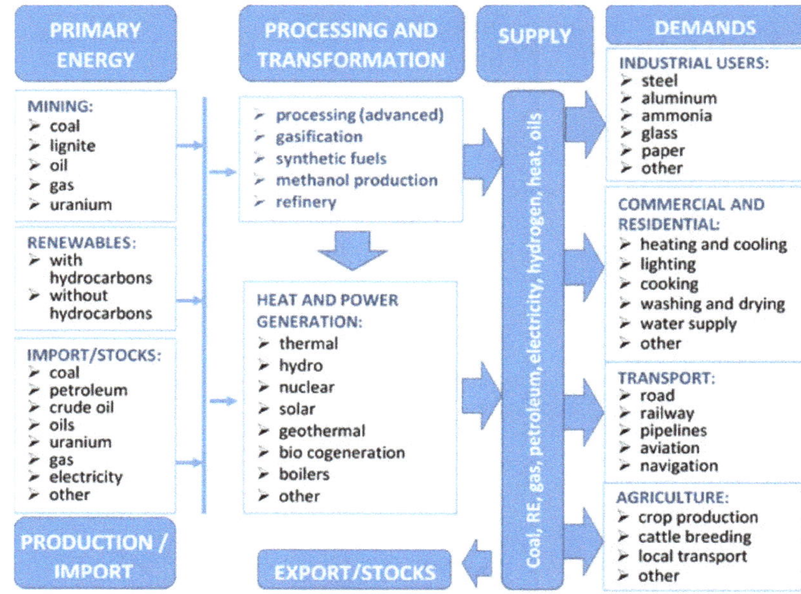

Figure 1. Main structural components of the TIMES-Ukraine model [45].

The most energy-intensive energy service demands determine energy consumption in the commercial and residential sectors. They include water heating, heating and cooling of dwellings, cooking, refrigeration, dish washing, washing and drying (ironing) of clothes, lighting, etc.

The agriculture sector in TIMES-Ukraine includes a representation of energy consumption for cattle breeding, crop production, local transport and other demands.

The modelling horizon in TIMES-Ukraine is defined until 2050 with most of the modelling periods comprising 5-year intervals. Energy service demand projections are included until the end of the modelling horizon and are based on aggregated macroeconomic indicators (gross domestic product (GDP) and real personal income), sector-specific production and economic performance indicators (gross value added, industrial production index, passenger and freight transportation), and demographic and social indicators (population categorised by place of living, number of dwellings and living conditions).

TIMES-Ukraine, and similar energy system models, are typically applied to analyse energy system development pathways in the long-term. Scenarios can be developed and analysed by changing the assumptions on, e.g., technologies, useful energy demands, prices or other exogenous variables. The result of the modelling is a least-cost solution for satisfying energy service demands of the entire energy system under given conditions and restrictions.

3. Analysed Scenarios

3.1. Main Scenarios

3.1.1. Frozen Policy

The Frozen Policy (FZP) scenario includes currently implemented policies only, which are based on the current (limited) level of implementation of the existing legislation, i.e., when legislation is not enforced, implemented only partially or with significant delays, or with limited scope and legislation targets. The FZP scenario is similar to the Business As Usual scenario developed for the updated NDC of Ukraine [44], but is based on updated macroeconomic, technical and other parameters.

This hypothetical scenario is useful in assessing the implications of the other policy options and scenarios considered in this study. The FZP scenario can also be considered as providing the upper limit of energy consumption and GHG emissions until 2050.

3.1.2. Least-Cost Development

The Least-Cost Development (LCD) scenario is an unconstrained scenario, free of targets and additional constraints, but with all decarbonisation measure options (energy efficiency, renewables, new technologies, etc.). The LCD scenario shows how the technological structure and energy mix may change through the competition of technologies to supply the demand, without implementing any significant policies (e.g., taxation).

3.1.3. Stated Policy

The Stated Policy (STP) scenario includes all stated future policies in addition to currently implemented ones. These are included to show the impact of timely implementation of existing climate-related legislation (Tables A1 and A2). Compared to the FZP scenario, this scenario provides insight into the effect of forthcoming policies and identifies the gains achieved through implementation of the new policies. The STP scenario is similar to the Reference Scenario developed for the updated NDC of Ukraine [44].

3.2. Alternative Scenarios

FZP is a very conservative representation of the energy system future. LCD and STP allow for more ambitious developments in renewable energy and GHG reductions, but cannot be characterised as disruptive or drastic. Therefore, we have analysed three alternative scenarios characterised by different methods for reducing GHG emissions: Climate Neutral Economy (CNE), Strong Carbon Tax (SCT) and Early Coal Phase-out (CPO) scenarios.

3.2.1. Climate Neutral Economy

The Climate Neutral Economy scenario is based on the Ukraine's recently updated NDC [5], which is consistent with a trajectory that achieves net zero GHG emissions by 2060. The CNE scenario includes only climate mitigation targets to model the least cost pathway to a low carbon society, with the GHG emissions in 2050 set to reduce to 14% of the 1990 level.

3.2.2. Strong Carbon Tax

Under the Strong Carbon Tax scenario, the carbon tax for industry, supply, power and heat sectors in Ukraine increases from the present level (€0.35 per ton) to about the average 2020 EU ETS carbon price (€26 per ton) by 2030 and reaches €140 per ton by 2050.

3.2.3. Early Coal Phase-Out

The Early Coal Phase-out scenario shows the cost and implications of a phase-out of coal from electricity and heat production by 2035. This scenario represents an example of a policy that has been seen in several European countries, e.g., Denmark and Finland, which have decided to phase out coal by 2028 and 2029, respectively.

4. Results

4.1. Power and Heat Sector

The total electricity production, share of renewables and share of generation from CHPs are presented in Figure 2. Electricity production increases between 2020 and 2050 in all scenarios, but the magnitude of the increase differs. In the FZP scenario, electricity production increases from 149 TWh in 2020 to 243 TWh in 2050. For the other scenarios, electricity demand more than doubles during the same period, reaching 311 TWh and 336 TWh in 2050 in the LCD and STP scenarios, respectively. Despite the lack of a renewable energy sources (RES) development policy in the FZP scenario, the share of RES in the structure of electricity generation will grow rapidly, and exceed the RES-targets of the current Energy Strategy of Ukraine to 2035.

The RES are consistently highest in the LCD scenario and lowest in the FZP scenario. In the FZP scenario, the RES share increases each decade and reaches 45% in 2050. However, in the LCD and STP scenarios, the RES values peak at 60% and 47%, respectively, in 2040. The RES then decrease in the subsequent decade, due to construction of additional nuclear units. Implementing measures of the STP scenario can also reduce electricity imports.

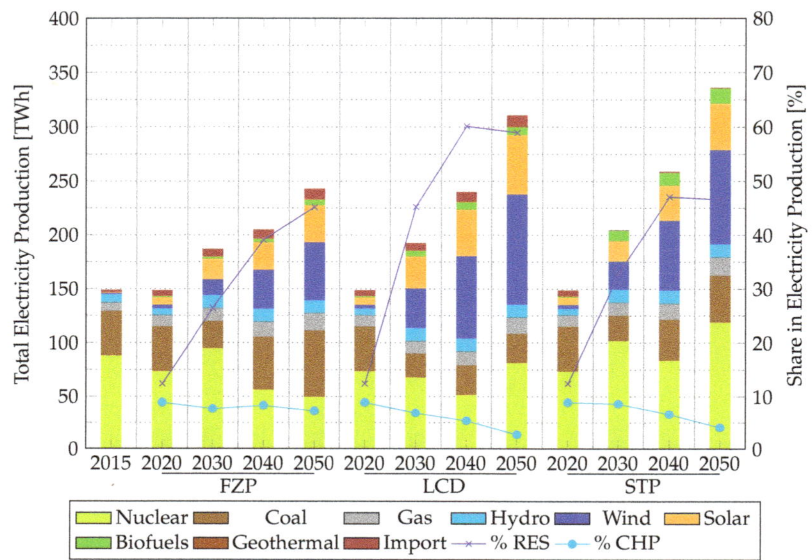

Figure 2. Total electricity production by fuel type for three main scenarios.

The district heating (DH) production in the FZP scenario will increase by more than 50% until 2050 compared to 2020, while the LCD and STP scenarios go in different direction (Figure 3). The DH production will decrease in these scenarios by around 5% in the same period. The reason for the decrease is cost-effective thermal insulation of buildings which is reducing heat production of all heat supply options, including DH.

In both the LCD and STP scenarios, DH production is converted from natural gas dominated production to biofuels as the primary energy source. In the FZP scenario, the majority of DH production originates from industrial boilers. At the same time, the share of CHP units in the DH production declines in all scenarios, resulting in shares of 26%, 22% and 30% in 2050, respectively, in the FZP, LCD and STP scenarios. The decreasing share is primarily due to the phase-out of gas-based autoproduction and only partial replacement by autoproduction using biofuels. CHPs with gas are almost completely replaced with an equal capacity of biofuel-based CHPs.

The share of renewables will reach almost 60% in 2050 in the LCD and STP scenarios, while only moderate increase in share is seen in the FZP scenario. Excess heat contributes with around 45 PJ in 2050 in both the LCD and STP scenarios, and 66 PJ in the FZP scenario. The utilisation of high temperature excess heat for DH does not grow significantly in the LCD and STP scenarios because a reduction of heating demand in buildings, due to thermal insulation, proves to be more cost-effective for the Ukrainian energy system. A significant increase in biofuel utilisation for DH from 2020 onwards will contribute to an increased RES share. Domestic biomass can thereby significantly replace imported gas in DH generation. However, this will require considerable improvement of the existing biofuels market. In all scenarios, there is a steady increase in excess heat utilisation. This the rate could potentially be increased through effective policy choices.

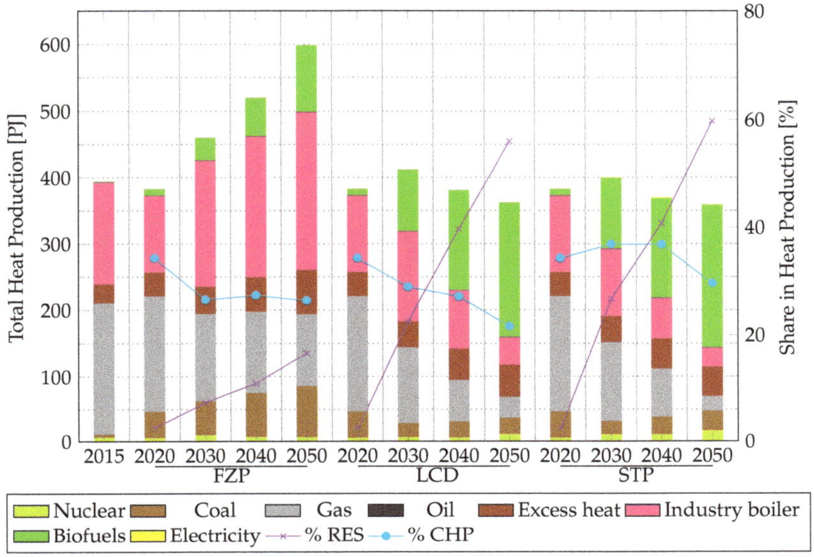

Figure 3. Total heat production by fuel type for three main scenarios.

4.2. Buildings

The use of electricity and fuels in buildings until 2050, in all main scenarios, is presented in Figure 4. In the FZP scenario, there is little change between 2020 and 2050, while the LCD and STP scenarios change noticeably from 2020 to 2050. Natural gas remains the dominant fuel for cooking in the FZP scenario even though its share drops from 69% in 2015 to 61% in 2050, while the use of electricity grows from 31% to 39% in the same period. The reason for using natural gas for cooking is the long lifetime of cooking devices and high cost of electricity relative to natural gas. On the other hand, there is a lack of policies that would give a stronger push to electric cookers. The LCD and STP scenarios develop in the opposing direction to the FZP scenario: the share of electric cooking grows to 93% in 2050.

The economic growth assumptions drive an increase of energy service demands from appliances. These assumptions, combined with absence of energy efficiency policies, translates in a 40% increase in electricity use for appliances between 2015 and 2050 in the FZP scenario. The policies introduced in STP scenario drive the development of energy efficient devices in buildings in the same direction as in LCD scenario, namely, the electricity use for appliances is 26% larger in the LCD and STP scenarios than in 2015 or 28% lower than in FZP scenario.

Space heating and sanitary hot water are produced mostly using natural gas boilers, DH and electric heating, in all scenarios. There is a minor change between the scenarios and over the analysed period: the minimum combined share of natural gas boilers, DH and electric heating in heat supply to buildings of 84% occurs in 2050 in LCD scenario, but most often it is between 87% and 92%. The shares of individual supply options are also stable: natural gas boilers vary between 48% and 56%, DH between 28% and 35%, while electric heating supplies between 5% and 7% of the demand. There are also two noticeable differences between the scenarios. Firstly, the impact of insulation measures on the space heating demand linearly grows from 3% in 2020 to 20%, 73% and 70% in 2050 in the FZP, LCD and STP scenarios, respectively. Therefore, space heating in the FZP scenario is responsible for two-thirds of the buildings' heating demands in 2050 and only ~40% in the LCD and STP scenarios. Secondly, solar heating is negligible in the STP scenario, while, being driven by decreasing technology costs and favourable policies, it supplies 10% of the heating demand in 2050 in LCD and STP scenarios.

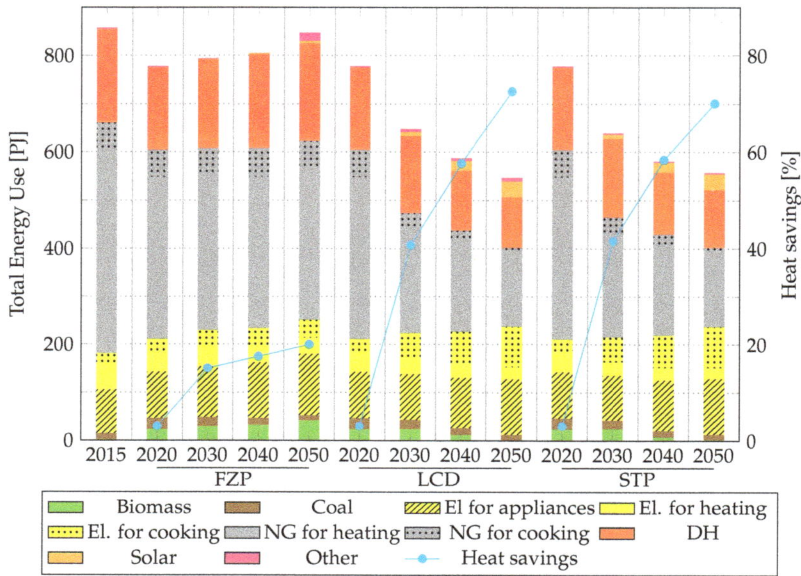

Figure 4. Final energy use in buildings by fuel type for three main scenarios.

4.3. Transport

The final energy use in the transport sector is shown in Figure 5. In all the main scenarios, energy consumption in transport is higher in 2050 compared to 2020, albeit quite differently, as a result of increased car ownership level. In the FZP scenario, energy use is almost doubled by 2050, compared to 2020. The LCD scenario shows a lower growth of ~80 PJ until 2030, followed by relatively unchanged levels of energy consumption to 2050, as the electric vehicle fleet undergoes expansion. Like in the LCD scenario, final energy use in transport in the STP scenario stays at a much lower level compared to FZP. Additionally, higher electric vehicle (EV) adoption rates bring the level further down. Contrary to the other scenarios, biofuels see a significant role as "drop-in" fuels in the period from 2030–2040 in the STP scenario.

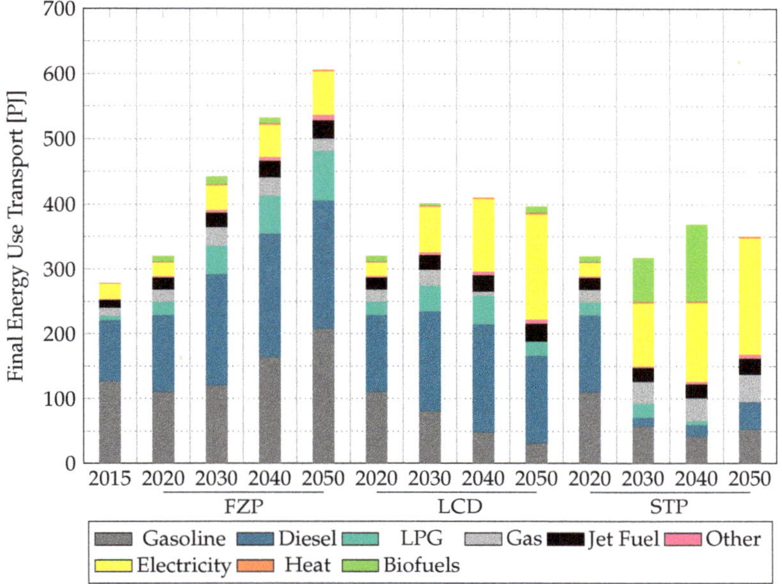

Figure 5. Final energy use in transport by fuel type for three main scenarios.

4.4. Industry

The final energy use and energy intensity of the industrial sector in Ukraine are shown in Figure 6. It can be seen that the energy use will more than double in 2050 compared to the baseline in 2015 in all scenarios. The energy intensity will however decline by more than 25% in the LCD and STP scenario. The STP will result in almost 300 PJ less final energy use than in the FZP scenario. Approximately 58% of energy use in the FZP scenario in 2050 is from direct use of fossil fuels, while this share is 55% in the LCD and STP scenario. The absolute amount of coal used is comparable in all scenarios, while the use of natural gas shows only a slight decrease in the LCD and STP scenarios. While the use of biofuels and oil is very small, electrification in industry will lead to considerably higher electricity use in the industry.

The energy intensity, as an indicator for energy efficiency of the industry, as well as the composition of the industry, will decline in all scenarios. The indicator describes the final energy use per USD of GDP from the industrial sector. The lowest energy intensity is reached in the STP scenario, which is slightly lower than that of LCD.

The heat used in industry does not increase in the LCD and STP scenario, while energy intensity reduces. This indicates the installation of efficient process equipment and electrification of processes in many parts of the industry.

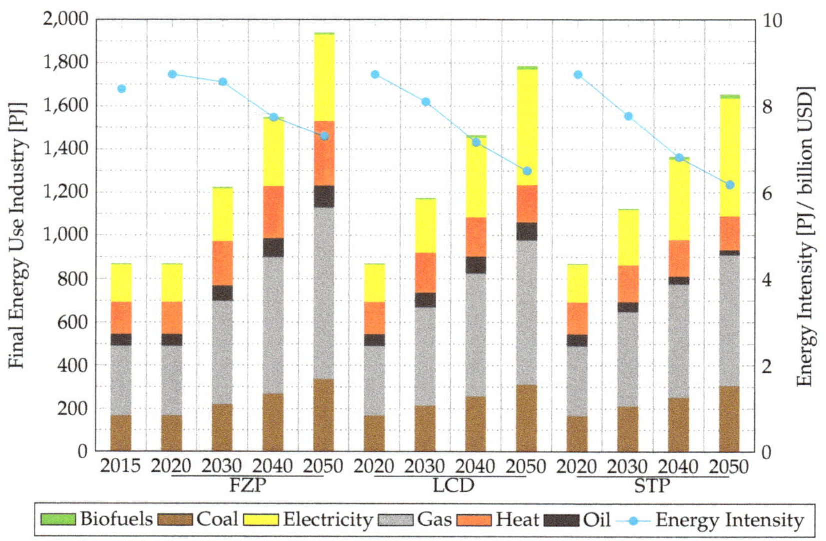

Figure 6. Final energy use in industry by fuel type for three main scenarios.

4.5. Supply

In the FZP scenario, development of the energy sector requires a significant increase in the consumption of energy resources, primarily carbon-intensive fuels, towards 2050 (Figure 7). Energy supply from renewable sources increases 4-fold between 2020 and 2050, from 157 to 622 PJ. The share of RES in TPES thereby increases to 12% by 2050. The carbon intensity of the TPES generally increases due to increased consumption of fossil fuels and reduced nuclear energy, despite increased renewable energy supply.

The TPES required in the respective STP and LCD scenarios is lower than the FZP scenario (Figure 8). The TPES in the LCD scenario will be less by 12% in 2030 and 16% in 2050, relative to the FZP scenario. Compared to the FZP scenario, both the STP and LCD scenarios are less dependent on fossil fuels and have increased RES share. The STP also has increased power supply from nuclear from 2030. The carbon-intensity of energy supply will decrease slowly, as additional demand will be supplied by RES in the LCD scenario, or both RES and nuclear in the STP scenario. The TPES results show that RES compete with other energy sources on market terms, without using policies or support mechanisms to stimulate investments.

The STP scenario has the highest share of nuclear power, due to commissioning of power plants in 2030 and 2050. Before 2050, the STP scenario has the highest RES share, but a decrease in RES share occurs in 2050 due to nuclear power plants. Nuclear power substitutes fossil fuels in the TPES compared to the FZP scenario, and RES and oil in the LCD scenario.

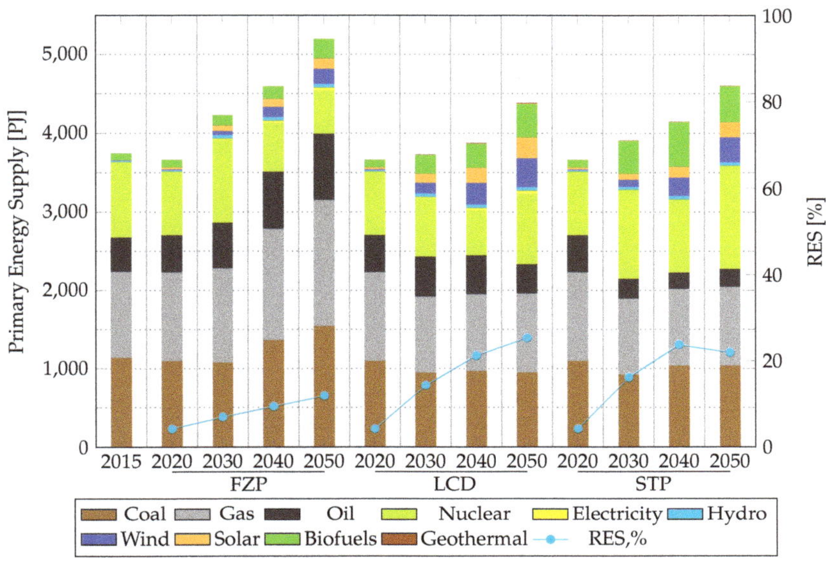

Figure 7. Total Primary Energy Supply for three main scenarios.

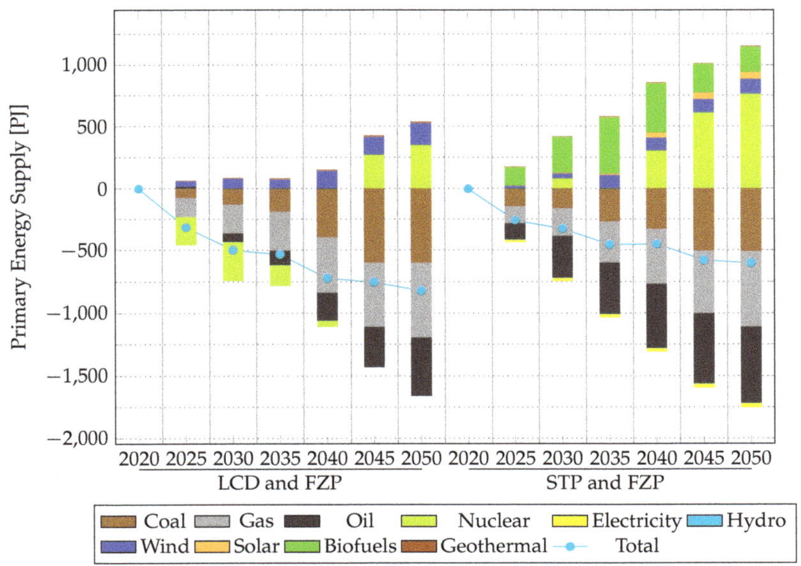

Figure 8. Difference between scenarios TPES.

4.6. Emissions

The total emissions in both the LCD and STP scenarios comply with the 2030 NDC for Ukraine, while they do not in the FZP scenario. In the LCD scenario, GHG emissions stabilise in the future at about the level of 2015 with the share of industry at around 50% (Figure 9). Compared to the FZP scenario, in the LCD scenario, the largest reduction of GHG emissions is in the electricity and heat generation and industry sectors. The transport sector reduces the emissions in 2050 by 23% compared to 2020, despite growth of demand,

and by 58% compared to the FZP scenario. In industry, the emissions in 2050 are more than 50% higher than in 2020, but still 20% smaller than in the FZP scenario. However, this is achieved together with an increase of production output and without structural changes.

In the STP scenario, GHG emissions decrease significantly until 2030, as a result of adopted targets and policies (see Table A1), but increase thereafter to the level specified in the Low Emission Development Strategy [4]. The FZP and LCD scenarios have no climate constraints, but GHG emission trajectories differ significantly due to diverse options for energy efficiency implementation, renewables and other measures (Figure 9). In order to maintain the trajectory of the GHG emission reductions achieved in the STP scenario before 2030, new and more ambitious policies relevant for the period beyond 2030 are required.

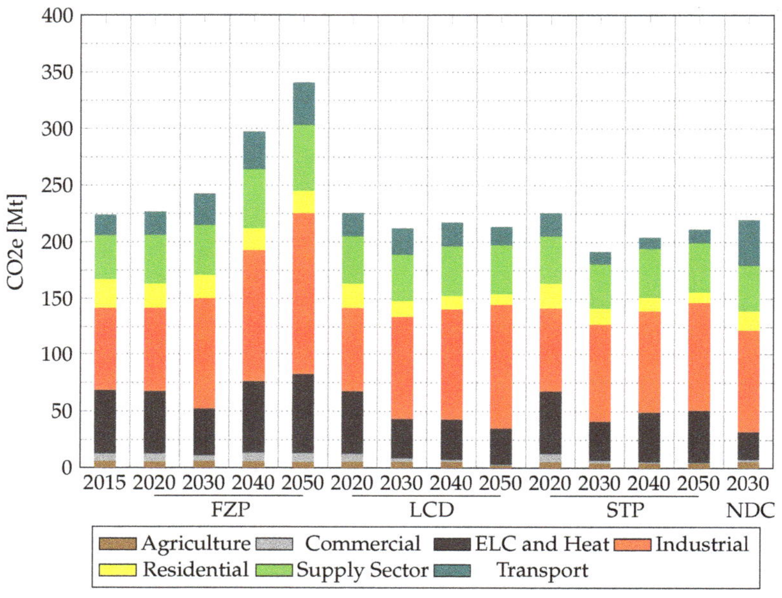

Figure 9. GHG emissions from Energy and Industrial process and product use for three main scenarios.

4.7. Costs

Figure 10 shows that, without any strong restrictions and targets, but with options to implement new technologies and measures, annual system costs in the LCD scenario are lower than in the FZP scenario. The STP scenario sees considerably higher investment in more efficient end use (demand) technologies and the power and heat sector leading to significantly lower fuel and O&M expenditures: the savings almost balance the increased costs by 2045 and outweigh them by 2050.

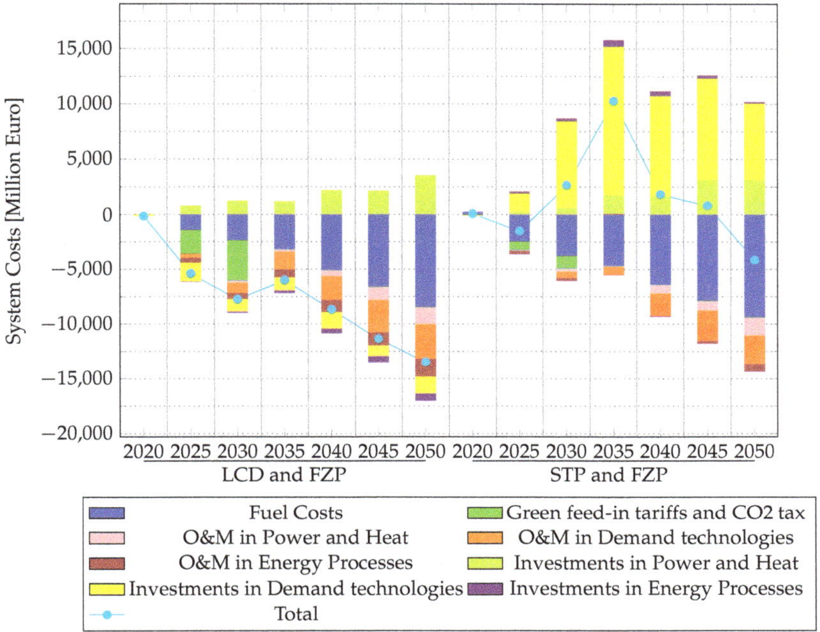

Figure 10. Cost difference between scenarios.

4.8. Alternative Scenarios

For each of the three alternative scenarios, the most significant variables are compared for the period 2020–2050 with their respective values in the FZP scenario. The summary of results for the alternative scenarios is presented in Table 1. They confirm a lack of ambition in the FZP scenario and present renewable alternatives to the development of the Ukrainian energy system in the main scenarios.

The alternative scenarios show similar development with regards to heat supply to buildings—heat savings in buildings grow drastically relative to the FZP scenario, while DH production decreases while increasing the share of renewables. The share of DH in the heat supply to buildings changes in a narrow range of [−2%, 7%], which means that buildings supplied by DH are equally affected by heat savings as are buildings supplied from individual heating sources.

The results of alternative scenarios for the transport sector are also uniform—the sector is switching towards electricity, leading to a decrease of fuel use due to the higher efficiency of electric vehicles compared to vehicles with internal combustion engines. The use of biofuels in the alternative scenarios remains insignificant (between 0.4% in the CPO to 1.3% in the CNE scenario), the large relative differences appear due to even smaller values in the FZP scenario.

In the industrial sector, the alternative scenarios point in the same direction— heavy industry in Ukraine reduces use of coal and oil considerably, while there is a slight decrease in the use of natural gas and DH. The use of renewables more than doubles. The use of electricity increases and thus provides twofold benefits—reduction of environmental emissions due to strong reduction of carbon content of electricity and a transition towards more energy efficient processes. This is accompanied by a reduction of fuel use in the industrial sector.

Electricity production increases in all the alternative scenarios relative to FZP due to increased electrification of the transport and industrial sectors. Fossil and nuclear-based generation reaches the end of its lifetime, at the time when the investment costs of photovoltaics (PV) and wind power are assumed to decrease. This brings a strong increase in electricity production from wind and PV and a decrease in fossil and nuclear production.

Decarbonised electricity and increased renewable energy use is responsible for reduction of emissions in the transportation and industrial sectors. At the same time, electricity use for heating decreases in the alternative scenarios, while the emissions from commercial and residential buildings decrease due to a large amount of heat saving measures (insulation), decarbonisation of DH supply and introduction of solar heating. The remaining use of electricity in buildings is limited to appliances, while natural gas boilers remain the dominant heat supply provider.

The alternative scenarios are also characterised by more than a doubling of renewable energy in the energy system. At the same time, the electrification of the industrial and transport sectors, along with energy efficiency measures in buildings and industry, lead to reduced TPES. Finally, the total system cost of the energy system is lower in the alternative scenarios than in the FZP scenario (between 3% in SCT and 8% in CPO). This means that major emissions reductions and minor cost reductions are possible for the future Ukrainian energy system.

Table 1. Summary of results for the alternative scenarios for the period 2020–2050.

			CNE	**SCT**	**CPO**
DH		Total Production	−21%	−25%	−21%
		Renewable Production	+157%	+191%	+163%
Buildings		DH share (%)	+7%	−3%	−2%
		Heat savings	+217%	+213%	+215%
Transport		Total Fuel Use	−19%	−19%	−19%
		Renewable fuels	+225%	+54%	+54%
		Electricity Use	+106%	+104%	+106%
Industry		Electricity Use	+19%	+28%	+15%
		Renewable	+84%	+107%	+55%
		Fuel Consumption	−7%	−11%	−5%
Electricity		Total Production	+12%	+16%	+11%
		Renewable Production	+108%	+124%	+107%
		Wind Production	+177%	+185%	+177%
		Solar Production	+61%	+89%	+60%
		Wind and Solar Production	+123%	+141%	+123%
		Nuclear Production	−22%	−19%	−21%
$CO_{2,eq}$		Total emissions	−31%	−38%	−27%
		Industry emissions	−20%	−31%	−13%
		Buildings emissions	−35%	−31%	−31%
		Transport emissions	−32%	−29%	−31%
		RE total	+113%	+124%	+106%
		TPES	−18%	−20%	−16%
		Objective function	−4%	−3%	−8%

5. Discussion

The results show that the Ukrainian energy system can reduce GHG emissions, increase renewable energy share and improve energy efficiency while not pressuring the total system costs. Increased electrification of industrial and transport sectors and buildings based on wind, solar and nuclear power is at the centre of this development. However, many assumptions influence the present analysis; they are discussed in this section.

Energy efficiency and electrification are the main pillars of the Ukrainian energy transition. The present analysis shows that the improved energy efficiency in the industrial, transportation and buildings sector is cost-beneficial for the energy system as a whole, but it does not analyse the business-economic perspective or behavioural aspects. The private-economic aspect is the most important for industries, while the decisions on energy efficiency improvements in buildings depend on both economic and behavioural factors. Finally, the choice of vehicles, especially in passenger transport depend both on economic and social factors. The behavioural aspects of the energy transition are not directly included in the present study.

In many studies, especially in countries with developed DH networks, DH is characterised by use of CHPs, utilisation of excess heat from industries and biofuels production, solar heating, large-scale heat pumps driven by renewable electricity and thermal storage. In these studies, DH is at the centre of the energy transition. The present study gives another view of the future—heating demands are reduced strongly (around 70%), individual natural gas boilers are supplying the majority of the demand, while DH, electric heating and solar heating (in alternative scenarios) are contributing in smaller shares. The focal point of insulation measures in the heating sector in ambitious renewable scenarios and unconstrained cost-optimised scenarios has twofold indication: (i) heat savings in Ukrainian buildings are less expensive in comparison with heat supply alternatives due to poor thermal standards of the buildings and (ii) stated policies regarding the energy efficiency of buildings in Ukraine are adequately designed from the energy system point of view.

CNE and SCT are the most ambitious emission-reduction scenarios in the study. The emissions reach 96 Mt and 86 Mt CO_{2eq} in 2050 or 18% and 28% reduction relative to 2020, respectively. However, as the world needs to reach net-zero carbon emissions by 2050 according to the Paris Agreement, even the alternative scenarios do not seem to be overly ambitious. Currently, the TIMES-Ukraine model lacks options such as use of hydrogen in industry or Direct Air Capture to reach net zero or net negative emission goals. On the other hand, the emission pathways presented in the present article are encouraging as they are achieved simultaneously with growth in energy service demands and reduced total system cost relative to the FZP scenario.

The CO_{2eq} emissions have different trajectories across scenarios. GHG emissions in FZP grow throughout the analysed period; they decrease in STP until 2030 and then grow again. In CNE and SCT emissions are stable to 2040 and decrease afterwards, while in CPO and LCD emissions grow to 2035 and then stay constant. Therefore, if reduction of GHG emissions is a prioritary policy, FZP should be avoided, STP could be followed in the short to medium term, CPO and LCD in the long term.CNE and SCT could be followed throughout the whole period. Instead of following a single scenario, a combination of measures behind the scenarios would be desirable.

Export and import of electricity is very limited in the present article. Export and import individually amount to approximately 2% of electricity production in Ukraine in 2020 in all scenarios. In all analysed scenarios, the import drops almost to zero in 2050, while in all scenarios except STP, the import reaches approximately 3% of domestic electricity production in 2050. Ukraine is in the process of joining the European Network of Transmission System Operators for Electricity (ENTSO-E) interconnection. After joining the network, larger exports and imports are expected as well as significant reduction of energy system costs.

All the issues discussed in this section deserve further consideration in further work.

6. Conclusions

The present article is among the only long-term analyses of the future Ukrainian energy system in the academic literature. The analysis is performed until 2050 with the

TIMES-Ukraine model, which covers all sectors of the Ukrainian energy system. We have analysed three main scenarios with moderate emission reduction ambitions and three alternative scenarios with higher ambitions.

The development of the Ukrainian energy system is based on energy efficiency improvements, electrification and renewable energy. Energy efficiency improvements are most pronounced in the buildings sector, while improved energy efficiency in industrial and transport sector is closely linked to electrification. Finally, renewable energy mostly enters the energy system through electricity production from wind and PV. In the buildings sector, space heating demand is reduced by approximately 70% in all scenarios except FZP. This means that in an average Ukrainian building in 2050, sanitary hot water could be responsible for approximately 60% of the heating demands, while the rest would go to space heating.

In the industrial sector, electrification is key to unlocking several benefits: improved energy efficiency, an increased share of renewable energy, and reductions in carbon and energy intensity. The improvement in energy efficiency results from the difference in efficiencies between combustion and electrical processes and leads to a reduction in final energy consumption in the industrial sector of 5% in LCD and 10% in STP scenario and between 5% and 11% in the alternative scenarios. As the overall industrial structure does not change in the presented scenarios (high share of energy intensive industries), improved energy efficiency leads to reduced energy intensity by 26–36% in 2050 relative to 2020 in all scenarios except FZP where it is reduced by 16%. Most of the increased renewable energy share comes indirectly from increased share of renewables in the electricity generation mix, even though the direct use of renewables in alternative scenarios reaches 11% to 14% in 2050. Reduced carbon emissions come from reduced use of fossil fuels and increased use of renewables and electricity.

Energy efficiency improvements and emission reductions in the transport sector are achieved through electrification. Average use of electricity in transport grows throughout the analysed period in all the scenarios except FZP; it reaches 51% in STP scenario and 40–42% in the other scenarios. Use of biofuels is negligible in all scenarios except STP where it is used as a "transitional fuel", namely, comprising 32–37% of final energy consumption in the transport sector.

The present analysis shows that the Stated Policy scenario presents a cost-efficient alternative for keeping Ukraine's GHG emissions at today's level. At the same time, the trajectory of the energy system under the STP scenario is very similar to the least-cost development (presented in LCD scenario). Finally, the SCT scenario results in 3% lower total GHG emissions compared to the STP scenario while reducing the total system costs.

Finally, large potential for cost-effective introduction of renewable energy in electricity production (mostly wind and PV), energy efficiency improvements in buildings and industries and electrification of transportation and industrial sectors should be the focus of policy measures in Ukraine as well as of future research.

Author Contributions: Conceptualisation, S.N.P., O.D., R.P. and O.B.; methodology, S.N.P., O.D., R.P. and O.B.; software, O.D., A.S. and O.B.; validation, S.N.P., O.D., A.S. and R.P.; formal analysis, S.N.P., O.D., A.S. and O.B.; investigation, S.N.P., O.D., A.S., R.P., F.B., R.G., M.B. and O.B.; data curation, O.D., A.S. and R.P.; writing—original draft preparation, S.N.P., O.D., A.S., F.B., R.G., M.B. and O.B.; writing—review and editing, S.N.P., O.D., F.B. and O.B.; visualisation, O.D., A.S., F.B. and O.B.; project administration, S.N.P.; All authors have read and agreed to the submitted version of the manuscript.

Funding: This research received no external funding.

Acknowledgments: The authors would like to acknowledge the contribution of Kenneth Karlsson and Mikkel Bosack Simonsen—support on the development of TIMES-Ukraine model; Galyna Trypolska—support on analysis of renewables policy and measures; Roman Yukhymets—support on analysis of energy efficiency policy and measures; and Tetiana Saprykina, Vladyslav Pekkoev and Valentyna Kozubra—support on data collection.

Conflicts of Interest: The authors declare no conflict of interest.

Abbreviations

The following abbreviations are used in this manuscript:

TPES	Total Primary Energy Supply
GHG	Greenhouse gas
LULUCF	Land Use, Land use Change and Forestry
NDC	Nationally Determined Contributions
DH	District heating
FZP	Frozen Policy
LCD	Least-cost Development
STP	Stated Policy
CNE	Carbon Neutral Economy
SCT	Strong Carbon Tax
CPO	Early Coal Phase-out

Appendix A

Table A1. Projections of GHG emissions in Energy and Industrial processes sectors in Ukraine 2050 Low-Emission Development Strategy [4].

	2015	2020	2025	2030	2035	2040	2045	2050
Share of GHG emission compared to 1990 in the most ambitious scenario,%	31	31	31	29	28	31	31	31

Table A2. Current adopted energy and climate targets in Ukraine.

Indicators	2015	2018	2020	2025	2030	2035
The Law of Ukraine on the Basic Principles (Strategy) of the State Environmental Policy of Ukraine for the period up to 2030 [8]						
Share of renewables (incl. hydro power plants) in TPES,%	4		8	12	17	
Primary Energy Intensity, toe/$1000 GDP (PPP)	0.28		0.2	0.18	0.13	
Share of GHG emission compared to 1990,%	37.8		<76	<60	<60	
Air pollutant emissions from stationary sources,% of 2015	100		<6	<16.5	<22.5	
Electric vehicles,% of new vehicles purchased			0.1	0.5	10	
Energy Strategy of Ukraine until 2035 [9]						
Primary Energy Intensity, toe/$1000 GDP (PPP) (constant 2011 US$)	0.28		0.20	0.18	0.15	0.13
Share of renewables (incl. big hydro) in TPES,%			8	12	17	25
Share of renewables (incl. hydro power plants) in power generation,%	5		7	10	>13	>25
Share of GHG emission compared to 1990,%			<60	<60	<60	<50
National transport strategy of Ukraine for the period up to 2030 [10]						
GHG emission and air pollutions from stationary sources,% of 1990					<60	
Share of alternatives fuels,%	10				50	
Share of electric transport in urban public transport,%					75	
National Renewable Energy Action Plan until 2020 [11]						
Share of renewables in cooling and heating systems,%	6.7	10	12.4			
Share of renewables in electricity production,%	8.3	10.4	11			
Share of renewables in transport,%	5	8.2	10			
Share of renewables in Gross Final Energy Consumption (GFEC),%	6.7	9.1	11			

Table A2. *Cont.*

Indicators	2015	2018	2020	2025	2030	2035
National Energy Efficiency Action Plan until 2020 [12]						
Share of retrofit residential buildings,%			25			
Share of retrofit public buildings,%			20			
Net-zero energy building,% per year			3			
Energy saving in 2020 from average FEC in 2005–2009,%			9			
Concept of implementation of the state policy of heat supply until 2035 [13]						
Heat production losses,%			8			
Transmission heat losses,%			12			10
Share of alternative energy in heat production,%				30	40	

References

1. IPCC. *Global Warming of 1.5 °C. An IPCC Special Report on the Impacts of Global Warming of 1.5 °C above Pre-Industrial Levels and Related Global Greenhouse Gas Emission Pathways, in the Context of Strengthening the Global Response to the Threat of Climate Change, Sustainable Development, and Efforts to Eradicate Poverty*; Masson-Delmotte, V., Zhai, P., Pörtner, H.-O., Roberts, D., Skea, J., Shukla, P.R., Pirani, A., Moufouma-Okia, W., Péan, C., Pidcock, R., et al., Eds.; 2018, in press.
2. UNFCCC. *Adoption of the Paris Agreement*; UNFCCC: Bonn, Germany, 2015.
3. MEPR. *Ukraine's Greenhouse Gas Inventory 1990–2019*; MEPR: Kyiv, Ukraine, 2021.
4. The Government of Ukraine. *Ukraine 2050 Low Emission Development Strategy*; Technical report; Governmentl of Ukraine: Kyiv, Ukraine, 2017.
5. MEPR. *Updated Nationally Determined Contribution of Ukraine to the Paris Agreement*; MEPR: Kyiv, Ukraine, 2021.
6. SSSU. *Energy Balance Data Time Series for the Period of 1990–2019*; SSSU: Kyiv, Ukraine, 2020.
7. IEA. *Ukraine Energy Profile*; IEA: Paris, France, 2020.
8. Verkhovna Rada of Ukraine. *The Law of Ukraine on the Basic Principles (Strategy) of the State Environmental Policy of Ukraine for the period up to 2030*; Verkhovna Rada of Ukraine: Kyiv, Ukraine, 2019.
9. The Government of Ukraine. *Energy Strategy of Ukraine till 2035*; Government of Ukraine: Kyiv, Ukraine, 2017.
10. The Government of Ukraine. *National Transport Strategy of Ukraine for the Period up to 2030*; Government of Ukraine: Kyiv, Ukraine, 2018.
11. The Government of Ukraine. *National Renewable Energy Action Plan till 2020*; Government of Ukraine: Kyiv, Ukraine, 2014.
12. State Agency on Energy Efficiency and Energy Saving of Ukraine. *National Energy Efficiency Action Plan till 2020*; State Agency on Energy Efficiency and Energy Saving of Ukraine: Kyiv, Ukraine, 2015.
13. The Government of Ukraine. *Concept of Realization of the State Policy of Heat Supply till 2035*; Government of Ukraine: Kyiv, Ukraine, 2017.
14. Lund, H.; Mathiesen, B.V. Energy system analysis of 100% renewable energy systems-The case of Denmark in years 2030 and 2050. *Energy* **2009**, *34*, 524–531. [CrossRef] [CrossRef]
15. Ćosić, B.; Krajačić, G.; Duić, N. A 100% renewable energy system in the year 2050: The case of Macedonia. *Energy* **2012**, *48*, 80–87. [CrossRef] [CrossRef]
16. Balyk, O.; Andersen, K.S.; Dockweiler, S.; Gargiulo, M.; Karlsson, K.; Næraa, R.; Petrović, S.; Tattini, J.; Termansen, L.B.; Venturini, G. TIMES-DK: Technology-rich multi-sectoral optimisation model of the Danish energy system. *Energy Strategy Rev.* **2019**, *23*, 13–22. [CrossRef] [CrossRef]
17. Capros, P.; Tasios, N.; Vita, A.D.; Mantzos, L.; Paroussos, L. Model-based analysis of decarbonising the EU economy in the time horizon to 2050. *Energy Strategy Rev.* **2012**, *1*, 76–84. [CrossRef] [CrossRef]
18. Dominković, D.F.; Bačeković, I.; Ćosić, B.; Krajačić, G.; Pukšec, T.; Duić, N.; Markovska, N. Zero carbon energy system of South East Europe in 2050. *Appl. Energy* **2016**, *184*, 1517–1528. [CrossRef] [CrossRef]
19. Sgobbi, A.; Nijs, W.; Miglio, R.D.; Chiodi, A.; Gargiulo, M.; Thiel, C. How far away is hydrogen? Its role in the medium and long-term decarbonisation of the European energy system. *Int. J. Hydrogen Energy* **2016**, *41*, 19–35. [CrossRef] [CrossRef]
20. Petrović, S.N.; Karlsson, K.B. Residential heat pumps in the future Danish energy system. *Energy* **2016**, *114*, 787–797. [CrossRef] [CrossRef]
21. Caldera, U.; Breyer, C. The role that battery and water storage play in Saudi Arabia's transition to an integrated 100% renewable energy power system. *J. Energy Storage* **2018**, *17*, 299–310. [CrossRef] [CrossRef]
22. Lund, H.; Möller, B.; Mathiesen, B.V.; Dyrelund, A. The role of district heating in future renewable energy systems. *Energy* **2010**, *35*, 1381–1390. [CrossRef] [CrossRef]
23. Mathiesen, B.V.; Lund, H.; Connolly, D. Limiting biomass consumption for heating in 100% renewable energy systems. *Energy* **2012**, *48*, 160–168. [CrossRef] [CrossRef]

24. Blesl, M.; Das, A.; Fahl, U.; Remme, U. Role of energy efficiency standards in reducing CO_2 emissions in Germany: An assessment with TIMES. *Energy Policy* **2007**, *35*, 772–785. [CrossRef] [CrossRef]

25. Fortes, P.; Simoes, S.G.; Gouveia, J.P.; Seixas, J. Electricity, the silver bullet for the deep decarbonisation of the energy system? Cost-effectiveness analysis for Portugal. *Appl. Energy* **2019**, *237*, 292–303. [CrossRef] [CrossRef]

26. Khanna, N.; Fridley, D.; Zhou, N.; Karali, N.; Zhang, J.; Feng, W. Energy and CO_2 implications of decarbonization strategies for China beyond efficiency: Modeling 2050 maximum renewable resources and accelerated electrification impacts. *Appl. Energy* **2019**, *242*, 12–26. [CrossRef] [CrossRef]

27. Gota, D.I.; Lund, H.; Miclea, L. A Romanian energy system model and a nuclear reduction strategy. *Energy* **2011**, *36*, 6413–6419. [CrossRef] [CrossRef]

28. Petrović, S.; Colangelo, A.; Balyk, O.; Delmastro, C.; Gargiulo, M.; Simonsen, M.B.; Karlsson, K. The role of data centres in the future Danish energy system. *Energy* **2020**, *194*. [CrossRef] [CrossRef]

29. Scott, I.J.; Carvalho, P.M.; Botterud, A.; Silva, C.A. Long-term uncertainties in generation expansion planning: Implications for electricity market modelling and policy. *Energy* **2021**, *227*. [CrossRef] [CrossRef]

30. Ozawa, A.; Kudoh, Y.; Murata, A.; Honda, T.; Saita, I.; Takagi, H. Hydrogen in low-carbon energy systems in Japan by 2050: The uncertainties of technology development and implementation. *Int. J. Hydrogen Energy* **2018**, *43*, 18083–18094. [CrossRef] [CrossRef]

31. Pizarro-Alonso, A.; Ravn, H.; Münster, M. Uncertainties towards a fossil-free system with high integration of wind energy in long-term planning. *Appl. Energy* **2019**, *253*. [CrossRef] [CrossRef]

32. Andersen, K.S.; Wiese, C.; Petrovic, S.; McKenna, R. *Exploring the Role of Households' Hurdle Rates and Demand Elasticities in Meeting Danish Energy-Savings Target*; Energy Policy; Elsevier Ltd.: Amsterdam, The Netherland, 2020; Volume 146. [CrossRef]

33. Capros, P.; Paroussos, L.; Fragkos, P.; Tsani, S.; Boitier, B.; Wagner, F.; Busch, S.; Resch, G.; Blesl, M.; Bollen, J. European decarbonisation pathways under alternative technological and policy choices: A multi-model analysis. *Energy Strategy Rev.* **2014**, *2*, 231–245. [CrossRef] [CrossRef]

34. Kato, E.; Kurosawa, A. *Evaluation of Japanese Energy System toward 2050 with TIMES-Japan—Deep Decarbonization Pathways*; Energy Procedia; Elsevier Ltd.: Amsterdam, The Netherland, 2019; Volume 158, pp. 4141–4146. [CrossRef]

35. Hedegaard, K.; Münster, M. Influence of individual heat pumps on wind power integration—Energy system investments and operation. *Energy Convers. Manag.* **2013**, *75*, 673–684. [CrossRef] [CrossRef]

36. Odenberger, M.; Kjärstad, J.; Johnsson, F. *Prospects for CCS in the EU Energy Roadmap to 2050*; Energy Procedia; Elsevier Ltd.: Amsterdam, The Netherland, 2013; Volume 37, pp. 7573–7581. [CrossRef]

37. Fernandes, L.; Ferreira, P. Renewable energy scenarios in the Portuguese electricity system. *Energy* **2014**, *69*, 51–57. [CrossRef] [CrossRef]

38. Amorim, F.; Pina, A.; Gerbelová, H.; Pereira da Silva, P.; Vasconcelos, J.; Martins, V. Electricity decarbonisation pathways for 2050 in Portugal: A TIMES (The Integrated MARKAL-EFOM System) based approach in closed versus open systems modelling. *Energy* **2014**, *69*, 104–112. [CrossRef] [CrossRef]

39. Solomon, A.A.; Bogdanov, D.; Breyer, C. Solar driven net zero emission electricity supply with negligible carbon cost: Israel as a case study for Sun Belt countries. *Energy* **2018**, *155*, 87–104. [CrossRef] [CrossRef]

40. Ichinohe, M.; Endo, E. Analysis of the vehicle mix in the passenger-car sector in Japan for CO_2 emissions reduction by a MARKAL model. *Appl. Energy* **2006**, *83*, 1047–1061. [CrossRef] [CrossRef]

41. Hugues, P.; Assoumou, E.; Maizi, N. Assessing GHG mitigation and associated cost of French biofuel sector: Insights from a TIMES model. *Energy* **2016**, *113*, 288–300. [CrossRef] [CrossRef]

42. Tattini, J.; Mulholland, E.; Venturini, G.; Ahanchian, M.; Gargiulo, M.; Balyk, O.; Karlsson, K. A long-term strategy to decarbonise the danish inland passenger transport sector. *Lect. Notes Energy* **2018**, *64*, 137–153. [CrossRef]

43. Chepeliev, M.; Diachuk, O.; Podolets, R. Economic assessment of low-emission development scenarios for Ukraine. *Lect. Notes Energy* **2018**, *64*, 277–295. [CrossRef]

44. Diachuk, O.; Chepeliev, M.; Podolets, R. *Support to the Government of Ukraine on Updating its Nationally Determined Contribution (NDC)*; Modeling report; Technical report; Institute for Economics and Forecasting of the National Academy of Sciences of Ukraine: Kyiv, Ukraine, 2020.

45. Diachuk, O.; Podolets, R.; Yukhymets, R.; Pekkoiev, V.; Balyk, O.; Simonsen, M. *Long-term Energy Modelling and Forecasting in Ukraine: Scenarios for the Action Plan of Energy Strategy of Ukraine until 2035*; Danish Energy Agency: Copenhagen, Denmark, 2019.

46. Diachuk, O.; Chepeliev, M.; Podolets, R. *Transition of Ukraine to the Renewable Energy by 2050*; Art Book Ltd.: Kyiv, Ukraine, 2018; p. 88.

47. Podolets, R.; Diachuk, O. *Strategic Planning in Fuel and Energy Complex Based on TIMES-Ukraine Model: Scientific Report*; Technical report; Institute for Economics and Forecasting: Kyiv, Ukraine, 2011.

48. ERI; CNREC. *China Renewable Energy Outlook 2016*; Danish Energy Agency: Copenhagen, Denmark, 2016.

49. ERI; CNREC. *China Renewable Energy Outlook 2017*; Danish Energy Agency: Copenhagen, Denmark, 2017.

50. ERI; CNREC. *China Renewable Energy Outlook 2018*; Danish Energy Agency: Copenhagen, Denmark, 2018.

51. ERI; NDRC. *China Renewable Energy Outlook 2019*; Danish Energy Agency: Copenhagen, Denmark, 2019.

52. ERI; NDRC. *China Renewable Energy Outlook 2020*; Danish Energy Agency: Copenhagen, Denmark, 2021.

53. SGNEC. *Indonesia Energy Outlook 2019*; Danish Energy Agency: Copenhagen, Denmark, 2019.

54. EREA; DEA. *Vietnam Energy Outlook Report 2019*; Danish Energy Agency: Copenhagen, Denmark, 2019.
55. Loulou, R.; Goldstein, G.; Kanudia, A.; Lettila, A.; Remme, U. *Documentation for the TIMES Model Part I: TIMES Concepts and Theory*; ETSAP: Paris, France, 2016.
56. Loulou, R.; Lehtila, A.; Kanudia, A.; Remme, U.; Goldstein, G. *Documentation for the TIMES Model Part II: Reference Manual*; ETSAP: Paris, France, 2016.
57. Goldstein, G.; Kanudia, A.; Lehtila, A.; Remme, U.; Wright, E. *Documentation for the TIMES Model Part III: the Operation of the TIMES Code*; ETSAP: Paris, France, 2016.
58. Bhattacharyya, S.C.; Timilsina, G.R. A review of energy system models. *Int. J. Energy Sect. Manag.* **2010**, *4*, 494–518. [CrossRef] [CrossRef]
59. Gargiulo, M.; Gallachóir, B.Ó. Long-term energy models: Principles, characteristics, focus, and limitations. *Wiley Interdiscip. Rev. Energy Environ.* **2013**, *2*, 158–177. [CrossRef] [CrossRef]
60. Simoes, S.G.; Dias, L.; Gouveia, J.P.; Seixas, J.; De Miglio, R.; Chiodi, A.; Gargiulo, M.; Long, G.; Giannakidis, G. InSmart—A methodology for combining modelling with stakeholder input towards EU cities decarbonisation. *J. Clean. Prod.* **2019**, *231*, 428–445. [CrossRef] [CrossRef]
61. Blanco, H.; Nijs, W.; Ruf, J.; Faaij, A. Potential for hydrogen and Power-to-Liquid in a low-carbon EU energy system using cost optimization. *Appl. Energy* **2018**, *232*, 617–639. [CrossRef] [CrossRef]
62. Føyn, T.H.Y.; Karlsson, K.; Balyk, O.; Grohnheit, P.E. A global renewable energy system: A modelling exercise in ETSAP/TIAM. *Appl. Energy* **2011**, *88*, 526–534. [CrossRef] [CrossRef]
63. IEA-ETSAP. *Etsap-TIMES/TIMES_Model: Current and Previous TIMES Versions*; ETSAP: Paris, France, 2020. [CrossRef]
64. State Statistics Service of Ukraine. *Methodological Recommendations for Drawing up the Energy Balance*; SSSU: Kyiv, Ukraine, 2011.

Article

Translating Global Integrated Assessment Model Output into Lifestyle Change Pathways at the Country and Household Level

Clare Hanmer [1,*], Charlie Wilson [1,2], Oreane Y. Edelenbosch [3] and Detlef P. van Vuuren [3,4]

[1] Tyndall Centre for Climate Change Research, University of East Anglia (UEA), Norwich NR4 7TJ, UK; charlie.wilson@uea.ac.uk

[2] International Institute for Applied Systems Analysis (IIASA), A-2361 Luxemburg, Austria

[3] Faculty of Geosciences, Copernicus Institute of Sustainable Development, Utrecht University, 3584 CS Utrecht, The Netherlands; o.y.edelenbosch@uu.nl (O.Y.E.); detlef.vanvuuren@pbl.nl (D.P.v.V.)

[4] PBL Netherlands Environmental Assessment Agency, 2594 AV The Hague, The Netherlands

* Correspondence: c.hanmer@uea.ac.uk

Abstract: Countries' emission reduction commitments under the Paris Agreement have significant implications for lifestyles. National planning to meet emission targets is based on modelling and analysis specific to individual countries, whereas global integrated assessment models provide scenario projections in a consistent framework but with less granular output. We contribute a novel methodology for translating global scenarios into lifestyle implications at the national and household levels, which is generalisable to any service or country and versatile to work with any model or scenario. Our 5Ds method post-processes Integrated Assessment Model projections of sectoral energy demand for the global region to derive energy-service-specific lifestyle change at the household level. We illustrate the methodology for two energy services (mobility, heating) in two countries (UK, Sweden), showing how effort to reach zero carbon targets varies between countries and households. Our method creates an analytical bridge between global model output and information that can be used at national and local levels, making clear the lifestyle implications of climate targets.

Keywords: integrated assessment; lifestyle; scenarios; climate change mitigation; LTES

Citation: Hanmer, C.; Wilson, C.; Edelenbosch, O.Y.; van Vuuren, D.P. Translating Global Integrated Assessment Model Output into Lifestyle Change Pathways at the Country and Household Level. *Energies* **2022**, *15*, 1650. https://doi.org/10.3390/en15051650

Academic Editor: Luigi Aldieri

Received: 13 November 2021
Accepted: 9 February 2022
Published: 23 February 2022

Publisher's Note: MDPI stays neutral with regard to jurisdictional claims in published maps and institutional affiliations.

1. Introduction

The Paris Climate agreement has set out goals of limiting global warming to well below 2 °C and requires each country to maintain nationally determined contributions to greenhouse gas reductions over time [1]. Stringent climate targets require major demand-side transformations [2–4]. As energy demand is directly related to energy used in everyday life, these pathways imply significant changes in lifestyle [5,6]. The model-based scenarios used to explore the implications of the Paris climate targets provide aggregated projections of energy demand. There is a gap between these abstract parameters and information about change at the household level consistent with the long-term targets. The high-level scenario output for global regions does not indicate how energy demand varies in different geographies or across heterogenous household types.

In this paper, we introduce a multi-step 5Ds method to translate energy demand for global regions as output from global integrated assessment models (IAMs) into information about lifestyle change at the household level in specific countries. The 5Ds stand for disaggregation (of sectoral final energy to specific energy services), downscaling (from region to country), decomposition (of service-specific final energy into activity, structure, intensity components), differentiation (into household archetypes), and description of detailed household-level lifestyle change. This novel combination of established techniques reveals differences between countries and household types, which are not visible in aggregated model output. Presenting implications for households provides a bridge between global scenarios and research on low-carbon lifestyles at the national and local levels.

125

Our method takes as a starting point IAM scenarios that describe the changes in energy and land-use systems required to meet the Paris targets. These scenarios play a key role in IPCC assessment reports and inform both international negotiations and target setting and national policies [7,8]. The scenarios are developed using modelling frameworks that represent interactions between human and environmental systems as well as between supply and demand-sides of the energy system. Typical reporting on the demand-side is at the level of final energy use for broad sectors such as transportation (passenger and freight) or buildings (residential and commercial). In terms of spatial resolution, global IAMs typically report results for 10–30 world regions, often resolving large countries such as China, India, and Brazil but otherwise reporting at continental or subcontinental scales. Public databases such as the IAMC 1.5 °C Scenario Explorer [9] make data at this level of granularity accessible across multiple models. Appendix A provides an overview of IAM models.

The gap filled by this research is to translate this high-level data into information relevant to households in specific countries in order to make clear the lifestyle changes implied by the aggregated scenario data. This provides a simple alternative to complex national energy system models. A wide range of energy models are used to support national planning and policy development [10,11]. National models enable detailed consideration of local context and policy priorities but do not provide the representation of energy prices, technology development and global carbon budgets available from integrated global models [12].

The novelty of our study is:

1. To develop and test a methodology for translating global IAM output into lifestyle change implications for households;
2. To recognise variation between countries and between different households in the effort required to reach net-zero targets;
3. To reveal differences between mobility and heating-related lifestyle changes within 1.5 °C scenario pathways.

2. Literature Review

2.1. The Demand Reduction Challenge

The Sixth Assessment Report of the IPCC demonstrates the urgency of reducing greenhouse gas emissions [13,14]. In 2019, global GHG emissions were dominated by the use of coal (42%), followed by oil (34%) and natural gas (22%) [15]. Ambitious scenarios for emissions reduction show the importance of reducing energy demand as well as decarbonizing energy supply [5,16]. Buildings were responsible for 25% and transport for 27% of global CO_2 emissions in that year [15]. Changes in the way people use energy in everyday life are required to reduce these emissions [2]. The literature describing the changes required at individual consumption level highlights the magnitude of the changes in behaviour required [17–19]. Our research provides a method to present transformations in energy demand at a household level, to support communication with the public about the changes necessary in their country [20,21].

2.2. Analysing Energy Demand

Our 5Ds approach provides a quantitative pathway of change over time in energy services per household, bringing together precedents from disparate sources within the literature: disaggregation (from energy-services and energy-systems analysis), downscaling (from spatially-explicit modelling), decomposition (from sectoral demand analysis), differentiation (from bottom-up energy demand modelling), and description (from lifestyle narratives). Multiple steps are required in order to analyse energy demand across different dimensions of energy use sector and spatial scale. Chen et al. in this issue [22] demonstrate the power of decomposition to analyse average per capita emissions for global regions on a sectoral basis. Our analysis combines decomposition with additional steps to focus in on emissions for specific services from a variety of household types in a particular country.

The energy service approach is an established entry point into energy systems analysis, since useful service provision is the ultimate purpose of the energy system [23–25]. Energy is used in everyday life to provide services to users such as mobility, heating, and cooling. Downscaling results from a larger to smaller geographic areas is common for many different types of spatially explicit analysis (see, for example, Hoskins et al. [26] on land use, and Byers et al. [27] on vulnerability to climate change).

Energy demand at the service level can be decomposed into activity (A), structure (S), and intensity (I) components that distinguish respectively the quantity, type, and efficiency of service provision. This approach draws on a long tradition of ASI decomposition in energy demand analysis and modelling, notably in transport where activity is quantified in passenger-km, structure is expressed as a mix of alternative modes, and intensity is related to fuel efficiency per mode [28,29]. ASI decompositions are used in sectoral demand analysis and energy efficiency market reports to understand the relative contributions of different factors to changes in energy demand [30–34].

There can be significant variation of energy service use across households. This heterogeneity is considered in granular sectoral energy models that resolve a variety of socio-demographic and physical characteristics. For example, building stock models based on a set of dwelling archetypes are commonly used for bottom-up analysis of residential energy demand. Physical characteristics such as building fabric properties are key drivers of heating energy consumption for each dwelling archetype [35–37]. Vehicle stock models are core components of energy demand projections for the transport sector. These models consider the effects of income change over time and may include heterogeneity in physical determinants of travel demand such as urban or rural location, and access to public transport [38].

Information about energy demand pathways for particular household types provides opportunities for communication with the public directly relevant to their way of life [39,40]. Narrative storylines with quantitative underpinnings are powerful tools for communication and public engagement [17,41]. Lifestyle change pathways can also be used in deliberative contexts to explore the perceived feasibility, appeal, and policy requirements for low carbon futures [20].

2.3. Extending Global IAM Analysis of Low-Carbon Lifestyles

Integrated Assessment Models combine knowledge from multiple scientific and economic disciplines to provide reproducible scenarios for future energy use and the impact of this on the climate [42]. This includes a detailed representation of energy supply (see for example [43,44]), however the focus in this analysis is the IAM scenario results for energy demand.

Our approach complements existing work to extend the scope of IAM scenarios. Many IAMs come from an energy supply optimisation or computable general equilibrium modelling tradition, so they have a relatively coarse representation of energy demand [45]. However, in recent years, more detail has been included in the representation of energy demand such that more models now include subsectoral detail (e.g., passenger mobility by mode) as well as activity levels describing the quantify of energy service provided (e.g., passenger-kms) [46,47]. ASI decompositions have also been applied to compare drivers of change in final energy across IAM pathways [47,48].

Various approaches have been used to downscale IAM scenario output for global regions to countries or smaller geographic areas. For example, van Vuuren et al. [49] describe the use of simple algorithms to downscale population, income and emissions to national territories and 0.5° grid squares. Sferra et al. [50] downscale regional emissions to the country level using a reduced complexity optimisation model that mimics the framework of the IAM.

A common critique of global IAMs is that there is a lack of heterogeneous consumer agents or actors explicitly represented in the models [51,52]. Some IAMs do include certain types of household heterogeneity important for analysing specific research questions such as access to energy services between urban and rural households [53] or for households at different income levels [54]. Rising concerns around inequality and just transitions [55–57] are drawing attention to the importance for global models to capture within-country variation in the opportunities and capacities of different household types.

Recent collaborative initiatives between global and national modelling teams such as the CD-Links project [58,59] and the COMMIT project [12,60] have encouraged consistency between national and global scenarios. The 5Ds approach complements these activities by offering a simple technique to derive national results consistent with global models, without the specialist modelling resource required to link detailed models at the global and country level.

We contribute to the growing body of research considering lifestyle aspects of global IAM scenarios. Van den Berg et al. [61] distinguish the two main approaches used in global IAMs to-date. The first describes lifestyle changes in qualitative terms in scenario narratives, and then 'translates' those narratives into exogenous inputs or modelling assumptions such as reduced levels of activity [16,62] or increased levels of service efficiency [5]. The second approach simulates lifestyle changes endogenously as a function of changing technology costs, availability, or preferences [63,64]. Both approaches consider lifestyle change ex ante as a focus of the scenario or modelling exercise. Our 5Ds approach provides a complementary ex post or post-processing step that can in principle be applied to the output of any IAM scenario modelling, whether or not lifestyle change is considered ex ante. The stepwise approach differentiates service-level changes across heterogenous households within a country and so provides a higher resolution perspective on lifestyle change.

In the next section, we outline the five steps of this 5Ds method before providing illustrative examples of its application to heating and mobility energy services derived from a 1.5 °C scenario by the IMAGE model (a widely-used global IAM [65]).

3. Generalisable Method for Translating IAM Regional Output into Household-Level Lifestyle Change

3.1. Overview and Principles

The 5Ds method has five calculation steps to process IAM scenario output and a final communication step (Figure 1). Energy service use is derived for a household based on IAM totals for the *base year* (in the recent past) and one or more *target years* (the future end date of interest). Figure 1 shows the five calculation steps (including decomposition at both region and country level) before a final communication step of describing lifestyle change pathways.

The endpoint of the calculations is an activity-structure-intensity (ASI) decomposition of energy demand. *Activity* is defined as the amount of energy service, *intensity* as the final energy consumed for each unit of activity, and *structure* as the different combinations of fuel and technology used to deliver the service (identifying the share of total activity for each). Table 1 shows the ASI dimensions for heating and mobility.

Equation (1) shows structure S defined in terms of activity A. Each form of service—delivered with technology j using fuel f—provides a proportion S_{fj} of the total amount of service, defined as the fraction of the total activity A_T.

$$S_{fj} = \frac{A_{fj}}{A_T} \tag{1}$$

Equation (2) shows the full decomposition, relating final energy E to total activity A_T, the fraction of activity S_{fj} for each combination of fuel f and technology j, and the intensity I_{fj} for each combination or form of the service.

$$E = A_T \sum_f \sum_j S_{fj} I_{fj} \tag{2}$$

The algorithm employed at each of the five calculation steps is selected based on the characteristics of the energy service. In each case, the algorithm is linked to an underlying assumption about how the variable of interest for the smaller unit (subsector, country or group of households) relates to that for the larger unit of which it is part (sector, region or country). Five basic types of algorithm are employed. These are summarized in Table 2 and are described further in Appendix B. The first algorithms correspond to the three types of generic downscaling algorithms described in van Vuuren et al. (2007) for translating data at large spatial scales to smaller spatial scales (country or grid level). These downscaling algorithms can be applied to "any process in which coarse-scale data is disaggregated to a finer scale while ensuring consistency with the original data set" [49]. Two additional types of algorithm are applied in cases where the calculation goes beyond applying a scaling ratio to the larger unit.

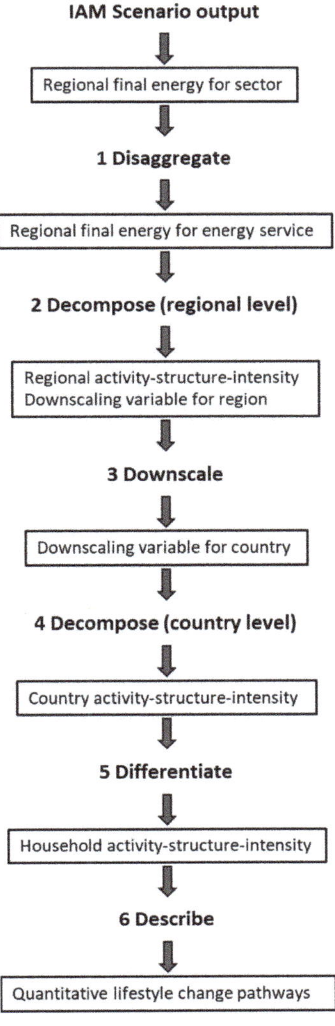

Figure 1. Flow diagram for generic 5Ds method. Initial scenario output from global IAMs is from publicly available repositories. Subsequent calculation steps are carried out for target year. Information passed from one step to the next is shown in boxes.

Table 1. ASI dimensions for mobility and heating.

	Activity	*Structure*	*Intensity*
General Definition	Amount of service used	Activity share of each form of the service	Energy use for each unit of activity
Heating	Building floor area heated (m^2)	Proportion of floor area heated by each combination of heating technology and fuel (e.g., natural gas boiler, electric heat pump, biomass boiler)	Final energy/floor area heated (MJ/m^2 yr)
Mobility	Distance travelled (passenger-km)	Proportion of distance travelled by each combination of mode and fuel (e.g., electric train, diesel bus, electric car: a switch from internal combustion engine to electric vehicles would be captured as a structural shift as fuel has changed even though mode is still private driving).	Final energy/passenger-km (MJ/p-km yr)

Table 2. Summary of algorithms.

Algorithm No	Description
1	Linear scaling (fixed proportion of larger unit)
2	Convergence (converges to mean for larger unit)
3	External input (apply ratios derived from detailed model)
4	Decompose into ASI components
5	Apply rule-based assumptions

3.2. Detailed Steps in 5Ds Method, with Illustration of Each Step for Mobility

In this section, we outline the series of steps to derive energy demand for a particular service at the household level in a specific country in the target year. The starting point is the final energy for the end use sector and region of interest reported in the IAMC data template [66]. This indicates the minimum information likely to be available from global IAM output.

We illustrate the application of the method to passenger mobility. The illustrative example uses data from a 1.5 °C scenario from the IMAGE IAM, referred to here by the abbreviated name '1.5C Total'. This is the 'All' deep mitigation pathway described by van Vuuren et al. [16], which incorporates both lifestyle change and rapid electrification based on renewable energy. We selected this pathway as one that shows more marked changes in certain energy-service demands, but, as we noted earlier, our methodology applies equally to scenarios in which lifestyle change is not explicitly considered. Output from the '1.5C Total' scenario for the Western Europe region is used to derive energy service use in the UK and Sweden. In Appendixs B through E, we provide further details of the calculations and the external data sources used for each step.

3.2.1. Step 1 Disaggregate

The first step is to disaggregate IAM sectoral final energy for a region to the level of a specific energy service (Figure 1). Linear scaling (Algorithm 1) is applied if the share of sectoral final energy for the energy service can be assumed to remain constant. The share is calculated from base year calibration data and then applied to the target year IAM sector total. For services where future energy use for the service is not expected to track overall sectoral trends, scaling based on a higher resolution sector model (Algorithm 3) is applied.

Illustration for Mobility

Figure 2 shows the results of the disaggregation step for the '1.5C Total' scenario. Final energy for transport is disaggregated between freight and passenger mobility. There is no reason to expect the passenger mobility share of energy for transport to stay constant so a detailed scenario model for the target year is required to provide scaling ratios to apply to IAM final energy for transport (Algorithm 3). The assumptions and equations to derive final energy for passenger transport modes when these are not directly available from IAM output are outlined in Appendix C.

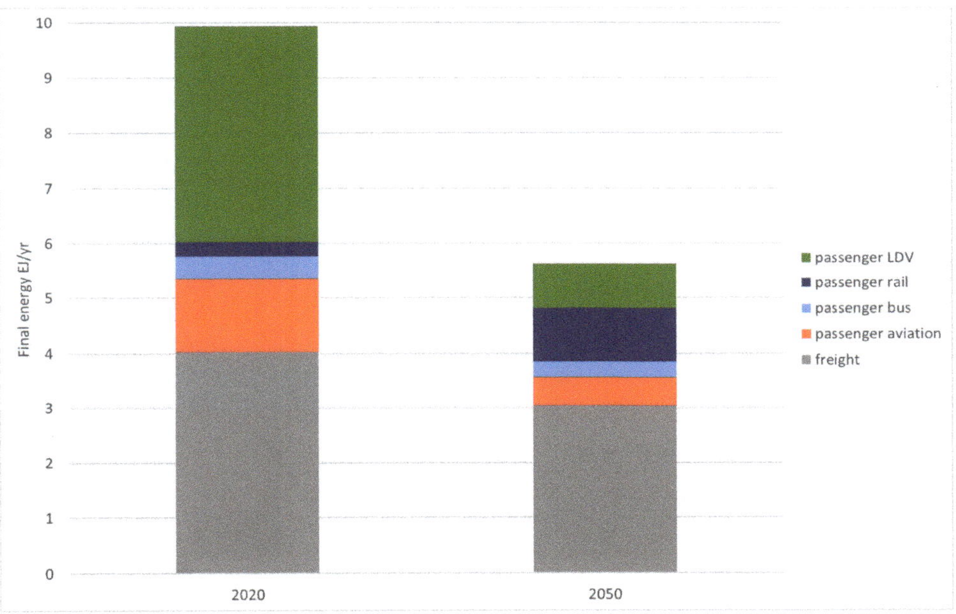

Figure 2. Transport disaggregation. This figure shows the division of final energy for transport in the IMAGE '1.5C Total' scenario for the Western Europe (WEU) region in 2020 and 2050 [16]. LDV = Light Duty Vehicle (predominantly cars). Although in this case data are directly reported by IMAGE, in models with less granular resolution, the disaggregation step would estimate the passenger mobility proportions of total final energy for transport.

3.2.2. Step 2 Decompose for Region

Final energy for the region is decomposed across each combination of technology and fuel, establishing activity and intensity for each element of the structure (Figure 1). If both activity and final energy data are available from IAM output, intensity can be derived directly from this and no further calculations are required. Otherwise, intensity for each technology and fuel combination (for base year, and projected for target year) is estimated based on the literature. The energy balance for each fuel (Equation (2)) is then solved for activity (see Section 2.3 for heating, for which activity is known and the unknown is the building fabric property H). The term fuel as used in this report includes all energy carriers (such as electricity, hydrogen and heat) in addition to primary fuels (such as coal and natural gas). In cases where a fuel maps onto more than one form of service, a set of assumptions must be made about the allocation of the fuel across different technologies (e.g., electricity for mobility could supply electric LDVs, trains, or buses).

Illustration for Mobility

The generic decomposition Equation (2) expressed for passenger mobility is shown in Equation (3):

$$E = d_T \sum_k \sum_f I_{kf}\, S_{kf} \tag{3}$$

Activity is expressed in terms of distance d (passenger-km) travelled. E is the final energy for passenger transport and I is the intensity for mode k using fuel f (MJ/passenger km). Different modes k in this equation are equivalent to different technology types in the generic decomposition Equation (2). The structure S_{kf} is the proportion of total distance d_T travelled by mode k using fuel f.

To reduce the complexity of the example, transport fuel options are grouped into two categories: electricity and liquid (which combines all liquid and gaseous fuels including petroleum, biofuels, hydrogen, compressed natural gas). This distinction preserves the ability to analyse the transition to electric vehicles projected in 1.5 °C scenarios.

The IMAGE scenario output includes final energy and distance travelled (activity) for each passenger transport mode and fuel combination for the region in the target year, so transport intensity in the example is derived from this data, and there is no need for further decomposition calculations. Figure 3 shows the results of the decomposition step for the regional level.

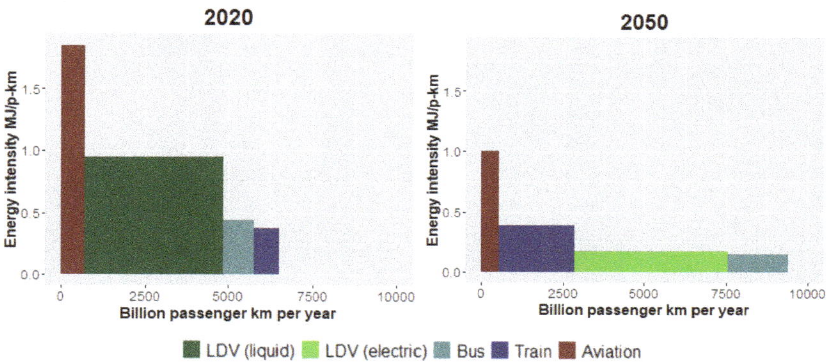

Figure 3. Decomposition of mobility for Western Europe region from '1.5C Total' scenario [16]. The *X*-axis shows activity (distance travelled) and the *Y*-axis shows intensity. Each coloured bar represents one form of mobility service, with the area of the bar proportional to final energy for that service. The widths of the coloured bars represent the structure (share of activity for each form of service).

Many IAMs report activity (distance travelled) and intensity (energy used per passenger km) figures for each mode of transport. In cases where the only output is aggregated across all modes of transport, so activity and intensity information for each mode are not directly available from the IAM, decomposition of mobility final energy for the target year involves solving a set of simultaneous equations. These are based on assumptions about the allocation of final energy for each fuel across different modes and matching the relative distances travelled by different modes with those from a detailed sector model. These assumptions and equations are detailed in Appendix C.

3.2.3. Step 3 Downscale

The next step is to downscale from region to country level (Figure 1). Household requirements for an energy service vary between countries in a region because of differences in factors such as climate, income, building stock characteristics, and typical travelling distances. The components of activity and intensity which vary by country are identified and the 'scaling variable' which represents these is established.

The downscaling step takes the total of the 'scaling variable' for the region and allocates this across countries in the region. The type of algorithm applied for downscaling is based on assumptions about whether the trends for energy use for the service in the country are likely to:

- follow the same trends as the region as a whole (apply Algorithm 1);
- converge on the regional mean (apply Algorithm 2);
- diverge or otherwise follow irregular trajectories (apply Algorithm 3 by using external input from a higher resolution analysis of the specific energy service).

This follows the 'simple algorithm' approach laid out in Van Vuuren et al. [52].

Illustration for Mobility

For mobility, the quantity which differs between countries is the distance travelled by each transport mode and this is identified as the 'scaling variable'. New vehicle technologies as well as efficiency standards are widely diffused, so it is assumed the intensity for each mode and fuel combination is constant across countries in a region.

Future mobility patterns will be influenced by the evolution of current travel practices, vehicle stocks and infrastructure. This means it is unlikely that all countries within a region will follow the same trends or have the same mix of transport modes in the target year. The approach taken for downscaling the distance travelled from region to country is to use external input from a higher resolution scenario modelling analysis reporting country level results (Algorithm 3). This provides the country to region ratio of distance travelled by particular transport modes. This ratio is used to downscale the regional distance by mode (in passenger-km) derived from the global IAM scenario. The external input in this illustration of the method uses the Directed Vision scenario [67] This is a scenario describing strong policy action at the European level to deliver on the EU's 2050 net-zero target; the scenario was interpreted by a suite of inter-linked sectoral and energy-system models, including ASTRA which resolves vehicle fleet and transport choices for 27 EU countries, which is broadly consistent with the IMAGE '1.5C Total' mitigation outcome. Figure 4 shows the results of the downscaling step.

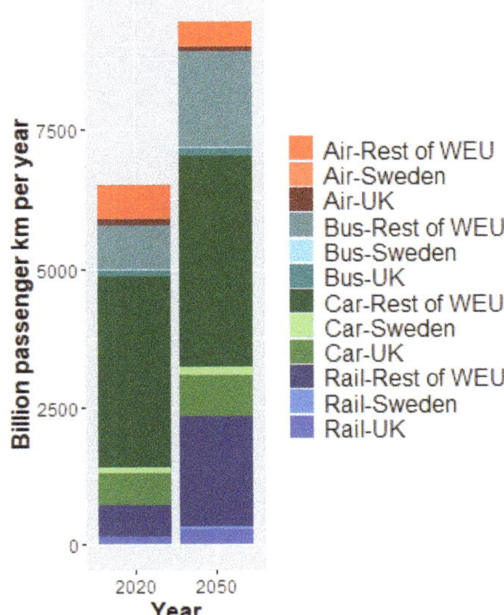

Figure 4. Downscaling mobility from WEU to the UK and Sweden. The total distances travelled for the Western Europe (WEU) region from the '1.5C Total' scenario [16] are downscaled to the UK and Sweden using ratios derived from a detailed scenario [67].

3.2.4. Step 4 Decompose for Country

Following downscaling, a second decomposition step at the country level is carried out (Figure 1). The unknowns and constraints for this decomposition are established based on the characteristics of the energy service considered. Local infrastructure constrains which forms of energy service are accessible. Energy infrastructure development is highly path dependent [68,69]. For services with infrastructure constraints, rules are applied to allocate future shares for each combination of technology and fuel (algorithm 5).

The country decomposition results can be used to assess the contribution of changes in activity, structure and intensity to the overall change in final energy. Ang [70] provides an overview of the development of 'index decomposition analysis' used by researchers to investigate trends in energy use. These techniques, developed for the analysis of historical energy data, have also been applied to emissions projections from IAM scenarios [47,48,71]. Appendix B.4 explains how the relative contribution of activity, structure and intensity effects can be calculated using the Sun method.

Illustration for Mobility

The mobility example illustrates how the country decomposition step draws on results from both the downscaling and regional decomposition steps. Mobility activity for the country is the sum of the distances by mode established in the downscaling step. As the 1.5 °C scenario used in this illustration sees rapid and pervasive electrification of the vehicle fleet, it is assumed that the electrified share of each mode converges to a regional average (established in the regional decomposition) by the target year of 2050. The structure is established by applying the regional electrified shares to the downscaled distances by mode for the country (e.g., splitting the distance travelled by LDV between electric and liquid fuel vehicles by applying the regional electrification ratio to the downscaled country distance travelled by LDV). As explained in the previous section it is assumed that regional intensities also apply at the country level. Infrastructure constraints are assumed not to apply.

Figure 5 shows the results of the decomposition step, expressed as changes in activity, structure and intensity for mobility for two countries in the 1.5 °C scenario illustration. The household travel patterns in 2020 are similar in both countries. In both countries to 2050, overall activity increases but with very significant improvements in intensity projected at the regional level. There is a shift from LDVs using liquid fuels to battery electric vehicles, distance travelled by air reduces, and distance travelled by train increases. Table 3 shows the relative contributions of each ASI effect to the change in final energy. In both countries in this scenario the intensity effect (UK 45%, Sweden 48%) is significant, but lower than the structure effect (UK 62%, Sweden 66%). The changes in structure caused by the transition from petroleum fuels to electric vehicles provide the greatest overall contribution to reduction in final energy.

Table 3. Percentage contributions of activity, structure and intensity effects to overall change in final energy between 2020 and 2050 for household mobility in Sweden and the UK derived from '1.5C Total' scenario. Negative figures indicate an increase in energy.

	Activity Effect	**Structure Effect**	**Intensity Effect**
Sweden	−14%	66%	48%
UK	−7%	62%	45%

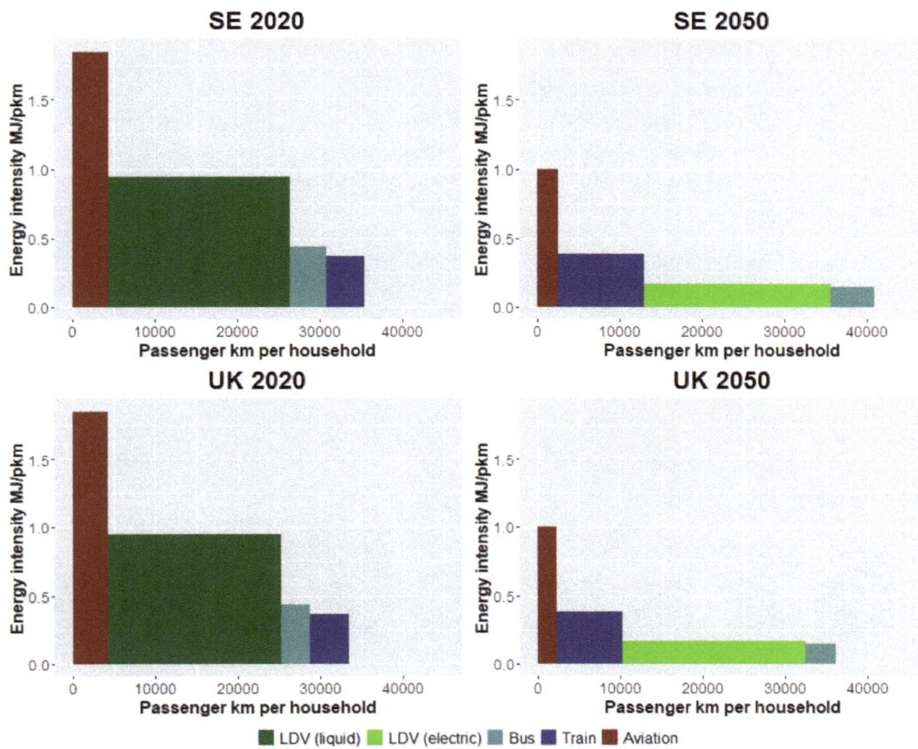

Figure 5. Decomposition of mobility in Sweden and the UK derived from '1.5C Total' scenario [16]. Coloured bars show activity, intensity, and structure as in Figure 3. X-axis shows activity levels for mean household in country.

3.2.5. Step 5 Differentiate

The last calculation stage, differentiation, draws on sectoral modelling and empirical analysis to identify the main causes of variation in energy services across households within a country. Archetypes (groups of households) with distinct characteristics that shape the activity, structure, and intensity of their energy service consumption are identified. The dimensions of variation can be socioeconomic (e.g., income), geographic (e.g., urban), or physical (e.g., building type). These are combined to create a simple set of household archetypes (e.g., eight archetypes along $2 \times 2 \times 2$ dimensions of variation). The share of national activity for each form of service for each household archetype in the base year is established by drawing on household surveys and other national data.

The target year final energy at the country level is differentiated across this set of household archetypes, with activity, structure and intensity established for each. The 'scaling variable' for the service (see step 3) indicates the components of activity and intensity that vary between archetypes. The algorithm to project the 'scaling variable' from base to target year is selected based on the characteristics of the service and decisions on whether differences between archetypes are likely to persist.

For some services, infrastructure or other physical constraints affect the suitability of different forms of service to a particular archetype (e.g., access to necessary infrastructure such as gas networks or electric vehicle charging points). In these cases, service-specific rules (Algorithm 5) are applied to establish the structure for the archetypes in the target year. In the mobility example, it is assumed there are no archetype-specific constraints; in other words, each archetype has equal access to electric vehicles and charging infrastructure. It is also assumed that the country electrification ratio for each mode can be applied for each archetype, thus establishing the structure for each.

Illustration for Mobility

For mobility, archetypes are differentiated based on household size and income, urban or rural location, all factors which are known to influence distance travelled. The distance travelled by mode (the 'scaling variable') by each archetype in the base year is established from survey data characterising the heterogeneous travel behaviour of a representative set of households, such as the National Travel Survey in the UK [72]. In the illustration it is assumed that income and location differences persist so that each archetype's share of the country total distance by mode is the same in the base and target years (algorithm 1). It is also assumed that average intensity for each mode and fuel combination is the same for each group of households. Figure 6 shows the different activity level and structure for eight UK household types in 2020 and 2050 from the 1.5 °C scenario illustration, with higher activity levels associated with larger households, higher incomes, and a rural location. Figure 7 shows distances travelled by four Swedish archetypes.

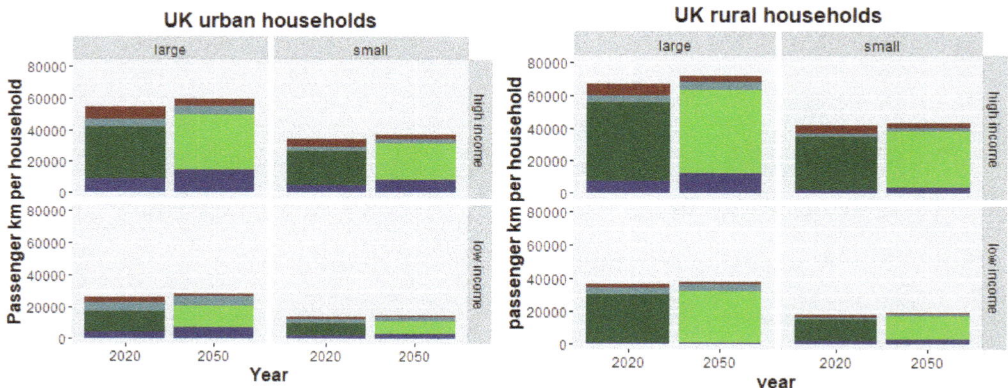

Figure 6. Shows results for mobility for UK households divided into four archetypes by income and household size.

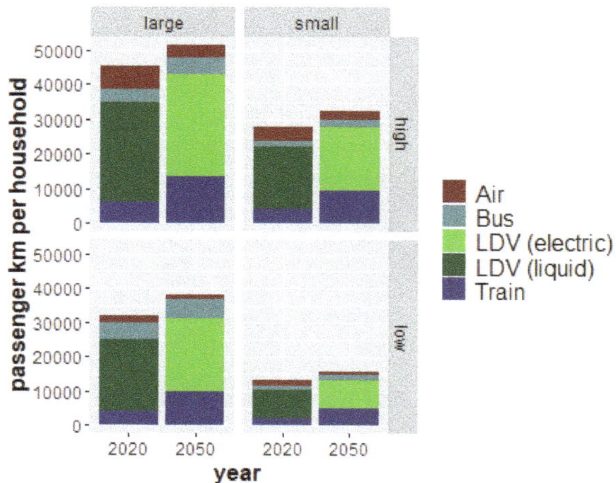

Figure 7. Mean distances travelled by four Swedish household archetypes, derived from '1.5C Total' scenario [16]. Large households > 2 people, high income > SEK 500,000 household income.

3.2.6. Step 6 Describe

In the final communication step, quantitative descriptions of changes in the activity, structure, intensity components of energy services between base and target year are developed for each household archetype (Figure 1). This enables audiences to compare the impact of the modelled scenario on different types of household and to identify the likely impact on lifestyle for households like their own. Narrative storylines derived from the IAM scenario provide context and interpretive detail.

The ASI decomposition results also enable messages about lifestyle changes to be positioned in terms of the Avoid-Shift-Improve framework, which has been widely used to characterize interventions and policies for changing energy demand [17,73]. Actions consistent with given warming outcomes can be described in three categories:

- avoid *activity* by reducing how much service is used;
- shift within *structure* by choosing a lower energy form of service provision;
- improve *intensity* by using a more efficient technology.

Illustration for Mobility

The use of the Avoid-Shift-Improve framework can be illustrated for the UK mobility results shown in Figures 5 and 6. In this case, rather than showing an 'avoid' story, the overall distance travelled (activity) per household increases by 8% between 2020 and 2050. The narrative of change is about major shifts between modes of transport, for both private and public modes of travel. In the case of private vehicles, the '1.5C Total' scenario shows a complete replacement by 2050 of cars fuelled by petrol and diesel with battery electric vehicles. The lifestyle implications of this change can be described, for example, by pointing out that households will need to integrate vehicle charging in their regular routines, rather than filling up their cars at petrol stations. Charging points will become important elements of local infrastructure [74]. For public transport, there is a significant increase in distance travelled by train (67% for the UK example) combined with a halving of travel by air. In 2050, holidays and business travel involving flights are less frequent. In lifestyle terms, a shift in long distance travel from plane to train is likely accompanied by changes in attitudes, with local holidays and virtual meetings perceived as satisfactory alternatives to trips to other countries.

The steps for the 5Ds method as applied to mobility are summarised in Table 4.

Table 4. Summary of data, assumptions, and algorithms for calculation steps applied to mobility. Steps in italics not carried out in illustrative example, as disaggregated data are available from IAM.

Step in 5Ds Method	External Data (Base Year)	External Data (Target Year)	Assumptions (Target Year)	Algorithm (Type)
Disaggregate	*Calibration data for sector.*	*Detailed sector scenario.*	*Personal mobility share of final energy for transport sector is same as for detailed scenario.*	*External input (3)*
Decompose (regional level)	*Intensity for each mode and fuel.*	*Intensity for each mode and fuel.*	*Assumptions about electrification level for each mode*	*Energy balance (4)*
Downscale to country	Distance by mode for country and region.	Distance by mode for country and region from higher resolution analysis.	Country-level modal shares of regional activity match external input from higher resolution analysis.	External input (3)
Decompose (country level)	*Intensity for each mode and fuel.*	*Intensity for each mode and fuel.*	*Proportion of each mode electrified for country is same as for region. Intensity same as region*	*Energy balance (4)*
Differentiate	National travel survey data: distance by mode for different household archetypes.	-	Ratio of archetype to national average distance travelled per mode stays constant.	Linear scaling (1)

3.3. Application of 5Ds Method to Heating

The same principles and series of steps illustrated for mobility can be applied to other energy services such as residential space heating, with differences in the characteristics of specific energy services leading to differences in implementation. In this second illustration, we show briefly how the 5Ds method is applied to residential space heating. Full details for each calculation step are provided in Appendix D.

For heating, the starting point for *disaggregation* is final energy for the commercial and residential sector (reported in the standard IAMC template). For *downscaling*, the space heating requirements in a country depend not only on the heated floor area but also the climate and the building fabric properties (a poorly insulated building stock requires more energy to heat than a well insulted one in the same climate).

The generic *decomposition* Equation (2) expressed for residential heating is:

$$E = a_T H \sum_j \sum_f \frac{S_{jf}}{\eta_{jf}} \tag{4}$$

E is final energy for space heating and S_{jf} is the fraction of total floor area a_T heated by technology j using fuel f. The amount of heating service received by building occupants from a fixed amount of energy depends on both the (active) efficiency of the heating conversion technology (η) and the (passive) efficiency of the building fabric [75].

For the *differentiation* step, the variation between archetypes of space available and access to infrastructure is considered. For example the economics of district heating mean that it is best suited for densely inhabited urban areas with large numbers of smaller homes, while costs to supply more widely spaced, larger homes would be higher. Analysis of space heating structure enables a *description* of how the proportion of each archetype which use a particular heating system changes over time and how this will affect the everyday life of the households involved.

Using the IMAGE '1.5C Total' scenario to illustrate the results of these steps, Table 5 shows the relative activity, structure and intensity (ASI) effects expressed as a percentage of the overall change in final energy. Activity at the household level does not change as it is assumed that floor area per household does not change from 2010 to 2050 in these two countries.

Table 5. Percentage contributions of activity, structure and intensity effects to overall change in final energy per household between 2010 and 2050 for heating in Sweden and the UK derived from '1.5C Total' scenario.

	Activity Effect	Structure Effect	Intensity Effect
Sweden	0%	−1%	101%
UK	0%	28%	72%

Figure 8 shows the change in heating types for the UK and Swedish housing archetypes. In Sweden low carbon heating options are currently in widespread use and district heating infrastructure is already in place. This is reflected in the very low percentage change projected for the structure component in Table 5. In the UK, much greater shifts in structure are apparent for all archetypes, in line with the national shift away from gas heating.

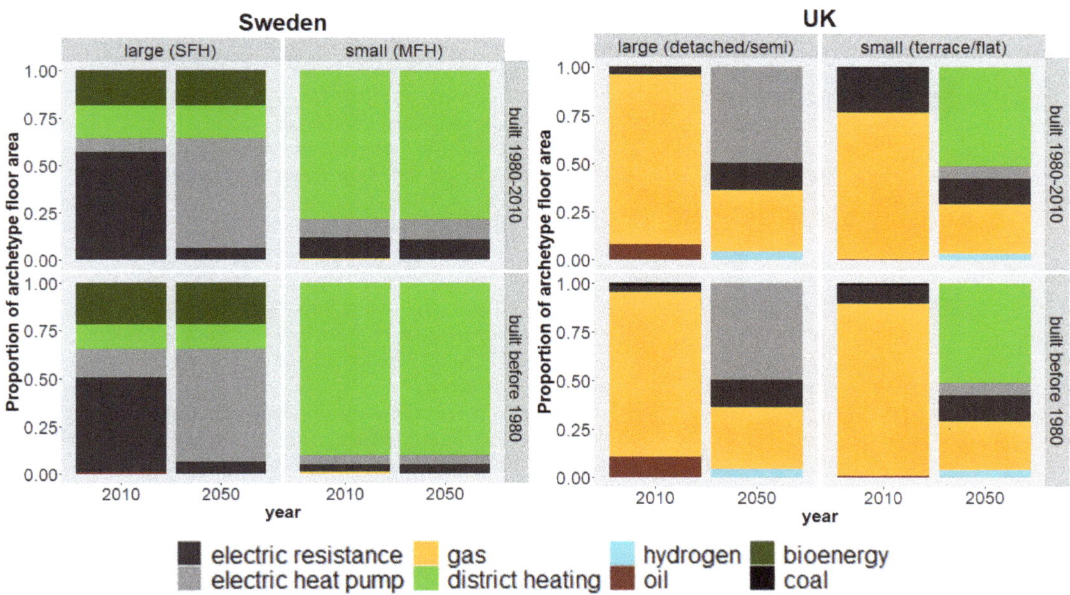

Figure 8. Share of floor area with each form of heating for UK and Swedish dwelling archetypes in 2010 and 2050 (authors' analysis).

Table 5 shows that the energy reduction in Sweden is dominated by the intensity effect, and intensity change also contributes 72% of the UK reduction, highlighting the importance of improvements in building fabric to improve intensity. Activity at the household level does not change as it is assumed that floor area per household does not change to 2050 in these two countries. Based on the UK archetypes shown in Figure 8, the description for each type of house introduces the heating systems projected for 2050, with resulting implications for lifestyle change. For example, in the majority of larger homes where a gas boiler is replaced with an electric heat pump, households are likely to notice changes in the patterns of heating in the home associated with lower radiator temperatures and requirements for demand management of electrical heating [76–78].

3.4. Uncertainties in Results

Since the principle of the method is to divide up the total emissions from the IAM scenario across households the uncertainties associated with the global scenario projections also apply to the 5Ds results. Wilson et al. [79] provide an overview of methods for the evaluation of IAM models. Sensitivity to scenario assumptions and uncertainties in calibration data for the IMAGE model have been the subject of detailed analysis [75,76].

4. Discussion

This section discusses the general applicability of the method and its relevance for national policy making and for lifestyle research. We reflect on the results of the illustrative example and the opportunities for communication with the public.

4.1. General Applicability

We have illustrated the 5Ds method here for two countries (UK, Sweden) within a single global region (Western Europe). However, the method can be generalised to any country in any global region, and is also flexible to work with any global modelling analysis reporting sectoral final energy at the regional level. The principles of the 5Ds method can also be applied to energy services beyond mobility and heating (see Appendix F for further discussion). Sectoral modelling of energy use in buildings often distinguishes hot water, cooling, cooking, lighting, and appliances as well as space heating [80–82]. The 5Ds method enables more detailed consideration of the energy services associated with these categories than is visible in standard IAM output. It allows lifestyle changes for countries and households implied by different scenarios and models to be compared in a standardised way, for example exploring the trade-offs between scenarios emphasising supply-side transformation (e.g., [83]) versus demand-side transformation (e.g., [5]).

4.2. Benefits for National Policy Analysis

The 5Ds method provides country-specific information within a consistent global context since it derives national energy service data from global IAM output. This is relevant for national policymakers and researchers interested in the local impact of global scenarios. The long-term viewpoint from global IAMs (which typically provide projections up to 2100) can also provide an alternative perspective to national modelling, which is typically concerned with shorter timescales.

The ASI analysis can support national policy development and planning. Policy to reduce or *avoid* activity focuses on behaviour change, with a combination of information, incentives and "nudges" to make the desired behaviour easier [84]. A range of policy options are open to encourage a *shift* the structure to lower carbon forms of energy service. Economic incentives can influence consumer choice (for example when selecting a replacement heating system) while regulation–for instance, banning the sale of fossil fuel heating boilers after a particular date–can remove high emissions options [85]. Infrastructure planning is a crucial element in enabling shifts to low emission energy services. Many shifts cannot be achieved without the development of new energy supply infrastructure (e.g., electric vehicle charging points, district heating networks). An important policy lever to *improve* the efficiency of service provision is regulation in the form of product standards, as well as financial incentives and R&D funding, to improve the energy performance of technologies and service-provisioning systems [86].

The results for activity, structure, and intensity contributions to overall reduction in final energy in Tables 3 and 5 illustrate how these may vary between countries, leading to different policy priorities. As an example, for heating the level of necessary infrastructure change in the UK is much greater than that in Sweden. There is less contrast between the countries in the components of energy reduction for mobility. In the '1.5C Total' global scenario, both countries share a key priority of encouraging a shift to electric LDVs.

4.3. Lifestyle Change

As a post-processing step, the 5Ds method requires no changes to IAM code or modelling approach in order to communicate lifestyle change implications of mitigation pathways in more detail than is possible from standard IAM output. The information about changes in energy services at the household level is relevant for civil society actors interested in lifestyle change to achieve emissions reductions. The results are accessible for those who are not involved in policy and scenario modelling discussions. For example, the stringent mitigation scenario in the illustrations above shows households in the UK and Sweden travelling further by train but flying less than they do today.

The 5Ds method enables qualitative storylines of change at global, national, and local scales to be linked. The IAM scenario narrative provides an over-arching storyline about what is happening elsewhere in the world, within which the description of changes at the household level in a particular country can be situated. Global scenarios draw on reference pathways, such as the widely used Shared Socioeconomic Pathways (SSPs) [87]. The narrative for each pathway describes trends in the world economy, demographics and technology development. This provides a global context for energy services at national and household levels. For example, the '1.5C Total' scenario (the starting point for the illustrations above) includes assumptions about rapid electrification in all end-use sectors and consumer changes in habits towards a lower greenhouse gas lifestyle [16]. These draw on an underlying narrative of co-operation between nations leading to rapid technology diffusion, and a shift in emphasis towards human well-being with less focus on economic growth. The 5Ds method adds stories about the changes in daily life that will be experienced by particular household types under the scenario.

The narrative can thus highlight both global and local conditions associated with the scenario. For example, the significant improvements in intensity for mobility seen in the '1.5C Total' scenario result from technology development encouraged by government policy and international cooperation. Intensity intersects with lifestyle at the level of individual decisions about purchasing vehicles with high energy efficiency (or choosing these among shared mobility options). Low-intensity options will be encouraged by high carbon prices in this stringent mitigation scenario, and the scenario narrative also shows these decisions influenced by society-wide preferences for options with low environmental impact. The significant changes in intensity of heating energy in both the UK and Sweden imply major overhauls of the building stock in both countries, but households in the UK will have to adapt to new forms of heating, while Swedish heating types will change much less because low-carbon heating technology is already prevalent.

4.4. Limitations of the Method

Each of the calculation steps in the 5Ds method requires a series of assumptions on data inputs and data processing algorithms, which are documented in this article. These assumptions require a degree of background information and an awareness of the statistical information on the energy service being analysed to understand the infrastructural, policy, and behavioural context of future lifestyle change, as well as to calibrate base year assumptions.

5. Conclusions

5.1. Practical Implications of the Research

By translating results from long term global scenarios into national level projections for energy demand, the method offers a technique for national planning, highlighting the national infrastructure changes and policy priorities implied by the global model results and enhancing interpretability and usefulness of IAM results. The approach reveals differences between countries and household types which are not visible in aggregated model output, showing variation between mobility- and heating-related lifestyle changes within net-zero pathways.

The method broadens the potential audience for IAM scenarios to members of the public interested in changes in lifestyle. The energy service results enable a focus on how aspects of daily life such as residential heating or passenger transport will change over time. The identification of changes in everyday lifestyles makes IAM results more transparent for citizens, offering new ways to communicate scenarios to the public in a way that resonates with people's lived experience. Details of the practical consequences for specific households are set within an overarching narrative about global emissions reductions and worldwide developments in population, economy, and technology.

5.2. Future Research Directions

The 5Ds method complements efforts to make global scenario modelling results publicly available. It can be applied to output from any IAM model, which reports energy demand, and to all types of mitigation and baseline scenarios. The purpose of the study was to develop, test, and demonstrate the methodology. The next steps for research are to apply the methodology to comparatively assess results across different scenarios and models as well as to extend the application to additional energy services such as illumination, cooking and cooling. The method could contribute to a multi-model comparison of IAM results, providing the basis to compare national and household level energy demand results (based on the same assumptions and calibration data) across an ensemble of different models. This would also allow systematic investigation of scenario uncertainty at the service level.

Author Contributions: Conceptualization, C.H. and C.W.; methodology, C.H., C.W., O.Y.E. and D.P.v.V.; software, C.H.; formal analysis, C.H.; data curation, O.Y.E. and C.H.; writing—original draft preparation, C.H. and C.W.; writing—review and editing, O.Y.E. and D.P.v.V.; visualization, C.H. and C.W. All authors have read and agreed to the published version of the manuscript.

Funding: This research was funded by the CAST Centre for Climate Change and Social Transformations (UK ESRC Grant ES/S012257/1), with additional funding from the Energy Demand changes Induced by Technological and Social innovations (EDITS) initiative coordinated by RITE, the Research Institute of Innovative Technology for the Earth (Japan), and IIASA, the International Institute for Applied Systems Analysis (Austria) and funded by the Ministry of Economy, Trade, and Industry (METI) in Japan. D.P.v.V. and O.Y.E. thank the KR foundation for funding the IMAGE-Lifystyle project, contributing to the paper. The APC was funded by IRENA.

Institutional Review Board Statement: Not applicable.

Informed Consent Statement: Not applicable.

Data Availability Statement: Data sources are listed in Appendix E.

Acknowledgments: Thanks to Vassilis Daioglou and David Gernaat, Netherlands Environmental Assessment Agency, for assistance supplying and interpreting IMAGE output data, and to Stephanie Heitel, Fraunhofer ISI, for her assistance.

Conflicts of Interest: The authors declare no conflict of interest.

Appendix A. An Overview of Integrated Assessment Models

Global integrated assessment models (IAMs) are tools for simulating the energy and land-use transformations that are necessary to limit global warming to meet climate targets. They represent linkages between energy, land use, climate, and [45,88]. IAMs are used to analyse long-term global climate outcomes under what-if assumptions about future drivers of change [79]. The Intergovernmental Panel on Climate Change (IPCC) fifth assessment report drew on 1134 scenarios from 30 global IAMs [42,89]. The 2018 IPCC Special Report on global warming of 1.5 °C drew on 411 scenarios from 10 global IAMs [90]. IAMs contribute directly to climate policy formulation, including UNFCCC international negotiations and national strategies and targets [91]. The models have been extensively peer reviewed and are increasingly open source.

IAMs typically have a high degree of resolution of energy supply. For example, the IMAGE IAM models 6 options for heating fuels, 17 different options for electricity generation, and 10 different forms of hydrogen production [92]. Efficiency and cost changes over time are modelled for energy demand. Transport demand is either based on top-down modelling (related to population and economic growth), or a hybrid with different technology options represented [47] (nine different passenger transport modes are represented in IMAGE [92]). Industrial demand sectors are represented in varying levels of detail (in the IMAGE model steel, cement and plastics production are each separately represented). Residential energy is the other major demand sector that is represented. Household income is a key driver for residential energy demand [54].

Figure A1 shows a schematic representation of the IMAGE IAM model illustrating how drivers are linked to impacts through human and earth systems. IAMs project a cost-optimized mix of energy supply given the scenario assumptions and climate target. A baseline scenario with assumptions for economic and population growth is chosen (frequently one of the Shared Socioeconomic Pathways [87]). Deep mitigation scenarios are implemented by introducing a uniform global carbon tax to meet the radiative forcing target associated with the specific climate target [16].

IMAGE 3.0 framework

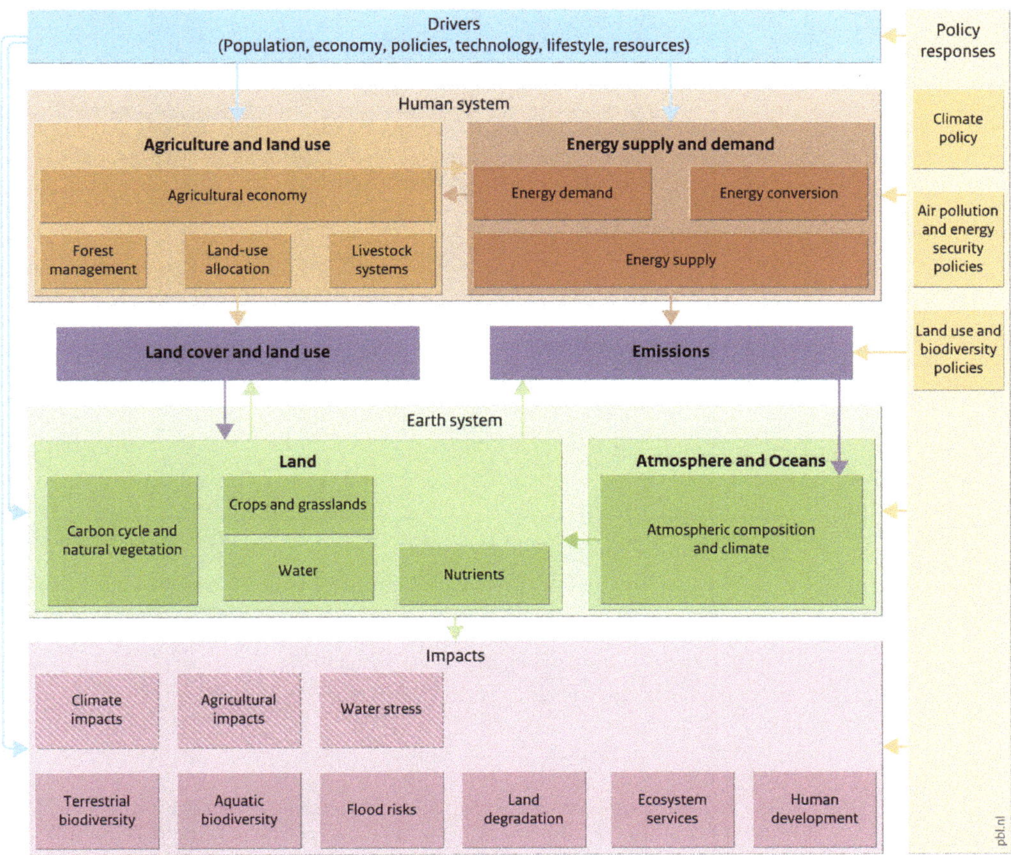

Source: PBL 2014

Figure A1. An overview of the IMAGE framework and its components, reproduced from https://models.pbl.nl/image/index.php/IMAGE_framework_summary (accessed 10 January 2021).

Appendix B. Principles of 5Ds Calculations

Table A1. Nomenclature for Appendices.

a	floor area
A	activity
B	final energy for bus
C	final energy for LDV
E	final energy
F	final energy for freight transport
H	building heating required per unit area
I	Intensity
L	total final energy from 'liquid' fuels (incorporates gaseous fuels)
P	total final energy from electricity
Q	useful space heat energy

Table A1. *Cont.*

R	final energy for passenger rail
T	total final energy for residential and commercial
S	Structure
V	final energy for (passenger) aviation
ε	electrification ratio–ratio of distance travelled by mode using electricity to total distance travelled by mode
ζ	ratio of total energy for transport from IAM to total from detailed model
η	efficiency of heating technology / vehicle
subscripts	
C	country
e	electric
f	fuel
g	fuel (excluding electricity)
I	derived from IAM
k	convergence year
l	liquid
M	derived from detailed scenario model
p	electric heat pump
r	electric resistive
R	region
T	total
α	archetype

Appendix B.1. Technology and Fuel Combinations

The structure to which mobility energy use is decomposed is a set of mode and fuel combinations. The share of activity for the region, country, or household for each combination is derived for each combination (form of service).

The modes considered for passenger mobility are:

- LDV (light duty vehicles, predominantly cars)
- Bus
- Rail
- Aviation

The fuel options are grouped into two categories: electricity and liquid. The liquid category combines all liquid and gaseous fuels (petroleum, biofuels, hydrogen, CNG), and a uniform efficiency is assumed for all vehicles in the same mode using liquid fuels.

The structure dimension for heating is the share of the activity that is attributed to each combination of heating technology and fuel. Eight combinations are considered in this analysis:

1. Electric resistance heater
2. Electric heat pump
3. Gas boiler
4. Heat from district heating network
5. Hydrogen boiler
6. Oil boiler
7. Biomass boiler
8. Coal boiler

Appendix B.2. Base Year Selection

The heating illustration takes 2010 as the base year. Calibration was found for dates between 2010 and 2013 (see Appendix D), so 2010 is the closest date with IAM data available (IMAGE IAM scenario output is reported at decade intervals).

For mobility, a base year of 2020 was chosen. Detailed model scenario data was used for this year, rather than calibration data from national and international statistics. This overcame difficulties in finding consistent calibration data across all transport modes.

Appendix B.3. Calculation Algorithms

Algorithm type 1, linear scaling, assumes that the smaller unit represents a constant proportion of the larger unit. Algorithm type 2, convergence, assumes values for the smaller unit converge to an average value for the larger unit. Algorithm type 3 is based on external input from an alternative model or scenario with greater resolution of energy service demands and does not assume a simple linear relationship between smaller and larger units. The scenario narrative from this external model, particularly the level of mitigation stringency, should be matched as closely as possible to that of the IAM. Algorithm type 4 refers to the decomposition of energy service demand into activity, structure and intensity (Equation (2)). Algorithm type 5, rule-based allocation, is applied when the options for the smaller unit are limited by physical constraints such as the availability of fuel supply infrastructure. In such cases, a decision tree set of questions is followed to allocate appropriate fuels to the smaller unit.

Appendix B.4. ASI Contributions to Change in Final Energy

The analysis of ASI contributions follows the Sun index decomposition method [85] to divide up the change in final energy ΔE between base year b and target year t.

$$\Delta E = E^t - E^b \tag{A1}$$

ΔE is expressed as the sum of the effects due to activity, intensity and structure:

$$\Delta E = EA_{effect} + EI_{effect} + ES_{effect} \tag{A2}$$

Each of these effects is expressed in terms of changes in overall activity A, intensity I_{fj} and structure S_{fj} summed across all combinations of fuel f and technology j.

$$EA_{effect} = \Delta A \sum_f \sum_j I_{fj}^b S_{fj}^b + \tfrac{1}{2}\Delta A \sum_f \sum_j \left(I^b \Delta S_{fj} + S_{fj}^b \Delta I_{fj}\right) + \tfrac{1}{3}\Delta A \sum_f \sum_j \Delta I_{fj}\Delta S_{fj} \tag{A3}$$

$$EI_{effect} = A^b \sum_f \sum_j S_{fj}^b \Delta I_{fj} + \tfrac{1}{2}\sum_f \sum_j \Delta I_{fj}\left(S^b \Delta A + A^b \Delta S_{fj}\right) + \tfrac{1}{3}\Delta A \sum_f \sum_j \Delta I_{fj}\Delta S_{fj} \tag{A4}$$

$$ES_{effect} = A^b \sum_f \sum_j I_{fj}^b \Delta S_{fj} + \tfrac{1}{2}\sum_f \sum_j \Delta S_{fj}\left(I^b \Delta A + A^b \Delta I_{fj}\right) + \tfrac{1}{3}\Delta A \sum_f \sum_j \Delta I_{fj}\Delta S_{fj} \tag{A5}$$

Appendix C. Additional Details of 5Ds Method Applied to Mobility

The main text provides an illustration of the 5Ds method applied to mobility using the '1.5C Total' scenario generated by the IMAGE IAM as an example. IMAGE reports mobility-specific data with more granularity than other IAMs. This appendix sets out the additional calculations needed if only aggregated sectoral IAM output is available.

Appendix C.1. Disaggregation

The starting point is regional energy used by mode from the detailed scenario output. This is used to disaggregate the IAM total energy for transport between four subsectors: freight, aviation, rail, and all road passenger transport (bus and LDV combined). It is assumed that the proportion of energy used by each sector is the same as that for the detailed scenario, and subsector totals are found by scaling detailed model amounts by ratio, ζ, of total energy from IAM to total energy from detailed model:

$$V_{RI} = \zeta V_{RM} \tag{A6}$$

$$C_{RI} + B_{RI} = \zeta(C_{RM} + B_{RM}) \quad \text{etc.} \tag{A7}$$

Appendix C.2. Decomposition

If data for final energy for each transport mode is not available, a number of simplifications are made to derive a set of equations to relate total transportation liquid fuel final energy L and electrical final energy P for the region from the IAM scenario and the energy for each fuel for each mode.

Balance of liquid across modes:

$$L_R = F_{IR} + C_{IR} + B_{IR} + V_{IR} \tag{A8}$$

Balance of electricity across modes (assumes all trains are electric in target year, and there is no electric freight or aviation):

$$P_{RI} = C_{eR} + B_{eR} + R_{eR} \tag{A9}$$

Two further assumptions are made:

- The ratio of distance travelled by bus to distance travelled by LDVs data is the same as that derived from the detailed sector model.
- The electrification ratio ε (of distance travelled using electric fuel to total distance travelled) is the same for LDV and for bus.

This leads to four equations in four unknowns (C_{IR}, B_{IR}, C_{eR}, B_{eR}—the liquid and electricity energy totals for bus and for LDV), which can be solved simultaneously.

Appendix D. 5Ds Method Applied to Heating

This appendix provides additional details of the application of the 5Ds method to heating.

Appendix D.1. Downscaling

The 'scaling variable' which varies between countries for space heating is the useful energy for space heating, Q (derived for the region in the previous step). The ratio of useful heating energy for country of interest, Q_C, to that for the whole region is established for the base year from calibration data. This ratio is then used to downscale the useful energy for the region in the target year (Algorithm 1). This linear scaling is based on the assumption that the country uses the same percentage of total regional useful heat as in the base year, i.e., ignoring changes in relative levels of population, floor area, and fabric heat loss among countries. Figure A2 shows the results of this step.

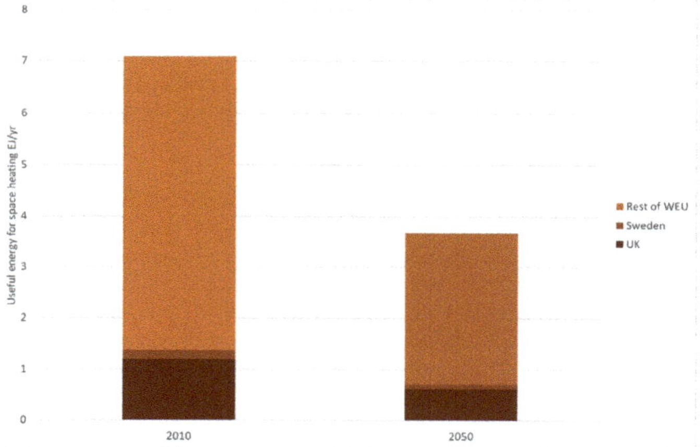

Figure A2. Downscaling of useful energy for heat. Total for Western Europe (WEU) derived from '1.5C Total' scenario [16]. Linear scaling from 2010 calibration data applied in 2050.

Appendix D.2. Decomposition

As stated in the main text, the generic decomposition Equation (2) expressed for residential heating is:

$$E = a_T \, H \sum_j \sum_f \frac{S_{jf}}{\eta_{jf}} \tag{A10}$$

E is final energy for space heating and S_{jf} is the fraction of total floor area a_T heated by technology j using fuel f. The amount of heating service received by building occupants from a fixed amount of energy depends on both the (active) efficiency of the heating conversion technology (η) and the (passive) efficiency of the building fabric [75]. Intensity can be expressed as H/η, where H is the mean useful space heating required for unit area. H is directly related to the heat lost from the building over the year. As fabric insulation is improved, this quantity will decrease.

The types of heating used in a country are strongly influenced by available infrastructure and established traditions so a set of rules (Algorithm 5) are applied to determine the mix of forms of service for the country in the target year. These rules to estimate the proportion of floor area heated with each fuel, which take into account policy ambitions and relate fossil fuel shares in the country to those in the region, are:

1. Fossil fuels (gas, coal, oil). Find the ratio of fraction of floor area heated by the fossil fuel in country to the fraction of floor area heated in the region by the fuel in the base year. Apply this ratio to the target year regional proportion. This linear rather than convergence relationship is based on the assumption that the existing infrastructure and installed equipment base will influence the share of future fossil fuel use for an extended period in the future.
2. Hydrogen. It is assumed that uptake of hydrogen will involve a conversion of a similar proportion of the existing natural gas infrastructure in each country. The area heated by hydrogen is derived by multiplying the area heated by gas in country by the regional ratio of area heated by hydrogen to area heated by gas.

For two low carbon options, biomass and district heating, the policy ambitions in the country are taken into account in a series of decision steps:

3. Bioenergy. If the current proportion is sustainable and economically likely to continue, assume bioenergy share of floor area heated is same as base year. If it is not, reduce in line with national policy forecasts.
4. District heating. If there are national policy targets to increase district heating, estimate the share in the target year based on these national ambitions. Otherwise, keep the current proportion constant.

Electric heating forms the balance once other fuel proportions have been estimated. An estimate of how this is divided between heat pumps and resistive heating is based on national policy aspirations.

Figure A3 illustrates the decomposition of heating energy in the UK and Sweden derived from the IMAGE '1.5C Total' scenario. Between 2010 and 2050, a significant shift in the UK away from gas heating and an increase in the share of total area heated by heat pumps and district heating is visible. In contrast, there is less change in the structure for Sweden, reflecting the high share of low carbon heating in 2010 (see Table 5). There is a substantial reduction in intensity in both countries. The reduction in useful space heat per m^2, H, (from 468 MJ/m^2 yr to 162 MJ/m^2 yr for Sweden and from 486 MJ/m^2 yr to 194 MJ/m^2 yr for the UK) represents a very significant improvement in building fabric in both countries.

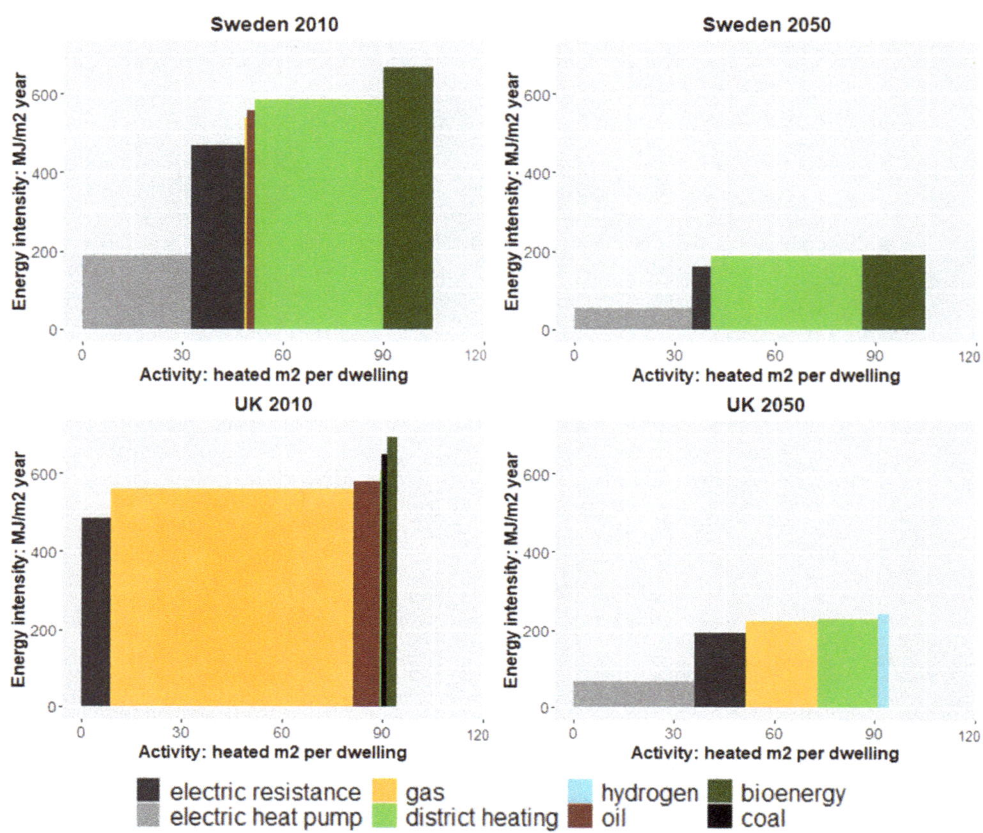

Figure A3. Heating decomposition for Sweden and the UK derived from '1.5C Total' scenario [16]. The *X*-axis shows activity and the *Y*-axis intensity. Each coloured bar represents one form of heating service. The widths of the coloured bars represent the share of floor area for each form of service (structure) expressed as the mean across all dwellings.

Appendix D.3. Differentiation

For the differentiation step, the variation between archetypes of space available and access to infrastructure is considered. For example, the economics of district heating mean that it is best suited for densely inhabited urban areas with large numbers of smaller homes, while costs to supply more widely spaced, larger homes would be higher. A set of rules is followed to allocate the national heating fuel totals across each archetype in the target year (Algorithm 5). These take into account relative shares for each archetype in the base year and the suitability of two low carbon-heating options (district heating and electric heat pumps) for archetypes with particular characteristics. The rules applied in the illustration are:

- Allocate country total for each fossil fuel pro rata to existing archetypes, which use that fuel in the base year (assume no fossil fuels are used in newbuild archetypes).
- Allocate hydrogen in proportion to gas use.
- Allocate district heating equally across small home archetypes based on its suitability for high density housing.
- Allocate biomass pro-rata based on initial proportions for each archetype in the base year.

The balance of floor area for each archetype, once all other fuels have been allocated, is allocated to electric heating. The heat pump and electric resistive heating totals for the country are divided across the archetypes based on an assumption about the ratio of heat pumps in large home archetypes to small home archetypes (larger homes are more likely to have the space required to install heat pumps).

Figure A4 repeats Figure 8 in the main text and shows the change in heating types for the UK and Swedish housing archetypes, which results from applying these rules to divide up the country total heating energy derived from the '1.5C Total' scenario. In Sweden, low-carbon heating options are currently in widespread use and district heating infrastructure is already in place; this is reflected in the vary low percentage change projected for structure in Table 5. The high prevalence of district heating in the smaller Swedish dwellings (MFH—multi-family homes) persists to 2050. Larger Swedish dwellings (SFH—Single Family Homes) have a different mix of heating in 2010, but, again, these are dominated by low carbon technologies, so there is little change in structure to 2050 apart from an increase in the share of electrical heating provided by heat pumps.

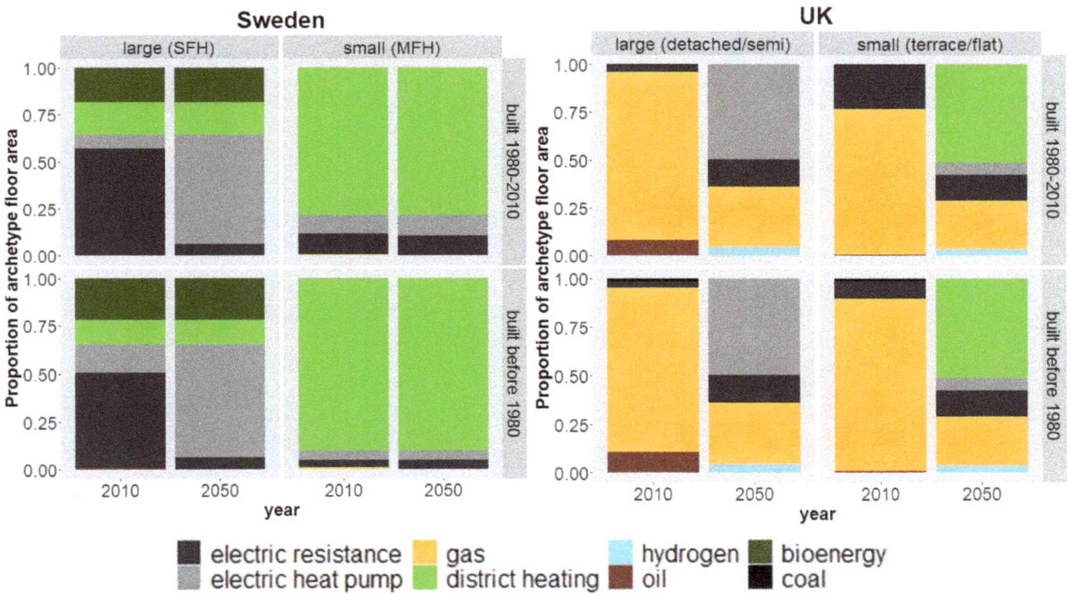

Figure A4. Share of floor area with each form of heating for the UK and Swedish dwelling archetypes in 2010 and 2050 (authors' analysis).

By contrast, in the UK, structure change contributes 28% of final energy reduction highlighting the importance of a shift away from gas heating. In the UK, much greater shifts in structure are apparent for all archetypes, in line with the national shift away from gas heating. The main replacement technology in small homes is district heating, which is particularly suitable for dense housing in urban areas, with electric heat pumps playing a significant role in larger homes, which are more likely to have the necessary space to install this technology.

Appendix D.4. Description

The analysis of space heating structure enables a description of how the proportion of each archetype which use a particular heating system changes over time and how this will affect the everyday life of the households involved. The intensity results indicate the improvements in building fabric implied by the scenario.

Based on the UK archetypes illustrated above, the description for each type of house would introduces the heating systems projected for 2050. The changes the household are likely to experience depend on the type of the new heating system. The heating service provided by district heating systems (the most common heating system in UK smaller homes in 2050) is very similar to that from a gas boiler, although there may be disruption outside the home associated with the installation of new heating infrastructure. In the majority of larger homes, where a gas boiler is replaced with an electric heat pump, households are likely to notice changes in the patterns of heating in the home associated with lower radiator temperatures and requirements for demand management of electrical heating [76–78].

The description would also include the impact of improvements in building fabric. Retrofitting of insulation to upgrade existing homes may be combined with ventilation improvements. Residents are likely to experience disruption during the installation phase, and a changed, more stable thermal environment following the upgrade [93].

Appendix E. Data Sources for Illustrations

Table A2. Data sources for mobility illustration.

Data	Level	Source	Notes
IAM passenger transport final energy by fuel	Region (WEU)	IMAGE Scenario LOWTOT_19	This is the "all" scenario in van Vuuren et al. [16]
Population, number of households	Region and country (UK/SE)	Eurostat [94] for base year. ONS [95] and Statistics [96] for target year	Target year scaled for population increases from IMAGE IAM output
Intensities for mode and fuel	Region	Derived from IMAGE data	Regional figures also applied for country
Comparator scenario with country data	Country	ASTRA Directed Vision scenario [67]	
Archetype household and distance travelled data	UK	National Travel Survey (NTS) 2002–19 [72]	Large: more than two people High income: >GBP 25,000 household income. Rural or urban based on NTS settlement classification
Archetype household and distance travelled data	SE	Swedish National Travel Survey 2011–16 [97]	Large: more than two people High income > SEK 500,000 annual income

Table A3. Data sources for heating illustration.

Data	Level	Source	Notes
IAM residential space heating final energy by fuel	Region (WEU)	IMAGE Scenario LOWTOT_19	This is the "all" scenario in [16]
Residential floor area, population, number of households	Region and country (UK/SE)	Eurostat [94] ONS [95]	Target year scaled for regional floor area and population increases from IMAGE IAM output
Space heating final energy by fuel (calibration data)	Region and country	Odyssee-mure [98]	Data for 14 countries available–scaled by population to match WEU region in IAM
Heating technology conversion efficiencies	Region and country	Compilation from the literature [19,99–102]	Regional figures also applied for country
Archetype heat loss rate	UK and SE	National typology brochures [103]	
Archetype floor area and fuel use	UK	English Housing Survey 2011–12 [104]	Separate analysis of survey dataset to derive mean for each archetype Old: built before 1980 New: built after 1980 Small: flat and terrace Large: Detached and semi detached
Archetype floor area and fuel use	SE	National building statistics [103,105,106]	Old: built before 1980 New: built after 1980 Small: flerbostadshus (multi-family home) Large: småhus (one and two family dwelling)

Appendix F. Generalizing the 5Ds Method to Other Energy Services

The article has focused on two energy services–mobility and heating. This appendix outlines how the principles of the 5Ds method can be applied to other energy services such as hot water, cooling, cooking, lighting, and appliances.

Established indicators and statistics in each sector indicate guide the ASI decomposition step for each energy service. For example, if illumination is considered, the lighting technology categories (LED, fluorescent and other) employed by the IEA [107] is a suitable starting point for the structure component. We have described how rules are applied in the decompose and differentiate steps to reflect infrastructure and other physical constraints that influence low carbon options available in a particular country. This is relevant for hot water and cooking, which have multiple fuel options dependant on specific local infrastructure (e.g., natural gas distribution network, availability of biofuels). These energy services contrast with cooling and illumination which are associated with a single fuel, electricity.

The key dimensions for differentiation between households depend on the characteristics of each service. Building on established traditions of bottom-up sector energy models, we differentiate households within a country based on their circumstances which influence energy demand (size, income, type of home etc.). These categories are straightforward to establish based on national statistics. It is also easy for members of the public to identify the group to which they belong. In the case of residential cooling, for example, cooling energy demand is strongly associated with household income [32]. Locational effects should also be considered–for example, in developing countries, the fuels used for cooking are likely to differ between urban and rural locations as well as by household income [54,108,109].

A potential limitation to generalising to different services in a country is the availability of suitable calibration data. Some sectors (such as transport) have well-established data collection protocols [110], but other services may fall outside the remit of national statistics agencies. Data resources for developing nations are typically less extensive than those in developed countries. In particular, more extensive data sets with which to differentiate households are likely to be available for developed countries, with few developing countries conducting large-scale surveys [111]. Detailed sector models may not be available for use in the disaggregation and downscaling steps.

We have described how expectations of patterns of change in energy demand are embedded in the selection of the algorithm to use at each step as the energy demand totals from the IAM scenario are allocated across countries and household archetypes. The method can be adapted to answer specific research questions. For example, an alternative algorithm choice would allow investigation of national differences within a region under different convergence assumptions. The household differentiation step offers opportunities to investigate equitable emissions reduction across different household groups [56,112,113].

References

1. Rose, S.K.; Richels, R.; Blanford, G.; Rutherford, T. The Paris Agreement and next Steps in Limiting Global Warming. *Clim. Chang.* **2017**, *142*, 255–270. [CrossRef]
2. IPPC. Global Warming of 1.5 °C. An IPCC Special Report on the Impacts of Global Warming of 1.5 °C above Pre-Industrial Levels and Related Global Greenhouse Gas Emission Pathways, in the Context of Strengthening the Global Response to the Threat of Climate Change, Sustainable Development, and Efforts to Eradicate Poverty. 2018. Available online: https://www.ipcc.ch/sr15/ (accessed on 11 February 2022).
3. Luderer, G.; Vrontisi, Z.; Bertram, C. Residual Fossil CO_2 Emissions in 1.5–2 °C Pathways. *Nat. Clim. Chang.* **2018**, *8*, 626–633. [CrossRef]
4. Mundaca, L.; Ürge-Vorsatz, D.; Wilson, C. Demand-Side Approaches for Limiting Global Warming to 1.5 °C. *Energy Effic.* **2019**, *12*, 343–362. [CrossRef]
5. Grubler, A.; Wilson, C.; Bento, N.; Boza-Kiss, B.; Krey, V.; McCollum, D.L.; Rao, N.D.; Riahi, K.; Rogelj, J.; De Stercke, S.; et al. A Low Energy Demand Scenario for Meeting the 1.5 °C Target and Sustainable Development Goals without Negative Emission Technologies. *Nat. Energy* **2018**, *3*, 515–527. [CrossRef]
6. Saujot, M.; Gallic, T.L.; Waisman, H. Lifestyle Changes in Mitigation Pathways: Policy and Scientific Insights. *Environ. Res. Lett.* **2020**, *16*, 015005. [CrossRef]

7. van Beek, L.; Hajer, M.; Pelzer, P.; van Vuuren, D.; Cassen, C. Anticipating Futures through Models: The Rise of Integrated Assessment Modelling in the Climate Science-Policy Interface since 1970. *Glob. Environ. Chang.* **2020**, *65*, 102191. [CrossRef]

8. Weyant, J. Some Contributions of Integrated Assessment Models of Global Climate Change. *Rev. Environ. Econ. Policy* **2017**, *11*, 115–137. [CrossRef]

9. Huppmann, D.; Kriegler, E.; Krey, V.; Riahi, K.; Rogelj, J.; Rose, S.K.; Weyant, J.; Bauer, N.; Bertram, C.; Bosetti, V.; et al. *IAMC 1.5 °C Scenario Explorer and Data Hosted by IIASA*; Integrated Assessment Modeling Consortium & International Institute for Applied Systems Analysis: Laxenburg, Austria, 2018.

10. Strachan, N. UK Energy Policy Ambition and UK Energy Modelling—Fit for Purpose? *Energy Policy* **2011**, *39*, 1037–1040. [CrossRef]

11. Süsser, D.; Ceglarz, A.; Gaschnig, H.; Stavrakas, V.; Flamos, A.; Giannakidis, G.; Lilliestam, J. Model-Based Policymaking or Policy-Based Modelling? How Energy Models and Energy Policy Interact. *Energy Res. Soc. Sci.* **2021**, *75*, 101984. [CrossRef]

12. Fragkos, P.; Laura van Soest, H.; Schaeffer, R.; Reedman, L.; Köberle, A.C.; Macaluso, N.; Evangelopoulou, S.; De Vita, A.; Sha, F.; Qimin, C.; et al. Energy System Transitions and Low-Carbon Pathways in Australia, Brazil, Canada, China, EU-28, India, Indonesia, Japan, Republic of Korea, Russia and the United States. *Energy* **2021**, *216*, 119385. [CrossRef]

13. Arias, P.; Bellouin, N.; Coppola, E.; Jones, R.; Krinner, G.; Marotzke, J.; Naik, V.; Palmer, M.; Plattner, G.-K.; Rogelj, J.; et al. *Climate Change 2021: The Physical Science Basis. Contribution of Working Group I to the Sixth Assessment Report of the Intergovernmental Panel on Climate Change*; Technical Summary; Cambridge University Press: Cambridge, UK, 2021.

14. Cartwright, E.D. "Code Red"—Recent IPCC Report Warns Time Is Running Out on Climate Change. *Clim. Energy* **2021**, *38*, 11–12. [CrossRef]

15. IEA Emissions by Sector–Greenhouse Gas Emissions from Energy: Overview–Analysis. Available online: https://www.iea.org/reports/greenhouse-gas-emissions-from-energy-overview/emissions-by-sector (accessed on 29 December 2021).

16. van Vuuren, D.P.; Stehfest, E.; Gernaat, D.E.H.J.; van den Berg, M.; Bijl, D.L.; de Boer, H.S.; Daioglou, V.; Doelman, J.C.; Edelenbosch, O.Y.; Harmsen, M.; et al. Alternative Pathways to the 1.5 °C Target Reduce the Need for Negative Emission Technologies. *Nat. Clim. Chang.* **2018**, *8*, 391–397. [CrossRef]

17. Lettenmeier, L.; Koide, R.; Toivo, V.; Amellina, A.; Akenji, L. *1.5-Degree Lifestyles: Targets and Options for Reducing Lifestyle Carbon Footprints*; Technical Report; Institute for Global Environmental Strategies: Hayama, Japan, 2019.

18. Goldstein, B.; Gounaridis, D.; Newell, J.P. The Carbon Footprint of Household Energy Use in the United States. *Proc. Natl. Acad. Sci. USA* **2020**, *117*, 19122–19130. [CrossRef] [PubMed]

19. Committee on Climate Change (CCC). *The Sixth Carbon Budget: The UK's Path to Net Zero*; Committee on Climate Change (CCC): London, UK, 2020.

20. Cherry, C.; Scott, K.; Barrett, J.; Pidgeon, N. Public Acceptance of Resource-Efficiency Strategies to Mitigate Climate Change. *Nat. Clim. Chang.* **2018**, *8*, 1007–1012. [CrossRef]

21. Capstick, S.; Demski, C.; Cherry, C.; Verfuerth, C.; Steentjes, K. Climate Change Citizens' Assemblies: CAST Briefing Paper 03. Available online: https://cast.ac.uk/wp-content/uploads/2020/03/CAST-Briefing-03-Climate-Change-Citizens-Assemblies.pdf (accessed on 11 February 2022).

22. Chen, H.-H.; Hof, A.F.; Daioglou, V.; de Boer, H.S.; Edelenbosch, O.Y.; van den Berg, M.; van der Wijst, K.-I.; van Vuuren, D.P. Using Decomposition Analysis to Determine the Main Contributing Factors to Carbon Neutrality across Sectors. *Energies* **2022**, *15*, 132. [CrossRef]

23. Cullen, J.M.; Allwood, J.M. Theoretical Efficiency Limits for Energy Conversion Devices. *Energy* **2010**, *35*, 2059–2069. [CrossRef]

24. Grubler, A.; Johansson, L.; Mundaca, L.; Nakicenovic, N.; Pachauri, S.; Riahi, K.; Rogner, H.-H.; Strupeit, L. Chapter 1-Energy Primer. In *Global Energy Assessment—Toward a Sustainable Future*; Cambridge University Press: Cambridge, UK, 2012.

25. TWI2050 Innovations for Sustainability. *Pathways to an Efficient and Post-Pandemic Future. Report Prepared by the World in 2050 Initiative*; International Institute for Applied Systems Analysis (IIASA): Laxenburg, Austria, 2020.

26. Hoskins, A.J.; Bush, A.; Gilmore, J.; Harwood, T.; Hudson, L.N.; Ware, C.; Williams, K.J.; Ferrier, S. Downscaling Land-Use Data to Provide Global 30" Estimates of Five Land-Use Classes. *Ecol. Evol.* **2016**, *6*, 3040–3055. [CrossRef]

27. Byers, E.; Gidden, M.; Leclère, D.; Balkovic, J.; Burek, P.; Ebi, K.; Greve, P.; Grey, D.; Havlik, P.; Hillers, A.; et al. Global Exposure and Vulnerability to Multi-Sector Development and Climate Change Hotspots. *Environ. Res. Lett.* **2018**, *13*, 055012. [CrossRef]

28. Creutzig, F.; Jochem, P.; Edelenbosch, O.Y.; Mattauch, L.; van Vuuren, D.P.; McCollum, D.; Minx, J. Transport: A Roadblock to Climate Change Mitigation? *Science* **2015**, *350*, 911–912. [CrossRef]

29. Schipper, L.; Saenger, C.; Sudardshan, A. Transport and Carbon Emissions in the United States: The Long View. *Energies* **2011**, *4*, 563–581. [CrossRef]

30. Ang, B.W.; Liu, N. Energy Decomposition Analysis: IEA Model versus Other Methods. *Energy Policy* **2007**, *35*, 1426–1432. [CrossRef]

31. IEA. *Energy Efficiency Indicators 2020*; International Energy Agency: Paris, France, 2020.

32. IEA. *The Future of Cooling in China*; International Energy Agency: Paris, France, 2019.

33. IEA. *Energy Efficiency Market Report*; International Energy Agency: Paris, France, 2018.

34. Marrero, G.A.; Ramos-Real, F.J. Activity Sectors and Energy Intensity: Decomposition Analysis and Policy Implications for European Countries (1991–2005). *Energies* **2013**, *6*, 2521–2540. [CrossRef]

35. Famuyibo, A.A.; Duffy, A.; Strachan, P. Developing Archetypes for Domestic Dwellings—An Irish Case Study. *Energy Build.* **2012**, *50*, 150–157. [CrossRef]

36. Kavgic, M.; Mavrogianni, A.; Mumovic, D.; Summerfield, A.; Stevanovic, Z.; Djurovic-Petrovic, M. A Review of Bottom-up Building Stock Models for Energy Consumption in the Residential Sector. *Build. Environ.* **2010**, *45*, 1683–1697. [CrossRef]

37. Mata, É.; Sasic Kalagasidis, A.; Johnsson, F. Building-Stock Aggregation through Archetype Buildings: France, Germany, Spain and the UK. *Build. Environ.* **2014**, *81*, 270–282. [CrossRef]

38. Brand, C.; Tran, M.; Anable, J. The UK Transport Carbon Model: An Integrated Life Cycle Approach to Explore Low Carbon Futures. *Energy Policy* **2012**, *41*, 107–124. [CrossRef]

39. Milne, S.; Chambers, K.; Elks, S.; Hussain, B.; McKinnon, S. *Living Carbon Free*; Energy Systems Catapult: Birmingham, UK, 2019.

40. ADEME. *We Demain Objectif 2030 10 Familles, 10 Scénarios Pour Un Mode de Vie Plus Durable*; ADEME: Paris, France, 2015.

41. Akenji, L.; Chen, H. *Framework for Shaping Sustainable Lifestyles: Determinants and Strategies*; United Nations Environment Programme: Nairobi, Kenya, 2016.

42. Clarke, L.; Jiang, K. Chapter 6: Assessing Transformation Pathways. In *Working Group III contribution to the IPCC 5th Assessment Report, Climate Change 2014: Mitigation of Climate Change*; Cambridge University Press: Cambridge, UK, 2014.

43. Gernaat, D.E.H.J.; Van Vuuren, D.P.; Van Vliet, J.; Sullivan, P.; Arent, D.J. Global Long-Term Cost Dynamics of Offshore Wind Electricity Generation. *Energy* **2014**, *76*, 663–672. [CrossRef]

44. de Boer, H.S.; van Vuuren, D. Representation of Variable Renewable Energy Sources in TIMER, an Aggregated Energy System Simulation Model. *Energy Econ.* **2017**, *64*, 600–611. [CrossRef]

45. Sathaye, J.; Shukla, P.R. Methods and Models for Costing Carbon Mitigation. *Annu. Rev. Environ. Resour.* **2013**, *38*, 137–168. [CrossRef]

46. Edelenbosch, O.Y.; van Vuuren, D.P.; Blok, K.; Calvin, K.; Fujimori, S. Mitigating Energy Demand Sector Emissions: The Integrated Modelling Perspective. *Appl. Energy* **2020**, *261*, 114347. [CrossRef]

47. Edelenbosch, O.Y.; McCollum, D.L.; van Vuuren, D.P.; Bertram, C.; Carrara, S.; Daly, H.; Fujimori, S.; Kitous, A.; Kyle, P.; Broin, E.Ó.; et al. Decomposing Passenger Transport Futures: Comparing Results of Global Integrated Assessment Models. *Transp. Res. Part Transp. Environ.* **2017**, *55*, 281–293. [CrossRef]

48. van den Berg, N.J.; Hof, A.F.; van der Wijst, K.-I.; Akenji, L.; Daioglou, V.; Edelenbosch, O.Y.; van Sluisveld, M.A.E.; Timmer, V.J.; van Vuuren, D.P. Decomposition Analysis of per Capita Emissions: A Tool for Assessing Consumption Changes and Technology Changes within Scenarios. *Environ. Res. Commun.* **2021**, *3*, 015004. [CrossRef]

49. van Vuuren, D.P.; Lucas, P.L.; Hilderink, H. Downscaling Drivers of Global Environmental Change: Enabling Use of Global SRES Scenarios at the National and Grid Levels. *Glob. Environ. Chang.* **2007**, *17*, 114–130. [CrossRef]

50. Sferra, F.; Krapp, M.; Roming, N.; Schaeffer, M.; Malik, A.; Hare, B.; Brecha, R. Towards Optimal 1.5° and 2 °C Emission Pathways for Individual Countries: A Finland Case Study. *Energy Policy* **2019**, *133*, 110705. [CrossRef]

51. Keppo, I.; Butnar, I.; Bauer, N.; Caspani, M.; Edelenbosch, O.; Emmerling, J.; Fragkos, P.; Guivarch, C.; Harmsen, M.; Lefèvre, J.; et al. Exploring the Possibility Space: Taking Stock of the Diverse Capabilities and Gaps in Integrated Assessment Models. *Environ. Res. Lett.* **2021**, *16*, 053006. [CrossRef]

52. Mercure, J.-F.; Pollitt, H.; Bassi, A.M.; Viñuales, J.E.; Edwards, N.R. Modelling Complex Systems of Heterogeneous Agents to Better Design Sustainability Transitions Policy. *Glob. Environ. Chang.* **2016**, *37*, 102–115. [CrossRef]

53. Krey, V.; O'Neill, B.C.; van Ruijven, B.; Chaturvedi, V.; Daioglou, V.; Eom, J.; Jiang, L.; Nagai, Y.; Pachauri, S.; Ren, X. Urban and Rural Energy Use and Carbon Dioxide Emissions in Asia. *Energy Econ.* **2012**, *34*, S272–S283. [CrossRef]

54. Daioglou, V.; van Ruijven, B.J.; van Vuuren, D.P. Model Projections for Household Energy Use in Developing Countries. *Energy* **2012**, *37*, 601–615. [CrossRef]

55. Jones, B.R.; Sovacool, B.K.; Sidortsov, R.V. Making the Ethical and Philosophical Case for "Energy Justice". *Environ. Ethics* **2015**, *37*, 145–168. [CrossRef]

56. Oswald, Y.; Owen, A.; Steinberger, J.K. Large Inequality in International and Intranational Energy Footprints between Income Groups and across Consumption Categories. *Nat. Energy* **2020**, *5*, 231–239. [CrossRef]

57. Sovacool, B.K.; Burke, M.; Baker, L.; Kotikalapudi, C.K.; Wlokas, H. New Frontiers and Conceptual Frameworks for Energy Justice. *Energy Policy* **2017**, *105*, 677–691. [CrossRef]

58. CD-Links CD-Links. Available online: https://www.cd-links.org/ (accessed on 12 August 2021).

59. Schaeffer, R.; Bosetti, V.; Kriegler, E.; Riahi, K.; van Vuuren, D. Climatic Change: CD-Links Special Issue on National Low-Carbon Development Pathways. *Clim. Chang.* **2020**, *162*, 1779–1785. [CrossRef] [PubMed]

60. COMMIT COMMIT. Available online: https://themasites.pbl.nl/commit/ (accessed on 12 August 2021).

61. van den Berg, N.J.; Hof, A.F.; Akenji, L.; Edelenbosch, O.Y.; van Sluisveld, M.A.E.; Timmer, V.J.; van Vuuren, D.P. Improved Modelling of Lifestyle Changes in Integrated Assessment Models: Cross-Disciplinary Insights from Methodologies and Theories. *Energy Strategy Rev.* **2019**, *26*, 100420. [CrossRef]

62. van Sluisveld, M.A.E.; Martínez, S.H.; Daioglou, V.; van Vuuren, D.P. Exploring the Implications of Lifestyle Change in 2 °C Mitigation Scenarios Using the IMAGE Integrated Assessment Model. *Technol. Forecast. Soc. Chang.* **2016**, *102*, 309–319. [CrossRef]

63. Edelenbosch, O.Y.; McCollum, D.L.; Pettifor, H.; Wilson, C.; Vuuren, D.P. van Interactions between Social Learning and Technological Learning in Electric Vehicle Futures. *Environ. Res. Lett.* **2018**, *13*, 124004. [CrossRef]

64. Li, F.G.N. Actors Behaving Badly: Exploring the Modelling of Non-Optimal Behaviour in Energy Transitions. *Energy Strategy Rev.* **2017**, *15*, 57–71. [CrossRef]

65. Stehfest, E.; van Vuuren, D.P.; Bouwman, L.; Kram, T.; Alkemade, R.; Bakkens, M.; Biemans, H.; Bouwman, A.; den Elzen, M.G.J.; Janse, J.; et al. *IMAGE 3.0*; PBL: Piscataway, NJ, USA, 2014.

66. IIASA IAMC Documentation. Available online: https://data.ene.iiasa.ac.at/database/ (accessed on 12 February 2021).

67. Crespo del Granado, P.; Resch, G.; Holz, F.; Welisch, M.; Geipel, J.; Hartner, M.; Forthuber, S.; Sensfuss, F.; Olmos, L.; Brenath, C.; et al. Energy Transition Pathways to a Low-Carbon Europe in 2050: The Degree of Cooperation and the Level of Decentralization. *Econ. Energy Environ. Policy* **2020**, *9*. [CrossRef]

68. Fouquet, R. Path Dependence in Energy Systems and Economic Development. *Nat. Energy* **2016**, *1*, 16098. [CrossRef]

69. Unruh, G.C. Escaping Carbon Lock-In. *Energy Policy* **2002**, *30*, 317–325. [CrossRef]

70. Ang, B.W. Decomposition Analysis for Policymaking in Energy: Which Is the Preferred Method? *Energy Policy* **2004**, *32*, 1131–1139. [CrossRef]

71. Zhang, R.; Fujimori, S.; Hanaoka, T. The Contribution of Transport Policies to the Mitigation Potential and Cost of 2 °C and 1.5 °C Goals. *Environ. Res. Lett.* **2018**, *13*, 054008. [CrossRef]

72. *Department for Transport National Travel Survey, 2002–2019*, 14th ed.; Data Collection; UK Data Service: Colchester, UK, 2020.

73. Creutzig, F.; Fernandez, B.; Haberl, H.; Khosla, R.; Mulugetta, Y.; Seto, K.C. Beyond Technology: Demand-Side Solutions for Climate Change Mitigation. *Annu. Rev. Environ. Resour.* **2016**, *41*, 173–198. [CrossRef]

74. Li, W.; Long, R.; Chen, H.; Geng, J. A Review of Factors Influencing Consumer Intentions to Adopt Battery Electric Vehicles. *Renew. Sustain. Energy Rev.* **2017**, *78*, 318–328. [CrossRef]

75. Cullen, J.M.; Allwood, J.M.; Borgstein, E.H. Reducing Energy Demand: What Are the Practical Limits? *Environ. Sci. Technol.* **2011**, *45*, 1711–1718. [CrossRef]

76. Caird, S.; Roy, R.; Potter, S. Domestic Heat Pumps in the UK: User Behaviour, Satisfaction and Performance. *Energy Effic.* **2012**, *5*, 283–301. [CrossRef]

77. Judson, E.P.; Bell, S.; Bulkeley, H.; Powells, G.; Lyon, S. The Co-Construction of Energy Provision and Everyday Practice: Integrating Heat Pumps in Social Housing in England. *Sci. Technol. Stud.* **2015**, *28*, 26–53. [CrossRef]

78. Parrish, B.; Hielscher, S.; Foxon, T.J. Consumers or Users? The Impact of User Learning about Smart Hybrid Heat Pumps on Policy Trajectories for Heat Decarbonisation. *Energy Policy* **2021**, *148*, 112006. [CrossRef]

79. Wilson, C.; Guivarch, C.; Kriegler, E.; van Ruijven, B.; van Vuuren, D.P.; Krey, V.; Schwanitz, V.J.; Thompson, E.L. Evaluating Process-Based Integrated Assessment Models of Climate Change Mitigation. *Clim. Chang.* **2021**, *166*, 3. [CrossRef]

80. Shi, J.; Chen, W.; Yin, X. Modelling Building's Decarbonization with Application of China TIMES Model. *Appl. Energy* **2016**, *162*, 1303–1312. [CrossRef]

81. Urge-Vorsatz, D.; Petrichenko, K.; Staniec, M.; Eom, J. Energy Use in Buildings in a Long-Term Perspective. *Curr. Opin. Environ. Sustain.* **2013**, *5*, 141–151. [CrossRef]

82. Zhou, N.; Khanna, N.; Feng, W.; Ke, J.; Levine, M. Scenarios of Energy Efficiency and CO_2 Emissions Reduction Potential in the Buildings Sector in China to Year 2050. *Nat. Energy* **2018**, *3*, 978–984. [CrossRef]

83. Kriegler, E.; Bauer, N.; Popp, A.; Humpenöder, F.; Leimbach, M.; Strefler, J.; Baumstark, L.; Bodirsky, B.L.; Hilaire, J.; Klein, D.; et al. Fossil-Fueled Development (SSP5): An Energy and Resource Intensive Scenario for the 21st Century. *Glob. Environ. Chang.* **2017**, *42*, 297–315. [CrossRef]

84. Tummers, L. Public Policy and Behavior Change. *Public Adm. Rev.* **2019**, *79*, 925–930. [CrossRef]

85. Bemelmans-Videc, M.-L.; Rist, R.C.; Vedung, E. *Carrots, Sticks & Sermons: Policy Instruments and Their Evaluation*; Comparative Policy Analysis Series; Transaction Publishers: Piscataway, NJ, USA, 1998; ISBN 978-1-56000-338-0.

86. Peñasco, C.; Anadón, L.D.; Verdolini, E. Systematic Review of the Outcomes and Trade-Offs of Ten Types of Decarbonization Policy Instruments. *Nat. Clim. Chang.* **2021**, *11*, 257–265. [CrossRef]

87. O'Neill, B.C.; Kriegler, E.; Ebi, K.L.; Kemp-Benedict, E.; Riahi, K.; Rothman, D.S.; van Ruijven, B.J.; van Vuuren, D.P.; Birkmann, J.; Kok, K.; et al. The Roads Ahead: Narratives for Shared Socioeconomic Pathways Describing World Futures in the 21st Century. *Glob. Environ. Chang.* **2017**, *42*, 169–180. [CrossRef]

88. von Stechow, C.; McCollum, D.; Riahi, K.; Minx, J.C.; Kriegler, E.; van Vuuren, D.P.; Jewell, J.; Robledo-Abad, C.; Hertwich, E.; Tavoni, M.; et al. Integrating Global Climate Change Mitigation Goals with Other Sustainability Objectives: A Synthesis. *Annu. Rev. Environ. Resour.* **2015**, *40*, 363–394. [CrossRef]

89. Krey, V. Global Energy-Climate Scenarios and Models: A Review. *Wiley Interdiscip. Rev. Energy Environ.* **2014**, *3*, 363–383. [CrossRef]

90. Rogelj, J.; Popp, A.; Calvin, K.V.; Luderer, G.; Emmerling, J.; Gernaat, D.; Fujimori, S.; Strefler, J.; Hasegawa, T.; Marangoni, G.; et al. Scenarios towards Limiting Global Mean Temperature Increase below 1.5 °C. *Nat. Clim. Chang.* **2018**, *8*, 325–332. [CrossRef]

91. Weitzel, M.; Vandyck, T.; Keramidas, K.; Amann, M.; Capros, P.; den Elzen, M.; Frank, S.; Tchung-Ming, S.; Díaz Vázquez, A.; Saveyn, B. Model-Based Assessments for Long-Term Climate Strategies. *Nat. Clim. Chang.* **2019**, *9*, 345–347. [CrossRef]

92. Reference Card-IMAGE-IAMC-Documentation. Available online: https://www.iamcdocumentation.eu/index.php/Reference_card_-_IMAGE (accessed on 7 January 2022).

93. Walliser, A.; Rajkovich, N.B.; Forester, J.; Friesen, C.; Malbert, B.; Nolmark, H.; Williams, J.; Wheeler, S.M.; Segar, R.B.; Utzinger, M.; et al. Exploring the Challenges of Environmental Planning and Green Design: Cases from Europe and the USA Renovating to Passive Housing in the Swedish Million Programme. *Plan. Theory Pract.* **2012**, *13*, 113–174. [CrossRef]
94. Eurostat Eurostat Database. Available online: https://ec.europa.eu/eurostat/web/main/data/database (accessed on 5 July 2021).
95. Office for National Statistics. *2016 Based Household Projections in England*; ONS: London, UK, 2021.
96. Statistics Sweden. *The Future Population of Sweden 2021–2070*; Statistics Sweden: Stockholm, Sweden, 2021.
97. RVU Sweden. *Transport Analysis Swedish National Travel Survey 2011–16*; RVU Sweden: Lindingö, Sweden, 2020.
98. Odyssee-mure. Available online: https://www.indicators.odyssee-mure.eu/energy-efficiency-database.html (accessed on 29 January 2020).
99. Dunbabin, P.; Charlick, H.; Green, R. *Detailed Analysis from the Second Phase of the Energy Saving Trust's Heat Pump Field Trial*; Department of Energy and Climate Change: London, UK, 2013.
100. Element Energy. *UCL IEDE Analysis on Abating Direct Emissions from 'Hard-to-Decarbonise' Homes, with a View to Informing the UK's Long Term Targets*; A Study for the Committee on Climate Change; Element Energy: Cambridge, UK, 2019.
101. HETAS. Available online: Hetas.co.uk (accessed on 12 May 2021).
102. Vesterlund, M.; Sandberg, J.; Lindblom, B.; Dahl, J. Evaluation of Losses in District Heating System, a Case Study. In Proceedings of the International Conference on Efficiency, Cost, Optimization, Simulation and Environmental Impact of Energy Systems, Guilin, China, 16–19 July 2013.
103. Episcope-TABULA. Available online: https://episcope.eu/ (accessed on 28 January 2020).
104. Department for Communities and Local Government. *English Housing Survey, 2011: Housing Stock Data*, 4th ed.; UK Data Service: Colchester, UK, 2017.
105. Swedish Energy Agency. *Swedish Energy Agency Energy Statistics for Multi-Dwelling Buildings 2019*; Swedish Energy Agency: Eskilstuna, Sweden, 2020.
106. Swedish Energy Agency. *Swedish Energy Agency Energy Statistics for One- and Two Dwelling Buildings 2019*; Swedish Energy Agency: Eskilstuna, Sweden, 2020.
107. IEA. *IEA Lighting: Tracking Report*; IEA: Paris, France, 2020.
108. Shan, M.; Wang, P.; Li, J.; Yue, G.; Yang, X. Energy and Environment in Chinese Rural Buildings: Situations, Challenges, and Intervention Strategies. *Build. Environ.* **2015**, *91*, 271–282. [CrossRef]
109. van Ruijven, B.J.; van Vuuren, D.P.; de Vries, B.J.M.; Isaac, M.; van der Sluijs, J.P.; Lucas, P.L.; Balachandra, P. Model Projections for Household Energy Use in India. *Energy Policy* **2011**, *39*, 7747–7761. [CrossRef]
110. ITF. *ITF Transport Statistics*; ITF: London, UK, 2020.
111. Ma, J.; Ye, X. Modeling Household Vehicle Ownership in Emerging Economies. *J. Indian Inst. Sci.* **2019**, *99*, 647–671. [CrossRef]
112. Oxfam. *Stockholm Environment Institute the Carbon Inequality Era: An Assessment of the Global Distribution of Consumption Emissions among Individuals from 1990 to 2015 and Beyond*; Oxfam: Stockholm, Sweden, 2020.
113. United Nations Environment Programme. Emissions Gap Report 2020. 2020. Available online: https://www.unep.org/emissions-gap-report-2020 (accessed on 11 February 2022).

Article

Zero-Emission Pathway for the Global Chemical and Petrochemical Sector

Deger Saygin and Dolf Gielen *

International Renewable Energy Agency (IRENA), Innovation and Technology Centre (IITC), 53113 Bonn, Germany; deger.saygin@shura.org.tr
* Correspondence: info@irena.org

Abstract: The chemical and petrochemical sector relies on fossil fuels and feedstocks, and is a major source of carbon dioxide (CO_2) emissions. The techno-economic potential of 20 decarbonisation options is assessed. While previous analyses focus on the production processes, this analysis covers the full product life cycle CO_2 emissions. The analysis elaborates the carbon accounting complexity that results from the non-energy use of fossil fuels, and highlights the importance of strategies that consider the carbon stored in synthetic organic products—an aspect that warrants more attention in long-term energy scenarios and strategies. Average mitigation costs in the sector would amount to 64 United States dollars (USD) per tonne of CO_2 for full decarbonisation in 2050. The rapidly declining renewables cost is one main cause for this low-cost estimate. Renewable energy supply solutions, in combination with electrification, account for 40% of total emissions reductions. Annual biomass use grows to 1.3 gigatonnes; green hydrogen electrolyser capacity grows to 2435 gigawatts and recycling rates increase six-fold, while product demand is reduced by a third, compared to the reference case. CO_2 capture, storage and use equals 30% of the total decarbonisation effort (1.49 gigatonnes per year), where about one-third of the captured CO_2 is of biogenic origin. Circular economy concepts, including recycling, account for 16%, while energy efficiency accounts for 12% of the decarbonisation needed. Achieving full decarbonisation in this sector will increase energy and feedstock costs by more than 35%. The analysis shows the importance of renewables-based solutions, accounting for more than half of the total emissions reduction potential, which was higher than previous estimates.

Keywords: chemical and petrochemical sector; decarbonisation; renewable energy; circular economy; electrification; material flow analysis

Citation: Saygin, D.; Gielen, D. Zero-Emission Pathway for the Global Chemical and Petrochemical Sector. *Energies* **2021**, *14*, 3772. https://doi.org/10.3390/en14133772

Academic Editor: Victor Manuel Ferreira Moutinho

Received: 28 April 2021
Accepted: 16 June 2021
Published: 23 June 2021

1. Introduction

The chemical and petrochemical sector is of vital economic importance. Global production amounted to 5.7 trillion United States dollars (USD) in 2017, including pharmaceuticals. Production is projected to quadruple by 2060 [1]. The sector's reliance on fossil fuels and fossil feedstocks results in the emissions of carbon dioxide (CO_2) during the production, use and end-of-life phases. As a result, the chemical and petrochemical sector is a major contributor to global industrial CO_2 emissions, ranking third behind iron and steel-making and cement production. Total direct emissions from production, product use and waste handling amounted to 1.6 gigatonnes (Gt) of CO_2 per year, while indirect emissions related to electricity supply accounted for 0.6 Gt of CO_2 per year. Production of chemicals results in around 1.1 Gt of energy and processing CO_2 emissions annually, accounting for about half of the full life cycle carbon footprint (estimated based on Ref. [2]). Emissions from the use of around 178 million tonnes (Mt) of urea fertiliser and decomposition/incineration of around 60 Mt of plastics per year result in an additional 0.3 Gt of CO_2 per year [3]. Another 0.2 Gt of CO_2 emissions arise from the use of solvents and surfactants.

The sector produces plastics, fibers, solvents, inorganic chemicals and hundreds of other types of products. Plastics production grew 20-fold over the past five decades to reach

360 Mt by the end of 2018 [4]. In addition, 115 Mt of other synthetic organic materials were produced. However, three-quarters of the total energy and non-energy use is accounted for by the manufacturing of certain products, such as: olefins (ethylene, propylene, butadiene) aromatics, ammonia, methanol and carbon black (see Figure 1). Plastics and fibers account for most of the product mix in volume terms, at around 400 Mt per year in 2018. Polyolefins (made from ethylene, propylene and butadiene) account for nearly half of all plastics production. Various polyethylene (PE) grades, polypropylene (PP) and polyamide (PA) account for 30%, 17% and 15% of all plastics production worldwide, respectively. Polyvinyl chloride (PVC) and polyethylene terephthalate (PET) combined account for another 19% of the total [5].

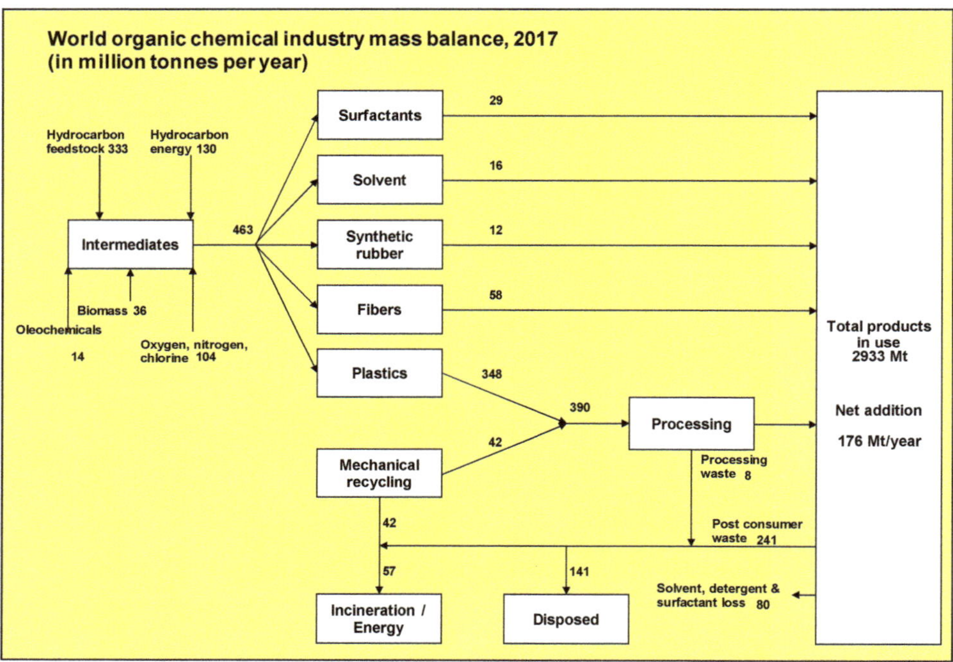

Figure 1. Estimated world petrochemicals production, processing and recycling balance, 2017. Source: updated to the year 2017 based on Ref [6].

Around 175 Mt of ammonia was produced in 2020 and was mainly used as nitrogen fertiliser. Annual methanol production amounted to more than 98 Mt in 2019. Methanol is used in the production of formaldehyde, acetic acid, di-methyl terephthalate, olefins and solvents. While ammonia and methanol are largely produced from coal in China, gas-based production dominates elsewhere. Figure 1 provides an overview of material flows in global petrochemicals production in 2017.

The chemical and petrochemical sector is the largest energy user in the manufacturing industry, with a total consumption of 46.8 exajoules (EJ) in 2017 (including non-energy use, NEU) [2]. Oil and gas dominated the sector's total consumption, with around 10% of global natural gas supply and 12% of all oil consumed by this sector. The chemical and petrochemical sector is unique, as significant amounts of fossil fuels are used as raw material (i.e., feedstock or NEU) [7]. This NEU reflects the energy content of the products that are sold. For products such as ammonia, methanol and plastics, the NEU exceeds the process energy use in their production [8]. This has profound consequences for strategies to abate emissions in the life cycle of the sector's products, which will be elaborated on below. The added carbon accounting complexity that results from NEU and carbon storage in materials and products is an aspect that warrants more attention in long-term energy scenarios [9]. As a result of this complexity, emission reductions in this sector constitute one of the main challenges for realising the Paris Agreement goals [10]. Moreover, as a large user of oil products, the sector's continued reliance on fossil fuels results in emissions outside of its boundaries in the petroleum sector [11,12]. This paper provides an assessment of 20 options that can be categorized on five main strategies to put the sector's life cycle CO_2 emissions on a pathway to net-zero by the mid-21st century. The analysis investigates each option's contribution to put the sector on a net-zero pathway and the respective CO_2 mitigation cost. The Supplementary Materials (Section A) provides a detailed overview of the status of low-carbon technologies worldwide.

We address two research questions in this paper:

- How can zero emissions be achieved, considering the full product life cycle?
- What is the potential contribution of renewables-based solutions?

This paper combines specific technology assessments to provide sector and life cycle level insights at the global level, with relevance in terms of future energy demand, location choices, plant siting and investment needs. The analysis covers direct emissions from production, as well as materials use and waste handling. The analysis accounts for emissions and carbon storage in products and their subsequent treatment in the waste management phase.

Section 2 provides a review of existing decarbonisation studies for the chemical and petrochemical sectors. Section 3 provides the details of the methodology and technology data. Section 4 presents results followed by a discussion of the opportunities and challenges of decarbonisation in Section 5, and the conclusion is Section 6.

2. Review of Literature on Decarbonisation of the Chemical and Petrochemical Sector

So far, the sector has made limited progress in reducing absolute CO_2 emissions levels at a global level, as demand growth has exceeded efficiency gains. Technical energy efficiency potentials have been exhausted. Multiple conversion processes are usually integrated in large, ageing industrial complexes that result in high energy efficiency on site, but that also limits achieving additional energy savings by switching to the best available technologies [8]. Around half of the sector's heat demand relates to high-temperature processes, which complicate renewable energy deployment [6]. Petrochemical production is increasingly integrated into refinery operations, with modern refinery designs yielding 50% petrochemicals in the product mix. Such plant design locks in fossil energy use going forward. The integration also complicates energy accounting for petrochemical products.

Plastics and other synthetic organic materials are currently produced from fossil fuel feedstocks. These can be replaced with biomass or synthetic feedstocks produced from CO_2 and renewables-based hydrogen. Bioplastics constitute less than 1% of current plastics production. The high cost of low-carbon alternatives acts as a major barrier [6,13–15] 35% of the emissions reduction potential lies with materials systems optimisation, while the remaining 65% is related to energy use in the materials production processes [16]. However, the circular economy is not well developed in this sector, despite decades of efforts. A majority share of post-consumer plastic and textiles is incinerated or dumped in landfills [17]. Low recycling rates and low energy recovery rates add to energy use and CO_2 emissions [18]; the future reconciliation of product demand growth and sustainability is therefore challenging.

Several studies have assessed the future CO_2 emissions reduction potentials in the sector. However, the conclusions regarding emissions reductions are not in line with the recent net-zero emissions objectives formulated by major economies [19]. For instance, a study for the Dutch chemical and petrochemical industry, which is representative of the global chemical and petrochemical sector, concluded that a 90% reduction in national sectorial emissions is feasible [20]. This would require 63 billion euros of investments (USD 75 billion), split into 26 billion for new chemical plants and 37 billion for energy supply.

Energy and feedstock supply cost would rise by 50%. The average emission mitigation cost would amount to 140 euros per tonne of CO_2 (USD 170/t CO_2). Annually, the industry would require 280 petajoules (PJ) of biomass and 11.4 gigawatts (GW) of offshore wind capacity. Biomass feedstock accounts for more than one-third, while renewable energy and CCS each account for one-sixth of the effort, and the remainder is accounted for by energy efficiency, closure of materials chains and nitrous oxide emission reductions.

Deep emissions reductions in Europe are technically possible through power supply decarbonisation and CCS integration with chemical processes in the 2030–2050 timeframe [21]. A range of current and future technologies can sustain Europe's track record of energy and emissions intensity improvements: final energy demand can be maintained at a constant level, and emissions could be virtually eliminated with energy efficiency (33% of the total emissions reductions), CO_2 capture and storage (CCS) (25%), renewable electricity (20%) and fuel switching and measures to reduce nitrous oxide emissions (22%). To enable continuous and competitive production, access to large amounts of affordable and reliable energy and feedstock will be necessary, which can be challenging for renewables [20]. Infrastructure will be crucial, including transmission grids for renewable power, pipelines for hydrogen, CO_2 and heat, and waste logistics and recycling.

According to the International Energy Agency's (IEA) Reference Technology Scenario (RTS), global sectorial CO_2 emissions would grow by around 40% globally from the current level, in line with a plastics demand growth of 600 Mt/yr [22]. The Clean Technology Scenario (CTS) estimates direct annual CO_2 emissions of 0.8 Gt by 2050, equivalent to a 60% reduction compared to the RTS. The IEA's analysis focuses on the reduction of emissions from direct energy use and processes that only cover two-thirds of the sector's total life cycle emissions. Energy efficiency therefore plays a key role in the IEA's analysis, contributing 25% to the mitigation effort. The role of alternative feedstocks and plastics recycling is limited to 15%. The IEA analysis suggests a continued importance of fossil fuel use in the sector, which is inconsistent with net zero by 2050.

According to Ref. [23], electrification of processes and new catalytic conversion routes can be listed as key options. Biomass and recycling are key strategies to reduce fossil feedstock use, while CO_2-based fuels and chemicals are unlikely to be significant contributors to global abatement in the next two decades. For energy supply, clean hydrogen, heat pumps and waste energy use, as well as energy management systems, are low-carbon options for decarbonisation.

Historical chemical and petrochemical sector energy efficiency trends have been assessed widely [24–26], but only a few studies have estimated the future efficiency potential [27–29]. More studies have focused on the assessment of renewable fuels and feedstocks and electrification potentials [30,31]. The sector's long-term decarbonisation potential is typically assessed as part of all energy-intensive industry sectors [32–35]. While such a broad perspective is useful from a general industrial policy perspective, gaining insights into the potential, investment needs and challenges of these options in isolation from other sectors is crucial to design sector-specific decarbonisation policies. According to Ref. [36] the industry focuses mainly on supply side mitigation options. Downstream options like material efficiency have received less attention due to the limited availability of material flows and supply chain data, as well as the insufficient understanding of potentials. (The industry often argues that its products reduce life cycle emissions compared to other materials for a range of specific products. A full life cycle analysis for the whole sector would require an assessment of the use stage of buildings, cars, and other type of complex products where plastics and other materials are deployed, which is beyond the scope of this paper).

According to Ref. [37] the lack of manufacturing experience, cost evaluations and proofs of concept of most mitigation measures on a large industrial scale. This is particularly the case for the hydrogen- and CO_2-based routes, but also for emerging biomass routes [38]. While technologies for all proposed production pathways are in principle available and demonstration plants are in operation, more efforts are needed to deploy these technologies on an industrial scale.

3. Materials and Methods: Prioritisation of Technology Options for Decarbonisation

A net-zero pathway has been developed for the global chemical and petrochemical sector to 2050, based on a detailed bottom-up technology approach. The full product life cycle emissions are covered in the analysis. Technology-specific mitigation costs have been collected to assess transformation impacts on the sector's total energy and feedstock cost. The results presented in the paper are part of IRENA's global energy system optimisation model. Thus, critical issues for the sector, such as competition for scarce biomass resources and the availability of renewable power for chemicals production, have been considered in technology choices.

In the case of the chemical and petrochemical sector, a large share of the energy inputs is used as feedstock, and around two-thirds of all carbon input is stored in chemicals. Moreover, as earlier analyses have shown, the sector's energy statistics include large uncertainties which require bottom-up methodologies that combine the production and energy use data of individual chemicals [26,29,39]. In this study, energy balances are thus combined with materials flow analysis and materials system optimisation, which includes various stages of the product life cycle. According to Ref. [36], there is a need to enhance the understanding of downstream mitigation options and their techno-economic potential for the proper modeling of impacts from varying efficiencies in material service provision. They also state that it would be important to include the relevant aspects of the MATerials Technologies for greenhouse gas Emission Reduction (MATTER) project, conducted in the late 1990s, which may have represented the peak of ambition with regards to integrated energy- and materials-related climate change mitigation research and other similar models in integrated assessment model frameworks [16].

The tracking of carbon flows from production to the waste management stage in this study helps to better understand the circular economy potential and its role in net-zero strategy development. Such bottom-up modeling can inform integrated assessment models in the representation of complex solutions, such as circular economy concepts.

The analysis covers the 2017–2050 period, and is based on a techno-economic assessment of technologies for decarbonising the global chemical and petrochemical sector, with a special focus on five particular strategies. Each strategy includes several technological options. The energy and emissions impact of each technology has been assessed to 2050, by gauging its potential under the 1.5 °C case compared to the Planned Energy Scenario (PES) [40]. Global results are estimated based on a bottom-up assessment of the energy use and emissions in China, India, Japan, 27 countries of the European Union (EU-27), the US, the remainder of the Group of 20 (G20) countries and the rest of the world. The Supplementary Materials (see Section B) provides further details regarding the scenario definitions and the additional data and assumptions used for the analysis.

As a first step, the production volumes of the major chemical production processes (i.e., high value chemicals, ammonia, methanol and carbon black), their respective specific energy consumption (for fuel and feedstock) values and the production process fuel mix were collected for the base year 2017. The combination of production volume, fuel mix and the specific energy consumption yields the total energy and non-energy use from the production of these chemicals for the base year 2017. These major chemical production processes account for more than 60% of the sector's total global energy and non-energy uses and related CO_2 emissions (Tables 1 and 2). The energy use related to the production of all other chemicals has been estimated with a country/region-specific coefficient. This share of energy use is attributed to the downstream processing of the chemical building blocks of plastics, fibres, solvents and hundreds of other types of products. Projections reflect the growth of this energy use in proportion with the rest of the sector. (The coefficient includes corrections for energy accounting in the process energy and non-energy use categories in the IEA energy balances, based on our bottom-up assessment of the non-energy use. The coefficient is estimated as a ratio of the bottom-up estimate of the process energy use based on the selected chemicals and the reported process energy use according to the IEA energy balances. While our bottom-up estimate covers 86% of the total NEU reported in the IEA energy balances, we assume that non-energy use is 100% covered by the production of the chemicals selected for this analysis. The 14% of the total reported global non-energy use according to the IEA energy balances is equivalent to 3.6 EJ in absolute terms [2] and a share of this is assumed to be consumed as process energy. Similar statistical accounting issues have been reported previously [8,39]).

Table 1. Global energy and non-energy use for petrochemical production according to the energy statistics, 2017.

[EJ/yr]	Energy	Non-Energy	Total	Total in This Analysis
Coal	4.5	0.1	4.7	4.7
Natural gas	5.7	7.7	13.5	14.0
Oil	2.6	18.9	21.6	18.6
Biomass and waste	0.1	1.0	1.1	1.1
Heat	2.4	-	2.4	2.4
Electricity	4.6	-	4.6	4.6
Total	20.0	27.8	47.8	45.4

Source: Ref. [2] and own analysis. Note: biomass for NEU has been included based on bottom-up information.

Table 2. Estimated global energy and non-energy use per type of product, 2017.

	[EJ/yr]
Ammonia	6.2
Methanol	2.7
High vale chemicals	21.2
Carbon black	1.0
Total	31.1

Source: own analysis.

To assess the total energy and carbon flows in waste management, additional data for the total volume of plastics production, demand and plastic waste generation have been collected for 2017. In a subsequent step, the energy demand in the PES in the year 2050 has been estimated by considering the growth in production of chemicals and plastics (see Table 3). Projections for future plastics demand growth range from 1% to 3% per year [6,41–43]. The higher end of this range was used for the PES, with lower demand in the 1.5 °C case due to greater circular economy efforts. In the PES 2050, the production fuel mix and the shares of waste management options are the same as in 2017 for each country/region, whereas the production growth varies depending on the regional dynamics. In the PES, the growth in energy demand is to some extent offset by improvements in energy efficiency. It is assumed that the specific energy consumption (excluding feedstock/NEU) of all chemicals would reach the level of current best practices, which results in a savings potential of 15% by 2050, compared to 2017 [8].

The net-zero pathway (1.5 °C case) takes five major strategies into account:

- improve energy efficiency in the production process by adopting best practices and breakthroughs, including substituting fossil fuels with direct renewable energy resources, electrification and other renewables for process heat generation (A)
- a switch to biomass and synthetic feedstocks based on renewable "green" hydrogen and CO_2 (B)

- a shift to circular economy to reduce primary materials demand by increasing reuse and recycling of plastics and by reducing per capita plastics and chemicals demand through changing consumer behavior and substitution with other materials (C)
- decarbonising production processes and waste handling by CCS (D)
- shifting power supply to carbon-free electricity, notably renewables (E)

Table 4 shows the technologies assessed in each pathway and the global cost of CO_2 mitigation (per tonne) for each decarbonisation technology. For each strategy, the 2050 country/region implementation potential in the 1.5 °C case relative to the PES has been estimated (see Supplementary Materials, Section B). In a subsequent step, the impacts of decarbonisation on the total energy and feedstock demand and CO_2 emissions have been estimated. Finally, a carbon flow analysis was conducted to assess the impact of the uptake of these technology options on the global plastics metabolism and to gain insight into the carbon storage in materials and products through the non-energy use emission accounting tables (NEAT) model for the calculation of carbon storage in petrochemical products [44]. NEAT calculates both CO_2 emissions and carbon storage resulting from the non-energy use of fossil fuels, independent from the energy statistics and the national GHG inventory, and complements energy statistics with material flow analysis [39]. Supplementary Materials (Section C) provides the details of the carbon flow analysis methodology.

Table 3. Technology options covered in the assessment.

Technology Option	Application	Rationale/Explanation and Key References Used to Estimate the Fossil Fuel Substitution Potentials	Cost	Unit	References for Costs
(A) Energy efficiency, renewable energy and process heat electrification					
Best practice technologies (1)	Improving energy efficiency to reduce process heat demand	Global energy saving potential of best practice technologies that are currently available in the market [8] would result in a continuation of the current average energy efficiency trends of 0.5%/yr if they are implemented in all production processes by 2050 [45]. [1] The rate of improvement is average over the period to 2050, and does not necessarily follow a linear path.	20–60	USD/t CO_2 in 2030	[46]
Breakthroughs and heat integration (1)		New technology options and cross-cutting technologies such as advanced membranes to reduce process heat demand by 2050 [47] would double the improvements to 1%/yr. [2] While pinch analysis for heat integration shows 50% and 30% savings for hot and cold utilities, respectively [48], actual potential could still be lower, since efficiency is typically assessed at site level where a high level of steam system integration reduces potential.	Up to 200	USD/t CO_2 in 2050	[47]
Solar process heat (2)	Fuel switching	Solar process heating systems can replace fossil fuels for process heat generation [49,50]. [3]	0–100	USD/t CO_2 in 2030	[49]
Biomass for process heat (4)		Biofuels produced from various biomass feedstocks can replace fossil fuels for process heat generation by 2030/50 [6]. [4]	0–75	USD/t CO_2 in 2030	[6]
Electrification of process heating combined with renewables (5)		Synthetic naphtha produced from renewable hydrogen can replace crude oil-based naphtha for HVC production [51]. [14]	−60–450	USD/t CO_2 in 2050	[52]
		Heat pumps can replace fossil fuels to supply low-temperature process heat [49,50,53,54]. [5]	0–50	USD/t CO_2 in 2030	[49]
(B) Switching from fossil fuel-based feedstocks to biomass and synthetic feedstocks					
Biomass for plastics (9)	Feedstock switching	Biomass can replace fossil fuels used as feedstock for plastics production [6]. [9]	0–500	USD/t CO_2 in 2009	[6]
Biomass for ammonia (10,19)	Feedstock switching	Biomass can replace fossil fuels used as feedstock for ammonia production [55]. [10]	250–400	USD/t CO_2	[22,56,57]
Biomass for methanol (10,19)	Feedstock switching	Biomass can replace fossil fuels used as feedstock for methanol production, either through gasification to methanol or by using biomethane in the traditional production route [58]. [11]	−150–450	USD/t CO_2	[37,56,57,59,60]
Renewable-hydrogen for ammonia (11,20)	Feedstock switching	Renewable hydrogen can replace fossil fuels used as feedstock for ammonia production [22]. [12]	0–150	USD/t CO_2	[37,60,61]
Renewable-hydrogen for methanol (11,20)	Feedstock switching	Renewable hydrogen can replace fossil fuels used as feedstock for methanol production [58]. [13]	−50–200	USD/t CO_2	[37,58,60]
Methanol for olefins (13)	Feedstock switching	Renewable hydrogen-based methanol can be used for olefins production, thereby reducing the need of fossil fuels feedstocks [58]. [14]	50–300	USD/t CO_2	[37]

Table 3. *Cont.*

Technology Option	Application	Rationale/Explanation and Key References Used to Estimate the Fossil Fuel Substitution Potentials	Cost	Unit	References for Costs
(B) Switching from fossil fuel-based feedstocks to biomass and synthetic feedstocks					
Synthetic fuels (naphtha) (12)	Feedstock switching	Synthetic naphtha produced from renewable hydrogen can replace crude oil-based naphtha for HVC production [51]. [15]	−60–450	USD/t CO_2 in 2050	[52]
CO_2 (14)	Feedstock switching	Electrocatalytic CO_2 production can replace fossil fuels used as feedstock in ethylene production. [16]	−30–80	USD/t CO_2 in 2050	[15]
(C) Circular economy concepts					
Demand reduction/Reuse (18)	Demand reduction	Plastics demand is reduced from high end of plastics production projections (3%/yr) to the average of the range found in literature (2%/yr). Reuse of plastics has been assessed as part of this demand reduction strategy.	N/A	N/A	-
Mechanical recycling (6)	End of life	Global mechanical recycling rate is assumed to grow around two-fold by 2030 [42] and triple by 2050. [6]	−140–200	USD/t CO_2 in 2015	[62–65]
Chemical recycling (7)	End of life	Chemical recycling rate is assumed to be commercialized and reach the level of mechanical recycling by 2050 [42,66]. [7]	80–500	USD/t CO_2 in 2015	[20,65–67] and industry sources
Incineration with highly efficient energy recovery (8)	End of life	All remaining post-consumer plastic waste is assumed to be incinerated with high efficiency combined with CCS [43]. [8]	−200–−50	USD/t CO_2 in 2020	Own estimate
(D) CCS					
Capture and storage (15)	Process emissions	All high-purity process CO_2 emissions can be captured by 2050. Three-quarters of all emissions from fuel combustion are assumed to be captured. It is assumed that a shift to energy recovery is meaningful from a climate perspective if only coupled with CCS. Biomass use is primarily for cogeneration of heat and power and all processes are assumed to be coupled with CCS. [17]	0–50	USD/t CO_2 in 2040	[20,60,68,69]
Capture and storage (16)	Emissions fossil fuel combustion from energy recovery		50–150	USD/t CO_2	
Capture and storage with biomass (3)	Emissions from biomass-based heat generation		150–200	USD/t CO_2	[70,71]
(E) Carbon-free electricity supply (17)					

Note: 20 technologies included in the analysis are numbered according to the order of introduction in the assessment of their emissions reductions impact. [1] Net additional cost of improving energy efficiency with best practice technologies for high value chemicals (HVC), ammonia, methanol and chlorine is estimated as USD 6 billion per year in 2030 worldwide. Avoided global CO_2 emissions are estimated to be between 100 and 300 Mt per year by 2030. The resulting CO_2 mitigation cost is estimated to range from USD 20 to USD 60 per tonne CO_2. [2] According to the BLUE scenarios of the IEA (where global energy-related CO_2 emissions are halved by 2050 compared to 2007), breakthrough energy efficiency technologies would require a carbon price of up to USD 200/t CO_2 for their deployment by 2050. [3] Cost of generating low-temperature process heat is estimated as USD 10–30 per gigajoule (GJ) from solar thermal systems by 2030. Fossil fuel-based process heat generation costs are USD 10–20/GJ. The generation cost difference results in CO_2 mitigation cost of USD 0–125/t CO_2, assuming around 100 kg CO_2 can be avoided per GJ of process heat generated. [4] Costs are estimated based on delivering one GJ of low- and medium-temperature process heat from agricultural and forest residues and dedicated energy crops compared to a mix of fossil fuels in eight world regions from various boiler and combined heat and power generation technologies. The referenced study provides a range for mitigation costs from as low as USD −150 to as high as USD 150 per tonne CO_2. This range has been narrowed for this study. [5] Cost of generating low-temperature process heat (<150 °C) is estimated as USD 10–25/GJ for heat pumps by 2030. Fossil fuel-based generation costs USD 10–20/GJ. The generation cost difference results in CO_2 mitigation cost of USD 0–50/t CO_2, assuming around 100 kg CO_2 can be avoided per GJ of process heat generated. [6] In the EU (e.g., Belgium, France, Germany, the Netherlands and Portugal), the costs for plastic collection, sorting and transport range between euro 200 and euro 1450 per tonne plastic waste (in some countries the green dot fee is used as a proxy for recycling costs). Treatment costs of recycled plastics in the

Netherlands are around euro 250–300/t of separated plastic. Hence the final production cost of recycled plastic could range from euro 500 to euro 2000 per tonne. Price of virgin plastics range from euro 750 to euro 1500 per tonne. [7] According to industry sources, plastics produced from chemical recycling have a green premium of between 10% and 30%, which adds from euro 75 to euro 450 per tonne over virgin plastics. Around 1 tonne of CO_2 is avoided per tonne of chemical recycling resulting in CO_2 mitigation cost of USD 75–450/t CO_2 in today's conditions. [8] Costs represent post-consumer plastic waste incineration in combined heat and power plant (total efficiency of 90% operating at 85% capacity factor) with USD 200–400/t waste price. Costs exclude CCS costs of capturing CO_2 from power plants at between USD 50–150/t CO_2. [9] Costs are estimated for drop-in and new bio-based chemicals and plastics based on their market value and production costs (assuming a sugar price of USD 400/t). [10] Costs are estimated for biomass gasification with a fuel price of USD 8/GJ compared to steam reforming of methane route without CCS. Additional cost of biomass-based ammonia production is USD 600–1000, which avoids a total of 2.3 tonnes CO_2 emissions per tonne of ammonia. [11] Costs are estimated for various biomass feedstocks such as wood and residues (ranging from euro 160 to 940 per tonne of methanol) compared to production costs of methanol in Europe from natural gas at USD 100–400/t methanol. Biomass-based methanol avoids around 1.7 tonnes of CO_2 per tonne of methanol. [12] Costs of renewable-hydrogen based ammonia production (between USD 460 and 800 per tonne of ammonia) are estimated compared to steam reforming of methane route in Europe (USD 450 per ton) for electricity prices of USD 28–63 per megawatt-hour (MWh) at full load hours of 4000–4500 h per year. Avoided CO_2 emissions per tonne of ammonia are assumed as 2.3 tonnes per tonne of ammonia. [13] Costs of renewable-hydrogen-based methanol production (for electricity prices of USD 12–58/MWh at full load hours of 3000–7000 h per year, between USD 350 and 1000 per tonne of methanol today that can go down to USD 70–630 per tonne by 2050) are estimated compared to steam reforming of methane route in Europe (USD 100–400 per tonne). Avoided CO_2 emissions of 1.7 tonnes per tonne of methanol are assumed. [14] Costs of producing olefins from renewable-hydrogen route could be up to by a factor of two times more expensive than the conventional naphtha-based route of producing ethylene and propylene. At least 1.89 tonnes of CO_2 can be avoided per tonne of high value chemical. [15] Costs are estimated based on the production of synthetic methane and liquids from renewable electricity in Iceland, North and Baltic seas and North Africa and the Middle East in 2050 that range from between euro 6.5 cents and euro 19 cents per kilowatt-hour (kWh) of final product. The reference price of liquid fuel is taken as euro 8 cents per kWh. [16] Costs are estimated for electrocatalysis route at energy conversion efficiency of 70%, Faradaic efficiency of 90%, grid emission intensity of 350 g CO_2/kWh and an electricity price of USD 40/MWh. Electrocatalysis route has an estimated ethylene production cost of USD 1100/t. The fossil fuel-based route cost ranges from USD 600 to 1300 per tonne. [17] Costs include transportation costs and electricity demand for CO_2 compression 100 kWh/t captured CO_2. Additional bioenergy demand of 2 GJ per tonne captured CO_2 is assumed for heat requirements of the solvent reboiler.

4. Transformation Scenario for the Chemical and Petrochemical Sector

In this section, we discuss the commodity and technology characteristics (Section 4.1), changes in the energy use (Section 4.2), CO$_2$ emissions and carbon flows (Section 4.3) and cost implications (Section 4.4) of decarbonisation, as well as its implications for the global energy system (Section 4.5).

4.1. Commodity and Technology Characteristics

Global plastics demand is projected to grow 2.5-fold in the PES. This growth of 3% per year is the high end of the literature projections (1000 Mt/yr by 2050) [42]. In the 1.5 °C case, demand reduction strategies reduce plastics demand by one third to 650 Mt/yr in 2050, or around 2% growth per year [72]. In the 1.5 °C case, ammonia and methanol production grow significantly as new market segments emerge for chemical building blocks, shipping fuels and power generation [22,73,74]. (In comparison, PES assumes a 2.5- and 2-fold growth in ammonia and methanol demand, respectively.) Figure 2 illustrates the changing material flows. Green hydrogen is treated as fuel and feedstock. Renewable electricity needed for hydrogen production is shown separately.

4.2. Energy Use

In the PES, total demand for plastics increases from 385 Mt in 2017 to 986 Mt by 2050. Sectorial demand for process heat and electricity more than doubles between 2017 and 2050, from 20.9 EJ to 44.5 EJ per year (see Figure 3). The PES includes autonomous energy efficiency improvements of 0.5%/yr, which result in 15% energy savings by 2050 (7.8 EJ/yr). The growing demand for plastics and other synthetic organic materials more than doubles NEU to 62.4 EJ in 2050. The process energy and NEU mix remains the same throughout the entire period; oil products represent more than 40% of the sector's total consumption, while gas represents about one-third. Electricity's share in the total process energy use is 20% (see Figure 4).

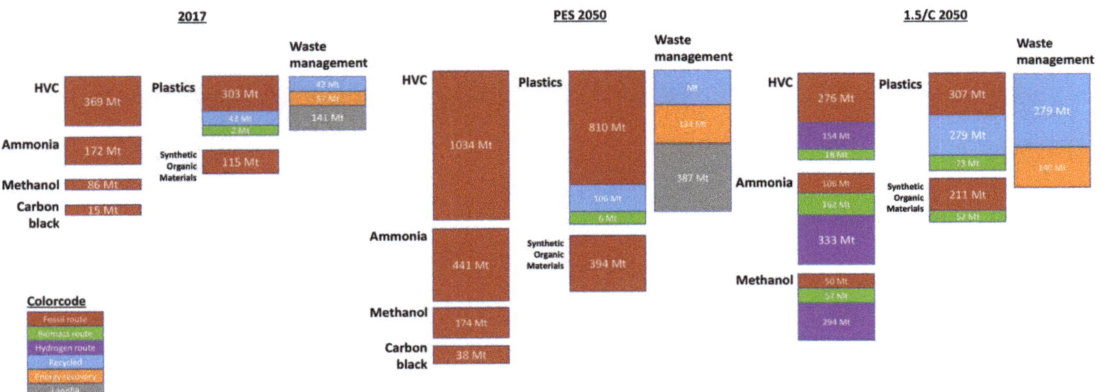

Figure 2. Estimated production volumes of the key chemicals, 2017–2050.

Table 4. Global production volumes and the specific fuel and feedstock use of the assessed chemicals in final energy terms, 2017 and 2050.

	2017			2050 PES			2050 1.5 °C		
	Production	Process Energy	Feedstock Use	Production	Process Energy [2]	Feedstock Use	Production	Process Energy	Feedstock Use
	(Mt/year)	(GJ/t)	(GJ/t)	(Mt/year)	(GJ/t)	(GJ/t)	(Mt/year)	(GJ/t)	(GJ/t)
Conventional routes [1]									
Ethylene (steam cracking)	135	16.3	45.0	379	13.9	45.0	80	12.6	45.0
Propylene (steam cracking)	50	16.3	45.0	135	13.9	45.0	27	12.6	45.0
Propylene (fluid catalytic cracking)	21	3.2	45.0	58	2.7	45.0	22	2.5	45.0
Benzene (steam cracking)	17	16.3	-	48	13.9	-	15	12.6	-
Benzene (naphtha extraction)	44	3.2	40.1	120	2.7	40.1	51	2.5	40.1
Toluene	26	3.2	20.3	75	2.7	20.3	36	2.5	20.3
Xylene	47	3.2	41.0	140	2.7	41.0	68	2.5	41.0
Butadiene (steam cracking)	16	16.3	-	44	13.9	-	14	12.6	-
Butadiene (C4 separation)	16	7.3	44.6	44	6.2	44.6	21	5.6	44.6
Butylene	30	3.2	45.0	84	2.7	45.0	41	2.5	45.0
Carbon black	15	9.0	32.8	38	7.7	32.8	-	-	-
Ammonia	172	15.0	20.7	440	12.8	20.7	106	11.6	20.7
Methanol	86	10.0	20.0	174	8.5	20.0	50	7.8	20.0
Alternative routes									
Plastics from biomass (excluding bio-ethylene) [3]	-	-	-	-	-	-	73	28.3	45.0
Synthetic organic materials from biomass [3]	-	-	-	-	-	-	52	23.3	45.0
Ethylene from biomass [4]	-	-	-	-	-	-	7	61.0	45.0
Ethylene from green hydrogen and captured CO_2 [6]	-	-	-	-	-	-	14	10.6	45.0
Methanol to olefins [5]	-	-	-	-	-	-	62	5.0	-
Steam cracking with synthetic naphtha (ethylene + propylene) [7]	-	-	-	-	-	-	65	42.2	45.0

Table 4. *Cont.*

	2017			2050 PES			2050 1.5 °C		
	Production	Process Energy	Feedstock Use	Production	Process Energy [2]	Feedstock Use	Production	Process Energy	Feedstock Use
	(Mt/year)	(GJ/t)	(GJ/t)	(Mt/year)	(GJ/t)	(GJ/t)	(Mt/year)	(GJ/t)	(GJ/t)
Alternative routes									
Steam cracking with synthetic naphtha (benzene + butadiene) [7]	-	-	-	-	-	-	12	24.6	-
Ammonia from green hydrogen [8]	-	-	-	-	-	-	330	5.8	20.7
Ammonia from biomass [9]	-	-	-	-	-	-	162	19.1	20.7
Methanol from green hydrogen [10]	-	-	-	-	-	-	117	11.1	20.0
Methanol from green hydrogen for olefins [10]							177	11.1	20.0
Methanol from biomass [11]	-	-	-	-	-	-	57	36.7	20.0

[1] See Ref. [8] for background assumptions related to the fuel and feedstock use of each chemical. [2] 0.5% autonomous efficiency improvements that lead to 15% less demand (savings achievable with today's best practice technologies) in fuel use by 2050 compared to 2017 have been assumed [45]. [3] Average of the 34–163 GJ biomass per tonne of chemical range for bio-based plastics and synthetic organic materials production. Feedstock demand is deducted to estimate fuel use [6]. [4] Bio-based ethylene production requires in total 100 GJ biomass per tonne of ethylene [75]. Ethylene's feedstock demand is deducted to estimate the fuel use. [5] See note 10 for details of methanol production from green hydrogen. An additional 5 GJ process electricity is required for conversion [37]. [6] The process requires 13.9 MWh electricity per tonne of ethylene at 100% Faradaic efficiency [15]. Ethylene's feedstock demand is deducted to estimate the electricity use as process energy. [7] The synthetic liquids production process requires 927 kWh of renewable electricity. As a first step, 800 kWh equivalent hydrogen is produced at 86% efficiency. Processing yields a mix of jet fuel (146 kWh), diesel (348 kWh) and naphtha (81 kWh). Based on energy allocation of outputs, electricity input per GJ naphtha is estimated as 1.6 GJ. Final product feedstock demand is deducted to estimate the hydrogen demand [76]. [8] Total electrolyser electricity demand is 52 MWh per tonne of hydrogen (including 2 MWh for hydrogen storage). One tonne of ammonia requires 27 GJ hydrogen and an additional 2.3 GJ for its synthesis. The process exports 2.5 GJ heat per tonne of ammonia which is assumed to be converted to electricity on site for process energy with a conversion efficiency of 50% [77]. Ammonia's feedstock demand is deducted to estimate the hydrogen demand. [9] 1 kg biomass yields 0.9 kg syngas. Ammonia production requires 2.1 m^3 syngas per kg of ammonia. In total, 38 GJ biomass is needed for 1 tonne of ammonia production. In addition, 2.4 GJ natural gas is partially oxidized per tonne of ammonia and 0.25 GJ electricity is needed for process energy [78]. Ammonia's feedstock demand is deducted to estimate the hydrogen demand. [10] 1 tonne of methanol requires 0.19 tonnes of hydrogen and 1.4 tonnes of CO_2. This requires 31.1 GJ hydrogen per tonne of methanol. In addition, 1.3 GJ fuel and 1.2 GJ electricity are needed in the process [59]. Methanol's feedstock demand is deducted to estimate the hydrogen demand. [11] Production of 1 tonne of methanol requires 34.3 GJ bagasse, 9.5 GJ steam and 5.9 GJ electricity. It is assumed that steam and electricity are produced on site from 22 GJ bagasse per tonne of methanol, resulting in total bagasse demand of 56.7 GJ [79]. Methanol's feedstock demand is deducted to estimate the fuel demand for process energy.

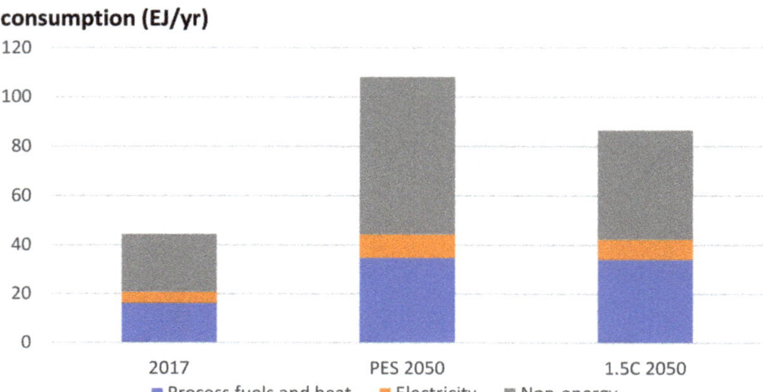

Figure 3. Development in the estimated total final consumption of the global chemical and petrochemical sector between 2017 and 2050.

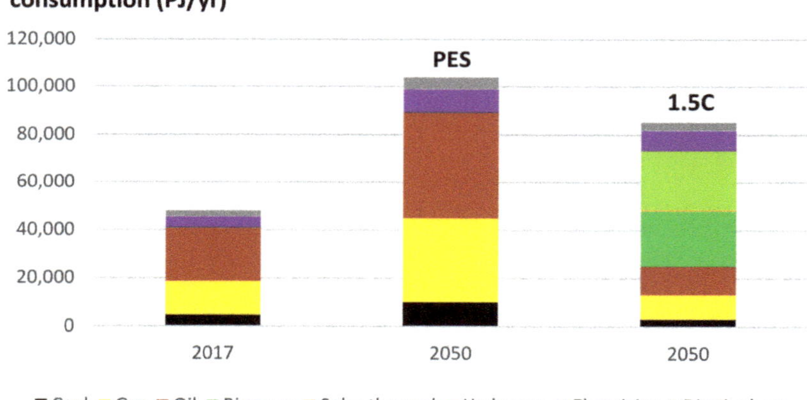

Figure 4. Breakdown of the estimated total final consumption of the global chemical and petrochemical sector by energy carrier, 2017 and 2050.

In the 1.5 °C case, efficiency breakthroughs and electrification limit process energy use to 42.9 EJ/yr in 2050, a doubling from the 2017 level. This is equivalent to a 5% reduction in total process energy demand compared to the PES, resulting from an annual 1% improvement in energy efficiency of processes, albeit an increase in energy use due to higher demand for ammonia and methanol. (This accounts for the changes in demand for chemicals in the 1.5 °C case compared to the PES: plastics demand decreases by 35% and the demand for ammonia and methanol increases by 47% and 82%, respectively, in 2050). NEU grows by 58% between 2017 and 2050 to reach 43.4 EJ/yr. The limited growth in NEU is driven by circular economy strategies for plastics, which include a combination of demand reduction (limiting demand to 657 Mt/yr by 2050) and higher mechanical and chemical recycling rates.

The final consumption mix changes in the 1.5 °C case (see Figure 4): the share of fossil fuel use in total process energy drops from 65% in 2017 to 24% in 2050, a reduction of 19.4 EJ/yr compared to PES in 2050. Direct use of renewables increases to 49% of process energy use including 28% bioenergy, 19% green hydrogen and 2% solar thermal. Electricity accounts for 20% of all process energy use (2320 terawatt-hours (TWh)). This excludes electricity for hydrogen production. If green hydrogen production is included, sector's electricity demand would increase fivefold.

Fossil fuels constitute nearly all NEU supply in the PES by 2050. Their share decreases to 36% in the 1.5 °C case, with gas and oil representing 15% and 18% of the total, respectively; coal's share drops to 3%. The remainder is a mix of biomass (25%) and green hydrogen (39%) feedstocks. Green hydrogen is the largest source of feedstock supply in the 1.5 °C case. It is used to produce HVCs, ammonia and methanol. It also is the basis for olefins production via renewable methanol, which accounts for 12% of the total HVC production in 2050. The introduction of renewables-based feedstocks and circular plastic economy strategies impact the use of oil feedstocks for HVC production. Compared to the PES, oil feedstock uses decline by 25 EJ/yr (equivalent to about 13 million barrels per day) to 8 EJ. Natural gas feedstock use is reduced by 70% in the 1.5 °C case (equivalent savings of 18.3 EJ/yr or 520 billion cubic meters per year) compared to the PES in 2050. Renewables, including renewable power and district heating, contribute to 68% of total final consumption.

Biomass demand for NEU increases from around 1 EJ in 2017 to more than 10.9 EJ in 2050. Another 12.2 EJ of biomass is needed for process energy, raising the total demand to around 23 EJ. Green hydrogen demand for process energy and NEU reaches 8.1 EJ and 16.9 EJ by 2050, respectively (in total around 210 Mt/yr, nearly twice today's global hydrogen demand).

Figure 5 provides a breakdown of biomass use. Apart from process heat, biomass feedstock is used to produce plastics (5.5 EJ/yr), ammonia, methanol and other chemicals (4.5 EJ/yr) and other high-value chemicals (0.9 EJ/yr). To meet the sector's total biomass demand, around 1.3 Gt of primary biomass would be needed each year, equivalent to 75% modern bioenergy use in 2017.

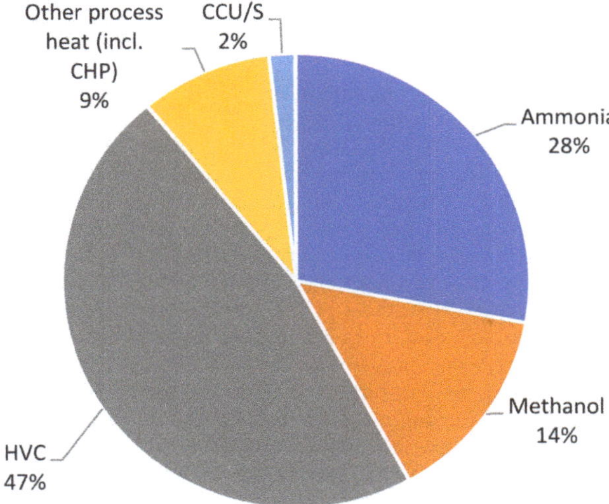

Figure 5. Estimated global use of biomass as fuel and feedstock in the 1.5 °C case, 2050.

The chemical and petrochemical sector's electricity demand is estimated to reach around 2640 TWh/yr in 2050 in the PES – equivalent to 80% growth from 2017. The sector's demand would be around 5% of the estimated total global gross electricity demand in 2050 [40]. In the 1.5 °C case, demand for electricity is slightly lower, at around 2320 TWh/yr (Figure 6). However, this excludes the electricity needed for green hydrogen production for ammonia, methanol and synthetic fuels, estimated at 9895 TWh/yr in 2050. Total electricity use in the sector equals 17% of global electricity demand in the 1.5 °C case. Compared to the PES, electricity efficiency improvements save 925 TWh/yr. Heat pumps for low-temperature process heat generation require another 320 TWh/yr electricity. Electrification has profound impacts; a total of 73 GW of (electric) heat pump capacity would be needed to supply low-temperature process heat. To meet the hydrogen demand for chemicals production, a total of 2435 GW of electrolyser capacity would be needed. (Assuming 350% heat pump efficiency and 65% electrolyser efficiency, with 50% capacity factor for both systems.)

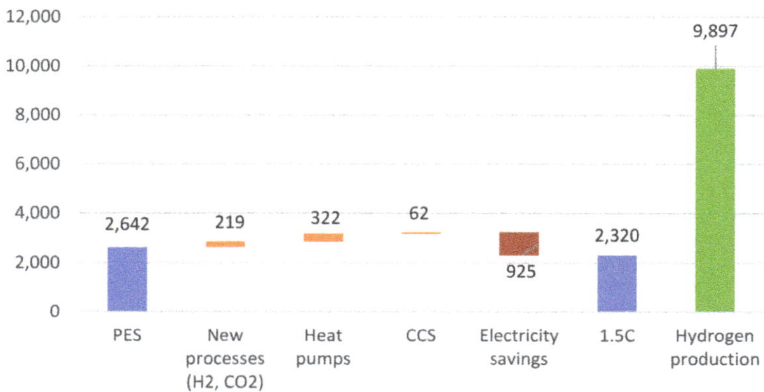

Figure 6. Changes in the estimated electricity demand in the global chemical and petrochemical sector between the PES and the 1.5 °C case, 2050.

4.3. Emissions Reductions and Carbon Flows

Renewable solutions, in combination with direct and indirect electrification, account for 40% of the emissions mitigation effort to go from 4.74 Gt in 2050 in the PES to zero emissions in the 1.5 °C case, including indirect electricity production emissions of 0.84 Gt and 0.11 Gt, respectively. These emissions reductions include all options (see Figure 7). Recycling rates increase six-fold, and this is coupled with deep demand reduction and CCS-retrofitted energy recovery (circular economy concepts account for 21% of the effort). However, all of this is still not enough: there is a need for 1.2 Gt per year of CCS to remove CO_2 from fuel combustion flue gases and the ammonia production process (26% of the total effort). 15% emission savings result from improving energy efficiency, and 8% from renewable-based process heat generation and feedstocks. The relatively small energy efficiency contribution is on top of the PES energy efficiency gains. Demand reduction including reuse of plastics contributes another 16% to total emissions reductions in the 1.5 °C case compared to the PES (350 Mt demand reduction yielding 0.56 Gt emissions savings). Higher mechanical and chemical recycling rates contribute another 5%: from 105 Mt in the PES to 276 Mt mechanical and chemical recycling in the 1.5 °C case. Around 16% of emissions reductions are related to switching to hydrogen-based feedstocks for methanol and ammonia (but also methanol and synthetic naphtha feedstocks for ethylene production). The implication is that part of industry will relocate to regions with lower cost renewable power sources. Finally, the contribution from power supply transformation is 15%, through a shift to renewable electricity.

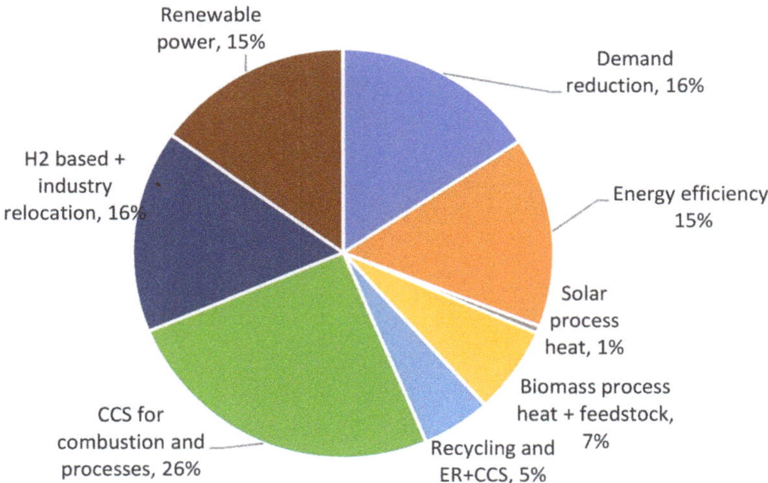

Figure 7. Breakdown of the estimated CO_2 emissions reductions in the 1.5 °C case compared to the PES, 2050. Note: the breakdown has been estimated by removing the technology penetration of each measure from the 1.5 °C case to arrive back at the estimated PES results in 2050 based on the following order: demand reduction, energy efficiency, renewable process heat, plastic waste treatment, renewable feedstocks, CCUS and renewable power/process heat electrification. The breakdown and the average mitigation costs may change somewhat if a different order is followed.

Analysis of the zero-emission pathway shows a 60% reduction potential in the sector's direct emissions (from 3.9 Gt in the PES to 1.58 Gt in the 1.5 °C case) from energy efficiency, renewable heat and feedstock, hydrogen-based routes and industry relocation and demand reduction (Figure 8). Reducing the remaining 40% relies on CCS integration with production processes and waste management of plastics, as well as through biomass carbon accounting practices. CO_2 emissions captured from fossil fuel-based production processes, process emissions and incineration total 0.94 Gt (0.83 Gt of which flows back for use with green hydrogen in the production of synthetic hydrocarbon feedstocks). Another 0.55 Gt is captured from biomass sources, which implies negative emissions. Finally, 0.14 Gt of biomass carbon is recycled back into plastic production. As a result, the sector's direct emissions become carbon neutral by 2050.

Figure 9 shows the sector's carbon flows in the PES and the 1.5 °C case (top and bottom, respectively). The graphs show the major changes that are required, with much more use of biomass carbon as well as carbon recycling and CO_2 capture and storage.

4.4. Costs of Emissions Reductions

We estimate the costs of decarbonising the chemical and petrochemical sector as the product of CO_2 emissions mitigation potential and the cost of each option considered in the analysis. On average, total mitigation costs amount to USD 310 billion per year in 2050; this results in an average mitigation cost of USD 64/t CO_2 (see Table 5). The total cost of mitigation equals more than 35% of the total energy and feedstock cost of the global chemical and petrochemical sector, estimated at around USD 860 billion per year in 2050. This is comparable with the findings of Ref. [72], which estimates an increase in production cost of 20–43% by 2050 for the deep decarbonisation of plastics production compared to business as usual (including energy, investment, operation and maintenance costs).

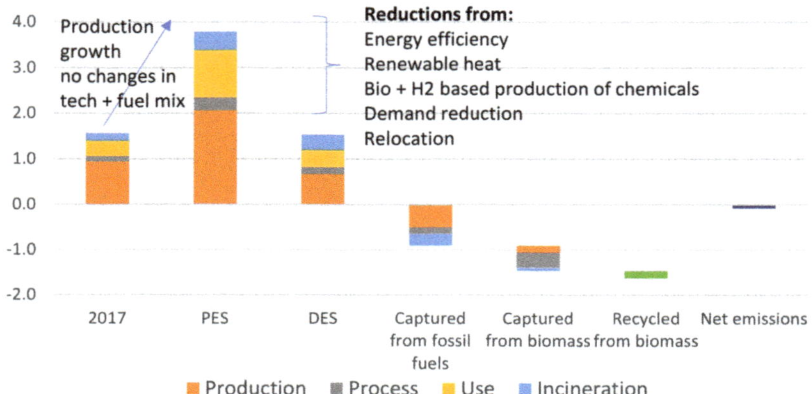

Figure 8. Changes in the sector's CO_2 emissions, 2017–2050.

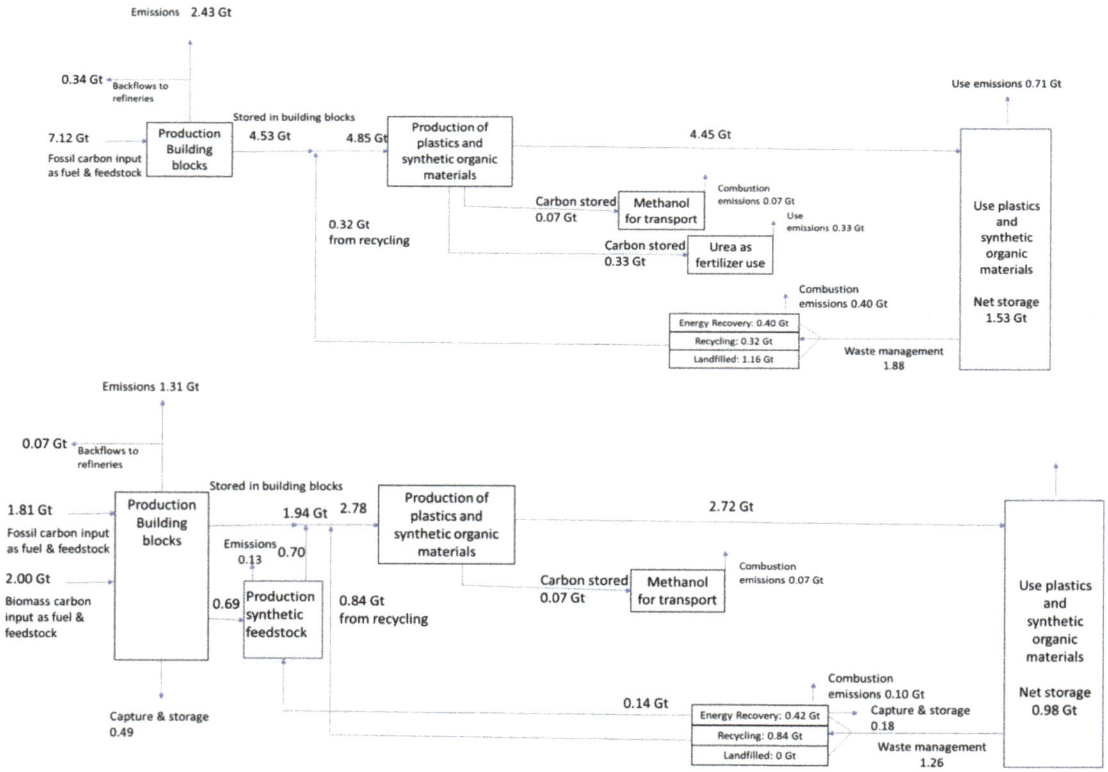

Figure 9. Embodied carbon flows for chemicals and petrochemicals in the PES and the 1.5 °C case in 2050. Figures refer to CO_2 equivalent flows.

Table 5. Estimated CO_2 mitigation cost in the 1.5 °C case, 2050.

	Emissions Mitigated	**Mitigation Cost Range**
	[Gt CO_2/yr]	[USD/t CO_2]
Demand reduction	0.76	0–50
Energy efficiency	0.72	25–125
Solar process heat	0.03	0–100
Biomass process heat	0.13	0–75
Recycling	0.24	−50–300
Energy recovery + CCS	0.31	−50–100
Biobased chemicals	0.13	−100–400
CCS for combustion and processes	1.18	0–200
H_2-based chemicals	0.54	−100–300
Industry relocation	0.05	0–50
Renewable power	0.73	−25–25
Total	4.79	−20–150

The 1.5 °C case technology portfolio identified requires a total investment of at least USD 4.5 trillion between 2018 and 2050 (on average USD 140 billion per year over the entire period) – an increase of USD 2.55 trillion compared to the PES (Figure 10). Low-carbon technologies require an additional USD 4.3 trillion, but fossil fuel-based production capacity investment needs are reduced by USD 1.8 trillion compared to the PES. Investments related to feedstock switching to biomass and hydrogen represent 61% of the total, followed by energy efficiency (18%), CCS (9%), recycling and energy recovery (8%) and direct use of renewables, including heat pumps (4%). Investments exclude infrastructure needs such as waste collection systems or hydrogen pipelines. Notably, investment cost for circular economy solutions is uncertain due to the complex supply chains and may be underestimated – more research is warranted. At the same time, such investments provide auxiliary environmental services, so their allocation to energy transition is a topic for debate.

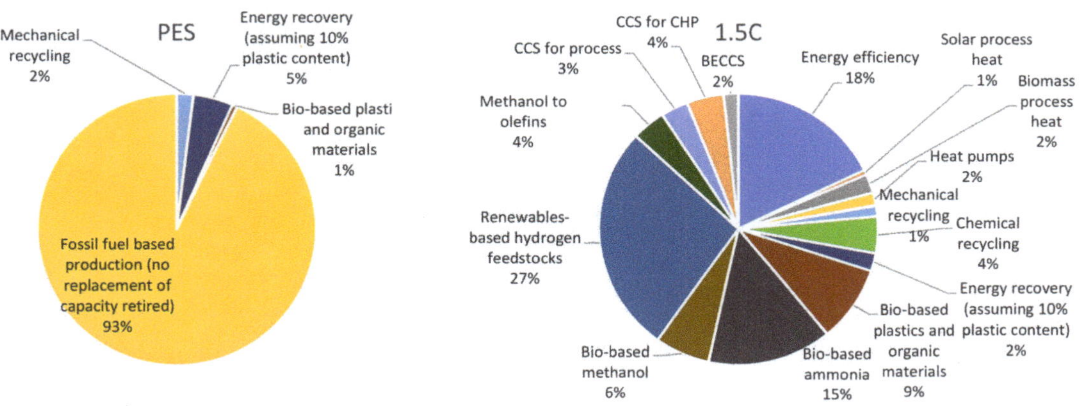

Figure 10. Estimated investment needs in the global chemical and petrochemical sector according to the PES and the 1.5 °C case, 2017–2050.

4.5. Implications for the Global Energy System

The chemical and petrochemical sector's role in the global energy system would grow substantially in the PES. Oil demand would double in absolute terms. In the 1.5 °C case, however, oil demand would decrease 30% between now and 2050 (Table 6). Still, the sector accounts for 60% of remaining oil demand in 2050, which is nearly a 6-fold growth from today's share. This shows the importance of the petrochemical industry energy and feedstock demand for total oil demand projections. A similar effect can be seen for natural gas, where demand in the PES grows 2.5-fold between now and 2050, while consumption is reduced in the 1.5 °C case. The difference in demand levels exceeds today's natural gas demand in Europe. The scenarios also differ markedly in biomass use, a six-fold increase in the 1.5 °C case compared to the PES. Total electricity demand is nearly five times higher in the 1.5 °C case when the needs for hydrogen production are accounted for, requiring more than 7000 GW of renewable power. Furthermore, in the future the chemical and petrochemical sector will remain deeply integrated with the energy sector, but the nature of the integration changes fundamentally. Despite the significant growth of renewables, the sector would rely on significant use of CCS for production processes and waste incineration, accounting for a quarter of total global CCS use. Around 1200 waste incinerators would require CCS deployment – up from four plants today.

Table 6. Indicators for energy systems relevance of the chemical and petrochemical sector.

	Unit	2017	2050 PES	2050 1.5 °C Case	1.5 °C case % World Demand 2050
Oil demand	[mbd]	7.1	18.1	5.2	60
Gas demand	[BCM]	525	1343	357	11
Biomass use	[Mt/yr]	5.1	9.0	1320	15
BECCS [1]	[Mt/yr]	0.0	0.0	550	6.5
Fossil CCS [2]	[Mt/yr]	0.0	0.0	940	11
Electricity demand [3]	[TWh/yr]	1278	2645	2307	3.2
Green hydrogen demand	[Mt/yr]	0.0	0.0	210	34
Hydrogen electrolyser capacity [4]	[GW]	0.0	0.0	2468	48
Heat pumps	[GW]			73	30
Solar thermal	[mln m^2]			190	5

[1] share of total global CCS. [2] share of total global CCS. [3] excludes green hydrogen production (10 PWh/yr). [4] chlorine production 1 Mt green hydrogen by-product today excluded.

4.6. Impact on Commodity Prices

Commodity price volatility that resulted from the Covid-19 crisis has been widely reviewed in the literature [80–83]. Oil and gas prices responded markedly, but have recovered since. Additionally, prices of commodities that are in demand because of the energy transition have risen substantially, as is the case for copper and lithium. Changing resource prices may also affect the cost effectiveness of transition strategies for the chemical and petrochemical industry. Fossil fuel prices are likely to decline, while prices of scarce biomass may rise; however, carbon pricing can still compensate wholly or partially for such developments. For renewable electricity and green hydrogen, as well as CCS, it is likely that economies of scale will overcome any scarcity effects. Longer term, the analysis indicates a 40% rise of energy and feedstock cost.

The impact of energy transition on product prices will vary. It will be most pronounced for the energy-and carbon-intensive products, and the effect will be moderate for more sophisticated products with higher value added. While prices will reflect the increased cost, the supply and demand balance will remain volatile, and prices will therefore continue to fluctuate.

5. Discussion of Decarbonisation Challenges

This analysis shows the technology needs for a zero-emission pathway and its impacts on sector's energy consumption, feedstock needs, carbon flows and investments. In the PES, the sector is responsible for a rising share of the global oil and gas demand. This trend can be reversed through a combination of biomass feedstock use, CCS, circular economy and renewable hydrogen. The implications of such a transition for the sector structure will be profound (Section 5.1). We also reflect on the robustness of the analysis (Section 5.2).

5.1. Discussion of Results

The analysis highlights a need for life cycle policies that encompass both energy and materials. The sector outlook is uncertain, and this poses a risk that investors must consider. The wrong investments in the coming years can result in billions of dollars of stranded assets. The uncertainty also creates a conundrum in terms of where to invest. The analysis suggests that global strategies cannot simply be applied equally at the country/region level without tailoring. At the country level, analysis shows a large potential for hydrogen in China, India and in the rest of the G20 countries (see Figure 11). Biomass share in total final consumption is high in the "Rest of the G20" and in the "Rest of the World" countries. Solar thermal use is higher in India and the United States compared to others. Coal would continue to represent the largest fossil fuel use in China, whereas Japan's sector would continue to rely on oil to a large extent. In other countries and regions, gas would comprise the majority of fossil fuel use.

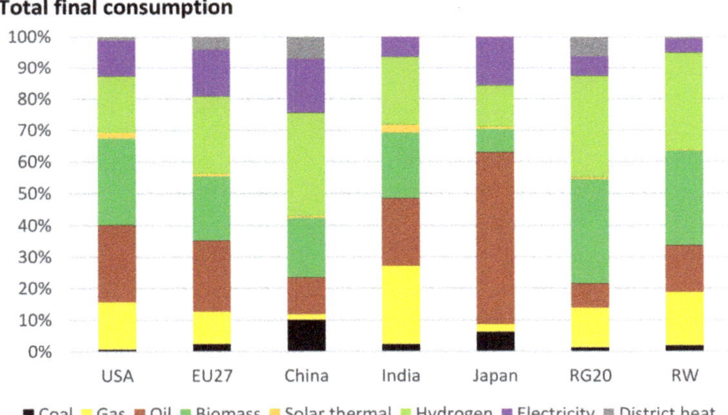

Figure 11. Breakdown of the estimated total final consumption by country/region in the 1.5 °C case, 2050. Note: RG20 = rest of G20; RW = rest of the world.

New petrochemical complexes are concentrated in the Middle East (based on cheap oil and gas) and China (driven by national product demand). However, future location choice may be determined by the access to low-cost renewable power, biomass feedstock and CO_2 storage potential. While it is possible to transport large amounts of biomass to central processing plants, the economics may favor smaller, decentralised plants close to biomass production sites. Such a decentralised structure is evident in existing biomass industries, such as sugar and ethanol plants, as well as pulp and paper mills, where access to fuel and raw material supply is crucial. Similar to existing sugar/ethanol biorefineries, new types of electricity/biofuel/biochemical biorefineries may emerge that can adjust their product mix based on market circumstances. North and South America, as well as South Asia, are potential locations for such a roll-out, given resource endowment and existing economic structure. Low-cost renewable-hydrogen production will be concentrated in remote desert locations, including in the Middle East and Africa as well as Australia and Chile, among others. Since hydrogen is a commodity that can be traded, it offers an opportunity for countries that already produce and export gas – a pillar of today's chemicals production – to switch to renewable exports. Manufacture of products such as ammonia, methanol and other chemicals should take place nearby to hydrogen production sites in order to reduce shipping costs, thereby highlighting the opportunity for new industrial activities that may result in a global relocation of chemicals production. Such developments are evident, with green ammonia projects in Australia, Chile, Oman and Saudi Arabia currently under development.

Demand for chemicals is currently concentrated in developed countries [46]. As developing countries catch up, demand for plastics could triple, as assumed in this analysis. However, concerted action to minimise consumption and maximise circular economy efforts may reduce plastics demand from nearly 1000 Mt in PES to 650 Mt in 1.5 °C case in 2050, with future fossil fuel-based production returning to the 2017 level of 350 Mt. Therefore, the growth potential is significant but uncertain. The outcome will depend on new product policy, waste management policies, innovation and R&D in material sciences, as well as logistics in end-of-life management of post-consumer plastic waste [13].

We demonstrated a single zero-emission pathway based on the 1.5 °C case. We argue the robustness of this finding, since the pathway is comparable with the findings of other studies, as reviewed in Section 2, whilst we added several new insights at the technology and material levels. The sectorial pathway assumes the rest of the global economy will join a zero-emission pathway, following the ambitious climate policy choices countries have started taking. However, key hurdles are present for the sector. In a nutshell, these are: (a) the availability, accessibility, and acceptance of CO_2 storage sites for the CCS route, but not their safety, which has been extensively proven; (b) the very high electricity and energy demand (due to the need to synthesise hydrogen via electrolysis and to energise CO_2 for the CCU route, with the associated strict requirement of very low carbon-intensity of the electricity mix); and (c) the availability of land for biomass production, as well as competition with other biomass and land use.

Decarbonisation solutions need to be developed by considering that a large share of the carbon is stored in products. This limits the contribution of traditional measures such as energy efficiency. New sources of carbon feedstock have been identified from CO_2 capture, biomass and recycling. The CCS route has two main advantages: (i) it can exploit the existing technology and the infrastructure of the current petrochemical and chemical industry, without the need for a complete reshaping of it; and (ii) CO_2 captured from point sources and/or from air plus permanent CO_2 storage in geological formations constitutes the key elements of the negative emissions technologies [84]. The analysis assumed massive use of CCS, to the tune of 1.5 Gt per year (including BECCS as a backstop, which is not yet deployed globally). In comparison, today's CCS use across all sectors is well below 50 Mt per year. It is likely that the incumbents will opt for CCS-based solutions that can be integrated into existing plants, while new players will aim for more innovative solutions. However, local acceptance and availability of suitable CO_2 storage sites could restrict its application. The key role of CCS in this sector is so far not fully appreciated and very few pilot projects exist beyond enhanced oil recovery. Therefore, the potential to ramp up CCS use is unclear. While there has been some progress on CCS for ammonia plants, other components are lagging. In the context of the life cycle of petrochemicals, emerging BECCS technologies and CCS for waste incinerators must be part of the solution. Our analysis suggests significant CCU use as captured carbon from chemical production processes would supply the carbon needed for green hydrogen-based routes. At the sector level, however, CCU must be combined with biomass use to fully replace primary fossil fuel feedstock and ensure carbon neutrality, as CCU for fossil CO_2 sources would yield only a 35–50% emissions reduction for the petrochemicals sector overall.

Around 1.3 Gt of biomass needs to be deployed. Large biobased industries will likely be located close to the biomass supply. Whereas harbors with large petrochemical activities, such as Rotterdam and Antwerp, are developing biorefineries based on imported biomass, the economic feasibility of such strategies is not yet evident. Another key uncertainty is the availability of sustainable biomass feedstock. To put the 1.2 Gt into perspective, that equals the potential bioenergy production of the United States [85]. This illustrates the land use implications and the logistical challenges of such a strategy [6]. More than one-fifth of all products should be biomass-based by 2050. Some studies suggest even more ambitious shares of 40% to 70% by 2050, but the progress in recent decades has been modest [55,86]. Food production and consumption requirements in a world with rising populations, sustainability concerns, changing consumption patterns and climate change effects result in an uncertain outlook for biomass-based production [87].

Synthetic feedstocks provide a technically feasible option, but our analysis suggests that given the high cost and small pilot plant deployment scale today, its growth will be too slow to have a significant impact by 2050. The same applies for green hydrogen; while demand may grow substantially on relative terms, the small capacity today means that a roll-out will take time. Around 0.3 GW of hydrogen electrolyser capacity is in place today, while the 1.5 °C case assumes 2435 GW of electrolyser capacity for the chemical and petrochemical sector alone by 2050. This equals nearly half of the total green hydrogen manufacturing capacity that is projected for 2050 in the 1.5 °C case [40]. Green ammonia production will represent an economically viable early opportunity for renewable hydrogen deployment, and the first commercial plants are expected in the coming years. Wider use of hydrogen and other synthetic feedstocks will require the availability of low-cost electricity, high utilisation of electrolyser capacities and improvements in the efficiency and costs of electrolysers. Any transition in this sector will have profound effects on the power system and its cost-effective development will depend on the availability of renewable power. Growth of green hydrogen production must be matched by the roll-out of massive renewable power generation capacity, on top of the necessary transition of the existing generation capacity. Supplying the necessary power to meet 25 EJ hydrogen by 2050 would require around 7000 GW of electricity generation capacity, roughly the level of total global power generation capacity today [40]. Assuming an average investment cost of USD 1000/kW for renewable power, this translates into a total investment of USD 7 trillion.

As the analysis shows, biomass-based feedstocks will be the key solution to stem the growing demand for fossil fuels for plastics and chemicals production. Production of 300–350 Mt of biomaterials (from less than 10 Mt per year today) requires a growth rate of 13% per year over the next three decades. This is compatible with the second scenario of the drawdown project (https://drawdown.org/, accessed on 1 June 2021), where bioplastics demand grows to 357 Mt, or 46% of the market in 2050. Plants in Brazil and India have already demonstrated that bio-ethylene can be produced at competitive prices if low-cost biomass is available. The next step is the accelerated expansion of bio-based chemicals to substitute petrochemical-based polymers, starting with high value-added opportunities. Early niche markets include beverage bottles and cosmetics packaging.

The analysis assumes stringent circular economy measures, including minimisation of product use, such as one-way packaging, new ecological product design and maximum recycling efforts. Increased waste recycling is also essential. Around half of all plastics should be recycled by mid-century (from around 10% today); this includes chemical and mechanical recycling. Higher mechanical recycling rates require industry innovations, notably in collection and sorting. A better collection infrastructure would lead to a larger supply of well-sorted, high-quality, post-consumer plastics. This would increase the scale and further improve the economics of mechanical recycling [88]. Chemical and feedstock recycling offer the possibility of operating at a larger scale with less pure feedstock. The various scenarios suggest rapid growth of pyrolysis, a technology that is not yet fully proven, and that would incur significant carbon and energy losses (see also Supplementary Materials).

5.2. Strategic Implications

Overall, the sector's development in the 1.5 °C case is very different from the sector's current investment trends. This is a cause for concern, as it points to a significant risk of stranded assets that need to be better understood.

In this analysis, several key aspects that pertain to the decarbonisation of the global chemical and petrochemical sector have been combined, namely the energy impacts, emissions reductions potential and the costs and investment needs of the key low-carbon emission technologies covering the life cycle of chemicals and plastics, the impacts of decarbonisation on location choices and plant siting, as well as on materials use and waste handling, and finally, the role of accounting carbon storage in products as a crucial step in the complete assessment of the sector's emissions. This combination helps to provide an overall picture for the sector's net-zero pathway, thereby complementing the many existing studies that individually focus on the various components of decarbonisation. At the same time, the analysis is subject to uncertainty, as it is based on a set of bold yet uncertain assumptions regarding technology uptake for a limited number for key chemicals. Higher product granularity and further regional granularity may affect the findings.

While thorough technological analysis was carried out for certain solutions (e.g., energy efficiency, renewable energy heating and feedstocks, hydrogen and CCU/S), the potentials of recycling and other circular plastic strategies are uncertain. Technology progress continuously changes the outlook for a zero-emission pathway. For instance, green hydrogen has only emerged in the last couple of years. New prospects of green ammonia production and the scale up of hydrogen use in the production of methanol and other HVCs brighten the outlook for energy transition in the sector. The emphasis of the sector's technology and emissions analyses has somewhat shifted strategy in recent years, from energy efficiency and biomass feedstocks to understanding the role of renewables for heating, hydrogen, CCUS and circular energy. It is unlikely that the technology transformation outlook will change fundamentally in the coming three decades, and we therefore regard our choice of five components as the key strategies. Specific to the chemical and petrochemical sector, most carbon is stored in products, and this limits the strategy scope; either carbon supply is carbon free, or CO_2 is stored after use. This aspect is not properly reflected in existing models, as they lack material flows, and few industry strategies properly account for such scope-3 emissions. For example, if urea continues to be used as nitrogen fertiliser (and CO_2 is released in the use phase), there is no way around biomass feedstock for ammonia production to ensure renewable CO_2 supply. At a regional level, the analysis suggests the need for targeted solutions, notably for China and the Middle East, which deserve attention in the coming years. More refined analysis will result in a higher accuracy regarding the 1.5 °C case's viability, and provide a better understanding regarding the steps needed from now until 2050.

6. Conclusions

A zero-carbon chemical and petrochemical industry is feasible by mid-21st century. Today, fossil hydrocarbon feedstocks are at the center of the chemical and petrochemical industry – this has profound implications for CO_2 mitigation strategies. A life cycle approach is needed to capture the full greenhouse gas emissions impact and all mitigation opportunities. A set of twenty options across five strategies have been identified that can be deployed for this purpose. Energy efficiency and renewables-based process heating, biomass feedstocks, circular economy concepts, synthetic hydrocarbons from green hydrogen and CO_2, CCS and the correct accounting of biomass carbon have key roles to play; together they can yield deep emissions reductions.

However, the product cost may increase by more than 35% compared to today's level. Such a cost increase implies that a premium must be paid for green products, or the negative environmental impacts of fossil fuels and feedstocks must be priced properly. Renewables would provide nearly 70% of final consumption of energy, and feedstock and renewable supply solutions – in combination with direct and indirect electrification – account for

40% of the emission mitigation effort. When BECCS is included, the role of renewables increases to more than half of all emissions reduction. The key role of renewables-based solutions represents a new insight that reflects the significant cost reduction and technology improvement witnessed in recent years. Investment needs amount to USD 4.5 trillion between now and 2050, and CO_2 mitigation would cost on average USD 64/t in 2050 – these costs are lower than previous estimates, yet this transition will not happen by itself. There is no significant experience with such an operating structure beyond a few scattered demonstration plants. The upscaling effort will be significant, and a certain lock-in of pathways may occur. The sector's structure must change fundamentally, and the implications for the global energy system can be significant, as well as the material flow and location choice effects. Significant uncertainty remains in terms of how quickly this transition will take place, and what direction it will take – this creates an important conundrum for investors today. A concerted global effort to transition the chemical and petrochemical sector seems unlikely. Front runners – consumers, governments and chemical and petrochemical clusters and companies alike – will need to force this change, and this will require attention for competitiveness issues and carbon leakage. For example, certification of green supply chains may be required, as well as the creation of market niches, including a mandatory share of green product supply. Governments must create the right enabling environment to allow for transition experiments and foster the necessary growth to achieve the required economies of scale and technology learning.

Supplementary Materials: The following are available online at https://www.mdpi.com/article/10.3390/en14133772/s1, Figure B-1: Estimated change in production of chemicals in the PES and 1.5 °C case by 2050 compared to the 2017 level; Table A-1: Potentials and cost of key mitigation options; Table B-1. Estimated production volume, specific energy consumption and fuel mix of the major products in selected country/regions, 2017; Table B-2. Feedstock consumption per unit of product; Table B-3: Technology penetration assumptions in the 1.5 °C case in 2050; Table B-4: Estimated plastics production, demand and waste generation in the PES and 1.5 °C case, 2017-2050; Table D-1. Investment cost assumptions of technology options; Table D-1. Global overview of completed, operating and operational carbon capture storage and utilization facilities.

Author Contributions: Conceptualisation, D.S. and D.G.; methodology, D.S. and D.G.; formal analysis, D.S. and D.G.; data curation, D.S. and D.G.; writing—original draft preparation, D.S. and D.G.; writing—review and editing, D.S. and D.G.; visualisation; funding acquisition, D.S. Both authors have read and agreed to the published version of the manuscript.

Funding: This research received no external funding.

Acknowledgments: We thank Ricardo Gorini, Rodrigo Leme, Gayathri Prakash and Sean Ratka (all IRENA) for their valuable comments, and Florian Ausfelder and Alexis Bazzanella (DECHEMA) who have reviewed earlier versions of this manuscript.

Conflicts of Interest: The authors declare no conflict of interest.

References

1. OECD. *Saving Costs in Chemicals Management: How the OECD Ensures Benefits to Society*; OECD: Paris, France, 2019.
2. IEA. *World Energy Balances 2019*; OECD/IEA: Paris, France, 2019.
3. Khemka, P. *Ammonia and Urea: Global Overcapacity to Continue to Subdue Global Operating Rates in the Medium Term*; Nexant: White Plains, NY, USA, 2019.
4. Plastics Europe. *Plastics—The Facts 2019. An Analysis of European Plastics Production, Demand and Waste Data*; Plastics Europe: Brussels, Belgium, 2019.
5. Grau, E. Bio-Sourced Polymers. Master's Thesis, University of Bordeaux, Bordeaux, France, 2019.
6. Saygin, D.; Gielen, D.; Draeck, M.; Worrell, E.; Patel, M. Assessment of the technical and economic potentials of biomass use for the production of steam, chemicals and polymers. *Renew. Sustain. Energy Rev.* **2014**, *40*, 1153–1167. [CrossRef]
7. Weiss, M.; Neelis, M.; Zuidberg, M.; Patel, M. Applying bottom-up analysis to identify the system boundaries of non-energy use data in international energy statistics. *Energy* **2008**, *33*, 1609–1622. [CrossRef]
8. Saygin, D.; Patel, M.; Worrell, E.; Tam, C.; Gielen, D. Potential of best practice technology to improve energy efficiency in the global chemical and petrochemical sector. *Energy* **2011**, *36*, 5779–5790. [CrossRef]
9. IRENA. *Global Renewables Outlook*; International Renewable Energy Agency: Abu Dhabi, United Arab Emirates, 2020.

10. Rogelj, J.; Shindell, D.; Jiang, K.; Fifita, S.; Forster, P.; Ginzburg, V.; Handa, C.; Kheshgi, H.; Kobayashi, S.; Kriegler, E.; et al. Mitigation Pathways Compatible with 1.5 °C in the Context of Sustainable Development. In *Global Warming of 1.5 °C. An IPCC Special Report on the Impacts of Global Warming of 1.5 °C above Pre-Industrial Levels and Related Global Greenhouse Gas Emission Pathways, in the Context of Strengthening the Global Response to the Threat of Climate Change, Sustainable Development, and Efforts to Eradicate Poverty*; Cambridge Press: Cambridge, UK, 2018.
11. Ike, G.N.; Usman, O.; Sarkodie, S.A. Testing the role of oil production in the environmental Kuznets curve of oil producing countries: New insights from Method of Moments Quantile Regression. *Sci. Total. Environ.* **2020**, *711*, 135208. [CrossRef]
12. Sadik-Zada, E.R.; Gatto, A. The puzzle of greenhouse gas footprints of oil abundance. *Socio-Econ. Plan. Sci.* **2021**, *75*, 100936. [CrossRef]
13. Chen, G.-Q.; Patel, M.K. Plastics Derived from Biological Sources: Present and Future: A Technical and Environmental Review. *Chem. Rev.* **2012**, *112*, 2082–2099. [CrossRef]
14. Karan, H.; Funk, C.; Grabert, M.; Oey, M.; Hankamer, B. Green Bioplastics as Part of a Circular Bioeconomy. *Trends Plant Sci.* **2019**, *24*, 237–249. [CrossRef]
15. De Luna, P.; Hahn, C.; Higgins, D.; Jaffer, S.A.; Jaramillo, T.F.; Sargent, E.H. What would it take for renewably powered electrosynthesis to displace petrochemical processes? *Science* **2019**, *364*. [CrossRef]
16. Gielen, D. Materializing Dematerialization. Ph.D. Thesis, Technical University Delft, Delft, The Netherlands, 1999.
17. Kümmerer, K.; Clark, J.H.; Zuin, V.G. Rethinking chemistry for a circular economy. *Science* **2020**, *367*, 369–370. [CrossRef]
18. Hatti-Kaul, R.; Nilsson, L.J.; Zhang, B.; Rehnberg, N.; Lundmark, S. Designing Biobased Recyclable Polymers for Plastics. *Trends Biotechnol.* **2020**, *38*, 50–67. [CrossRef]
19. ECIU. Net Zero Tracker. Energy & Climate Intelligence Unit. 2021. Available online: https://www.eciu.net/netzerotracker (accessed on 24 April 2021).
20. Stork, M.; de Beer, J.; Lintmeijer, N.; den Ouden, B. *Chemistry for Climate: Acting on the Need for Speed. Roadmap for the Dutch Chemical Industry towards 2050*; Ecofys and Berenschot: Utrecht, The Netherlands, 2018.
21. CEFIC. *European Chemistry for Growth. Unlocking a Competitive, Low Carbon and Energy Efficient Future*; CEFIC: Brussels, Belgium, 2013.
22. IEA. *The Future of Petrochemicals. Towards More Sustainable Plastics and Fertilisers*; OECD/IEA: Paris, France, 2018.
23. Rissman, J.; Bataille, C.; Masanet, E.; Aden, N.; Morrow, W.R.; Zhou, N.; Elliott, N.; Dell, R.; Heeren, N.; Huckestein, B.; et al. Technologies and policies to decarbonize global industry: Review and assessment of mitigation drivers through 2070. *Appl. Energy* **2020**, *266*, 114848. [CrossRef]
24. Economidou, M. *Assessing the Progress towards the EU Energy Efficiency Targets Using Index Decomposition Analysis*; No. EUR 28710 EN; Publications Office of the European Union: Luxembourg, 2017.
25. Reuter, M.; Patel, M.K.; Eichhammer, W. Applying ex post index decomposition analysis to final energy consumption for evaluating European energy efficiency policies and targets. *Energy Effic.* **2019**, *12*, 1329–1357. [CrossRef]
26. Saygin, D.; Worrell, E.; Tam, C.; Trudeau, N.; Gielen, D.; Weiss, M.; Patel, M. Long-term energy efficiency analysis requires solid energy statistics: The case of the German basic chemical industry. *Energy* **2012**, *44*, 1094–1106. [CrossRef]
27. Boulamanti, A.; Moya, J.A. *Energy Efficiency and GHG Emissions: Prospective Scenarios for the Chemical and Petrochemical Industry*; No. EUR 28471 EN; Publications Office of the European Union: Luxembourg, 2017.
28. Kesicki, F.; Yanagisawa, A. Modelling the potential for industrial energy efficiency in IEA's World Energy Outlook. *Energy Effic.* **2015**, *8*, 155–169. [CrossRef]
29. Saygin, D.; Wetzels, W.; Worrell, E.; Patel, M.K. Linking historic developments and future scenarios of industrial energy use in the Netherlands between 1993 and 2040. *Energy Effic.* **2013**, *6*, 341–368. [CrossRef]
30. McLellan, B.; Corder, G.; Giurco, D.; Ishihara, K. Renewable energy in the minerals industry: A review of global potential. *J. Clean. Prod.* **2012**, *32*, 32–44. [CrossRef]
31. Schiffer, Z.J.; Manthiram, K. Electrification and Decarbonization of the Chemical Industry. *Joule* **2017**, *1*, 10–14. [CrossRef]
32. Åhman, M.; Nilsson, L.J.; Johansson, B. Global climate policy and deep decarbonization of energy-intensive industries. *Clim. Policy* **2017**, *17*, 634–649. [CrossRef]
33. Fais, B.; Sabio, N.; Strachan, N. The critical role of the industrial sector in reaching long-term emission reduction, energy efficiency and renewable targets. *Appl. Energy* **2016**, *162*, 699–712. [CrossRef]
34. Gerres, T.; Ávila, J.P.C.; Llamas, P.L.; Román, T.G.S. A review of cross-sector decarbonisation potentials in the European energy intensive industry. *J. Clean. Prod.* **2019**, *210*, 585–601. [CrossRef]
35. Gielen, D.; Taylor, M. Modelling industrial energy use: The IEAs Energy Technology Perspectives. *Energy Econ.* **2007**, *29*, 889–912. [CrossRef]
36. Levi, P.G.; Cullen, J.M. Mapping Global Flows of Chemicals: From Fossil Fuel Feedstocks to Chemical Products. *Environ. Sci. Technol.* **2018**, *52*, 1725–1734. [CrossRef] [PubMed]
37. Bazzanella, A.M.; Ausfelder, F. *Low Carbon Energy and Feedstock for the European Chemical Industry*; Technical Study; Dechema E.V.: Frankfurt am Main, Germany, 2017.
38. Serrano-Ruiz, J.C.; West, R.M.; Dumesic, J.A. Catalytic Conversion of Renewable Biomass Resources to Fuels and Chemicals. *Annu. Rev. Chem. Biomol. Eng.* **2010**, *1*, 79–100. [CrossRef] [PubMed]

39. Patel, M.; Neelis, M.; Gielen, D.; Olivier, J.; Simmons, T.; Theunis, J. Carbon dioxide emissions from non-energy use of fossil fuels: Summary of key issues and conclusions from the country analyses. *Resour. Conserv. Recycl.* **2005**, *45*, 195–209. [CrossRef]
40. IRENA. *World Energy Transitions Outlook: 1.5 °C Pathway (Preview)*; IRENA: Abu Dhabi, United Arab Emirates, 2021.
41. Bourguignon, D. *Plastics in a Circular Economy. Opportunities and Challenges*; Briefing; European Parliament: Brussels, Belgium, 2017.
42. Hundertmark, T.; Mayer, M.; McNally, C.; Simons, T.J.; Witte, C. *How Plastics-Waste Recycling Could Transform the Chemical Industry*; McKinsey & Company: New York, NY, USA, 2018.
43. WEF. *The New Plastics Economy. Rethinking the Future of Plastics*; World Economic Forum: Geneva, Switzerland, 2016.
44. Gielen, D.; Patel, M.K. The NEAT model, non-energy use emission accounting tables. In Proceedings of the First NEU-CO_2 workshop, Paris, France, 22–23 September 1999.
45. Zuberi, M.J.S.; Patel, M.K. Cost-effectiveness analysis of energy efficiency measures in the Swiss chemical and pharmaceutical industry. *Int. J. Energy Res.* **2019**, *43*, 313–336. [CrossRef]
46. Broeren, M.; Saygin, D.; Patel, M. Forecasting global developments in the basic chemical industry for environmental policy analysis. *Energy Policy* **2014**, *64*, 273–287. [CrossRef]
47. IEA. *Energy Technology Transitions for the Industry. Strategies for Next Industrial Revolution*; OECD/IEA: Paris, France, 2009.
48. Parvatker, A.G.; Eckelman, M.J. Simulation-Based Estimates of Life Cycle Inventory Gate-to-Gate Process Energy Use for 151 Organic Chemical Syntheses. *ACS Sustain. Chem. Eng.* **2020**, *8*, 8519–8536. [CrossRef]
49. IRENA. *Renewable Energy in Manufacturing. A Technology Roadmap for REmap 2030*; International Renewable Energy Agency: Abu Dhabi, United Arab Emirates, 2014.
50. Taibi, E.; Gielen, D.; Bazilian, M. The potential for renewable energy in industrial applications. *Renew. Sustain. Energy Rev.* **2012**, *16*, 735–744. [CrossRef]
51. Lechtenböhmer, S.; Nilsson, L.J.; Åhman, M.; Schneider, C. Decarbonising the energy intensive basic materials industry through electrification—Implications for future EU electricity demand. *Energy* **2016**, *115*, 1623–1631. [CrossRef]
52. Perner, J.; Unteutsch, M.; Lövenich, A. *The Future Cost of Electricity-Based Synthetic Fuels*; Agora Energiewende and Agora Verkehrswende: Berlin, Germany, 2018.
53. Jung, O. Industrial Heat Pumps; Svenska Kyl &, Värmepumpdagen, Mölndal, Sweden, 17 October 2019. Available online: https://www.kvpdagen.se/wp-content/uploads/2019/11/Sal%201/20191017%20Industrial%20heat%20pumps%20-Oliver%20Jung.pdf (accessed on 11 June 2021).
54. Van Kranenburg, K.; Schols, E.; Gelevert, H.; de Kler, R.; van Delft, Y.; Weeda, M. *Empowering the Chemical Industry. Opportunities for Electrification*; Whitepaper; TNO and ECN: Utrecht, The Netherlands; Petten, The Netherlands, 2016.
55. Daioglou, V.; Faaij, A.P.C.; Saygin, D.; Patel, M.K.; Wicke, B.; Van Vuuren, D.P. Energy demand and emissions of the non-energy sector. *Energy Environ. Sci.* **2014**, *7*, 482–498. [CrossRef]
56. IEA. *The Future of Hydrogen. Seizing Today's Opportunities*; OECD/IEA: Paris, France, 2019.
57. Karka, P.; Papadokonstantakis, S.; Kokossis, A. Cradle-to-gate assessment of environmental impacts for a broad set of biomass-to-product process chains. *Int. J. Life Cycle Assess.* **2017**, *22*, 1418–1440. [CrossRef]
58. IRENA; Methanol Institute. *Production of Renewable Methanol*; IRENA: Abu Dhabi, United Arab Emirates, 2021.
59. Matzen, M.; Alhajji, M.; Demirel, Y. Technoeconomics and Sustainability of Renewable Methanol and Ammonia Productions Using Wind Power-based Hydrogen. *J. Adv Chem Eng.* **2015**, *5*, 128. [CrossRef]
60. IRENA. *Accelerating the Energy Transition through Innovation*; International Renewable Energy Agency: Abu Dhabi, United Arab Emirates, 2017.
61. Philibert, C. Direct and indirect electrification of industry and beyond. *Oxf. Rev. Econ. Policy* **2019**, *35*, 197–217. [CrossRef]
62. Da Cruz, N.F.; Ferreira, S.; Cabral, M.; Simões, P.; Marques, R.C. Packaging waste recycling in Europe: Is the industry paying for it? *Waste Manag.* **2014**, *34*, 298–308. [CrossRef] [PubMed]
63. Ferreira, S.; Cabral, M.; da Cruz, N.; Simões, P.; Marques, R.C. The costs and benefits of packaging waste management systems in Europe: The perspective of local authorities. *J. Environ. Plan. Manag.* **2016**, *60*, 773–791. [CrossRef]
64. Gradus, R.H.; Nillesen, P.H.; Dijkgraaf, E.; van Koppen, R.J. A Cost-effectiveness Analysis for Incineration or Recycling of Dutch Household Plastic Waste. *Ecol. Econ.* **2017**, *135*, 22–28. [CrossRef]
65. Pilz, H. *Criteria for Eco-Efficient (Sustainable) Plastic Recycling and Waste Management*; Background Report for Associated Presentation No. Version 1.2; Denkstatt: Vienna, Austria, 2014.
66. Bergsma, G. *Chemical Recycling and Its CO_2 Reduction Potential*; CE Delft: Delft, The Netherlands, 2019.
67. Devasahayam, S.; Raju, G.B.; Hussain, C.M. Utilization and recycling of end of life plastics for sustainable and clean industrial processes including the iron and steel industry. *Mater. Sci. Energy Technol.* **2019**, *2*, 634–646. [CrossRef]
68. Kuramochi, T.; Ramírez, A.; Turkenburg, W.; Faaij, A. Comparative assessment of CO_2 capture technologies for carbon-intensive industrial processes. *Prog. Energy Combust. Sci.* **2012**, *38*, 87–112. [CrossRef]
69. Saygin, D.; Broek, M.V.D.; Ramírez, A.; Patel, M.; Worrell, E. Modelling the future CO_2 abatement potentials of energy efficiency and CCS: The case of the Dutch industry. *Int. J. Greenh. Gas Control* **2013**, *18*, 23–37. [CrossRef]
70. Bhave, A.; Taylor, R.H.; Fennell, P.; Livingston, W.R.; Shah, N.; Mac Dowell, N.; Dennis, J.; Kraft, M.; Pourkashanian, M.; Insa, M.; et al. Screening and techno-economic assessment of biomass-based power generation with CCS technologies to meet 2050 CO_2 targets. *Appl. Energy* **2017**, *190*, 481–489. [CrossRef]

71. Pröll, T.; Zerobin, F. Biomass-based negative emission technology options with combined heat and power generation. *Mitig. Adapt. Strat. Glob. Chang.* **2019**, *24*, 1307–1324. [CrossRef]

72. Material Economics. *Industrial Transformation 2050—Pathways to Net-Zero Emissions from EU Heavy Industry*; Material Economics: Stockholm, Sweden, 2019.

73. Brown, T. What Drives New Investments in Low-Carbon Ammonia Production? One Million Tons per Day Demand. Ammonia Industry. 2018. Available online: https://ammoniaindustry.com/what-drives-new-investments-in-low-carbon-ammonia/ (accessed on 11 June 2021).

74. IRENA. *Global Energy Transformation: A Roadmap to 2050*, 2019th ed.; IRENA: Abu Dhabi, United Arab Emirates, 2019.

75. Haro, P.; Ollero, P.; Trippe, F. Technoeconomic assessment of potential processes for bio-ethylene production. *Fuel Process. Technol.* **2013**, *114*, 35–48. [CrossRef]

76. Fasihi, M.; Bogdanov, D.; Breyer, C. Techno-Economic Assessment of Power-to-Liquids (PtL) Fuels Production and Global Trading Based on Hybrid PV-Wind Power Plants. In *Energy Procedia, Proceedings of the 10th International Renewable Energy Storage Conference, IRES 2016, Düsseldorf, Germany, 15–17 March 2016*; Elsevier: Amsterdam, The Netherlands, 2016; pp. 243–268.

77. Ikäheimo, J.; Kiviluoma, J.; Weiss, R.; Holttinen, H. Power-to-ammonia in future North European 100 % renewable power and heat system. *Int. J. Hydrogen Energy* **2018**, *43*, 17295–17308. [CrossRef]

78. Gilbert, P.; Thornley, P. Energy and carbon balance of ammonia production from biomass gasification. In Proceedings of the Bioten Conference on Biomass, Bioenergy and Biofuels, Birmingham, UK, 21–23 September 2010; CPL Press: Birmingham, UK, 2010.

79. Renó, M.L.G.; Lora, E.E.S.; Palacio, J.C.E.; Venturini, O.J.; Buchgeister, J.; Almazan, O. A LCA (life cycle assessment) of the methanol production from sugarcane bagasse. *Energy* **2011**, *36*, 3716–3726. [CrossRef]

80. Uddin, M.; Chowdhury, A.; Anderson, K.; Chaudhuri, K. The effect of COVID-19 pandemic on global stock market volatility: Can economic strength help to manage the uncertainty? *J. Bus. Res.* **2021**, *128*, 31–44. [CrossRef]

81. Ezeaku, H.C.; Asongu, S.A.; Nnanna, J. Volatility of international commodity prices in times of COVID-19: Effects of oil supply and global demand shocks. *Extr. Ind. Soc.* **2021**, *8*, 257–270. [CrossRef]

82. Bakas, D.; Triantafyllou, A. Commodity price volatility and the economic uncertainty of pandemics. *Econ. Lett.* **2020**, *193*, 109283. [CrossRef]

83. Gaudenzi, B.; Zsidisin, G.A.; Pellegrino, R. Measuring the financial effects of mitigating commodity price volatility in supply chains. *Supply Chain Manag. Int. J.* **2020**, *26*, 17–31. [CrossRef]

84. Gabrielli, P.; Gazzani, M.; Mazzotti, M. The Role of Carbon Capture and Utilization, Carbon Capture and Storage, and Biomass to Enable a Net-Zero-CO_2 Emissions Chemical Industry. *Ind. Eng. Chem. Res.* **2020**, *59*, 7033–7045. [CrossRef]

85. Langholtz, M.H.; Stokes, B.J.; Eaton, L.M. *2016 Billion-Ton Report: Advancing Domestic Resources for a Thriving Bioeconomy*; Economic Availability of Feedstocks (No. ORNL/TM-2016/160); US Department of Energy: Oak Ridge, TN, USA, 2016; Volume 1.

86. Mandley, S.; Daioglou, V.; Junginger, H.; van Vuuren, D.; Wicke, B. EU bioenergy development to 2050. *Renew. Sustain. Energy Rev.* **2020**, *127*, 109858. [CrossRef]

87. Daioglou, V.; Doelman, J.C.; Wicke, B.; Faaij, A.; van Vuuren, D.P. Integrated assessment of biomass supply and demand in climate change mitigation scenarios. *Glob. Environ. Chang.* **2019**, *54*, 88–101. [CrossRef]

88. IHS Markit. Plastics Recycling. In *Chemical Economics Handbook*; IHS Markit: London, UK, 2019.

 *energies*

MDPI

Article

Preparing the Ecuador's Power Sector to Enable a Large-Scale Electric Land Transport

Janeth Carolina Godoy [1,*], Daniel Villamar [2,†], Rafael Soria [3], César Vaca [4], Thomas Hamacher [1] and Freddy Ordóñez [2]

[1] Renewable and Sustainable Energy Systems, Technical University of Munich, Lichtenbergstr. 4a, 85748 Garching bei München, Germany; thomas.hamacher@tum.de
[2] Departamento de Ingeniería Mecánica, Escuela Politécnica Nacional, Ladrón de Guevara E11-253, Quito 170525, Ecuador; dvillamar@tech.epn.edu.ec or daniel.villamar@etudiant.univ-perp.fr (D.V.); freddy.ordonez@epn.edu.ec (F.O.)
[3] Department of Mechanical Engineering, Universidad San Francisco de Quito, Diego de Robles y Vía Interoceánica, Campus Cumbayá, Quito 170901, Ecuador; rsoriap@usfq.edu.ec
[4] b4Future, Guangüiltagua N37-266, Quito 170528, Ecuador; cesar.vaca@b4future.com
* Correspondence: carolina.godoy@tum.de
† Département Sciences Pour L'ingénieur, Université de Perpignan, École Doctorale 305, 52 Avenue Paul Alduy, 66100 Perpignan, France.

Abstract: The Ecuador's expansion plans for the power sector promote the exploitation of hydro power potential, natural gas and a small share of alternative renewable energies. In 2019, electricity generation reached 76.3% from hydroelectric power, 21.9% from thermal plants and 1.8% from other renewable resources. Although the power energy mix is mainly based on renewable technologies, the total energy demand is still dependent on fossil fuels, which is the case of the transport sector that alone accounted for 50% of the total primary energy consumed in the country. This paper analyzes the pathway to develop a clean and diversified electricity mix, covering the demand of three specific development levels of electric transportation. The linear optimization model (*urbs*) and the Ecuador Land Use and Energy Netwrok Analysis (ELENA) are used to optimize the expansion of the power system in the period from 2020 to 2050. Results show that reaching an electricity mix 100% based on renewable energies is possible and still cover a highly electrified transport that includes 47.8% of land passenger, and 5.9% of land freight transport. Therefore, the electrification of this sector is a viable alternative for the country to rely on its own energy resources, while reinforcing its future climate change mitigation commitments.

Keywords: hydropower; electric transport; energy modeling; ELENA; *urbs*; Ecuador

Citation: Godoy, J.C.; Villamar, D.; Soria, R.; Vaca, C.; Hamacher, T.; Ordóñez, F. Preparing the Ecuador's Power Sector to Enable a Large-Scale Electric Land Transport. *Energies* **2021**, *14*, 5728. http://doi.org/10.3390/en14185728

Academic Editor: Dolf Gielen

Received: 19 July 2021
Accepted: 6 September 2021
Published: 11 September 2021

1. Introduction

To prevent the worst climatic events, all countries of the world must contribute to the reduction of greenhouse gases (GHG) emissions as was agreed on the Paris Conference in 2015. The Paris Agreement set the goal of keeping the global warming well below 2 °C above pre-industrial levels and even more ambitious to 1.5 °C until the end of this century. To accomplish this, national efforts and pledges are established and published in the so-called National Determined Contributions (NDCs). The emissions gap for 2030-defined as the difference between global total GHG emissions from least-cost scenarios that keep global warming to 2 °C, 1.8 °C, or 1.5 °C, and the estimated global total GHG emissions resulting from a full implementation of the NDCs shows that current unconditional NDCs falls short 15 GtCO$_2$eq by 2030 compared with the 2 °C scenario, and by about 32 GtCO$_2$eq compared with the 1.5 °C scenario. Despite the emissions reduction by about 7% in 2020 compared with 2019 due to the COVID-19 outbreak, the GHG atmospheric concentration continues to rise, which means that the pandemic offered only a short-term reduction with negligible effect by 2030 unless countries pursue a strong post-pandemic recovery process, including a long term deep decarbonization towards net-zero GHG emissions [1].

The transport sector is a major contributor to GHG emissions, as it was responsible for 23% of global energy-related CO_2 emissions in 2014, 72% of which were produced by road transport as was reported by the 5th IPCC Assessment Report [2]. Additionally, it is the fastest-growing sector in terms of emissions and the least diversified energy end-use sector, consuming 65% of global oil in 2018 [3,4]; however, only 8 of the 47 revised NDC submissions for the period 2020–2025 include specific emissions targets for the transport sector [5].

Some of the decarbonization indicators proposed by the IPCC mitigation pathways are: (i) reduction of the carbon intensity of electricity generation, and (ii) increase the electrification rate in final energy consumption sectors. These pillars are strategic for the energy transition in the land transport sector. Even though in several developing countries there is a significant share of renewable energy for electricity generation, the use of electricity in the transport sector is still minimal [6–8]. This is especially true in the Latin American and Caribbean (LAC) region where the stock of battery electric vehicles (BEV) in 2020 represented less than 1% of its global fleet, where Mexico, Brazil and Chile stand out [8]. The LAC region is experiencing the highest growth in car ownership in the world—more than twice the global average of 27% [7]. On the other hand, LAC has the world's highest per capita bus use and also leads in the implementation of bus rapid transit, with systems present in 54 cities as of 2019 [7]. In LAC there are 2000 electric buses in 2020, this is less than 1% of the region's fleet. The city of Santiago (Chile) has the largest number of electric buses in the region, followed by Bogotá (Colombia) [6]. In addition, there are specific pilot projects that have implemented small fleets of institutional electric vehicles or taxis, and free electric chargers in shopping centers or parks in the main cities of the region [7–9].

Following the broad trends of rapid urbanization and increase of private car share in Latin-American countries [10], Ecuador has witnessed a growth of 161% of road transport vehicles (including freight vehicles) between 2008 to 2018 [11]. According to the most recent official national GHG inventory (2012), this sector contributed with 21% of total national emissions [12], and has historically been the most energy-consuming sector in Ecuador. In 2019, it represented almost 50% of the 94 million barrels of oil equivalent (BOE) of the total energy consumed in the country. The road transport is by far the most used mode of transport, accounting for almost 95% of the sectorial energy consumption. In the same year, the use of electricity in the transport sector accounted for 0.02% of final energy consumption [13], which is explained by the operation of a trolley BRT system in Quito. There is much expectation for the massive penetration of electric buses in compliance with the Energy Efficiency Law, which mandates that all new urban and inter-municipal buses from 2025 will be electric [14]. Rail transport is virtually nonexistent, and due to the size of the country (283,560 km^2) air transport is not a viable option for local freight. Heavy freight and passenger vehicles consumed 47% and 27% of the land transport energy demand respectively [13]. In the same year, it consumed 83% of the diesel and 76% of gasoline required in the country; moreover, in the last decade (2009–2019), the consumption of diesel and gasoline in the transport sector has increased in 74% and 119% respectively [13]. Although these data show the weight of the transport sector on the energy consumption and GHG emissions of the country, it has been disregarded in energy projections and plans, while there are no specific commitments in the NDC [15]. Understanding the implications of the future large-scale development of land electric transportation in Ecuador is a challenge that has been little explored [16].

According to the Ecuadorian Electrification Master Plan (PME), the deployment of hydropower will be the priority to supply the future electricity demand, complemented with natural gas for the dry season; whereas the solar, wind, biomass and geothermal deployment will continue at minimal levels [17]. In 2019, the electricity generation reached 76.3% from hydropower, 21.9% from thermal plants (diesel, natural gas, heavy oil), and 1.8%

from solar, wind, and biomass resources. In the last ten years (2009–2019), the hydropower installed capacity increased from 2.1 GW to 5.1 GW, whereas, the installed capacity of other renewable energy technologies just increased from 109 MW to 193 MW in the same period [13]. However, this strategy does not take fully into account the vulnerability to climate change due to the possible high or low hydropower availability scenarios [18–24].

At global scale, there are several academic studies discussing the impacts of stringent decarbonization scenarios with large participation of electric land transport [25–27]. There are less studies focusing on developing countries. For example for LAC region, Gils et al. [28] studied the feasibility of a 100% renewable energy power system in Brazil through sector coupling and regional development; Sauer et al. [29] analyzed the strengths and opportunities of developing in Bolivia and Paraguay a large industry of electric vehicles with Li-ion battery by taking advantage of the mineral potential and the availability of hydroelectricity of both countries, respectively; Meza et al. [30] discussed about the transformation of the Nicaraguan energy mix towards 100% renewable to support the electric mobility development; Lallana et al. [31] presented the required transformation of the productive matrix in Argentina to achieve decarbonization goals with large share of electric vehicles; and, finally, Godínez-Zamora et al. [32] assessed decarbonization scenarios and electrification of the transport sector in Costa Rica. The development of electric transportation in developing countries should also analyze the reliability in the operation of the whole transport system [33], innovation to develop smart urban electric transport systems based in electric car sharing [34], participation of citizens to guarantee governance and transparency [35,36], conditions to attract foreign direct investments and its macroeconomic impacts in terms of job creation and participation of local industries [37–39], and other Political Economy related topics [40,41].

From the literature review, we did not identify scientific publications that analyse the impact on the operation of the Ecuadorian national interconnected power system (SNI) due to a large development of electric land transportation. Although there are few publications presenting analysis of the long-term expansion of the energy system in Ecuador [16–18] and its economic and social impacts [42], there are no scientific publications where detailed power system operating criteria is considered for the calculation of the long-term expansion of the national power system in scenarios with large use of electric mobility. This work helps fill this gap in two ways: (i) it analyses long-term scenarios of massive participation of electric vehicles in Ecuador; (ii) the modelling framework considers detailed power system operating criteria, which are considered for the calculation of the long-term expansion of the national power system. This without counting on the soft-link with an integrated model of the entire energy sector that provides final energy demands for the transportation sector.

In line with the IPCC report on Mitigation Pathways Compatible with 1.5 °C, in this study we adopted a systemic approach for analyzing the inter-dependencies between end-use sectors and energy-supply [3]. In this sense, an electrified, low-emission transport sector could be achieved only if structural changes are applied to the energy matrix at a proper pace, balancing the introduction of renewable energy technologies and the phasing out of fossil-based power generation. Therefore, we analyze three scenarios with different degrees of electrification in the transport sector by 2050, which adds to the electricity demand of the other sectors. Additionally, we also included technology-focused measures (energy efficiency and fuel switching), as well as structural changes on the transport activity; the former contributes to the reduction of CO_2 emissions, while the latter to the reduction of energy consumption [3].

Although Ecuador contributes with a minimal part of the global GHG emissions, it cannot remain on the side-lines of the economic, technological and social opportunities that arise from sustainable and low carbon strategies for a post-pandemic and post-extractivist future [43,44]. This study seeks to understand to what extent and under what conditions renewable energy can suply electricity demand until 2050 in a context of transport decarbonization in Ecuador, and, at the same time, how they can complement each other to generate reliable and affordable electricity.

The document is structured as follows: Section 2 details the methods including the description of ELENA and *urbs* models, a presentation of the three analyzed scenarios, the detailed modeling of the Ecuadorian power sector using *urbs*, and finally, the model validation. Section 3 presents the results and discussion about the expansion of the power system by 2050 for the three scenarios and provides recommendations for future research works. Section 4 contains the main conclusions of this work. Finally, Section 5 presents the future works lied to the results.

2. Methods

2.1. Modeling Tools: urbs And Elena

2.1.1. Linear Optimization Model for Distributed Energy Systems (*urbs*)

The Ecuadorian power system was modelled using *urbs*, which is an open-source linear optimization-modeling framework for capacity expansion and unit commitment analyses developed by the Chair of Renewable and Sustainable Energy Systems at Technical University of Munich (ENS-TUM). It minimizes the annual energy system costs, including all investment costs by their annualized depreciation, as well as the operational and environmental costs. Furthermore, it allows the integration of multiple input and output commodities resulting in detailed representations of the energy conversion processes. *urbs* has a high temporal resolution (8760 h per year) that allows the visualization of the chronological behavior of supply and demand. This model ensures that the required demand is covered by the input commodities and technological processes at every time step. Energy and power capacities are expanded independently; however, a linear dependence between them is integrated [45–47]. The *urbs* toolchain is presented in Figure 1.

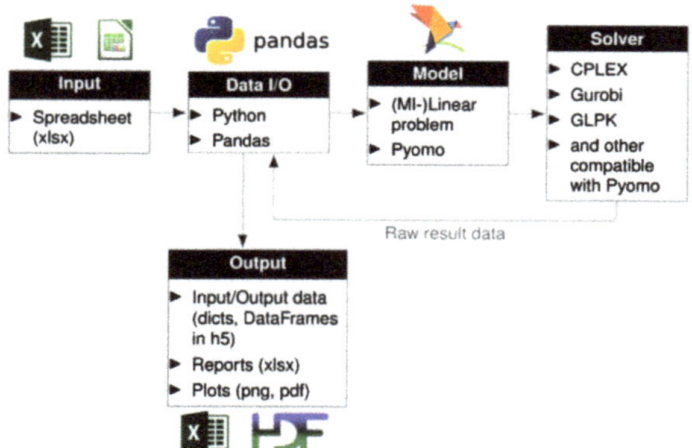

Figure 1. *urbs* toolchain [48].

2.1.2. Ecuador Land Use and Energy Network Analysis Model (Elena)

The Ecuador Land Use and Energy Network Analysis (ELENA) model is an application of the MESSAGE platform [49] for Ecuador [16]. ELENA model assesses the expansion of the energy and land occupancy systems, and its greenhouse gases (GHG) emissions evolution up to 2050. It is a partial equilibrium, integrated, perfect foresight, linear programming optimization model. The objective function is the total cost of the energy-land system expansion up to 2050. The analysis period is 2015 to 2050 (8 steps), the year has 60 time steps (monthly with 5-periods daily). The model considers three geographical regions in Ecuador: Coast, Andes and Amazon. To model the energy sector, ELENA considers the whole energy conversion chain, from primary energy to useful energy in five economic sectors, including transport [16]. For the land occupancy system, it calculates the land use changes according to the food demand and deforestation/reforestation scenarios up to 2050. Useful energy, food demands and deforestation/reforestation scenarios are exogenously calculated. The model includes a wide variety of technologies, each with its capital cost, O&M cost, efficiency, lifetime and other operational information. ELENA was developed during the Deep Decarbonization Pathways in Latin America and the Caribbean (DDP-LAC) project [9], with the support of the Cenergia Lab from the Federal University of Rio de Janeiro (COPPE/UFRJ) [50]. A detailed description of the model structure, mathematical approach and its data base is available in [16].

2.1.3. *urbs* and Elena Interaction

The ELENA and the *urbs* models are soft linked. In the present work the integrated assessment model ELENA was used to evaluate several scenarios with different commitment levels for decarbonization of the energy matrix of Ecuador. Each scenario was supported with a narrative that include not only environmental and technological restrictions but also behavioural changes from the demand perspective. *urbs* uses the transport final energy demand calculated with ELENA as input for computing the expansion of the power system until 2050. On the one hand, the scenario with no restrictions (minimal cost scenario) developed in [16] was used as input for the least cost scenario (LC) in the *urbs* model. On the other hand, the scenario with a stringent emission cap to maintain the country development in the path of 1.5 °C above pre-industrial levels determined by the IPCC (deep decarbonization scenario), presented in [16] was used to build the moderate (Mod) and the deep decarbonization pathway (DDP) scenarios. To ensure that the scenarios are compatible with an emission reduction trend of 1.5 °C, the ELENA model uses as constraint a carbon budget calculated with the COFFEE model. The methodological details for this calculation can be found in [16]. The three scenarios will be described in the next section. A scheme showing the interaction between the models is depicted in Figure 2.

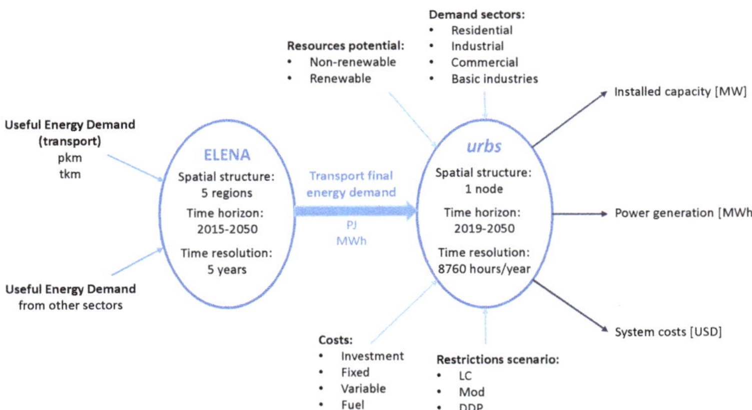

Figure 2. ELENA and *urbs* relationship in the context of this work.

2.2. Scenarios Definition

The scenarios used in this work are based on the National Energy Forecast [51], which assesses the Ecuadorian energy development until 2050 following a set of policies described in the Ecuadorian Master Electricity Plan (PME), the National Energy Agenda and the National Plan of Energy Efficiency (PLANEE). The set of premises used are presented in Table 1.

First, the National Energy Agenda proposes to use the hydropower potential to make it the main electricity source with at least 70% of the total power generation by 2040 [52]. Second, following that direction, the PME has planned by 2027 to incorporate 360 MW from hydropower, 410 MW of PV, wind, and geothermal projects, and 187 MW of combined cycle power plants [17]. Finally, the National Plan of Energy Efficiency (PLANEE) introduces a number of actions to improve energy consumption by incorporating energy efficiency measures in the energy supply and demand sectors. The detailed description of these measures can be found in [53].

Table 1. Premises for the LC, Mod, and DDP scenarios.

Variable	LC Scenario	Mod Scenario	DDP Scenario
GDP growth	4%	4%	4%
Changes in demand behavior	Small	Medium	High
Basic industries as electricity consumer	No	Yes	Yes
LPG replacement by induction stoves in the residential sector	0%	0%	3 million by 2025
Natural gas for transportation	Yes	Yes	No
Private mobility share	+36%	−55%	−55%
Public mobility share	−29%	Maintained	Maintained
Rail in public transportation	Reaches 1%	Reaches 3%	Reaches 3%
Non motorized mobility share	Maintained in 1%	Reaches 10%	Reaches 10%
Average travel distance for cars	+13%	−25%	−25%
Average travel distance for buses	Maintained	−36%	−36%
Occupancy in private cars	Maintained	+6%	+6%
Occupancy in buses	Maintained	+25%	+25%
Power generation	Expansion of the power system according to the least cost technology	Expansion of the power system following the PME guidelines by 2027. Diesel and heavy oil plants replaced by CCGT	Expansion of natural gas power plants allowed until 2030 and their phase out by 2050

Sources: [16,17,51].

2.2.1. Least Cost Scenario (LC)

From the demand side, the premise that characterized the LC scenario is the small change on the electricity consumption behavior in the residential, industrial, commercial, and transport sectors compared the base year 2019. The energy efficiency policies detailed in the PLANEE are not considered. Regarding to the transport sector, it presents an increase in the private mobility participation, no changes in the non-motorized share, and a minimal electrification share. In the supply side, the expansion plan proposed by the PME until 2027 is not considered. The objective of this scenario is looking for the minimum cost power system expansion that satisfies the future electricity demand.

2.2.2. Moderate Scenario (Mod)

The Mod scenario considers the same demand in the residential, industrial, and commercial sector as in the LC scenario; however, it includes the development of basic industries such refineries, shipbuilding, petrochemicals, and metallurgy (iron and steel) as a new electricity demand sector [54]. The transport sector presents significant reductions in the private mobility, and an increase in the non-motorized mobility share compared with the LC scenario. Moreover, the energy sources for the land transport are diesel, gasoline, natural gas, and electricity. In the supply side, this scenario is in line with the current energy policies for the power system expansion considered in the PME by 2027. It follows the same path for the expansion power system until 2050 with big participation of hydropower complemented with natural gas, and small participation of non conventional renewable technologies.

2.2.3. Deep Decarbonization Pathways Scenario (DDP)

The DDP scenario considers a bigger electrification rate in the consumer sectors compared with the LC and Mod scenarios due to the inclusion of the energy efficiency policies detailed in the PLANEE. The private, and public mobility shares, and therefore the energy consumption of the transport sector are the same in the Mod and DDP scenarios, neverthe-less, natural gas is no longer a fuel option for transport, and it is replaced with electricity. In the supply side, this scenario gives the opportunity to deploy other renewable technolo-gies through the natural gas constraint. The DDP scenario looks for the diversification of a clean energy mix, and a high rate of electrification in the transport sector.

2.3. Modeling the Ecuadorian Power Sector with urbs and ELENA

2.3.1. Model Structure

urbs consists of several model entities, such as commodities, processes, transmission, and storage. This work considers as commodities the fluctuating natural resources such solar radiation, wind velocity, and basins' flow rate, each of them with its own set of hourly time series of 8760 time steps. On the other hand, natural gas, diesel, heavy oil, biogas, geothermic, and electricity are considered as stock commodities (not fluctuating in time). Additionally, *urbs* needs the conversion processes which in the Ecuadorian case are hydropower plants, thermal plants, PV systems, wind farms, geothermal plants, biogas, and bagasse plants, whereas water reservoirs represent the stored commodity. This model considers Ecuador as one node, so we do not take into account the internal transmission lines.

The model needs the following inputs: (i) Total usable area for each specified site; (ii) Energy resources that includes renewable (solar, wind, geothermal, biomass, and biogas) and non-renewable (heavy oil, diesel, and natural gas) resources, as well as the imported electricity and transaction prices; (iii) Technical specifications of each type of power plant such as the installed and the maximum capacities, lifetime, minimum load fraction, max-imal power gradient, investment costs, fixed and variable costs; (iv) Electricity demand represented by a set of time series for each analyzed year; and (v) Scenarios described in Section 2.2. The outputs include (i) Database and plots of the power system profile with one hour resolution; (ii) Total installed capacity for every studied year; (iii) Costs of the system during the analyzed period of time; (iv) Detailed data of commodities consumption on each time step. The *urbs* model scheme for the Ecuadorian power system used in this work appears in Figure 3.

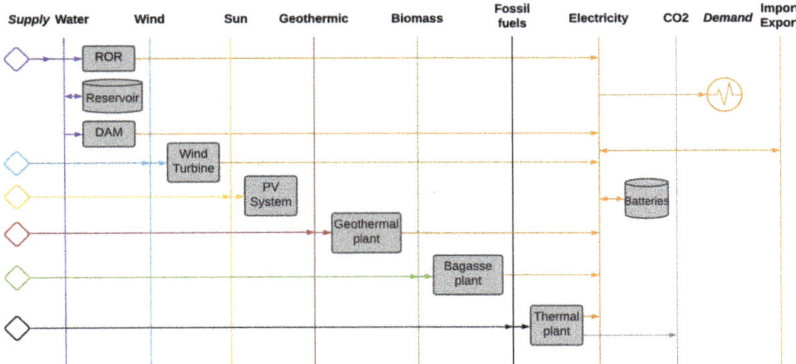

Figure 3. Ecuadorian power system representation in *urbs*.

2.3.2. Supply Side Modeling

The time period 2019–2050 is simulated as a cascade set-up using the years 2019, 2030, 2040 and 2050. The input data are updated for each simulated year, in this way, the evolution of the Ecuadorian power system for the whole time period is projected. The capacity expansion by technology, the addition of transmission capacity, the integration of storage technologies, the overall generated clean energy, and the total system costs for the representative years are simulated under the three different scenarios.

To start modeling the Ecuadorian power sector, we select 2019 as the base year. The actual electricity delivered to the Interconnected National System (SNI) in this year came from 76.3%

of hydroenergy, 21.9% of thermal power generation based on diesel, natural gas, and heavy oil, and 1.8% of renewable resources like biomass, solar, wind, and biogas. Total installed capacity reached 8512 MW, mostly hydropower plants located in the Highlands and in the Amazon region, while thermal plants are mainly located in the Amazon and Coast regions (see Figure 4) [55,56]. For the other modelled years in the Mod, and DDP scenarios, we use the power plant portfolio detailed in the PME by 2027 (see Table 2), and the feasible areas for wind parks and PV systems shown in Figure 5, whereas for the LC scenario the expansion of the power system follows the least cost criteria used by the model.

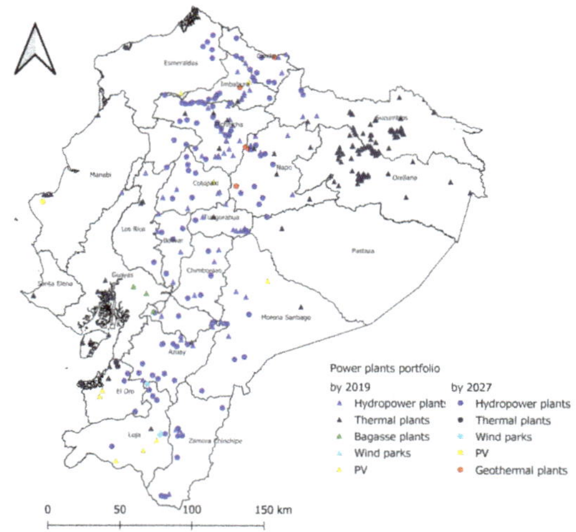

Figure 4. Power plants location by 2019 and by 2027 [55,56].

Table 2. Type of technology, installed capacity, and resource potential included in *urbs*.

Resource	Technology	Installed Capacity in 2019 [MW]	Potential [MW]
Water	Large DAM (>450 MW)	1075	6975.6
	Medium DAM (50–450 MW)	616	369.13
	Large ROR (>450 MW)	1987	1920
	Medium ROR (50–450 MW)	748	2229.8
	Small ROR (<50 MW)	653	1365.25
	Not defined	-	4061
Solar	PV-US	25	67,500
	PV-DG	0	n.a.
Wind	Wind park onshore	16.5	1000
Geothermal	Geothermal plants	0	900
Biomass	Bagasse plants	144.3	n.a.
Biogas	Municipal Solid Waste Biogas	7.26	n.a.
Natural gas	OCGT	19.42	n.a.
	CCGT	644.18	n.a.
Diesel	ICE	1216.42	n.a.
Heavy oil	ICE	1359.91	n.a.

PV-US: Photovoltaic utility scale, PV-DG: Photovoltaic distributed generation, OCGT: Open cycle gas turbine, CCGT: Combined cycle gas turbine, ICE: Internal combustion engine. Sources: [17,57].

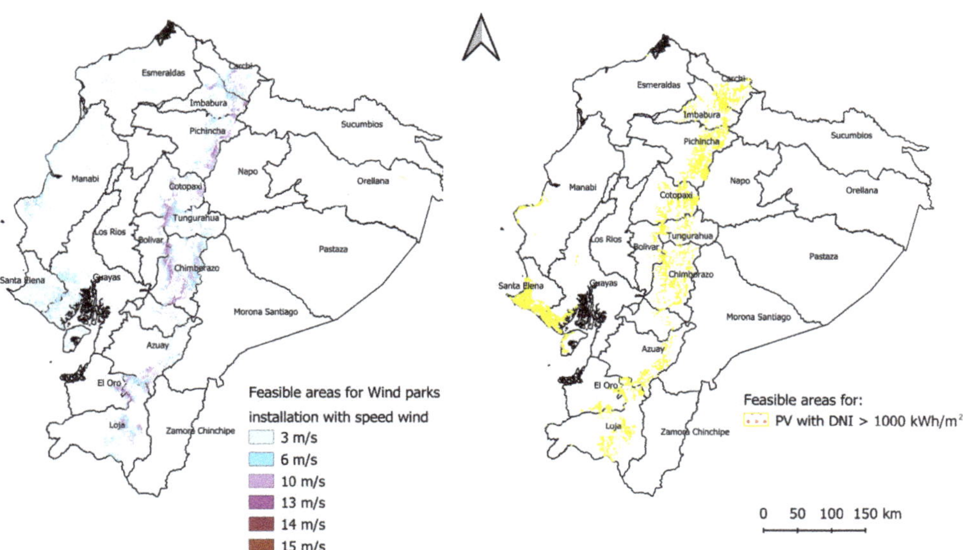

Figure 5. Feasible areas for wind parks with wind speed higher than 3 m/s. and for photovoltaic plants with DNI higher than 1000 kWh/m² [57].

The *urbs* model also requires economic data for every technology specified. Figure 6 shows the fluctuation of natural resources such as solar radiation, and wind velocity which are used as time series of availability factors obtained from the online tool Renewables.ninja developed by Pfenninger and Staffell [58]. The availability factors for hydropower plants are based on the average flow rates in the reservoirs of the Ecuadorian hydropower plants.

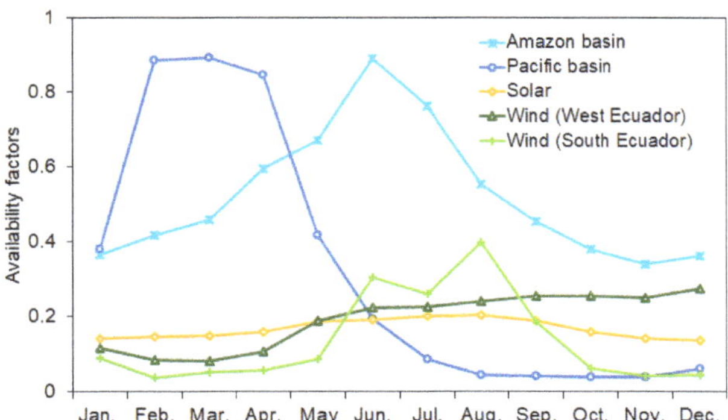

Figure 6. Monthly normalized availability factors for hydropower plants in the Amazon and Pacific basins, wind parks in Western and Southern Ecuador, and photovoltaics [58–61].

2.3.3. Demand Side Modeling

In the base year 2019, the residential, industrial, commercial, public lighting, and construction sectors consumed together 21.91 TWh of electricity. In Ecuador, these sectors are known as public service demand; the non-public service demand, mostly represented by oil companies (3.78 TWh consumption), is not considered as demand sector in this work [13]. As *urbs* needs an exogenous energy demand to be satisfied through the optimization process, we used the transport demand delivered by ELENA, which uses as input the passenger-kilometer (pkm) and ton-kilometer (tkm) data. The details of the transport demand projection are shown in Section 2.3.4, whereas the electricity demand of the other sectors is taken from the National Energy Forecast [51].

2.3.4. Transport

For modeling the transport sector, 2015 is the selected base year. The information for characterizing the vehicle fleet is available in the Statistical Yearbook of Transportation [62]. The total energy consumed by land passengers transport was 79.54 PJ (63.3 PJ of gasoline, and 16.24 PJ of diesel), and the consumption of the freight transport was 82.33 PJ (100% diesel). The activity level of the sector could be represented with the amount of energy and fuels consumed available from [13,63]. Some of the characteristics of the freight and passenger transport are presented in Tables 3 and 4 respectively.

Table 3. Freight Transport Characteristics.

Fleet Categorised by Size	LDV	MDV	HDV
No. of vehicles ×1000	138	189	97
Average traveling distances (km)	17,000	27,000	30,000
Average load transported (tons)	0.3	1.8	8.7

Table 4. Passenger Transport Characteristics.

Road Fleet	Cars	Motorbikes	Buses
No. of vehicles ×1000	1290	431	47
Average traveling distances (km)	14,200	6000	55,000
Occupancy rate	1.7	1.1	20

For the base year, passengers demand is the sum of all the vehicles multiplied by the year-average distance travelled for vehicle type multiplied by the average occupancy rate. The calculation for freight demand is similar, replacing the occupancy rate by the year-average load transported. To calculate the future pkm demand, a projection for the population [64] is used as driver, while the tkm demand is forecasted using the GDP as driver, both are presented in Table 5.

Table 5. Passenger and freight transport demand and drivers.

Year	2015	2020	2030	2040	2050
GDP(billion USD)	71.7	74.6	86.7	115.8	154.5
Population(millions)	16.3	17.5	19.8	21.8	23.4
Freight demand (Gtkm)	35.2	36.6	42.6	56.9	75.9
Passenger demand (Gpkm)	94.5	101.4	114.7	126.3	135.6

These main drivers are the same for the different scenarios, but the narratives considered for the scenarios are different. In the Mod and DDP scenarios, a reduction of the individual transport is considered in favour of public mass transport modes, whilst the actual growing tendency of private cars is maintained for the LC scenario. Tables 3–5 were the inputs for the transport demand projection calculated and shown in [16], which is used as exogenous demand in the present work.

Table 6 shows the energy consumption in PJ for the passengers land transport sector for the analyzed years. The LC scenario shows consistently the biggest energy demand for the analyzed period. Gasoline was the predominant consumed fuel with almost 85 PJ that represents 72.5% of the total energy consumption, far followed by diesel and natural gas with 14.2% and 12.5% respectively. Electricity demand specified for this scenario just represented 0.8% of the total energy consumption for passenger transportation by 2050.

In the Mod scenario, the natural gas consumption reached 9.63 PJ, followed by gasoline with 8.9 PJ and diesel with 7.84 PJ. These three fuels represented 82.3%, and electricity 17.7% of the total energy consumption by 2050. For the DDP scenario, natural gas was not considered as fuel for transportation, since it was replaced by electricity. It represented 47.8% of the total energy consumed for passenger land transport by 2050, followed by gasoline and diesel with 27.8% and 24.5% respectively.

Table 6. Passengers land transport demand in PJ.

	LC Scenario			Mod Scenario			DDP Scenario		
Source	**2030**	**2040**	**2050**	**2030**	**2040**	**2050**	**2030**	**2040**	**2050**
Electricity	0.43	0.68	0.95	1.92	3.97	5.68	21.48	18.66	15.31
Gasoline	75.29	74.97	84.74	20.86	7.06	8.90	20.86	7.06	8.90
Diesel	17.34	17.28	16.62	11.35	9.86	7.84	11.35	9.86	7.84
Natural gas	6.27	13.48	14.61	19.55	14.69	9.63	-	-	-
Total	99.33	106.42	116.92	53.69	35.58	32.06	53.69	35.58	32.06

Table 7 refers to the freight land transport demand in PJ. Electricity and diesel are the energy sources used in the three scenarios, with no participation neither natural gas nor gasoline. In the LC scenario, diesel represented almost 100% of the total energy consumed by freight land transport by 2050. For the Mod and DDP scenarios there is a small participation of electricity as energy source for the freight transport with 5.9% of the total energy required, the remaining 94.1% corresponds to diesel.

Considering the electricity consumption in Tables 6 and 7, the total electricity demand of the land transport sector for the LC, Mod, and DDP scenarios are shown in Table 8. The ELENA results presented in Table 8 are added to the electricity demand from the other consumption sectors for each scenario.

Table 7. Freight land transport demand in PJ.

	LC Scenario			Mod Scenario			DDP Scenario		
Source	**2030**	**2040**	**2050**	**2030**	**2040**	**2050**	**2030**	**2040**	**2050**
Electricity	0.03	0.06	0.12	4.18	8.98	15.42	4.18	8.98	15.42
Diesel	221.1	279.2	348.7	184.5	208.5	247.1	184.5	208.5	247.1
Total	221.1	279.2	348.8	188.7	217.5	262.5	188.7	217.5	262.5

Table 8. Land transport electricity demand in TWh.

	LC Scenario			Mod Scenario			DDP Scenario		
Source	**2030**	**2040**	**2050**	**2030**	**2040**	**2050**	**2030**	**2040**	**2050**
Electricity	0.13	0.21	0.30	1.70	3.60	5.86	7.13	7.68	8.54

Figure 7 shows the total electricity demand used as external input for *urbs*, from 2020 to 2050. Land transport has a notorious participation in the electricity consumed in Mod and DDP scenarios with 6.3% and 8% respectively by 2050 compared with the 0.4% in the LC. The Mod and DDP scenarios present a marked increase in electricity consumption due to the participation of the so-called basic industries and higher levels of electrification in the residential, industrial, commercial and transport sectors.

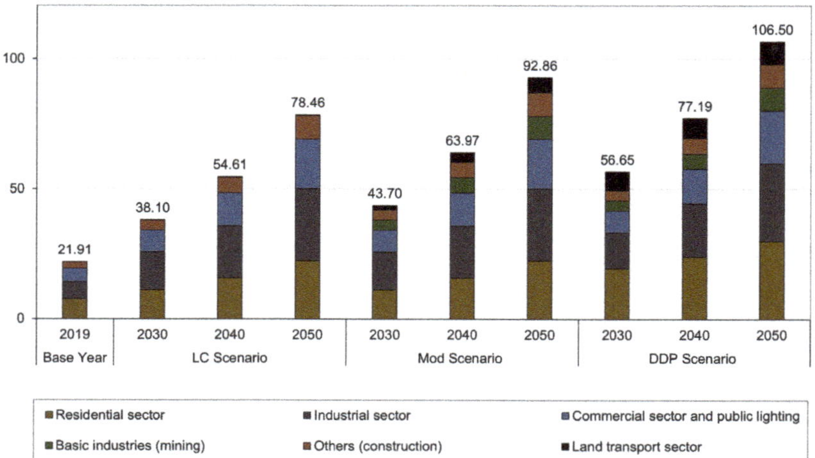

Figure 7. Electricity demand per sector in TWh for the LC, Mod, and DDP scenarios.

2.4. Model Validation

In 2019, the electricity generated by the Ecuadorian power system reached 31.72 TWh [55]. The installed capacity of this year, shown in Table 2, was modelled in *urbs* and the cost-optimal operation of the system, i.e., how much electricity is generated with each technology at every time step. In this section, there is a comparison between the electricity generation simulation (Base 2019) and the official data from ARCONEL (see Figure 8).

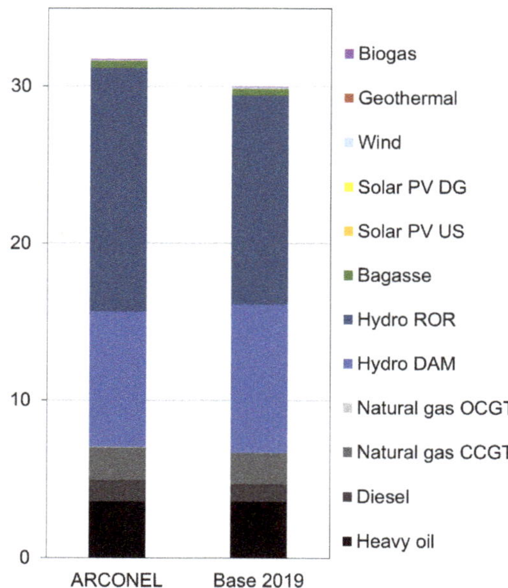

Figure 8. Annual electricity generation in TWh according to the *urbs* model Base 2019 and the ARCONEL statistics of 2019.

The electricity generation in the *urbs* model reaches 29.99 TWh, and properly represents the participation of the different technologies compared with the real data. In both mixes, hydropower has the biggest participation in the total power generation with 76.1% and 75.8% in ARCONEL and Base 2019, respectively. Thermal technologies run with heavy oil, diesel, and natural gas, in both mixes, account for 22% of the generation. Power generation from renewable resources (solar, wind, biomass, and biogas) represents 1.8% according to ARCONEL and 2% according to Base 2019. A relative error of 5.4% is observed for the total generation in the *urbs* model.

3. Results and Discussion

3.1. Analysis of the Transport Sector Energy Demand

The LC scenario shows a low electrification rate in the land transport sector, reaching only 0.30 TWh by 2050. It follows the current consumption behavior with a big participation of fossil fuels. In the Mod scenario, the electrification of the transport sector reaches 5.86 TWh by 2050, compared with the 0.011 TWh in 2019, but still contains an important share of natural gas as fuel for transport vehicles. The DDP scenario presents an electricity consumption of 8.54 TWh by 2050, which is 2.68 TWh more than the Mod scenario for the same year, with the particularity that natural gas used in the Mod scenario is totally replaced by electricity. Besides the increase of the electrification rate in the Mod and the DDP scenarios, these consider a substantial reduction in the land transport energy demand from 116.92 PJ in the LC scenario to 32.06 PJ by 2050 in the Mod and DDP scenarios (please refer to Table 6).

The results show that electrification of the land passengers transport sector at the levels proposed for each scenario is indeed possible. However, a high electrification of this sector by non-conventional renewable energies is only possible if at the same time final energy demand is reduced, as is the case in the Mod and DDP scenarios. The LC scenario depicts the trend growth of transport, derived from the expected increase in GDP and population, but without considering any measures to restrict the use of private motor vehicles and increase energy efficiency, so that its final consumption is more than three times the demand of the Mod and DDP scenarios. The electrification of the whole LC energy demand for land passengers transport (116.92 PJ) implies that renewable energies would have to be massively deployed and should include battery-based storage systems, which would increase investment costs even above the costs of the other scenarios. Analyzing the electrification of a constantly growing land passengers transport as the only measure of decarbonization was not the scope of this study.

In addition, electricity represents a small part of the total final energy consumption of land freight transport. This is because there are no credible assumptions to consider that electric trucks could enter the Ecuadorian market in the coming decades on a large scale. However, it is observed that in the Mod scenario, the total electricity consumption of land freight transport in 2050 is three times higher than the electricity consumption of passenger transport; while their consumption in the DDP scenario are almost equal, so the contribution of the freight land transport to the electricity demand and its impact on the energy mix cannot be disregarded (see Tables 6 and 7).

3.2. Installed Capacity and Electricity Generation

Electricity generation in Ecuador is already highly renewable, with more than 85% generated by hydropower plants; however, the deep electrification of the transport sector is still a challenge, not fully considered in the Ecuadorian energy policies or GHG emission reduction measures.

The installed capacity, presented in Figure 9, shows that for the three scenarios, hydropower continues as the least cost clean technology; complemented in the LC and Mod scenarios with natural gas, and in the DDP scenario with a mix of other renewable technologies. In the LC scenario, the installed capacity grows from 7.93 GW by 2030 to 14.25 GW by 2050. During the studied period, natural gas is used to complement hydropower generation, with minimal participation of other renewable technologies. By 2050, hydropower reaches 10.4 GW of installed capacity, followed by 3.08 GW of natural gas, other fuels such as diesel and heavy oil are no longer used by 2030, and the deployment of renewables such as solar, wind, geothermal, biomass, and biogas is minimal reaching in total 0.77 GW.

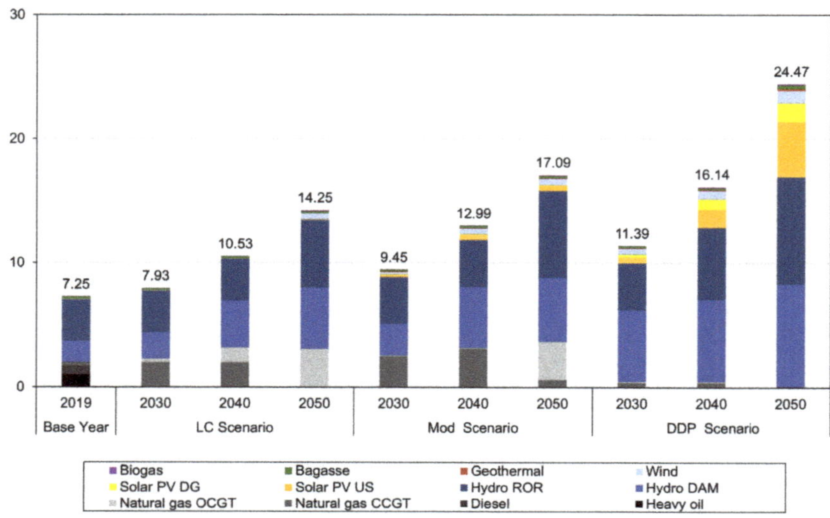

Figure 9. Installed capacity in GW for the Ecuadorian power sector by 2019, 2030, 2040, and 2050 per scenario.

In the Mod scenario, as the demand increases, the installed capacity also grows from 9.45 GW by 2030 to 17.09 GW by 2050. Natural gas complements hydropower during the whole analyzed period, and the participation of other renewables increases compared with the LC scenario. By 2050, hydropower has an installed capacity of 12.14 GW, followed by 3.66 GW of natural gas. Other non-conventional renewables reach 1.29 GW, which is in line with current national policies for the power system expansion stated in the Ecuadorian Master Electricity Plan.

The deployment of renewable energies, especially solar energy, in the DDP scenario replaces the expansion of natural gas power plants since 2030 and phases them out by 2050. PV-US technology reaches 4.5 GW, PV-DG 1.50 GW, wind farms 1 GW, bagasse 0.3 GW, geothermal 0.15 GW, and biogas 0.1 GW, all together represent 30.86% of the total installed capacity in the Ecuadorian power system; hydropower has the remaining 69.14%. In this scenario, by 2030 the energy mix reaches 95.94% based on renewable energies, 97.13% by 2040, and 100% by 2050.

Figure 10 shows that for the LC scenario, hydropower reaches more than 70% of the total generation, which is a goal stated in the National Energy Agenda [52] (78.95% by 2030, 82.96% by 2040, and 92.11% by 2050). Natural gas decreases its participation from 19.42% by 2030 to 3.12% by 2050. As the deployment of other renewable energies (solar, wind, bagasse, geothermal, and biogas) is minimal, it contributes with a small participation into the mix reaching 4.77% of the total generation by 2050.

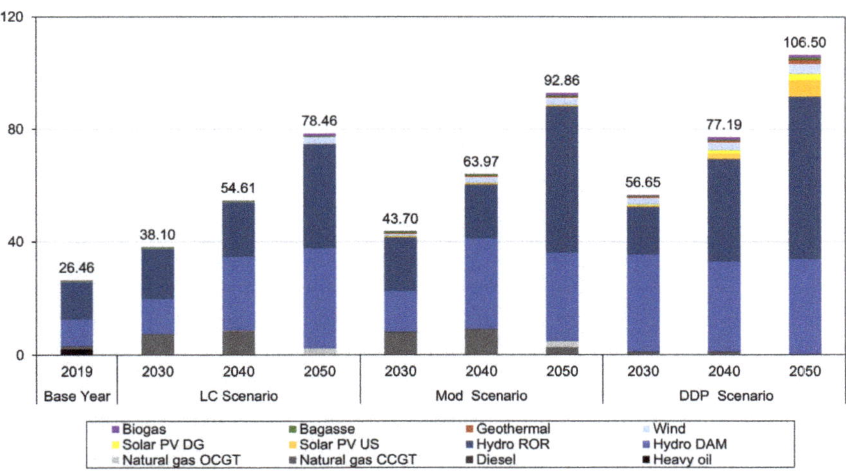

Figure 10. Power generation in TWh for the Ecuadorian power sector by 2019, 2030, 2040, and 2050 per scenario.

The Mod scenario also presents a high participation of hydropower into the energy mix. It represents 76.84% by 2030, 79.82% by 2040, and 89.52% by 2050 of the total electricity generation. Besides, the natural gas participation decreases from 18.39% in 2030, 14.34% in 2040, and 5.3% in 2050. The power generation from other renewable technologies reaches 5.17% of the total generation by 2050. This scenario presents a power mix mostly based on renewable resources during the whole period (81.61% by 2030, 85.66% by 2040, and 94.69% by 2050), but fossil fuels such natural gas still has a participation into the supply mix, and at the same time is used as fuel in the passenger land transport (please refer to Table 6).

The results for the DDP scenario show that, by 2050, is possible to reach a high electrification rate in the passengers land transport (47.8%) and at least 5.87% in the freight transport (see Tables 6 and 7), while the power mix is completely based on renewable resources. In 2050, the electricity generation matrix is composed of 85.98% hydropower, 7.65% solar, 3.40% wind, 1.24% geothermal, 0.86% bagasse, and 0.82% biogas. A 100% renewable mix is possible with the limitation of natural gas usage that allows the deployment of alternative renewable technologies.

Although the scope of this study was not the calculation of the GHG emission reductions, the need for a deep decarbonization of the transport sector is framed within the context of climate change mitigation. In this sense, all scenarios can be seen, at first glance, as clean due to the large share of hydropower in the present and projected future. However, in order to reduce GHG emissions through electrification, both supply-side and demand-side measures must be implemented. This is remarked in the case of the LC scenario, in which the electricity demand maintains the same behavior as today, therefore, the energy mix is mainly based on hydropower, but the transport sector remains highly dependent on fossil fuels. This reliance on hydropower is also seen in the other scenarios, as it is the most stable low-emission technology. Despite Ecuador's vast water resources and experience with hydropower plants, relying so heavily on a single technology can pose disadvantages. First, it has been demonstrated that water reservoirs can emit significant amounts of GHGs, especially in flooded tropical soils, as is the case of Ecuador [65]. Secondly, available studies show the vulnerability of hydroelectric projects to climate change in Ecuador, as water availability (high or low water scenarios) can induce a variation in electricity generation of between 29% and 86% [18], thus causing a significant risk of electricity shortages for demand sectors. Finally, large hydropower plants can be seen as a form of the classical extractivism model that encourage the exploitation of enormous quantities of natural resources causing socio-environmental conflicts, that have been well reported throughout Latin America [66–69].

In order to understand how an electrified transport sector in combination with other electricity demand sectors can be reliably supplied through renewable national resources, we analyzed in detail the DDP scenario. It contains the greatest effort in terms of diversifying demand sectors, achieving high levels of electrification in each of them. With this regard, the results obtained for this scenario fulfill the two selected decarbonization indicators of the IPCC mitigation pathways: the reduction in the carbon intensity of electricity, and the increase in the share of final energy provided by electricity. Note that this scenario could be reached only if technology-focused measures (energy efficiency and fuel switching), as well as structural changes to avoid or shift transport activity are implemented at the same time, which were integrated in the premises of this scenario (Table 1). For ease, Figure 11 shows just the load curves for the DDP scenario.

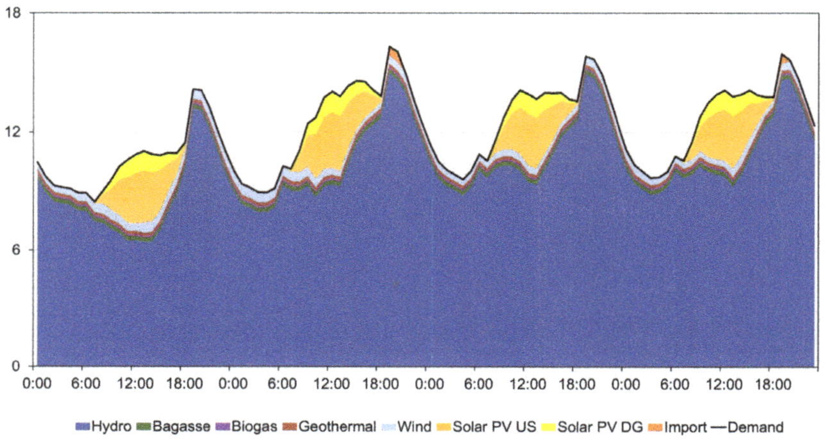

Figure 11. Electricity dispatch per hour in TWh during four days in October 2050 for the DDP scenario.

The load curves show that the diversification of natural resources for electricity generation makes it possible to use each resource according to its hourly availability. During the day (from 06:00 to 18:00), solar energy contributes to the energy mix, while water is stored in reservoirs for its usage at peak hours (from 19:00 to 22:00). All renewable technologies in combination are able to supply the whole demand without the necessity of fossil fuels; nevertheless, possible high energy peaks during the dry season (October to March) could require importing electricity from neighbour countries Colombia and Peru. An electrical interconnection network with these countries already exists, but this could lead to a delocalization of GHG emissions. It must be said that these three countries share the same time zone and would have similar peak demand times, thus is probable that the electricity purchased from these countries comes from non-renewable sources.

From 1:00 to 8:00 there is a valley in the electricity demand curve. This low consumption time slot could become, through the implementation of a low electricity tariff, an ideal period for recharging the batteries of electric vehicles. This kind of incentive would increase the appeal of this vehicles and ensure that an increasing fleet does not represent an extra load during peak time. Nowadays in Ecuador, electricity subsidies are determined according with the overall consumption level, switching to a time-based cost of electricity can also reduce the consumption at peak time which is one of the major concerns from the generation side. This kind of electricity price analysis should be considered as a topic for future research.

3.3. Costs of The System

The total costs of the system during the whole period vary among the three scenarios. For the LC scenario the total cost reaches USD 31.71 billion, whose investment component (USD 14.88 billion) is the highest, representing 47% of the total, followed by fixed costs (USD 8.83 billion), the fuel costs (USD 7.48 billion), variable costs (USD 0.41 billion), and import costs (USD 0.12 billion), which represent the import of electricity that is still needed in this scenario.

As the installed capacity increases for the Mod scenario, also the cost of the system (USD 44.32 billion) which is 1.4 times higher than the LC because of the increased deployment of alternative renewable technologies. This leads to a 58% share of the investment costs with respect to the whole system costs (USD 25.58 billion). As this scenario still has thermal electricity generation with natural gas, the costs for the fuel represents almost 19% (USD 8.30 billion), while the fixed costs reaches 22.54% (USD 9.99 billion), the variable costs depicts 0.77% (USD 0.34 billion), and the costs for electricity imports represents 0.24% (USD 0.10 billion).

The DDP scenario requires the strongest effort in terms of total costs (USD 59.44 billion), which are 1.9 and 1.3 times bigger than the LC and Mod scenarios, respectively. It is clear that the investment costs represent by far the largest component of the total system costs due to the new renewable technologies. The fixed costs (USD 12.15 billion) are higher than the fuel costs (USD 2.73 billion) due to the 100% renewable energy mix by 2050. There are still fuel costs in this scenario due to the presence of thermal electricity generation until 2040 that is then completely replaced by renewable technologies. The variable costs are USD 0.13 billion, and the import of electricity costs reaches USD 1.65 billion. All these results can be seen in the Figure 12.

Given that the three scenarios depict different installed capacities, a cost comparison can be misleading. It is important to remember that the DDP scenario is purposely designed to show an important national economic growth coupled with behavioral changes in the demand side and a strong effort to reach a totally renewable energy mix. This is the reason why DDP represents the highest cost, as it reflects the effort of a change towards sustainability in both demand and supply sides of the energy system. In contrast, the LC scenario represents the trend growth without major changes in both energy demand and supply. It reflects factors as techno-economic characteristics of the technological components, the infrastructure at the system level, and the institutional characteristics that favor one technology and act as barriers for others, and therefore promote technological lock-ins [70]. This is the Ecuadorian case, which shows a trend trajectory of deployment of the cheapest and most mature energy technologies (hydropower and thermal generation) as can be seen in the national energy policies and plans that consider only a small participation of alternative renewable generation technologies, even though in the future these are expected to become cheaper [71].

The cost of a fully renewable electricity matrix for Ecuador has to be analyzed also under the perspective that the country's oil era is likely coming to an end within the next decade or, according to the most optimistic estimates, within the next two decades [72]. If the country is no longer an oil producer but its technological dependency on this product continues, his energy and transport sectors will be vulnerable to the fluctuating oil market. A planned and gradual transition from fossil fuels to a clean energy mix would be less costly than a forced adoption of new technologies that could result in many stranded assets, so the depletion of oil reserves is an important factor to consider in the energy planning.

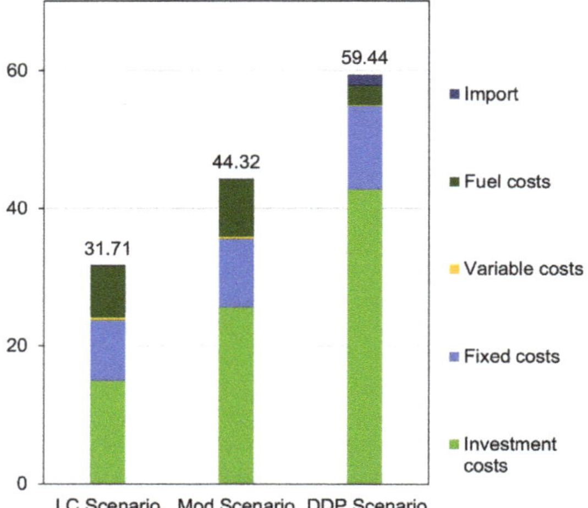

Figure 12. Costs of the system in USD billion for the LC, Mod, and DDP scenarios.

In this context, a most comprehensive cost analysis should be taken into account that can serve as inputs for policy and planning recommendations. Although this is not the objective of this work, some insights in this direction can be mentioned. First the presented cost analysis does not take into account the economic costs such as environmental and social ones. Many authors show the impact and related social costs of big energy projects, that can lead to several social issues even to the movement or disappearing of entire communities [44,66–69], with associated governmental costs needed to supply people with new houses, infrastructure, employment and so on. Most of the time these are not considered because are hidden behind the side effects of an infrastructure project and are not visualized as costs. Moreover, the environmental costs are also disregarded in most cases, especially because of the difficulty of assigning a cost to environmental degradation, and a weak consideration of the impacts of environmental services loss on other activities located in the places where the big energy projects are built, such as agriculture, tourism, and cultural values for people and communities, as is widely the case in Ecuador [73]. For coping with this, a multi criteria analysis can be used to visualize and take into consideration the social, environmental, technological, and political aspects of the energy projects, especially those of big scale, that help to redefine the energy portfolios and reflect the potential advantages of a more diversified and non-centralized energy mix. This is a field of further research to improve our analysis of the Ecuadorian energy mix transition.

4. Conclusions

This study analyzes whether a diversified and clean electricity mix can supply the demand of a highly electrified transport sector without neglecting the demand of other consumption sectors. The second inherent research question is how renewable energy technologies can be integrated to supply this demand taking into account technical and cost criteria of each type of technology. This work adds new information to previous studies on the decarbonization of the Ecuadorian demand sectors, as here we visualize different levels of electrification of the transport sector, and analyze the energy supply in the context of current national plans and also potential measures.

High electrification of passengers land transport can be achieved with the renewable resources available in the country, while a significant portion of land freight transport can also be electrified. For this, Ecuador could reject the use of fossil fuels, due to its vast renewable resources, especially water, however, the disadvantages of over dependence on this resource should be avoided through holistic energy planning, seeking for the best balance between available resources and technologies.

This work demonstrates that clean electrification of demand sectors, especially land transport is a viable alternative for the country to reinforce its future climate change mitigation goals. Finally, the results show that the deployment of renewable technologies has high costs in terms of investment, maintenance and operation, however, this should be seen as the cost of a necessary transition towards a sustainable, low-emission energy supply and demand, which can deliver large potential benefits to the national economy.

5. Future Work

Some refinements to the methodology could be proposed for future work. They include improving the characterisation of the fleet and its use to better understand the end uses of energy in the transport sector. Also, it is suggested to incorporate the hydrogen industrial chain into the modelling, which in long term could play an important role in the decarbonisation of the transport sector. More, a detailed modelling of large-scale energy storage technologies is suggested. These refinements should give tools to suggest energetic policies to encourage a sustainable and just energy transition, especially in developing countries. Such policies could include the implementation of differentiated electricity tariffs depending on the time of day, and policies related to mobility efficiency where mass rather than individual transport is prioritised. Finally, in order to further elaborate on the effects that transport electrification may have in developing countries such as Ecuador, the formulation of decarbonisation scenarios should include social, environmental, macroeconomic, governance and economic policy criteria. A green solution in the energy sector must foster sustainable development and reduce economic inequalities.

Author Contributions: Conceptualization, J.C.G., D.V., R.S., F.O. and T.H.; methodology, J.C.G., D.V. and R.S.; software, J.C.G. and D.V.; validation, J.C.G. and D.V.; formal analysis, J.C.G., D.V. and C.V.; investigation, J.C.G. and D.V.; resources, J.C.G., D.V. and R.S.; data curation, J.C.G. and D.V.; writing—original draft preparation, J.C.G., D.V. and R.S.; writing—review and editing, C.V. and J.C.G.; visualization, J.C.G. and D.V.; supervision, R.S. and T.H.; project administration, J.C.G.; funding acquisition, Bayerisches Hochschulzentrum für Lateinamerika (BAYLAT). All authors have read and agreed to the published version of the manuscript.

Funding: This research was funded by Bayerisches Hochschulzentrum für Lateinamerika (BAYLAT).

Institutional Review Board Statement: Not applicable.

Informed Consent Statement: Not applicable.

Data Availability Statement: No new data were created or analyzed in this study. Data sharing is not applicable to this article.

Acknowledgments: Gratitude to the Ecuadorian Secretariat of Higher Education, Science, Technology and Innovation (SENESCYT) for providing monetary support to Janeth Carolina Godoy for her doctoral studies at the ENS-TUM. Special thanks to the b4future team for their support during this work and their valuable inputs about initiatives for sustainable development for Ecuador. The author, Daniel Villamar, would like to thank the FSPI-Doctoral Schools Project of the French Embassy in Ecuador, financed by the Ministry of Europe and Foreign Affairs.

Conflicts of Interest: The authors declare no conflict of interest.

References

1. United Nations Environment Programme. *Emissions Gap Report 2020*; UNEP Publications: Nairobi, Kenya, 2020.
2. Sims, R.; Schaeffer, R.; Creutzig, F.; Cruz-Núñez, X.; D'Agosto, M.; Dimitriu, D.; Figueroa Meza, M.; Fulton, L.; Kobayashi, S.; Lah, O.; et al. Transport. In *Climate Change 2014: Mitigation of Climate Change. Working Group III Contribution to the IPCC Fifth Assessment Report*; Edenhofer, O., Pichs-Madruga, R., Sokona, Y., Farahani, E., Kadner, S., Seyboth, K., Adler, A., Baum, I., Brunner, S., Eickemeier, P., et al., Eds.; Cambridge University Press: Cambridge, UK; New York, NY, USA, 2015; pp. 599–670.

3. Rogelj, J.; Shindell, D.; Jiang, K.; Fifita, S.; Forster, P.; Ginzburg, V.; Handa, C.; Kheshgi, H.; Kobayashi, S.; Kriegler, E.; et al. Mitigation Pathways Compatible with 1.5 °C in the Context of Sustainable Development. In *Special Report on Global Warming of 1.5 °C*; Masson-Delmotte, V., Zhai, P., Pörtner, H.-O., Roberts, D., Skea, J., Shukla, P.R., Pirani, A., Moufouma-Okia, W., Péan, C., Pidcock, R., et al., Eds.; IPCC: Geneva, Switzerland, 2018; pp. 93–174.

4. IEA. World Oil Final Consumption by Sector. 2018. Available online: https://www.iea.org/data-and-statistics/charts/world-oil-final-consumption-by-sector-2018 (accessed on 5 August 2021).

5. Changing Transport. Facilitating Climate Actions in Mobility. Available online: https://www.changing-transport.org/updated-ndcs-transport/ (accessed on 5 July 2021).

6. REN21. *Renewables 2021 Global Status Report*; REN21 Secretariat: Paris, France, 2021.

7. SLOCAT, Tracking Trends in a Time of Change: The Need for Radical Action Towards Sustainable Transport Decarbonisation, Transport and Climate Change Global Status Report. 2021. Available online: https://tcc-gsr.com (accessed on 19 August 2021).

8. IEA. Global EV Data Explorer. Available online: https://www.iea.org/articles/global-ev-data-explorer (accessed on 24 August 2021).

9. Bataille, C.; Waisman, H.; Briand, Y.; Svensson, J.; Vogt-Schilb, A.; Jaramillo, M.; Delgado, R.; Arguello, R.; Clarke, L.; Wild, T.; et al. Net-zero deep decarbonization pathways in Latin America: Challenges and opportunities. *Energy Strategy Rev.* **2020**, *30*, 100510. [CrossRef]

10. UN Habitat for a Better Urban Future. Available online: https://unhabitat.org/node/142554 (accessed on 20 July 2021).

11. Información Estadística de Transporte. Available online: https://www.ecuadorencifras.gob.ec//transporte/ (accessed on 6 July 2021).

12. Ministerio del Ambiente. *Tercera Comunicación Nacional del Ecuador sobre Cambio Climático*, 1st ed.; MAE: Quito, Ecuador, 2017; ISBN 976-994-22-145-2.

13. MERNNR. *Balance Energético Nacional 2018*; MERNNR: Quito, Ecuador, 2020.

14. Asamblea Nacional de la República de Ecuador. *Ley Orgánica de Eficiencia Energética*; Editora Nacional: Quito, Ecuador, 2019.

15. Ministerio del Ambiente. *Primera Contribución Determinada a Nivel Nacional para el Acuerdo de París Bajo la Convención Marco de Naciones Unidas Sobre Cambio Climático*; MAE: Quito, Ecuador, 2019.

16. Villamar, D.; Soria, R.; Rochedo, P.; Szklo, A.; Imperio, M.; Carvajal, P.; Schaeffer, R. Long-term deep decarbonisation pathways for Ecuador: Insights from an integrated assessment model. *Energy Strategy Rev.* **2021**, *35*, 100637. [CrossRef]

17. MERNNR. *Plan Maestro de Electricidad 2018–2027*; Ministerio de Energía y Recursos Naturales No Renovables: Quito, Ecuador, 2020.

18. Carvajal, P.E.; Li, F.G.; Soria, R.; Cronin, J.; Anandarajah, G.; Mulugetta, Y. Large hydropower, decarbonisation and climate change uncertainty: Modelling power sector pathways for Ecuador. *Energy Strategy Rev.* **2019**, *23*, 86–99. [CrossRef]

19. Blackshear, B.; Crocker, T.; Drucker, E.; Filoon, J.; Knelman, J.; Skiles, M. *Hydropower Vulnerability and Climate Change a Framework for Modeling the Future of Global Hydroelectric Resources*; Middlebury College: Middlebury, VT, USA, 2011.

20. Gaudard, L.; Gabbi, J.; Bauder, A.; Romerio, F. Long-term Uncertainty of Hydropower Revenue Due to Climate Change and Electricity Prices. *Water Resour. Manag.* **2016**, *30*, 1325–1343. [CrossRef]

21. Vulnerabilidad al Cambio climáTico de los Sistemas de Producción Hidroeléctrica en Centroamérica y sus Opciones de Adaptación. Available online: https://publications.iadb.org/es/publicacion/15676/vulnerabilidad-al-cambio-climatico-de-los-sistemas-de-produccion-hidroelectrica (accessed on 22 July 2021).

22. Teotónio, C.; Fortes, P.; Roebeling, P.; Rodriguez, M.; Robaina-Alves, M. Assessing the impacts of climate change on hydropower generation and the power sector in Portugal: A partial equilibrium Approach. *Renew. Sustain. Energy Rev.* **2017**, *74*, 788–799. [CrossRef]

23. Ruffato-Ferreira, V.; da Costa Barreto, R.; Júnior, A.; Silva, W.; de Berrêdo, D.; Sena, J.; Vasconcelos, M. A foundation for the strategic long-term planning of the renewable energy sector in Brazil: Hydroelectricity and wind energy in the face of climate change scenarios. *Renew. Sustain. Energy Rev.* **2017**, *72*, 1124–1137. [CrossRef]

24. Understanding the Impact of Climate Change on Hydropower: The Case of Cameroon. Available online: http://hdl.handle.net/10986/18243 (accessed on 15 August 2021).

25. IEA. *Global EV Outlook 2021—Accelerating Ambitions Despite the Pandemic*; IEA: Paris, France, 2021.

26. International Renewable Energy Agency. *Global Energy Transformation: A Roadmap to 2050*; IRENA: Abu Dhabi, United Arab Emirates, 2019.

27. International Energy Agency. *Energy Technology Perspectives 2020*; OECD Publishing: Paris, France, 2020.

28. Gils, H.C.; Simon, S.; Soria, R. 100% Renewable Energy Supply for Brazil—The Role of Sector Coupling and Regional Development. *Energies.* **2017**, *10*, 1859. [CrossRef]

29. Sauer, I.L.; Escobar, J.F.; da Silva, M.F.P.; Meza, C.G.; Centurion, C.; Goldemberg, J. Bolivia and Paraguay: A beacon for sustainable electric mobility? *Renew. Sustain. Energy Rev.* **2015**, *51*, 910–925. [CrossRef]

30. Meza, C.G.; Amado, N.B.; Sauer, I.L. Transforming the Nicaraguan energy mix towards 100% renewable. *Energy Procedia* **2017**, *138*, 494–499. [CrossRef]

31. Lallana, F.; Bravo, G.; Le Treut, G.; Lefèvre, J.; Nadal, G.; Di Sbroiavacca, N. Exploring deep decarbonization pathways for Argentina. *Energy Strategy Rev.* **2021**, *36*, 100670. [CrossRef]

32. Godínez-Zamora, G.; Victor-Gallardo, L.; Angulo-Paniagua, J.; Ramos, E.; Howells, M.; Usher, W.; De León, F.; Meza, A.; Quirós-Tortós, J. Decarbonising the transport and energy sectors: Technical feasibility and socioeconomic impacts in Costa Rica. *Energy Strategy Rev.* **2020**, *32*, 100573. [CrossRef]

33. Auhadeev, A.E.; Litvinenko, R.S.; Fandeev, V.P.; Pavlov, P.P.; Butakov, V.M.; Litvinenko, A.R. Research of reliability of urban electric transport system. *E3S Web Conf.* **2019**, *124*, 05078. [CrossRef]
34. Mingrone, L.; Pignataro, G.; Roscia, M. Smart urban electric transport system: An innovative real model. In Proceedings of the 2015 International Conference on Renewable Energy Research and Applications (ICRERA), Palermo, Italy, 22–25 November 2015; pp. 1457–1462. [CrossRef]
35. Crețan, R.; O'Brien, T. Corruption and conflagration: (In)justice and protest in Bucharest after the Colectiv fire. *Urban Geogr.* **2019**, *41*, 368–388. [CrossRef]
36. Schaffitzel, F.; Jakob, M.; Soria, R.; Vogt-Schilb, A.; Ward, H. Can government transfers make energy subsidy reform socially acceptable? A case study on Ecuador. *Energy Policy* **2020**, *137*, 111120. [CrossRef]
37. Crețan, R.; Guran-Nica, L.; Platon, D.; Turnock, D. Foreign Direct Investment and Social Risk in Romania: Progress in Less-Favoured Areas. In *Foreign Direct Investment and Regional Development in East Central Europe and the Former Soviet Union*; Turnock D., Ed.; Routledge: London, UK, 2005; pp. 305–348.
38. Saget, C.; Vogt-Schilb, A.; Luu, T. *El Empleo en un Futuro de Cero Emisiones Netas en América Latina y el Caribe*, 1st ed.; Banco Interamericano de Desarrollo y Organización Internacional del Trabajo: Washington, DC, USA, 2020.
39. Pérez Urdiales, M.; Yépez-García, A.; Tolmasquim, M.; Alatorre, C.; Rasteletti, A.; Stampini, M.; Hallack, M. *El Papel de la Transición Energética en la Recuperación Sostenible de América Latina y el Caribe*; BID: Washington, DC, USA, 2021.
40. Crețan, R.; Vesalon, L. The Political Economy of Hydropower in the Communist Space: Iron Gates Revisited. *Tijdschr. Voor Econ. Soc. Geogr.* **2017**, *108*, 688–701. [CrossRef]
41. Vogt-Schilb, A.; Hallegatte, S. Climate policies and nationally determined contributions: Reconciling the needed ambition with the political economy. *Wires Energy Environ.* **2017**, *6*, e256. [CrossRef]
42. Ramirez, A.D.; Boero, A.; Rivela, B.; Melendres, A.M.; Espinoza, S.; Salas, D.A. Life cycle methods to analyze the environmental sustainability of electricity generation in Ecuador: Is decarbonization the right path? *Renew. Sustain. Energy Rev.* **2020**, *134*, 110373. [CrossRef]
43. International Atomic Energy Agency. *Transitions to Low Carbon Electricity Systems: Key Economic and Investments Trends. Changing Course in a Post-Pandemic World*; IAEA: Viena, Austria, 2021.
44. Gudynas, E. Caminos para las transiciones postextractivistas. In *Transiciones. Postextractivismo y Alternativas al Extractivismo en el Perú*, 2nd ed.; Alayza, A., Gudynas, E., Eds.; CEPES: Lima, Peru, 2010; pp. 165–190.
45. A Linear Optimisation Model for Distributed Energy Systems. Available online: https://github.com/tum-ens/urbs (accessed on 4 June 2021).
46. *urbs*: A Linear Optimisation Model for Distributed Energy Systems—*urbs* 1.0.0 Documentation. Available online: https://urbs.readthedocs.io/en/latest/ (accessed on 4 June 2021).
47. '*urbs*' Module Description. Available online: https://urbs.readthedocs.io/en/latest/api.html (accessed on 4 June 2021).
48. Candas, S.; Odersky, L.; Stüber, M. *urbs*-open source energy system modelling framework. In Proceedings of the Open Energy Modelling Workshop, Berlin, Germany, 15–17 January 2020.
49. MESSAGE. A Modeling Framework for Medium- to Long-Term Energy System Planning, Energy Policy Analysis, and Scenario Development. Available online: https://iiasa.ac.at/web/home/research/researchPrograms/Energy/MESSAGE.en.html (accessed on 30 March 2020).
50. Modelling Tool, Cenergia Lab. Available online: http://www.cenergialab.coppe.ufrj.br/tools (accessed on 30 March 2020).
51. Instituto Nacional de Eficiencia Energética y Energías Renovables. *Escenarios de Prospectiva Energética Para Ecuador a 2050*, 1st ed.; INER: Quito, Ecuador, 2016; ISBN 978-9942-8620-4-4.
52. Ministerio Coordinador de Sectores Estratégicos. *Agenda Nacional de Energía 2016–2040*; MICSE: Quito, Ecuador, 2016.
53. Ministerio de Electricidad y Energía Renovable. *Plan Nacional de Eficiencia Energética 2016–2035*, 1st ed.; Manthra Comunicación: Quito, Ecuador, 2017; ISBN 978-9942-22-148-3.
54. Secretaría Nacional de Planificación y Desarrollo. *Transformación de la Matriz Productiva*, 1st ed; Ediecuatorial: Quito, Ecuador, 2012.
55. Agencia de Regulación y Control de Electricidad. *Estadística Anual y Multianual del Sector Eléctrico Ecuatoriano 2019*; ARCONEL: Quito, Ecuador, 2019; ISBN 978-9942-07-946-6.
56. CENACE. *Informe Anual 2019*; CENACE: Quito, Ecuador, 2019.
57. Agencia de Regulación y Control de Electricidad. Estadística del Sector EléCtrico. Available online: https://www.regulacionelectrica.gob.ec/estadistica-del-sector-electrico/ (accessed on 8 March 2019).
58. Renewables.ninja. Available online: https://www.renewables.ninja/ (accessed on 15 June 2021).
59. Producción en Línea Detalle—CELEC EP. Available online: https://generacioncsr.celec.gob.ec/graficasproduccion/ (accessed on 15 June 2021).
60. Corporación Eléctrica del Ecuador. Available online: https://www.celec.gob.ec/ (accessed on 15 June 2021).
61. Biblioteca Operador Nacional de Electricidad CENACE. Available online: http://www.cenace.gob.ec/biblioteca/ (accessed on 15 June 2021).
62. INEC Anuario Estadístico de Transporte 2016. Availabe online: https://www.ecuadorencifras.gob.ec/documentos/web-inec/Estadisticas_Economicas/Estadistica%20de%20Transporte/2016/2016_AnuarioTransportes_%20Principales%20Resultados.pdf (accessed on 12 July 2021).

63. Instituto Nacional de Eficiencia Energética y Energías Renovables. *Manual Para el Procesamiento de Información del Balance Energético*; INER: Quito, Ecuador, 2015.
64. INEC. *Proyección Poblacional Nacional 2010–2050*; INEC: Quito, Ecuador, 2012.
65. Briones, A.; Uche, J.; Martínez-Gracia, A. Accounting for GHG net reservoir emissions of hydropower in Ecuador. *Renew. Energy* **2017**, *112*, 209–221. [CrossRef]
66. De Echave, J.; Diez, A.; Huber, L.; Revesz, B.; Lanata, X.; Tanaka, M. *Minería y Conflicto Social*, 1st ed.; Instituto de Estudios Peruanos: Lima, Peru, 2009.
67. Bebbington, A. The new extraction: Rewriting the political ecology of the Andes? *NACLA Rep. Am.* **2016**, *42*, 12–20. [CrossRef]
68. Acosta, A. *La Maldición de la Abundancia*, 1st ed.; Abya-Yala: Quito, Ecuador, 2009.
69. Global Atlas of Environmental Justice. Available online: https://ejatlas.org/ (accessed on 15 June 2021).
70. Del Río, P.; Unruh, G. Overcoming the lock-out of renewable energy technologies in Spain: The cases of wind and solar electricity. *Renew. Sustain. Energy Rev.* **2007**, *11*, 1498–1513. [CrossRef]
71. IRENA. *Renewable Power Generation Costs in 2019*, 1st ed.; IRENA: Masdar City, United Arab Emirates, 2020; ISBN 978-92-9260-244-4.
72. Espinoza, V.S.; Fontalvo, J.; Martí-Herrero, J.; Ramírez, P.; Capellán-Pérez, I. Future oil extraction in Ecuador using a Hubbert approach. *Energy* **2019**, *182*, 520–534. [CrossRef]
73. Godoy, C. Desde Paute Hasta Coca Codo Sinclair 40 Años de Hidroenergía en el Ecuador: Discurso Alrededor de Cambio de Matriz Energética. Master's Thesis, Facultad Latinoamericana de Ciencias Sociales, Quito, Ecuador, 2013.

Article

Contribution of Road Transport to the Attainment of Ghana's Nationally Determined Contribution (NDC) through Biofuel Integration

Peerawat Saisirirat [1,*], Johannex Fefeh Rushman [1,2], Kampanart Silva [1] and Nuwong Chollacoop [1]

[1] National Energy Technology Center (ENTEC), Renewable Energy and Energy Efficiency Research Team, National Science and Technology Development Agency (NSTDA), Bangkok 12120, Pathum Thani, Thailand; johannex.frushman@kstu.edu.gh (J.F.R.); kampanart.sil@entec.or.th (K.S.); nuwong.cho@entec.or.th (N.C.)

[2] Faculty of Engineering and Technology, Kumasi Technical University, Kumasi P.O. Box 854, Ashanti, Ghana

* Correspondence: peerawat.sai@entec.or.th; Tel.: +66-2-564-6500 (ext. 4747)

Abstract: Since the Paris Agreement in COP21, many countries around the world, including Ghana and Thailand, have established a Nationally Determined Contribution (NDC) to reduce greenhouse gas (GHG) emissions, with first update recently in COP26. With Ghana's ongoing effort at COP26 to change its baseline to 2019, this study established a detailed Ghana vehicle ownership model with necessary transport parameters to construct an energy demand model to provide insight for reducing GHG emission contributions from road transport through biofuel (both bioethanol and biodiesel) potential by recourse to a Low Emission Analysis Platform (LEAP), with two scenarios of development from Thailand's best practice for policy recommendation, which are alternative (ALT), with up to E20/B20, and extreme (EXT), with up to E85/B50, for new vehicles. In each case, energy demand and GHG emissions were analyzed from detailed data on Ghana's transport sector to show potential benefit from biofuel usages. From Ghana's transport sector contribution to NDC, 8.4% and 11.1% of GHG emission reduction in 2030 can be achieved with a 0.13% and 0.27% additional arable land requirement from ALT and EXT scenarios. Policy recommendation and implication were also discussed.

Keywords: greenhouse gas emissions; Ghana road transport; energy demand model; biofuel integration; arable land requirement

Citation: Saisirirat, P.; Rushman, J.F.; Silva, K.; Chollacoop, N. Contribution of Road Transport to the Attainment of Ghana's Nationally Determined Contribution (NDC) through Biofuel Integration. *Energies* **2022**, *15*, 880. https://doi.org/10.3390/en15030880

Academic Editors: Dolf Gielen and Kun Mo Lee

Received: 21 November 2021
Accepted: 18 January 2022
Published: 26 January 2022

Publisher's Note: MDPI stays neutral with regard to jurisdictional claims in published maps and institutional affiliations.

1. Introduction

Global warming has become a one of the most important issues of all economies in the world. Greenhouse gas (GHG) emissions contributes to the increase of global average temperature and, hence, is life-threatening for a number of species [1]. This has alerted the world to managing emissions and the consequent global warming potential while maintaining the energy-dependent ongoing development. The 3rd Conference of the Parties (COP) in 1997 resulted in the Kyoto Protocol [2], which was superseded by the Paris Agreement [3] adopted in COP21. The Paris Agreement requires the countries to commit the Nationally Determined Contribution (NDC) and update it every five years. COP26 in the United Kingdom welcomed the first update of the NDC, where most countries announced an ambitious emission reduction target to keep the global temperature increase under 1.5 °C [4]. The majority of the developed countries committed to over 50% GHG emission reduction by 2030 (compared to the 2005 level) in order to pave the pathway toward climate neutrality by 2050. As for Ghana, the intended NDC announced in 2015 aimed to *unconditionally lower its GHG emissions by 15 percent (11.1 MtCO$_2$e) relative to a business-as-usual (BAU) scenario emission of 73.95 MtCO$_2$e by 2030*, and to additionally reduce emissions by 30 percent (22.2 MtCO$_2$e) *on the condition that external support is made available* (Figure 1) [5]. These targets were replaced by more stringent ones in the updated NDC presented at COP26 [6]. Unconditional mitigation measures in all relevant sectors would result in 8.5 $MtCO_{2e}$ GHG reductions by 2025, and 24.6 $MtCO_{2e}$ by 2030. Additional conditional measures have the potential to achieve an increment of 16.7 $MtCO_{2e}$ by 2025, and 39.4 $MtCO_{2e}$ by 2030, if financial support is made available. The BAU scenario emission is being recalculated with 2019 as the base year.

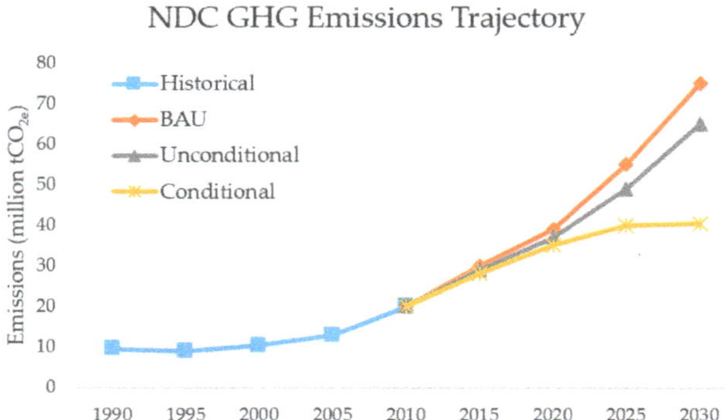

Figure 1. Ghana's GHG emission reduction trajectory in the NDCs [5].

The transportation sector makes a significant contribution to global GHG emissions and, thus, owns large potential for emission reduction. The transportation sector accounts for 14.3% of the total GHG emissions as of 2010, with 14% and 0.3% as direct and indirect emissions, respectively [7]. The statistics are similar in Ghana, where the transportation sector is the third largest contributor to the GHG emissions, making up to 17% of the total emissions [8]. Transportation in Ghana includes road, rail, marine, and aviation, and communication networks are centered in the southern region. The road transport contributes to 95% of freight and passenger carriers [9]. About 84% of passenger trips are made with public transport, mainly by low occupancy mini-buses and modified passenger vans, which are imported used cars. These vehicles have poor fuel economy and high GHG emissions [10]. This is projected to increase exponentially as a result of vehicle fleet accumulation that corresponds with the rapid GDP growth [11], as shown in Figure 2 [12]. Appropriate planning for GHG emission reduction for new and existing vehicles is inevitable to be able to achieve the updated NDC. In addition, even though Ghana has set ambitious targets to reduce GHG emissions by 4439.4 kTOE and 1338.4 kTOE by low carbon electricity generation and scale-up renewable energy penetration, there has been no plan on vehicle electrification, which could facilitate further decarbonization.

However, the only transportation-related policy action that appeared in the annex of the updated NDC is the *expansion of inter-and-intra-city transportation modes*, which contributes to 109.9 $ktCO_2e$ emission reduction. This aligns with *scaling up sustainable mass transportation*, which was mentioned as a policy action in the initial NDC. This policy must be coupled with public communication measures to incentivize people to use more efficient mass transport systems, rather than their own private cars or low occupancy vehicles. This action is supposed to involve the adoption of an urban mass transport system in terms of roads and rails to reduce vehicle traffic and cut down energy demand and, consequently, reduce GHG emissions [13,14]. There was no mention of policy actions contributing to alteration of fuel types to lessen GHG emissions.

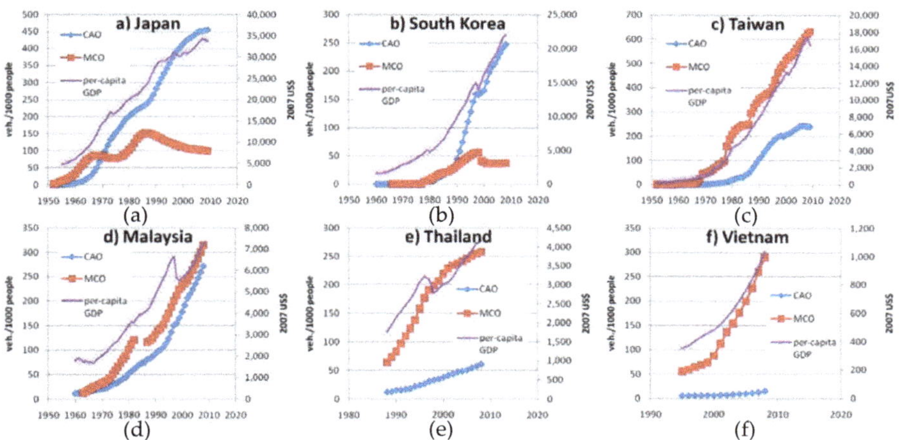

Figure 2. Historical ownership trends of car (CAO) and motorcycle (MCO) and GDP per capita [12] (**a**) Japan, (**b**) South Korea, (**c**) Taiwan, (**d**) Malaysia, (**e**) Thailand, and (**f**) Vietnam.

There have been various alternative fuels studied and used around the world. Biofuels, such as ethanol and biodiesel, are widely used to blend with gasoline and diesel in order to lower the reliance on fossil fuels. The United States [15] and the European Union [16] have various policy actions to accommodate biofuel substitution in the transportation sector in their territories. Usage of biofuels is primarily determined by biofuel crops available in the country. The United States and Brazil use corn and sugarcane to produce bioethanol [17], while China uses corn, wheat, and sweet sorghum [18]. In contrast, Malaysia relies heavily on palm-based biodiesel [19]. Sub-Saharan Africa also has large potential of biofuel usage in the region, though it needs a careful plan to balance between food production and GHG emission reduction [20]. Electrification can be another means for decarbonization, especially for non-agricultural-based countries. Battery electric vehicle is seen as a promising technology for a light duty vehicle in Europe, though it needs to be more competitive in terms of cost and social acceptance in order to achieve a large-scale penetration [21]. A study in the United States showed that government incentives are still needed to increase the market share of electric vehicles [22]. Similarly, China [23] and Japan [24] are also implementing sets of climate change-related policies to promote vehicle electrification. Though an electric vehicle may not be financially feasible at a first glance, which halts its penetration in most countries, it can be cheaper than an internal combustion engine vehicle with the consideration of GHG emissions throughout its life cycle, as well as carbon credit being implemented [25]. Apart from biofuel and electric vehicles, natural gas (NG) can be another alternative fuel with lower carbon emission, especially during transition to renewable fuels [26]. Since the initial investment on NG-based vehicles is not significantly different from oil-based vehicles, it can be easier to encourage the public to use them [27]. Appropriate policy and strategy to adopt these alternative fuels in Ghana will lead to far more reduction of GHG emissions from the transportation sector and, consequently, contribute to the targets committed in the NDC.

Thailand has been doing very well in promoting biofuel utilization in the transportation sector, in both ethanol in light duty vehicles and biodiesel in heavy duty vehicles [28]. This is due to its potential in cassava and molasses for ethanol production, and in palm oil for biodiesel production [29]. Electric vehicles have also been included in its Energy Efficiency Plan since 2015, as one of the measures to improve energy efficiency [30]. A study by the three electricity authorities revealed a goal to increase the number of electric vehicles up to 1.2 million by 2036 [31]. NG used as vehicle fuel has been successfully completed with government subsidies since 1993, sharing with major NG consumption in power sector [32]. Currently, NG consumption declines with NG price lifting to the actual market without government subsidies. Comparing between Thailand and Ghana, both countries have a similar tropical climate (located in tropical-equator zone, warm temperature, and high humidity of 26–27 °C and 70–85%RH) population density (about 750 thousand per square kilometer), and high ratio of arable area per capita (ranked at 61st and 82nd of 205 countries) [33,34]. Therefore, Thailand can serve as a good role model for biofuel integration and electrification in the transportation sector with an aim to reduce GHG emissions. Ghana's agro-ecological characteristics are suitable for planting biofuel crops, e.g., cassava, maize, sorghum [35], and a recent study showed potential of using the current energy surplus for electric vehicle charge [36]. As, in Ghana, new vehicles are generally imported used car from developed European countries, the number of electric vehicles can possibly expand following the European trend. NG vehicles are considered to have cleaner fuel [10], especially for large cities with traffic congestion issue, i.e., Accra, Kumasi. Ghana will have more NG potentials than Thailand as it has NG supported by the West Africa Gas Pipeline (WAGP). Ghana can refer to successful policy actions to promote these alternative fuels in Thailand to make use of its biofuel potential and excessive electricity production capacity. As both are developing countries, forecast in growth of vehicle fleet would also follow a similar trend, which is rather different from developed countries [37].

This study aims to indicate pathways to emission reduction in Ghana's transportation sector, particularly through biofuel integration, along with vehicle electrification and NG utilization, referring to good practices in transportation policy and planning of Thailand. Vehicle ownership models are developed to quantify and project the vehicular fleet until 2036 (from Thailand Integrated Energy Blueprint (TIEB)) and input to the Low Emission Analysis Platform (LEAP) [38] to assess the total energy demands and resulting GHG emissions. Two different scenarios, namely alternative scenario (ALT), which is based on the practices of Thailand, and extreme biofuel integration scenario (EXT), are used for the comparison with the business-as-usual (BAU) case. A policy recommendation is to be made following the outcome of the scenarios to promote additional contribution from the transportation sector to the updated NDC.

2. Methodology

The energy demand and GHG emissions were analyzed using the bottom-up approach due to its capability in indicating pathways of policy impacts. The calculation was performed based on the LEAP's algorithm, defined by a simple engineering relationship, as follows (1):

$$ED_{ij} = NV_i \times VKT_i \times SF_{ij} \times FE_{ij} \times HV_j, \tag{1}$$

where ED_{ij} is the energy demand (MJ) of vehicle category 'i' using fuel 'j', NV_i is the number of vehicle, VKT_j is the vehicle kilometer of travels (km), SF_{ij} is the fuel share (%), FE_{ij} is the fuel economy (liter or kg/km for fuel or kWh/km for electric vehicles), and HV_j is the fuel lower heating value (energy unit/physical unit of fuel). Then, the road transport energy demand can be integrated from all vehicle types and fuels, so that the considered measures or focusing policy can be analyzed and tracked. Afterward, GHG emissions can be calculated by multiplying the energy demand with the Emission Factor (EF), giving emission quantity per unit consumed energy. All analyses in this research were done using the Low Emission Analysis Platform (LEAP), a commercial software tool which is widely used for energy planning and climate change mitigation assessment [38].

2.1. Data Collection

Ghana's road transport is mainly made up of both public and private passenger vehicles and freight vehicles [39]. Private vehicles account for the majority of road vehicles, while public vehicles mostly belong to institutions. Historical data acquired from the Driver and Vehicle Licensing Authority (DVLA) of Ghana indicates that the majority of road transport is made up of private non-commercial vehicles. These include sedans, SUVs, and vans used for personal purposes. For data entry, vehicle data obtained from

DVLA were grouped into seven main categories, based on vehicle function and technical characteristics [16]. Table 1 shows vehicle categorization adopted for this research, and historical record of vehicle registration is shown in Figure 3.

Table 1. Categories of vehicles for road transport in Ghana.

Vehicle Category	Abbreviation	Description	Uses
Motorcycle and tricycle	MC	2- and 3-wheelers	Private passenger (lately tricycles) used for commercial purposes
Private Vehicles	PC	Include all sedan, SUVs, and Vans. Engine size up to 3.5 L	For personal/private passenger and freight non-commercial.
Taxi-Commercial vehicles up to 2 L	Taxi	All sedans with engine capacity up to 2 L	Used as Taxi for public commercial passenger transport
Mini-Buses and Vans-Commercial vehicles above 2 L	miniBus & Van	Smaller capacity vans and buses.	For commercial passenger and freight transport
Buses & Coaches	Bus & Coach	Larger capacity vehicles	Passenger and freight
Heavy-duty trucks	HD Truck	Without trailers (Capacity 16–22 tons)	For fright and construction purposes
Articulated trucks *	ARTICS	Mostly with trailers (Capacity from 24–32 tons)	Freight transport

Note: * Articulated trucks is a truck which has a permanent or semi-permanent joint in its construction, allowing the vehicle to turn more sharply.

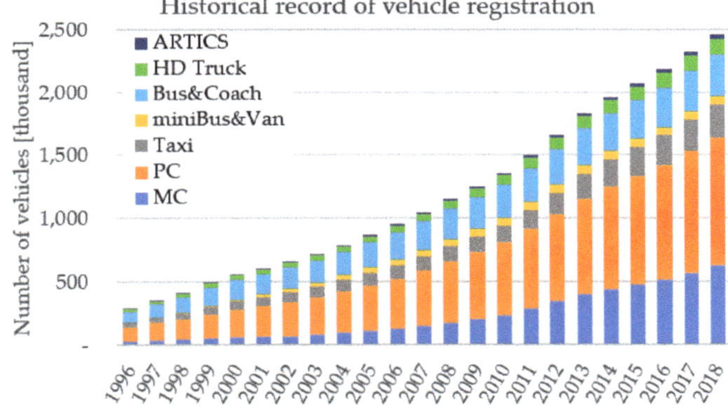

Figure 3. Historical record of vehicle registration [8].

Available historical data (1996–2018) in Figure 3 enables validation of a vehicle number model which is used to project vehicle number development between 2019–2036. The projected vehicle numbers are assembled with 'car survival' and vintage profile', to define vehicle stock-turnover rate in the LEAP program.

Besides, historical fuel consumption in Ghana which was reported from the National Petroleum Authority (NPA) of Ghana [40] was used as validation data for fuel demand calculation from LEAP. It must be emphasized that this data is the whole country's fuel demand. Nevertheless, the majority of Ghana's fuel consumption depends on the road transport sector, so it can be used to validate calculated results in this study. Other necessary data, such as Vehicle Kilometer Traveled (VKT) and Fuel Economy (FE), were collected from literature [13,14,39,41–43]. Note that the corrected averaged values of VKT and FE were assumed for all vehicles in each category, which can vary significantly, i.e., private vehicles include different vehicle segments: from small sedans up to large SUVs (with various engine size: 1.3–3.5 L), while commercial vehicles include small pickup-trucks and station wagons.

2.2. Model Development

Historically, economic development has been strongly associated with the transportation demand, particularly in the road vehicle numbers [11,37,44]. In this study, the vehicle ownership models are described in two functions, e.g., logistic and logarithmic, following literature [45,46]. In brief, the logistic vehicle-ownership (VO, vehicle number per capita) model is defined with maximum saturation level of car per capita (S), simplified from Button et al. [37], as follows Equation (2):

$$\ln\left(\frac{VO}{S - VO}\right) = B + a.ln(GDPpCap),\tag{2}$$

where *GDPpCap* is gross domestic product per capita, and a and b are model coefficients. Otherwise, logarithmic function is used for the public low occupancy vehicles (miniBus & Van) and the heavy-duty trucks (HD Trucks and ARTICS). Logarithmic function was chosen because the number of these vehicles is not related to the saturation level (*S*), and the function will not over-predict in long-term projection, as used in References [45,46]. The developed models of vehicle population in vehicle categories according to Table 1 are shown with adjusted coefficient of determination (adjusted R-squared, R^2_{adj}) in Table 2.

Table 2. Developed models of vehicle population.

Vehicle Category	Model	R^2_{adj}
MC	$\ln\left(\frac{VO_{MC}}{600-VO_{MC}}\right) = 2.656\ln(GDPpCap) - 23.450$	0.993
PC	$\ln\left(\frac{VO_{PC}}{812-VO_{PC}}\right) = 1.237\ln(GDPpCap) - 12.477$	0.984
TAXI	$\ln\left(\frac{VO_{TAXI}}{812-VO_{TAXI}}\right) = 1.616\ln(GDPpCap) - 16.671$	0.981
miniBus & Van	$VO_{miniBus\&Van} = -0.349\ln(GDPpCap) + 4.946$	0.873
Bus & Coach	$\ln\left(\frac{VO_{Bus\&Coach}}{812-VO_{Bus\&Coach}}\right) = 0.254\ln(GDPpCap) - 6.171$	0.964
HD Truck	$VO_{HD\ Truck} = 3.859\ln(GDPpCap) - 24.884$	0.994
ARTICS	$VO_{ARTICS} = 1.112\ln(GDPpCap) - 7.308$	0.966

VKT and FE represent transport activity and energy intensity for energy demand and GHG emission calculations. Table 3 outlines the base year values for these two variables, equivalent for each considered scenario. The data was extracted from limited sources [39,43] and analyzed according to involved parameters, i.e., vehicle size, technology, emission regulation level. The fuel economy data of EV has not yet been surveyed in Ghana. Therefore, the energy consumption of EV is assumed to be equal to 30% of conventional vehicles. This is estimated from the relationship between fuel economy of conventional vehicles and that of the battery EV, found in Reference [47].

Table 3. Average VKT and FE for the base year [39,43].

Vehicle Category	Average VKT (km)	Average FE (Lge */100 km)		
		Gasoline	Diesel	NG
Motorcycle and tricycle	12,500	3.7	-	-
Private Vehicle	25,000	9.7	8.8	10.2
Taxi-Commercial vehicle	30,000	9.7	8.8	10.2
Mini-Bus and Van	30,000	9.8	9.2	11
Bus & Coach	15,000	-	30.1	-
Heavy-duty truck	15,000	-	33.1	-
Articulated truck	12,000	-	33.1	-

Note: * Lge means liter of gasoline equivalent.

2.3. Validations

The developed model was validated by comparing calculated results obtained from LEAP [38] with historical records for both vehicle population and energy demand. As mentioned above, vehicle registration and fuel demand records were taken from DVLA [14] and NPA [42], respectively. Calculated vehicle numbers from the logistic and logarithmic functions fit well with the historical data, as shown in Figure 4. On the other hand, the calculated energy demands are lower for both diesel and gasoline fuels, as shown in Figure 5. The calculated values tend to be higher than the historical records. Noticeable difference in diesel fuel indicates that some portion of diesel fuel is consumed by other vehicles and equipment, i.e., agricultural and industrial sectors, which is slightly different from the assumption that the entire fuel usage is attributed to transport sector. Yet, the transport sector remains the largest contributor to fossil fuel usage [48], and the model is adequately robust for the calculation. Better agreement between actual and calculated values in gasoline fuel indicates that gasoline fuel demand depends heavily on the road transport sector.

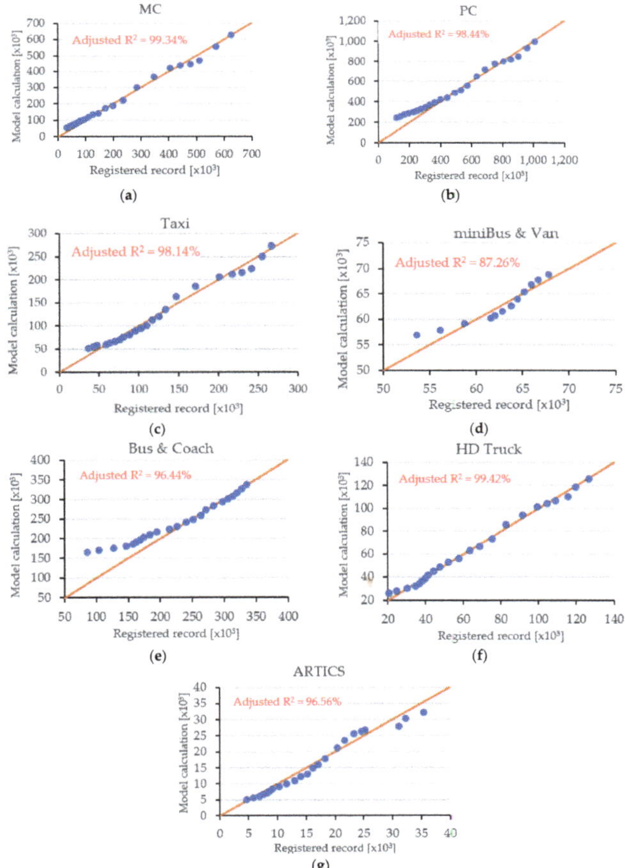

Figure 4. Validation of vehicle population models (**a**) Motorcycle, (**b**) Private vehicle, (**c**) Taxi, (**d**) Mini bus and Van, (**e**) Bus and Coach, (**f**) Heavy-duty truck, and (**g**) Articulated truck.

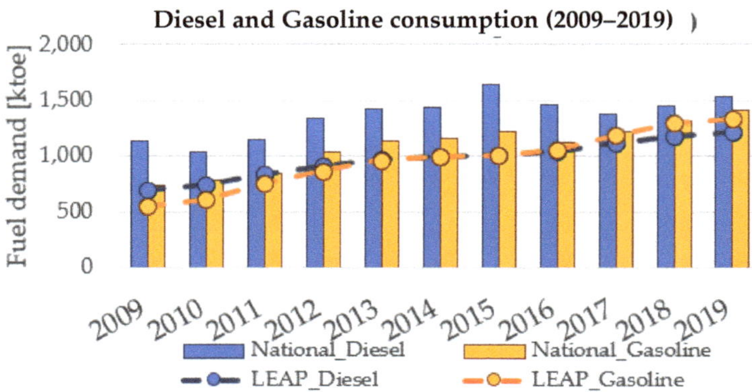

Figure 5. Validation of energy demand calculation.

2.4. Greenhouse Gas Emission Calculation

In this study, GHG emissions were calculated according to the Tier-I Intergovernmental Panel on Climate Change (IPCC) methodology [49]. This study focuses on the use phase GHG emission calculation (Tank-to-Wheel) to be comparable with the NDC target. Using the LEAP feature of Technology Environmental Database (TED), the amounts of GHG emissions can be calculated from the quantity of fuel consumption in Equation (1) by described vehicle technology level in various vehicle types and segments. The GHG emissions from consumed fuels are carbon-dioxide (CO_2), methane (CH_4), and nitrous oxide (N_2O). Non-CO_2 emissions were considered according to Ghana's NDC [6], as well as because a use of NG has its specific emission of methane. IPCC recommended that all GHG emissions should be reported in a mass unit of carbon dioxide equivalent (CO_{2e}), by multiplying the emission quantity with its Global Warming Potentials (GWP, equaled to 1 for CO_2 [49]). GHG emissions are calculated as followed Equation (3):

$$GHG_{ijk} = ED_{ij} \times EF_{ijk} \times GWP_k,$$ (3)

where GHG_{ijk} is the GHG emission type 'k' (kg CO_2-equivalent) produced from vehicle category 'i' using fuel 'j', EF_{ijk} the emission factor (kg/MJ), and GWP_k the global warming potentials (kg CO_{2e}/kg of emission 'k'). The GWP_k are shown in Table 4.

Table 4. Global warming potential of consumed fuel.

GHG Emission	GWP$_k$ (kg CO$_2$-Equivalent/kg of Emission 'k')
CO_2	1
CH_4	25
N_2O	298

2.5. Scenario Definitions

There are three scenarios considered in this study. First, the Business as Usual (BAU) scenario was defined from the current situation, as a baseline trend of energy consumption and GHG emissions in the road transport sector. The other two scenarios were developed as the guideline measures for biofuel integration in two possible levels of GHG mitigation, namely the alternative (ALT) and extreme (EXT) scenarios. In both the ALT and EXT scenario, the share of NG vehicles was included for the transport sector in the same approach to increase more cleaner gaseous fuels as in the power generation sector [10,50]. On the other hand, electric vehicles were also included for motorcycles, private passenger cars, taxis, mini vans, and buses, followed the "Drive Electric Initiative (DEI)" supported by the Energy Commission, Ministry of Energy of Ghana [51]. Electric vehicle penetration will indicate the impact of vehicle fuel economy improvement. Gas vehicles and electric vehicles will each achieve 10% by 2030 in both ALT and EXT scenarios by replacing shares of gasoline and diesel vehicles, proportionally (Figure 6).

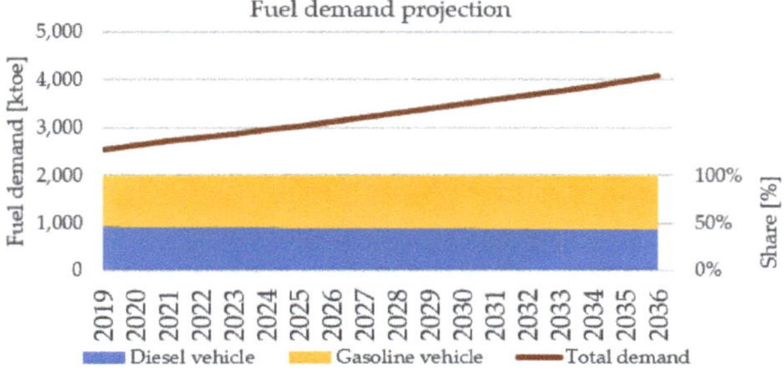

Figure 6. Fuel demand projection (BAU scenario).

Biofuel integration was described by considering the case study of Thailand. Anhydrous ethanol (less than 1% water) has been proven to replace some portion of gasoline in various blending fractions, namely gasohol E5, E10, E20, and E85. The descried numbers represent ethanol blending fraction in fuels, i.e., E20 is the blending of gasoline and ethanol fuels, with a fraction of 80% and 20%, respectively. In general, vehicles will have a maximum applicable limit of gasohol fuel specified in the vehicle handbook. Besides, the retrofit kits are widely available to increase this applicable limit, beyond the vehicle specification. However, the use of high blended gasohol, i.e., E20 and E85, is limited with ethanol production capacity; nevertheless, the Flex Fuel Vehicle (FFV), which can use wide types of gasohol fuels, is supported in many countries [52]. On the other hand, the methyl-ester of fatty acids, namely biodiesel, which is derived from vegetable oils or animal fats, is the fossil fuel replacement for a diesel vehicle. In the contrary, the limit of biodiesel blending fractions has stringent manufacturer's cautions in that the use of biodiesel without being essentially careful may cause severe damage on many vehicle parts. Currently, the biodiesel limit varies in different countries, depending on different national policies and available biodiesel resources, as well as regional weather. Biodiesel is not good in cold flow properties, so it is more favored in warm regions. Today, the maximum blending fraction of biodiesel B20 succeed as voluntary program in some tropical countries, e.g., Indonesia, Thailand, Brazil [53]. In contrast, there is an ambitious target to push forward biodiesel blending fraction, achieving B50 as a voluntary measure used in Indonesia [54,55].

According to the aforementioned, ALT and EXT scenarios were defined with probable and ambitious targets of new gasoline and diesel vehicles which will annually replace registered stock vehicles, according to a stock-turnover mechanism. Blending ratios of biofuel in both scenarios were adopted according to those that have been implemented in Thailand, i.e., gasohol E5, E10, E20, and E85 and biodiesel B5, B10, and B20. In addition, the biodiesel B50 was added for the ambitious measure in the EXT scenario. New vehicles were defined to gradually switch from fossil fuel to using biofuel every five years, starting from 2020 until 2031. The scenario starts from 2020, the same period as the Ghana's Strategic Program at COP26. Figure 7 and Table 5 show the scenario definitions and its timelines.

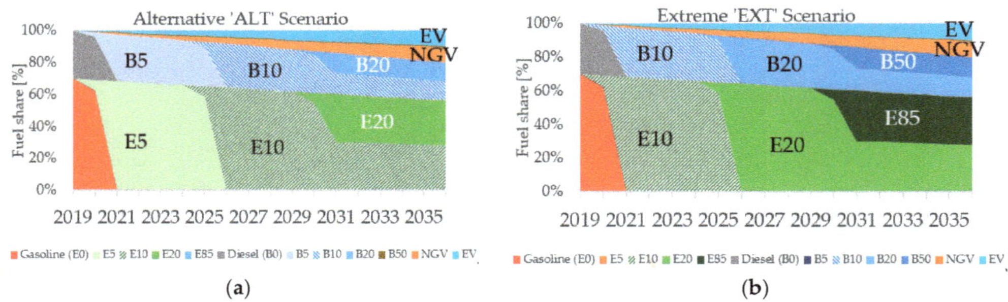

Figure 7. Graphical chart of biofuel integration in the considered scenarios for new light duty vehicles: (**a**) alternative scenario (ALT) and (**b**) extreme scenario (EXT).

Table 5. Scenario definitions for biofuel integration (Ethanol, EX; and Biodiesel, BX).

		2020 (Base-Year)	2021	2025	2026	2030	2031
BAU		There are no gasohol or biodiesel measures (E0 & B0).					
ALT	Ethanol (gasoline replacement)	E0: E5 = [90%:10%]	E5 [100%]	E5: E10 [90%:10%]	E10 [100%]	E10: E20 [90%:10%]	E10: E20 [50%:50%]
	Biodiesel (diesel replacement)	B0: B5 = [90%:10%]	B5 [100%]	B5: B10 [90%:10%]	B10 [100%]	B10: B20 [90%:10%]	B10: B20 [50%:50%]
EXT	Ethanol (gasoline replacement)	E0: E10 = [90%:10%]	E10 [100%]	E10: E20 [90%:10%]	E20 [100%]	E20: E85 [90%:10%]	E20: E85 [50%:50%]
	Biodiesel (diesel replacement)	B0: B10 = [90%:10%]	B10 [100%]	B10: B20 [90%:10%]	B20 [100%]	B20: B50 [90%:10%]	B20: B50 [50%:50%]

In the ALT scenario, new gasoline vehicles will switch to gasohol in the final share of E10 and E20, 50% each, and new diesel vehicles are defined to switch from using diesel fuel to the biodiesel share of B10 and B20, likewise with a shared 50% each. Besides, the ambitious target was defined in the EXT scenario. The final 50% share of new vehicles is specified for gasohol E20 and E85, on new gasoline vehicles, and biodiesel B20 and B50, on new diesel vehicles.

3. Results and Analyses

3.1. Impacts on Energy Demand

Figure 8 shows the impacts of biofuel integration, NG vehicle, and electric vehicle penetration on energy demand of the road transport sector. The calculated results show a contrast between BAU scenario and the two others. Besides, the total energy demands are similar in the ALT and EXT scenarios. This result indicates that the electric vehicle penetration can reduce total energy demand in road transport due to its higher energy conversion efficiency compared to all combustion engine vehicles. On the other hand, the biofuel integration and NG vehicles cannot help in reducing total energy demand but replaced conventional fossil fuel (gasoline and diesel fuels). Figure 9 shows the fuel switching comparing between conventional fossil fuels (gasoline and diesel) and alternative fuels (ethanol, biodiesel, electricity, and NG). By gradually increasing alternative fuels with better vehicle fuel economy, forecasted conventional fossil fuels will be reduced more than alternative fuel demand under the same assumption on number of vehicle and vehicle kilometer of travels, leading to net energy reduction from fuel switching.

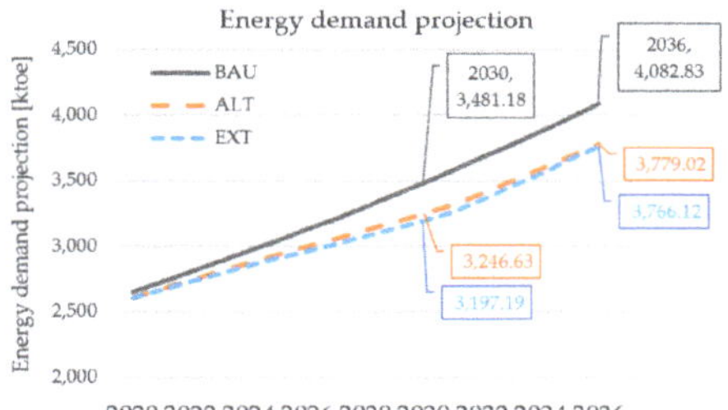

Figure 8. Energy demand projection.

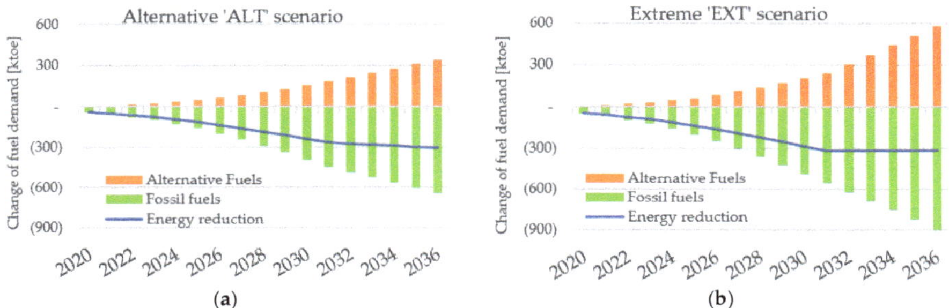

Figure 9. Fuel switching and energy reduction: (**a**) alternative scenario and (**b**) extreme scenario.

3.2. GHG Emission Mitigation

Figure 9 shows the projection of GHG emissions from the road transport sector. The results show that the road transport sector will contribute to 14.2% of the whole country's GHG emissions (Figure 10) in 2030. ALT and EXT scenarios can tear down the GHG emissions by introducing the electric vehicle technology (by improving fuel economy) and biofuel integration (apply carbon neutral fuels). GHG emission of ALT and EXT scenarios was lower than BAU scenario by 8.4% and 11.1% in 2030, and by 11.0% and 16.7% in 2036. The results show that biofuel integration measures offer only a moderate effect on GHG emission mitigation because the considered measures applied on new vehicles. The conventional vehicles in the road transport system require a period of replaced time. On the other hand, the calculated results confirm that GHG mitigation measures should be diversified. In addition, the results indicate that the impact of biofuel integration has higher potential on GHG mitigation than the impact on reducing energy demand.

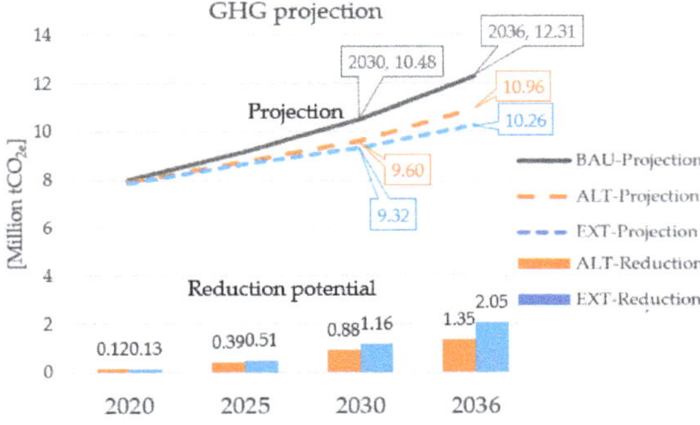

Figure 10. GHG emission projection.

3.3. Change in Biofuel Demand

This section indicates the calculated results of biofuel demand projection which must be prepared for sustainable biofuel development. Figure 11 shows the projection of ethanol and biodiesel demand according to ALT and EXT scenarios. The results indicate that ethanol demand will be higher than biodiesel demand for two reasons. First, the projection of energy demand from gasoline vehicle is higher than diesel vehicles, as shown in Figure 6. Otherwise, gasohol technology has more technology readiness than biodiesel in that the maximum blending fraction can be higher (85% of E85 compared to 50% of B50).

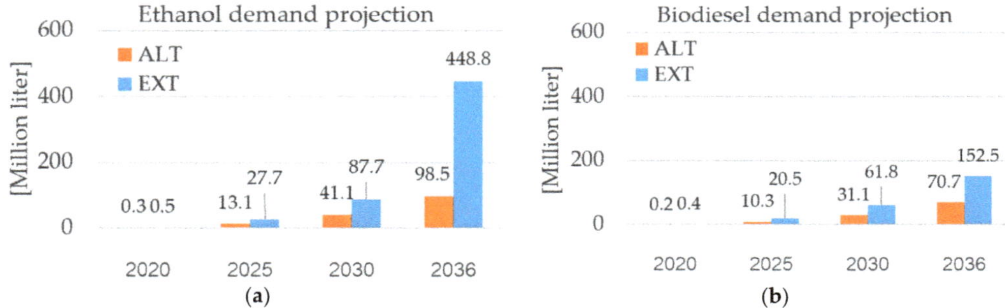

Figure 11. Projection of biofuel consumption in Alternative and Extreme scenario: (**a**) ethanol and (**b**) biodiesel.

On the other hand, the results show that biofuel demand has increased exponentially over the last five years. Biofuel promotion measures in this study are applied on the fuel share of new vehicle; therefore, biofuel demand will be gradually expanded. The stock vehicles which are not applicable for using biofuel will take a replacement period after endorsing measures. This behavior is good for the overall biofuel supply chain because the biofuel production can prepare to support the expansion of biofuel demand. This adjustment period includes preparation of farmland, growing periods of the biofuel crops, and construction periods for biofuel conversion industries, as well as the logistics of all biofuel supply chains.

4. Discussion

4.1. Land Preparation for Biofuel Crop Cultivation

Land preparation for biofuel crop cultivation is discussed in this section. The biofuel production yield by cultivation area (million liter/hectare, ML/ha) was taken from Ghana data surveyed by F. Kemausuor et al. [35]. Ethanol can be produced from cassava feedstock in the ratio to cultivation land of 18.75 ton/ha, or the ethanol yield is accounted for at 3000 L/ha. Besides, biodiesel can be produced from palm fruit feedstock in the ratio to cultivation land of 4.00 ton/ha, or biodiesel yield is accounted for at 6000 L/ha. Therefore, the cultivation area must be further prepared and give time for the biofuel crops to be ready for cultivation. The percentage of Ghana's arable lands are also determined by comparing to a total area of 147,827.40 ha [56]. Tables 6 and 7 show the biofuel resource preparation for ALT and EXT scenarios, respectively, up to about 1% arable land.

Table 6. Biofuel resource preparation for Alternative 'ALT' scenario.

Year	Ethanol			Biodiesel			Summary	
	Demand [Million liter]	Feed Stock [Million ton]	Land Required [ha]	Demand [Million liter]	Feed Stock [Million ton]	Land Required [ha]	Required Arable Area [ha]	Percentage of Total Arable Land [%]
2020	0.25	1562.5	83.3	0.20	133.3	33.3	116.66	0.00%
2025	13.13	82,062.5	4376.7	10.30	6866.0	1716.5	6093.16	0.04%
2030	41.12	257,000.0	13,706.7	31.09	20,724.6	5181.1	18,887.82	0.13%
2036	98.54	615,875.0	32,846.7	70.69	47,122.0	11,780.5	44,627.16	0.30%

Table 7. Biofuel resource preparation for Extreme 'EXT' scenario.

Year	Ethanol			Biodiesel			Summary	
	Demand [Million liter]	Feed Stock [Million ton]	Land Required [ha]	Demand [Million liter]	Feed Stock [Million ton]	Land Required [ha]	Required Arable Area [ha]	Percentage of Total Arable Land [%]
2020	0.52	3250.0	173.3	0.39	260.0	65.0	238.33	0.00%
2025	27.70	173,125.0	9233.3	20.53	13,685.3	3421.3	12,654.66	0.09%
2030	87.73	548,312.5	29,243.3	61.79	41,189.2	10,297.3	39,540.64	0.27%
2036	448.80	2,805,000.0	149,600.0	152.51	101,663.2	25,415.8	175,015.79	1.18%

4.2. Policy Considerations for Biofuel Integration

It can be seen from the results above that biofuel usage gradually increases with time until 2030 and sharply increases in 2036, in both ALT and EXT scenarios. This is due to the synergistic effects of large-scale replacement of vehicle fleets after 2030, and the increase in blending ratio of ethanol and biodiesel in gasoline and diesel, along with the increase in technology maturity of both fuels. In this study, rather than strictly follow the policy actions in Thailand by implementing the nationwide mandatory blending, obligation of biofuel usage was applied only to new vehicle fleets, even if it led to much slower penetration of biofuels in the country and, consequently, much less GHG emission reduction. There are two main reasons for this rather conservative assumption. One is that Ghana is a net fossil fuel exporter [40], where fossil fuels can be produced and used domestically, as well as exported to generate national income. Reduction in domestic production will affect the national economics, and it has to be done with caution. The other is the fact that the price of biofuel can be higher than gasoline or diesel, requiring adequate subsidy to gain public acceptance as the majority of the public will not be willing to pay significantly more for a cleaner fuel [57]. However, a preceding study showed that biofuel can be more economical than fossil fuels if its social benefits are taken into account [58]; hence, the two scenarios took into account the balance among national economic growth, public acceptance, and reduction of GHG emissions. This synchronizes well with the fact that, even though Ghana has made significant progress in reducing fossil fuel subsidies, it still continues to face public pressure to reinstate subsidies [59], especially when the international oil price is high [60]. Furthermore, as demonstrated above, that the land needed for biofuel crops is less than 0.3% and 1.2% of the total cultivation area in 2036 for ALT and EXT scenarios, respectively, competition between biofuel crops and food or feed crops would not happen. Farmers will be supported in the production of the biofuel feedstocks, and markets will be readily available. Since the availability of large continuous land has been identified as a barrier to the success of oil palm production in Ghana, small-holder plantations would be a better alternative. A good combination of policies for carbon-neutral fuel integration, management of fuel price structure, public communication, and good agricultural practices will result in gradual integration of biofuels, which can contribute to 11.0–16.7% of the *total* GHG emissions in the transportation sector compared to the BAU scenario within 15 years. Furthermore, additional national plan to decarbonize the entire transport system shall be considered, such as transport energy efficiency improvement (i.e., avoid-shift-improve measures) and scaling up the carbon-neutral fuel promotion, as well as the low emission electric vehicles.

4.3. Policy Considerations for NG Utilization and Vehicle Electrification

In both the ALT and EXT scenario, NG and electric vehicles are also planned to play significant roles in road transportation, with 10% share of all vehicles for NG, and 10% share of light duty vehicles for electric vehicles by 2036. Since Ghana has NG reserve and can import NG through the West Africa Gas Pipeline [61], the acquisition of NG should not be a big deal. The main issue would be regarding the technology acquisition and infrastructure development for NG vehicles. This issue is also applicable to the case of electric vehicles. The government could start developing infrastructure in urban area in the initial phase, and eventually expand to other areas. Another important concern regarding vehicle electrification is the energy sources for the electricity. At present, approximately 60% of the electricity is produced by thermal power plants. The government needs to gradually increase the share of renewable energy in electricity generation in order to ensure GHG emission reduction by shifting from an internal combustion engine vehicle to an electric vehicle.

5. Conclusions

This study indicated the two possible pathways, namely ALT and EXT scenarios, for GHG emission reduction in the transportation sector through biofuel integration, along with vehicle electrification and NG utilization, in order to achieve Ghana's updated NDC at COP26. The pathways were developed based on vehicular fleets projected by vehicle ownership models for developing countries, as well as energy demand and GHG emissions calculated by the Low Emission Analysis Platform (LEAP). The ALT scenario adopted the biofuel blending ratios implemented in Thailand, though it required biofuel blending with only a new vehicle fleet to maintain national economic competitiveness and to assure public acceptance. It also introduced NG and electric vehicles to follow the global trend. The EXT scenario maintained the assumptions of the ALT scenario, except for changes in blending ratios to the highest possible ones. Biofuel integration, along with the introduction of NG and electric vehicles, significantly reduced energy demand, though the difference between ALT and EXT scenarios was not notable. Energy demand of ALT and EXT scenarios was lower than the BAU scenario by 6.7% and 8.2% in 2030, and by 7.4% and 7.8% in 2036, respectively. On the other hand, it can be seen from the difference in GHG emissions in ALT and EXT scenarios that biofuel integration appreciably contributed to the reduction in GHG emissions. GHG emissions of ALT and EXT scenarios was lower than the BAU scenario by 8.4% and 11.1% in 2030, and by 11.0% and 16.7% in 2036, respectively. This amount of emission reduction could be achieved even if the biofuel mandates were only applied to new vehicle fleets due to the new vehicle technology compatibility with biofuel blends. The comparison between the land requirement for biofuel feedstock and the total cultivation area also indicated that the issue of land use competition with agricultural produces is unlikely to happen. NG would also play a significant role in both scenarios since Ghana has no challenge in NG acquisition, though the share of renewable energy in electricity generation would need to be carefully monitored. This is to ensure that electricity for charging EV will not emit significant GHG emissions when considering entire well-to-wheel emissions. The study demonstrated that a good combination of policy actions for clean fuel integration, management of fuel price structure, public communication, and good agricultural practices is necessary to achieve successful biofuel integration and, consequently, GHG emission reduction in Ghana's transportation sector. Furthermore, a transport decarbonization plan should be considered for the entire transport system, and the policy impact appraisal must be continuously updated, for effective GHG mitigation policy. These could be required to strengthen the GHG reduction capacity of the transport sector.

Author Contributions: Conceptualization, J.F.R. and N.C.; methodology, P.S.; software, J.F.R. and P.S.; validation, J.F.R., N.C., P.S. and K.S.; formal analysis, J.F.R. and P.S.; investigation, J.F.R.; resources, N.C. and K.S.; data curation, P.S.; writing—original draft preparation, J.F.R., P.S. and K.S.; writing—review and editing, N.C., P.S. and K.S.; visualization, P.S.; supervision, N.C.; project administration, N.C.; funding acquisition, N.C. All authors have read and agreed to the published version of the manuscript.

Funding: This research was funded by the National Science and Technology Development Agency (NSTDA) under the Postdoctoral Fellowship Program, contract number: NCR-MT-2563-10660-EN.

Institutional Review Board Statement: Not applicable.

Informed Consent Statement: Not applicable.

Data Availability Statement: Not applicable.

Conflicts of Interest: The authors declare no conflict of interest.

References

1. Yearbook of Global Climate Action 2020: Marrakech Partnership for Global Climate Action. Available online: https://unfccc.int/sites/default/files/resource/2020_Yearbook_final_0.pdf (accessed on 21 November 2021).
2. Breidenich, C.; Magraw, D.; Rowley, A.; Rubin, J.W. The Kyoto protocol to the United Nations framework convention on climate change. *Am. J. Int. Law* **1998**, *92*, 315–331. [CrossRef]
3. UNFCCC. Adoption of the Paris Agreement. In Proceedings of the UNFCCC Conference 21st Session, Paris, France, 30 November–11 December 2015.
4. Masson-Delmotte, V.; Zhai, P.; Pörtner, H.-O.; Roberts, D.; Skea, J.; Shukla, P.R.; Pirani, A.; Moufouma-Okia, W.; Péan, C.; Pidcock, R.; et al. *Global Warming of 1.5 °C. An IPCC Special Report Impacts of Global Warming*; IPCC: Geneva, Switzerland, 2018.
5. Republic of Ghana. *Ghana's Intended Nationally Determined Contribution (INDC) and Accompanying Explanatory Note*; GH-iNDC: Accra, Ghana, 2015; pp. 1–16.
6. MESTI. *Ghana: Updated Nationally Determined Contribution under the Paris Agreement (2020–2030)*; Environmental Protection Agency, Ministry of Environment, Science, Technology and Innovation: Accra, Ghana, UNFCCC, 2021; pp. 1–27.
7. IPCC. Summary for policymakers. In *Climate Change 2014, Mitigation of Climate Change. Contribution of Working Group III to the Fifth Assessment Report of the Intergovernmental Panel on Climate Change*; IPCC: Geneva, Switzerland, 2014.
8. EPA. *Ghana's Fourth National Greenhouse Gas Inventory Report*; UNFCCC: Bonn, Germany, 2019; pp. 1–318.
9. Owusu-Bio, M.; Frimpong, J.M.; Duah, G.P. The state of road transport infrastructure and ensuring passenger safety in Ghana. *Eur. J. Logist. Purch. Supply Manag.* **2016**, *4*, 79–85.
10. UNEP. Roadmap for the Promotion of Cleaner Buses in Accra, Ghana. 2017. Available online: https://wedocs.unep.org/handle/20.500.11822/31213 (accessed on 21 November 2021).
11. Dargay, J.; Gately, D.; Sommer, M. Vehicle ownership and income growth, worldwide: 1960–2030. *Energy J.* **2007**, *28*, 143–170. [CrossRef]
12. Tuan, V.A. Dynamic interactions between private passenger car and motorcycle ownership in Asia: A cross-country analysis. *J. East Asia Soc. Transp. Stud.* **2011**, *9*, 6.
13. Annan, J.; Arthur, Y.; Quanah, E. Modelling Transport Energy Demand in Ghana: The Policy Implication on Ghanaian Economy. *Br. J. Econ. Manag. Trade* **2015**, *10*. [CrossRef]
14. Agyemang-Bonsu, K.W.; Dontwi, I.K.; Tutu-Benefoh, D.; Bentil, D.E.; Boateng, O.G.; Asubonteng, K.; Agyemang, W. Traffic-data driven modelling of vehicular emissions using COPERT III in Ghana: A case study of Kumasi. *Am. J. Sci. Ind. Res.* **2010**, *1*, 32–40.
15. Hossain, A.K.M.N.; Serletis, A. Biofuel substitution in the U.S. transportation sector. *J. Econ. Asymmetries* **2020**, *22*, e00161. [CrossRef]
16. Drabik, D.; Venus, T. EU biofuel policies for road and rail transportation sector. In *EU Bioeconomy Economics and Policies: Volume II*; Palgrave Macmillan, Springer: Cham, Switzerland, 2019; pp. 257–276.
17. Mekonnen, M.M.; Romanelli, T.L.; Ray, C.; Hoekstra, A.Y.; Liska, A.J.; Neale, C.M.U. Water, energy, and carbon footprints of bioethanol from the U.S. and Brazil. *Environ. Sci. Technol.* **2018**, *52*, 14508–14518. [CrossRef]
18. Liu, H.; Huang, Y.; Yuan, H.; Yin, X.; Wu, C. Life cycle assessment of biofuels in China: Status and challenges. *Renew. Sustain. Energy Rev.* **2018**, *97*, 301–322. [CrossRef]
19. Petinrin, J.; Shaaban, M. Renewable energy for continuous energy sustainability in Malaysia. *Renew. Sustain. Energy Rev.* **2015**, *50*, 967–981. [CrossRef]
20. Gasparatos, A.; von Maltitz, G.; Johnson, F.; Lee, L.; Mathai, M.; de Oliveira, J.P.; Willis, K. Biofuels in sub-Sahara Africa: Drivers, impacts and priority policy areas. *Renew. Sustain. Energy Rev.* **2015**, *45*, 879–901. [CrossRef]
21. Andwari, A.M.; Pesiridis, A.; Rajoo, S.; Martinez-Botas, R.; Esfahanian, V. A review of battery electric vehicle technology and readiness levels. *Renew. Sustain. Energy Rev.* **2017**, *78*, 414–430. [CrossRef]

22. Soltani-Sobh, A.; Heaslip, K.; Stevanovic, A.; Bosworth, R.; Radivojevic, D. Analysis of the electric vehicles adoption over the United States. *Transp. Res. Procedia* **2017**, *22*, 203–212. [CrossRef]

23. Tambo, E.; Duo-Quan, W.; Zhou, X.-N. Tackling air pollution and extreme climate changes in China: Implementing the Paris climate change agreement. *Environ. Int.* **2016**, *95*, 152–156. [CrossRef] [PubMed]

24. Xue, M.; Lin, B.-L.; Tsunemi, K. Emission implications of electric vehicles in Japan considering energy structure transition and penetration uncertainty. *J. Clean. Prod.* **2020**, *280*, 124402. [CrossRef]

25. Miotti, M.; Supran, G.J.; Kim, E.J.; Trancik, J.E. Personal vehicles evaluated against climate change mitigation targets. *Environ. Sci. Technol.* **2016**, *50*, 10795–10804. [CrossRef]

26. Murshed, M.; Ali, S.R.; Banerjee, S. Consumption of liquefied petroleum gas and the EKC hypothesis in South Asia: Evidence from cross-sectionally dependent heterogeneous panel data with structural breaks. *Energy Ecol. Environ.* **2020**, *6*, 353–377. [CrossRef]

27. Tseng, C.-H.; Chen, L.-L.; Shih, Y.-H. Quantifying the environment and health co-benefits of low-emission vehicles in Taiwan. *J. Integr. Environ. Sci.* **2014**, *12*, 39–48. [CrossRef]

28. Kumar, S.; Salam, P.A.; Shrestha, P.; Ackom, E.K. An assessment of Thailand's biofuel development. *Sustainability* **2013**, *5*, 1577–1597. [CrossRef]

29. Department of Alternative Energy Development and Efficiency. Guidelines for development of renewable energy and alternative energy. In *Alternative Energy Development Plan 2018-2037 (AEDP2018)*; DEDE: Bangkok, Thailand, 2020; pp. 15–22.

30. Department of Alternative Energy Development and Efficiency. Guidelines for energy efficiency Plan 2018–2037. In *Energy Efficiency Plan 2018-2037 (EEP2018)*; DEDE: Bangkok, Thailand, 2020; pp. 10–12.

31. Electric Generating Authority of Thailand. Metropolitan electricity authority of Thailand, provincial electricity authority of Thailand, electric vehicle projection in Thailand. In *Electrical Infrastructure Development Plan to Support Electric Vehicle in Thailand*; EGAT: Bangkok, Thailand, 2016; pp. 57–68.

32. Foster, N.; Bloyd, C. *Thailand Alternative Fuels Update 2016*; DE-AC05-76RL01830; Department of Energy US: Washington, DC, USA, 2016. Available online: https://www.pnnl.gov/main/publications/external/technical_reports/PNNL-25884.pdf (accessed on 29 December 2021).

33. The UNdata – A World of Information. Available online: http://data.un.org/Host.aspx?Content=About (accessed on 29 December 2021).

34. Arable Land (Hectares Per Person). The NationMaster 2003–2021. Available online: https://www.nationmaster.com/country-info/stats/Agriculture/Arable-land/Hectares-per-capita (accessed on 29 December 2021).

35. Kemausuor, F.; Akowuah, J.O.; Ofori, E. Assessment of feedstock options for biofuels production in Ghana. *J. Sustain. Bioenergy Syst.* **2013**, *3*, 119–128. [CrossRef]

36. Ayetor, G.; Quansah, D.A.; Adjei, E.A. Towards zero vehicle emissions in Africa: A case study of Ghana. *Energy Policy* **2020**, *143*, 111606. [CrossRef]

37. Button, K.; Ngoe, N.; Hine, J. Modelling vehicle ownership and use in low income countries. *J. Transp. Econ. Policy* **1993**, *27*, 51–67.

38. Heaps, C.G. *Low Emission Alternative Planning*; Stockholm Environment Institute: Somerville, MA, USA, 2020.

39. Faah, G. Road Transportation Impact on Ghana's Future Energy and Environment. Ph.D. Thesis, Freiberg University of Mining and Technology, Freiberg, Germany, 7 November 2008.

40. National Petroleum Consumption from 1999–2019. Available online: http://www.npa.gov.gh/downloads (accessed on 21 November 2021).

41. Estimation of Fuel Consumption and CO_2 Emissions in Ghana. Methodology and Results. Available online: https://wedocs.unep.org/bitstream/handle/20.500.11822/25374/CalculationofFuelConsumptionandCO2Emissions.pdf (accessed on 21 November 2021).

42. Ackom, G. United Nations Environment Programme, Vehicle Emission Enforcement, Driver and Vehicle Licensing Authority, Ghana. Available online: https://wedocs.unep.org/20.500.11822/21480 (accessed on 21 November 2021).

43. Asante, K. Improving Fuel Economy in Ghana: A Socio-Economic Analysis of Policy Options, Global Fuel Economy Inisitative, 2018, 13 January. Available online: https://www.globalfueleconomy.org/media/596960/ghana-costbenefitanalysissocioeconomicanalysis.pdf (accessed on 21 November 2021).

44. Nagai, Y.; Fukuda, A.; Okada, Y.; Hashino, Y. Two-wheeled vehicle ownership trends and issues in the ASIAN Region. *J. East. Asia Soc. Transp. Stud.* **2003**, *5*, 11.

45. Pongthanaisawan, J.; Sorapipatana, C. Greenhouse gas emissions from Thailand's transport sector: Trends and mitigation options. *Appl. Energy* **2013**, *101*, 10. [CrossRef]

46. Chollacoop, N.; Saisirirat, P.; Sukkasi, S.; Tongroon, M.; Fukuda, T.; Fukuda, A.; Nivitchanyong, S. Potential of greenhouse gas emission reduction in Thai road transport by ethanol bus technology. *Appl. Energy* **2013**, *102*, 112–123. [CrossRef]

47. Fuel Economy. All Electric Vehicles. Available online: https://www.fueleconomy.gov/feg/evtech.shtml (accessed on 29 December 2021).

48. Mensah, J.; Annan, J.; Andoh-Baidoo, F. Assessing the impact of vehicular traffic on energy demand in the Accra metropolis. *J. Manag. Policy Pract.* **2014**, *15*, 10.

49. Intergovernmental Panel on Climate Change (IPCC). *Revised 1996 IPCC Guidelines for National Greenhouse Gas Inventories*; IGES: Kanagawa, Japan, 1996.

50. Urban Electric Mobility Initiative (UEMI). Ghana Policy Environment Paper. Available online: http://www.uemi.net/uploads/4/8/9/5/48950199/uemi_ca_ghana.pdf (accessed on 21 November 2021).
51. Energy Commission, Ministry of Energy of Ghana. Drive Elecrtric Initiative (DEI). Available online: https://energycom.gov.gh/efficiency/drive-electric-initiative (accessed on 21 November 2021).
52. Sorda, G.; Banse, M.; Kemfert, C. An overview of biofuel policies across the world. *Energy Policy* **2010**, *38*, 6977–6988. [CrossRef]
53. Yusoff, M.N.A.M.; Zulkifli, N.W.M.; Sukiman, N.L.; Chyuan, O.H.; Hassan, M.H.; Hasnul, M.H.; Zulkifli, M.S.A.; Abbas, M.M.; Zakaria, M.Z. Sustainability of palm biodiesel in transportation: A review on biofuel standard, policy and international collaboration between Malaysia and Colombia. *Bioenergy Res.* **2020**, *14*, 43–60. [CrossRef]
54. Indonesian Palm Oil. Indonesia Leading in Biodiesel Production. Available online: https://thepalmscribe.id/indonesia-to-start-testing-biodiesel-mixtures-higher-than-b30/ (accessed on 21 November 2021).
55. SAWIT. President Wants B50 in Use by End 2020. Available online: https://www.bpdp.or.id/en/President-Wants-B50-in-Use-by-End-2020 (accessed on 21 November 2021).
56. The World Bank. World Bank Open Data: Historical Record of Arable Land (sq.km)—Ghana 1961–2018. Available online: https://data.worldbank.org/ (accessed on 21 November 2021).
57. Savvanidou, E.; Zervas, E.; Tsagarakis, K.P. Public acceptance of biofuels. *Energy Policy* **2010**, *38*, 3482–3488. [CrossRef]
58. Silalertruksa, T.; Bonnet, S.; Gheewala, S.H. Life cycle costing and externalities of palm oil biodiesel in Thailand. *J. Clean. Prod.* **2012**, *28*, 225–232. [CrossRef]
59. Vagliasindi, M. *Implementaing Energy Subsidy Reforms: Evidence from Developing Countries: Ghana*; World Bank Group—e-Library: Washingon, DC, USA, 2012; pp. 39–56.
60. Crawford, G. Ghana's Fossil-Fuel Subsidy Reform. Available online: https://www.ids.ac.uk/download.php?file=files/dmfile/LHcasestudy09-FossilfuelsGhana.pdf (accessed on 21 November 2021).
61. Energy Commission. National Energy Statistics 2006–2015. Available online: http://energycom.gov.gh/files/National%20Energy%20Statistics_2016.pdf (accessed on 21 November 2021).

Article

Using Decomposition Analysis to Determine the Main Contributing Factors to Carbon Neutrality across Sectors

Hsing-Hsuan Chen [1], Andries F. Hof [1,2,*], Vassilis Daioglou [1,2], Harmen Sytze de Boer [1], Oreane Y. Edelenbosch [2], Maarten van den Berg [1], Kaj-Ivar van der Wijst [1] and Detlef P. van Vuuren [1,2]

[1] PBL Netherlands Environmental Assessment Agency, 2594 AB The Hague, The Netherlands; hsinghsuan.chen@pbl.nl (H.-H.C.); vassilis.daioglou@pbl.nl (V.D.); Harmen-Sytze.deBoer@pbl.nl (H.S.d.B.); bergvdma@pbl.nl (M.v.d.B.); wijstvdk@pbl.nl (K.-I.v.d.W.); vuurenvd@pbl.nl (D.P.v.V.)

[2] Copernicus Institute of Sustainable Development, Faculty of Geosciences, Utrecht University, 3584 CB Utrecht, The Netherlands; o.y.edelenbosch@uu.nl

[*] Correspondence: Andries.Hof@pbl.nl

Abstract: This paper uses decomposition analysis to investigate the key contributions to changes in greenhouse gas emissions in different scenarios. We derive decomposition formulas for the three highest-emitting sectors: power generation, industry, and transportation (both passenger and freight). These formulas were applied to recently developed 1.5 °C emission scenarios by the Integrated Model to Assess the Global Environment (IMAGE), emphasising the role of renewables and lifestyle changes. The decomposition analysis shows that carbon capture and storage (CCS), both from fossil fuel and bioenergy burning, renewables and reducing carbon intensity provide the largest contributions to emission reduction in the scenarios. Efficiency improvement is also critical, but part of the potential is already achieved in the Baseline scenario. The relative importance of different emission reduction drivers is similar in the OECD (characterised by relatively high per capita income levels and emissions) and non-OECD (characterised by relatively high carbon intensities of the economy) region, but there are some noteworthy differences. In the non-OECD region, improving efficiency in industry and transport and increasing the share of renewables in power generation are more important in reducing emissions than in the OECD region, while CCS in power generation and electrification of passenger transport are more important drivers in the OECD region.

Keywords: net-zero emission; decomposition analysis; mitigation; integrated assessment; shared socioeconomic pathways; climate change; Paris Agreement; scenarios

Citation: Chen, H.-H.; Hof, A.F.; Daioglou, V.; de Boer, H.S.; Edelenbosch, O.Y.; van den Berg, M.; van der Wijst, K.-I.; van Vuuren, D.P. Using Decomposition Analysis to Determine the Main Contributing Factors to Carbon Neutrality across Sectors. *Energies* **2022**, *15*, 132. https://doi.org/10.3390/en15010132

Academic Editor: Dolf Gielen

Received: 15 November 2021
Accepted: 22 December 2021
Published: 24 December 2021

Publisher's Note: MDPI stays neutral with regard to jurisdictional claims in published maps and institutional affiliations.

1. Introduction

The Paris Agreement calls for achieving a balance between anthropogenic emissions and removals by sinks of greenhouse gases in the second half of this century (also called net-zero emissions) in order to achieve the objective of holding the global average temperature increase to well below 2 °C and preferably 1.5 °C. In line with this, many countries have set net-zero emissions targets. Scenarios developed by Integrated Assessment Models (IAMs) have shown that achieving these targets requires a fast transition in energy and land-use systems on a global scale [1]. There are, however, important differences in the strategies applied to reduce emissions to net-zero. Such strategies differ, for instance, in terms of choices made in the power system (focusing on renewables, nuclear power or carbon capture and storage (CCS)), the importance of bio-energy, the role of negative emissions (e.g., reforestation and bio-energy-CCS (BECCS)), the role of non-CO_2 greenhouse gases, the role of lifestyle change and the timing of reductions (e.g., [2,3]). Understanding the relative importance of these drivers in mitigating emissions is crucial to support and accelerate the energy transition, which is the main aim of IRENA's Long-Term Energy Transition (LTES) initiative. Decomposition analysis can identify the importance of each of these factors. This technique has been applied frequently to explain historical emission trends.

Decomposition has been extensively used to at trends in scenarios, but mostly at the aggregated level of Kaya indicators [4–6]. However, also more detailed level decomposition schemes have been used, such as those applied by van den Berg [7], who focused on the residential sector and passenger transport; Sharmina, et al. [8] and Edelenbosch et al. [9], who focused on passenger transport and industry; Llop [10], who did a structural decomposition analysis of gross energy output; Karmellos, et al. [11], who applied decomposition analysis on the EU power sector; Marcucci and Fragkos [12], who focused on long-term drivers of alternative net-zero pathways in China, India, Europe, and the USA; Guan et al. [13], who conducted decomposition analysis to investigate the driving forces of CO_2 emission in China; Jiang, Su, and Li [14], who conducted a decomposition analysis of the Chinese power grid. However, a thorough decomposition of all important emission sources globally has not yet been conducted, and none of the above studies included carbon capture and storage as decomposition factor.

This study applies decomposition analyses on the newly developed IMAGE 3.2 [15] 1.5 °C scenarios to analyse the largest contributing factors to emission reductions. This study uses recently published scenarios based on the Shared Socio-economic Pathways (SSPs), looking at different mitigation routes. The latter include the default strategy and two alternatives: one focusing strongly on a rapid expansion of renewables and electrification and one focusing strongly on lifestyle changes. In the study here, we use these scenarios. We analyse the results at the global level and for the OECD and non-OECD region (The OECD region includes Canada, the USA, Mexico, Western Europe, Central Europe, Turkey, Korea region, Japan, and Oceania. The non-OECD region covers Central America, Brazil, South America, Africa, Ukraine region, Central Asia, Russia region, Middle East, India, China, South-eastern Asia, and South Asia (https://models.pbl.nl/image/index.php/Region_classification_map), as these show very different characteristics. Emissions per capita and income levels per capita are relatively high in the OECD region, whereas the carbon intensity of the economy is relatively high in the non-OECD region.

Unlike previous studies, which only applied decomposition analysis on one or two energy sectors or for specific regions, this study uses the decomposition method to analyse changes in CO_2 emissions in the three sectors with the highest emission levels globally: power generation, industry, and transport (including passenger travel and freight transport). For all other emission sources (residential, services, land use, waste, and other non-CO_2 greenhouse gas emissions), we discuss the changes in emissions without applying a decomposition analysis. The paper aims to identify the main contributing factors to emission reductions in existing long-term 1.5 °C scenarios globally and in OECD and non-OECD regions.

2. Materials and Methods

While the decomposition method applied in this paper is generically applicable to scenario output of IAMs, we apply decomposition analysis here to existing scenarios developed by the IMAGE 3.2 Integrated Assessment model. We focus specifically on the changes in emissions from power generation, industry and transport for the SSP2 baseline and two 1.5 °C mitigation scenarios. Below, we explain the decomposition analysis and present the framework.

2.1. Decomposition Analysis

Decomposition analysis is used to identify the key contributions to changes in GHG emissions in scenarios. This decomposition analysis is applied to CO_2 emissions from energy use in industry, power generation, and transport. The decomposition is not applied to all other sources of GHG emissions as most of these sources are either relatively small or modelled in less detail. We focus on 2050 as total GHG emissions are very close to net-zero by this year in the mitigation scenarios.

Several decomposition methods have been developed to study the impacts of structural change on energy use in industry or on energy-related gas emissions. In literature, two forms of decomposition methods are mostly used: the Divisia based method and the Laspeyres based method [16,17]. As the scenarios used include negative emission values (via carbon dioxide removal technologies), we use the Laspeyres method instead of the Divisia method. However, the Laspeyres-based method leads to residuals, which means that the sum of each contributing factor to changes in emissions does not equal the total changes in emissions. Therefore, we use the Shapley/Sun method, which is based on the Laspeyres method but does not lead to residual values [9,16,18,19].

Due to the structural differences between the sectors, we have derived different formulas to show the contribution of the factors to emission changes for each sector. For transport, we have used an adapted version of the method used by van den Berg [7], who identified activity, mode shift, intensity, and fuel mix. We have added population as a driving factor to show the impact of increasing population on emissions explicitly. The resulting formula is given in Table 1.

Table 1. Decomposition formula for transport.

Population		Activity		Mode Shift		Efficiency		Carbon Intensity
Pop	$\times$	$\frac{Pkm\ or\ Tkm}{Pop}$	$\times$	M	$\times$	$\frac{FE}{Pkm}$	$\times$	$\frac{CO_2}{FE}$

Pop: population, Pkm: passenger-kilometer, Tkm: tonne-kilometer, M: mode share (%), FE: final energy use (TJ), CO_2: CO_2 emissions from transport (Gt).

We have applied this decomposition to both passenger travel and freight transport. We use the same equation for these subsectors; the only difference is that for passenger travel activity, we use passenger distance (Pkm) and for freight transport tonne-kilometres (Tkm). The travel modes also differ between passenger travel and freight transport. In passenger travel, we consider four modes: bus, train, car, and airplane. In freight transport, there are five modes: national shipping, international shipping, train, medium truck, heavy truck, and air cargo.

Decomposition of emissions from power generation and industry has been applied only in a few previous studies. In Karmellos et al. [11], emissions from power generation are decomposed by the following factors: the activity effect, defined as changes in Gross Domestic Product (GDP); the electricity intensity effect, defined as changes in the ratio of total electricity consumption to total GDP; the electricity trade effect, defined as changes in the ratio of electricity production to electricity consumption; the energy efficiency effect, defined as changes in the ratio of fuel input to the respective electricity output; and the fuel mixture effect, defined as changes in the share of a fuel in the total energy input of the power sector of the country. In Edelenbosch et al. [9], population growth, final energy use per capita, the share of electricity and hydrogen, and direct emissions of non-electric fuels are used as decomposition factors for industry.

The above studies excluded CCS as decomposition factors. Since CCS is potentially an important factor for emission reduction, we include this as an additional factor. This means that we define carbon intensity as the sum of CO_2 emissions and CO_2 captured divided by the energy use of fossil fuels (primary in the case of power generation and final in the case of industry). CCS includes all carbon captured, so both from resulting from the burning of fossil fuels and bioenergy. Furthermore, switching to renewables is a crucial mitigation strategy for power generation and has been included in the decomposition. For industry, we have selected electrification as a separate factor following Edelenbosch et al. [9]. For both industry and power generation, carbon intensity only relates to fossil fuels and biomass carbon intensity (i.e., fuel switch between coal, oil, gas, and biomass). The formulas for industry and power generation are given in Tables 2 and 3, respectively.

Table 2. Decomposition formula for industry.

Population		Activity		Electrification		Efficiency		Carbon Intensity		CCS
Pop	$\times$	$\frac{Act}{Pop}$	$\times$	$(1 - \%Elc)$	$\times$	$\frac{FE}{Act}$	$\times$	$\frac{CO_2 + CCS}{FE(1 - \%Elc)}$	$-$	CCS

Pop: population; *Act*: industry production value-added (US\$/cap); *Elc*: electricity share in energy use (%); *FE*: total final energy use of industry (TJ); *CO$_2$*: CO_2 emissions from industry (Gt CO_2); *CCS*: carbon capture and storage in industry, including BECCS (Gt CO_2).

Table 3. Decomposition formula for power generation.

Population		Activity		Renewables		Efficiency		Carbon Intensity		CCS
Pop	$\times$	$\frac{elec\ prod}{pop}$	$\times$	$elec\ prod\ (1 - \%nonfos)$	$\times$	$\frac{PE\ (1 - \%nonfos)}{elec\ prod\ (1 - \%nonfos)}$	$\times$	$\frac{CO_2 + CCS}{PE(1 - \%nonfos)}$	$-$	CCS

elec prod: electricity production (GWh); *%nonfos*: share of renewables and nuclear in primary energy use of power generation (%); *PE*: total primary energy use of power generation (TJ); *CO$_2$*: CO_2 emissions from power generation (Gt CO_2); *CCS*: carbon capture and storage in power generation, including BECCS (Gt CO_2).

2.2. Scenarios

The analysed scenarios were developed by the Integrated Assessment model IMAGE 3.2 [15]. IMAGE is a comprehensive ecological–environmental model framework that simulates the environmental consequences of human activities worldwide. A detailed description of the model can be found in the online documentation [15].

Three IMAGE scenarios are analysed: a baseline scenario for reference and two 1.5 °C mitigation scenarios (Table 4). The Baseline scenario is an updated version of the SSP2 baseline by van Vuuren et al. [20], calibrated to 2020 data where possible. SSP2 is based on middle-of-the-road socio-economic projections (e.g., population growth, economic growth, technology development, and lifestyle change). The recent update includes model updates, new insights in technology development and the impact of COVID-19 and its recovery measures (see [15]).

Table 4. Description of scenarios.

Scenario	Main Assumptions
Baseline	The Baseline scenario follows the SSP2 baseline assumptions; main drivers are updated to 2015–2020 data, including near-term projections up to 2025 for GDP following IMF (to account for COVID-19). Drivers follow relative growth rates of original SSPs from 2025 onwards. See Van Vuuren et al., 2021 [15].
1.5 °C Renewable scenario	A carbon tax is introduced to reach an end-of-century radiative forcing of 1.9 W/m^2 compared to preindustrial times. High electrification rates in all end-use sectors are possible.
1.5 °C Lifestyle scenario	A carbon tax is introduced to reach an end-of-century radiative forcing of 1.9 W/m^2 compared to preindustrial times. consumers change their habits towards a lifestyle that leads to lower GHG emissions.

Two 1.5 °C scenarios are based on the SSP2 baseline but assume climate policy to reach 1.5 °C. In the 1.5 °C Renewable scenario, it is assumed that high electrification rates in all end-use sectors are possible due to optimistic assumptions about the integration of variable renewable energy technologies and costs of transmission, distribution, and storage. As a result, the 1.5 °C Renewable scenario has a higher electrification rate and a higher renewable share. In the 1.5 °C Lifestyle scenario, consumers change their habits towards a lifestyle that leads to lower GHG emissions. This includes a less meat-intensive diet conforming to health recommendations, less CO_2-intensive transport modes (following the current modal split in Japan), less intensive use of heating and cooling (change of 1 °C in heating and cooling reference levels) and a reduction in the use of several domestic

appliances. These assumptions lead to more structural changes at the sector level, leading to less energy consumption [7,21–23] (also see [15]).

3. Results

In the first part of the results section, we show the total emissions by source under the three scenarios. In the second part, the results of the decomposition analysis are shown.

3.1. Total Emissions by Source

The three highest-emitting sources for which we apply decomposition are responsible for about half of total CO_2, CH_4, and N_2O emissions in the Baseline scenario, both by 2030 and 2050 (Figure 1). Based on the output from the IMAGE model, 11 other emission sources can be distinguished. The sum of the emissions from these sources equals the total of CO_2, CH_4, and N_2O emissions. Figure 1 shows global greenhouse gas emissions by 2030 and 2050 for these sources and separately for OECD countries and non-OECD countries.

In the Baseline scenario, GHG emissions continue to increase, especially in the non-OECD region. By 2050, the power sector will be responsible in the SSP2 scenario for the largest share of GHG emissions. Other large emitting sources are transport, CO_2 emissions due to land use, industry, and non-CO_2 energy. By 2050, non-OECD countries will be responsible for 77% of global GHG emissions in the baseline.

Total GHG emissions are net-negative in the OECD region in the mitigation scenarios by 2050. In the non-OECD region, emissions in 2050 are still positive, mostly from land-use emissions and transport (due to less efficient and higher carbon-intensive fuel usage than the OECD region). GHG emissions are reduced strongly across all sources in the 1.5 °C scenarios, especially CO_2 emissions from power generation and land use, where emissions are negative by 2050. For most other sources, emissions are close to zero. However, there are still considerable remaining CH_4 and N_2O emissions from animal husbandry and other land-use activities (agriculture and forestry), and, to a lesser degree, CO_2 emissions from transport. The emissions from these sources are more difficult to reduce in the IMAGE model (consistent with most other scenarios). Therefore, negative CO_2 emissions through BECCS and reforestation and afforestation are needed to achieve overall net-zero GHG emissions.

The clearest difference between the 1.5 °C Lifestyle scenario and the 1.5 °C Renewable scenario is the difference in non-CO_2 emissions from animal husbandry and land-use CO_2 emissions. This is largely caused by changing diets in the 1.5 °C Lifestyle scenario. The lower consumption of meat directly reduces emissions from animal husbandry. At the same time, less grazing land is needed, which frees up land that can be used for reforestation, leading to substantial negative land-use CO_2 emissions. By 2050, this also leads to lower emissions in the Lifestyle scenario—although initially, emissions are reduced more strongly in the Renewable scenario in power, industry, and transport sectors due to a faster switch to renewable energy.

The next section analyses the CO_2 emission reductions of power generation, transport, and industry in more detail using the decomposition analysis.

3.2. Main Mitigation Drivers

Here, we analyse how different factors contribute to mitigation in different strategies. For power generation, industry, and transport, we first show the waterfall charts that provide insight into the emission changes, then present the energy mix in 2015 and 2050 for the three scenarios to provide more detail on some of these factors.

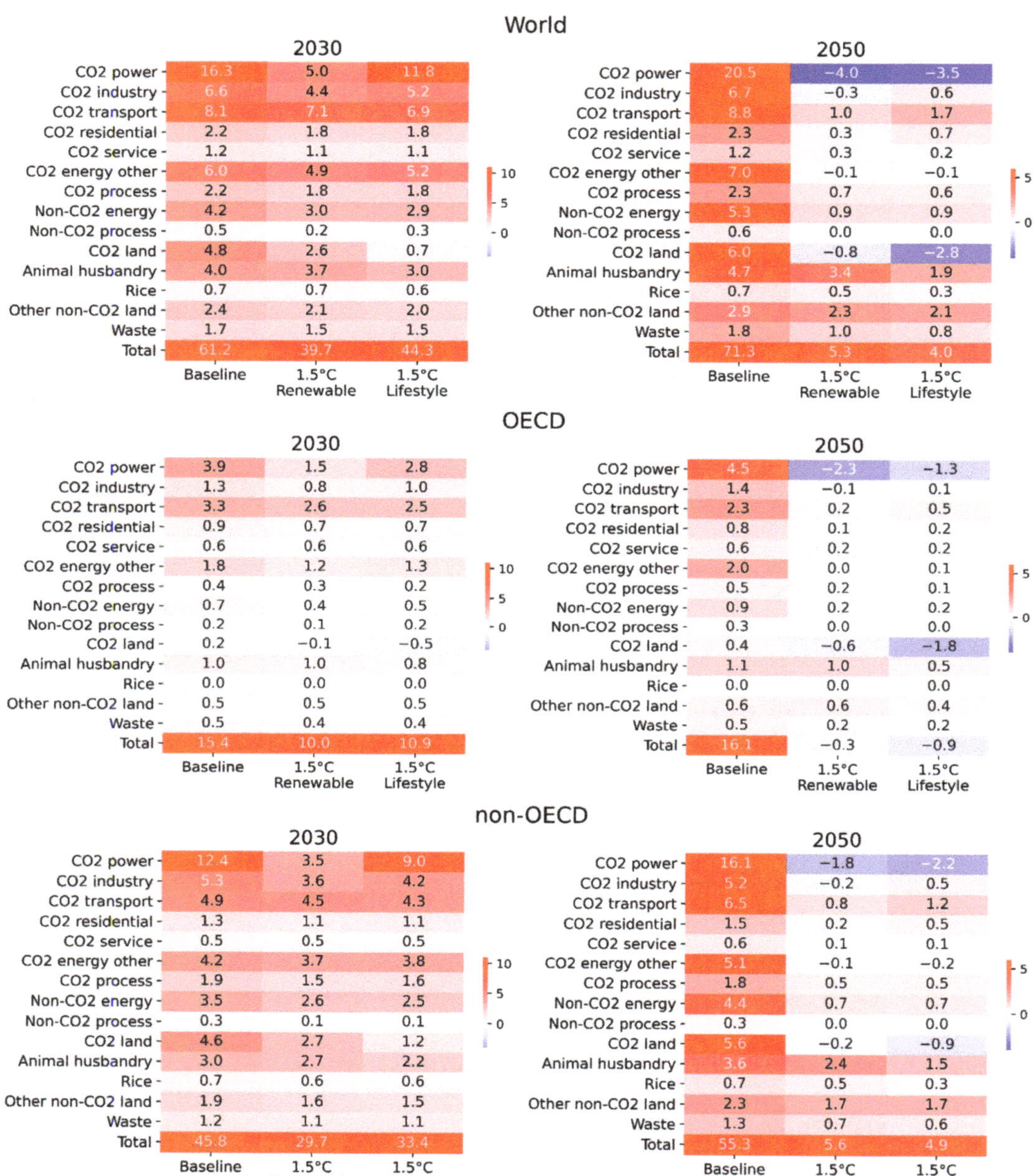

Figure 1. Total greenhouse gas emissions by emission source in 2030 and 2050 in Gt CO_2-eq. (using IPCC AR4 100-year global warming potentials [24]). The emission sources combined account for around 97% of global GHG emissions in recent years [25].

3.2.1. Power Generation

In power generation, we decompose emission trends into population change, activity change (power generation per capita), changes in efficiency (power generation divided by energy use in power generation), changes in non-fossil share (renewable and nuclear power generation), changes in non-renewable CO_2 intensity, and CCS.

In the Baseline scenario, CO_2 emissions from power generation strongly increase between 2015 and 2050, from 13.5 Gt CO_2/y to 20.5 GtCO_2/y (Figure 2). This is mainly due to higher electricity consumption per capita (9.8 GtCO_2/y), followed by population growth (4.1 GtCO_2/y). Efficiency improvements have a downward impact of 4.8 Gt CO_2 on emissions in the Baseline, followed by 2.1 Gt CO_2 reduction due to a reduction in carbon intensity. The latter is a result of changes in the energy mix: the share of coal decreases from 45% to 35%, and the share of natural gas increases from 18% to 29% (Figure 2).

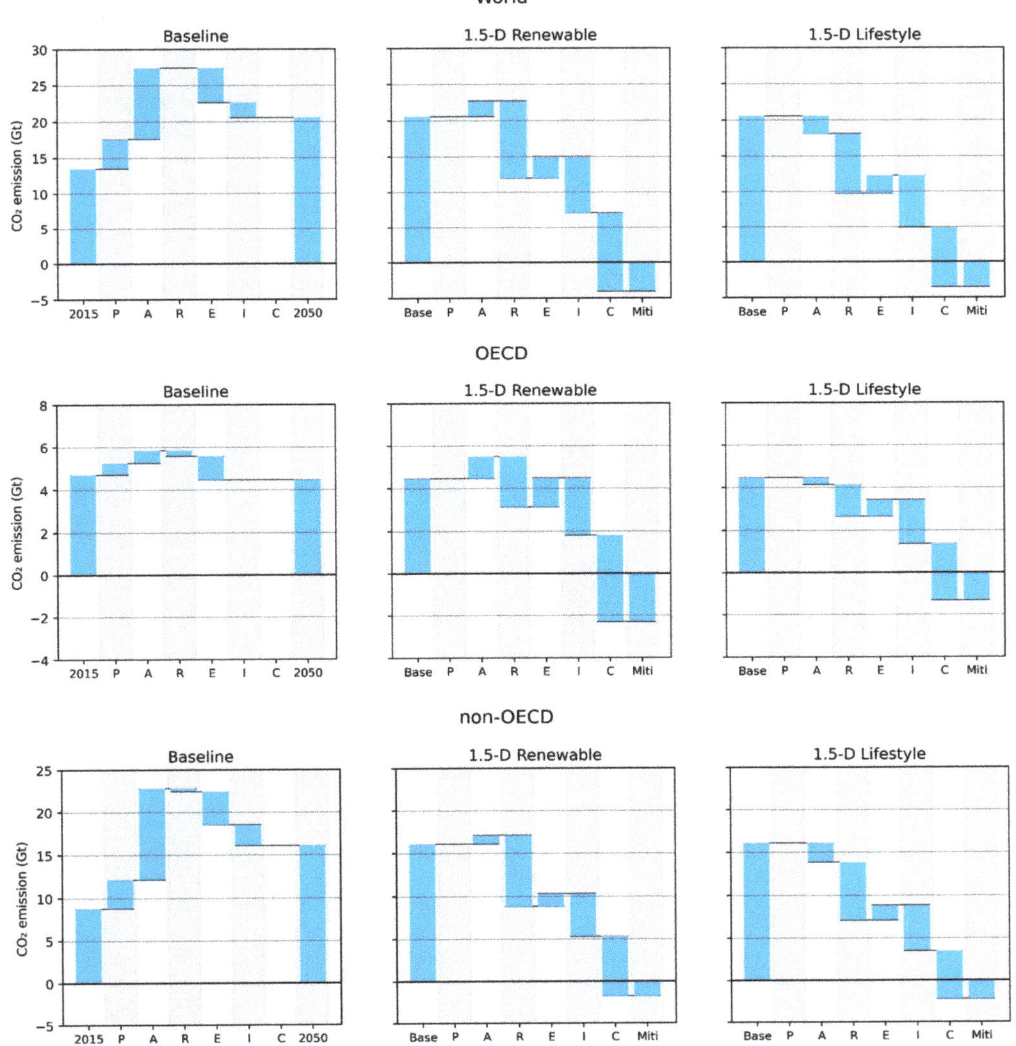

Figure 2. Changes in power generation emissions between 2015 and 2050 (Baseline) and between Baseline and the mitigation scenarios (in 2050) due to different drivers. P: population growth, A: activity changes, R: renewable and nuclear share, E: efficiency, I: carbon intensity, C: CCS.

Figure 2 also shows the decomposition analysis between the Baseline scenario and the 1.5 °C scenarios. The 1.5 °C Renewable and Lifestyle scenarios show substantial (−4 and −3.5 Gt CO_2) net-negative CO_2 emissions by 2050, made possible by the use of BECCS. This is reflected in the decomposition analysis, which shows that CCS is the most important driver for emission reduction in both 1.5 °C scenarios, leading to 8 Gt CO_2 reduction in the 1.5 °C Lifestyle scenario and 11 Gt CO_2 reduction in the 1.5 °C Renewable scenario. CCS is applied both on natural gas and bioenergy power plants.

Increasing the share of non-fossil energy (renewables and nuclear) has a similar impact on reducing emissions as CCS in both 1.5 °C scenarios. Indeed, the share of renewables and nuclear power in the energy mix increases from 14% in 2015 to 45–48% in 2050 in the 1.5 °C scenarios (Figure 3). CO_2 intensity improvements also contribute strongly to reducing emissions. This is mainly a consequence of carbon-intensive coal being phased out almost completely. In contrast, natural gas still has a significant share in the energy mix (largely with CCS).

The main difference between the two 1.5 °C scenarios is the contribution of the activity factor. In the 1.5 °C Renewable scenario, there is even an upward impact on emissions compared to baseline, resulting from higher electrification rates in end-use sectors. In contrast, in the Lifestyle scenario, some of the assumed changes reduce electricity demand increase somewhat. The stronger impact of especially renewables and CCS in reducing emissions means that overall, CO_2 emissions from power generation are slightly lower in the Renewable scenario than in the Lifestyle scenario.

The non-OECD region has higher population and activity growth, causing a significant increase in emissions in the Baseline scenario. In the non-OECD region, renewables play a more important role in reducing emissions than in the OECD region, while CCS plays a more important role in the OECD region.

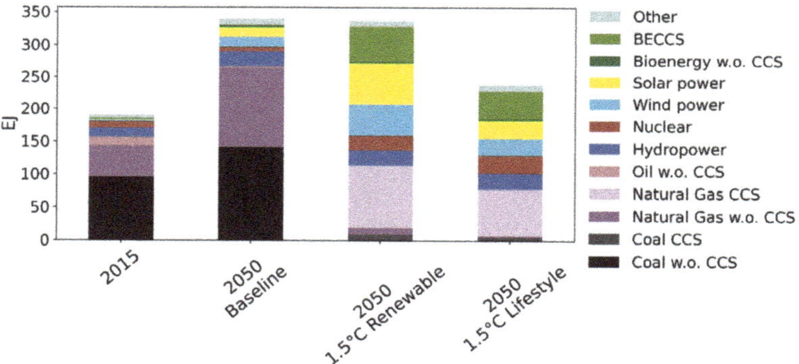

Figure 3. The global primary energy mix of power generation.

3.2.2. Industry

For industry, we decompose emission trends into the contribution of population change, activity change (industrial value-added per capita), changes in production efficiency, electrification, changes in non-renewable CO_2 intensity, and CCS.

Strong growth in activity is the most important contribution to increasing industrial emissions in the Baseline, followed by population growth (Figure 4). A strong improvement in energy efficiency partly offsets the increase in activity and population growth. In total, industrial emissions increase by 23% between 2015 and 2050 in the Baseline scenario.

In both 1.5 °C scenarios, CCS is the most important factor for reducing emissions from industry. In the 1.5 °C Renewable scenario, electrification and improving overall CO_2 intensity lead to higher emission reductions than in the 1.5 °C Lifestyle scenario. This leads to lower total industrial emissions in the 1.5 °C Renewable scenario than in the Lifestyle scenario, despite the higher impact of efficiency improvements in the 1.5 °C Lifestyle scenario.

In the baseline, activity and population growth leads to a much stronger emission increase in the non-OECD region than the OECD region. Efficiency improvement contributes more to emission reduction in the non-OECD region than in the OECD region.

The energy mix of the two 1.5 °C scenarios is similar by 2050: coal and oil are replaced by electricity and bioenergy (Figure 5). The main differences are (i) much lower energy use in the Lifestyle scenario, which was reflected by the stronger impact of energy efficiency in reducing emissions, and (ii) a much stronger electrification rate in the Renewable scenario.

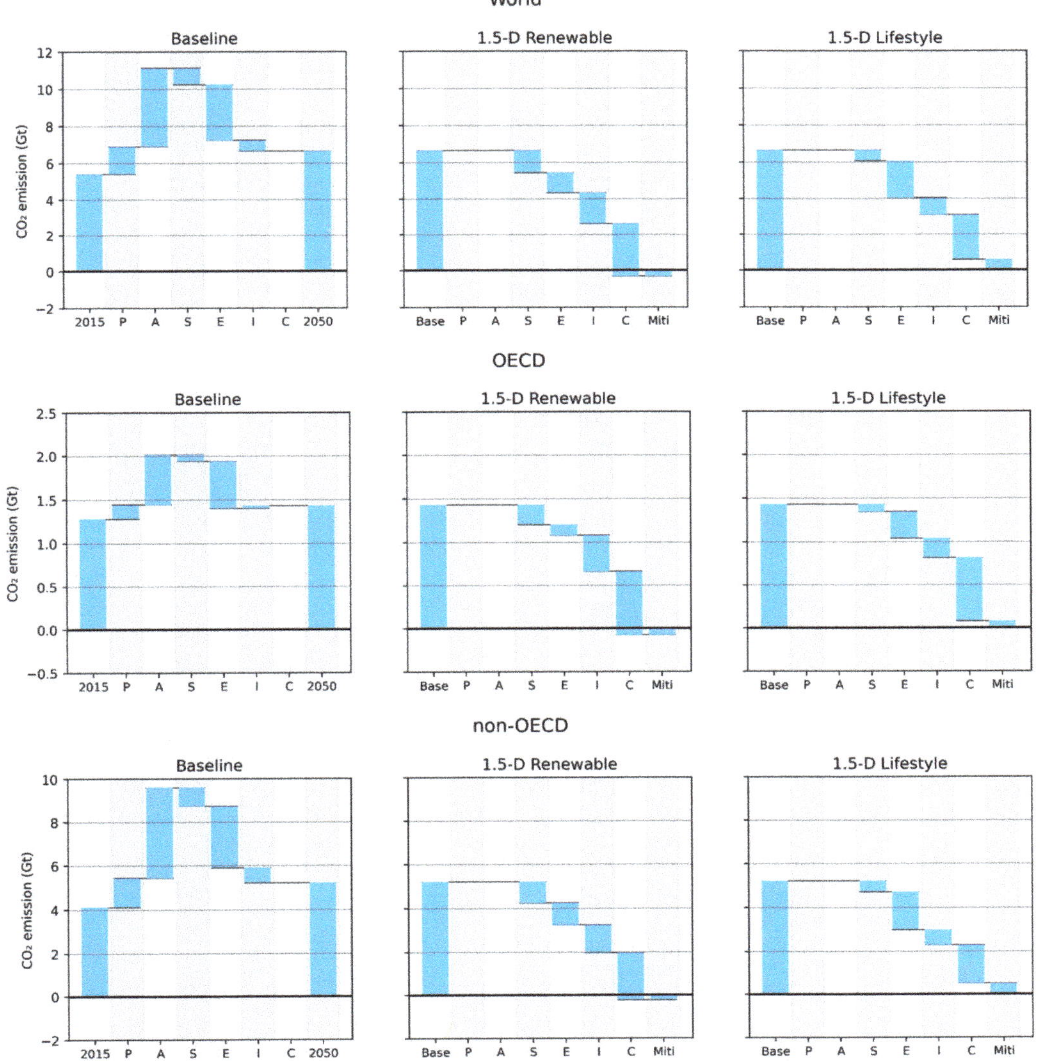

Figure 4. Changes in industry emissions between 2015 and 2050 (Baseline) and between Baseline and the mitigation scenarios (in 2050) due to different drivers. P: population growth, A: activity changes, S: electrification, E: efficiency, I: carbon intensity of non-renewable fuels, C: CCS.

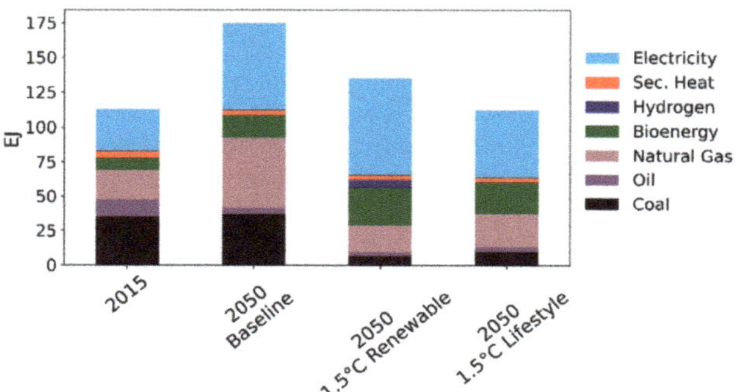

Figure 5. The global energy mix of industry in the baseline and the two 1.5 °C scenarios. (sec. heat = secondary heat).

3.2.3. Transport

IMAGE divides the emission sources into three sectors in the transport sector: passenger travel, freight transport, and bunkers (international aviation and international shipping). We apply the decomposition analysis to passenger travel and freight transport, including bunkers in these two subsectors. Figure 6 shows the decomposition results for passenger travel, and Figure 7 illustrates the energy mix. Also, Figure 8 shows the decomposition results for freight transport, with Figure 9 shows its energy mix.

Passenger Travel

In the Baseline scenario, CO_2 emissions of passenger travel increase from 4.3 Gt in 2015 to 7.6 Gt by 2050, due to the increasing travel distance per capita (contributing to 2.4 Gt emission increase), population growth (1.4 Gt emission increase), and switching to more carbon-intensive travel modes (1.1 Gt emission increase). The latter is due to a shift from public transport to car and air travel: the share of distance travelled by car increases from 39% to 47% and by air from 9% to 12% between 2015 and 2050. Efficiency improvements in cars and a shift to more electric cars partly offset the increases in emissions due to the above factors, contributing to a 1.7 Gt emission reduction.

There is a large difference between the OECD and non-OECD regions. In the OECD region, CO_2 emissions decrease in the Baseline scenario as the impact of the activity increase and mode shift on emissions is much smaller, also the impact of efficiency improvements is much higher than in the non-OECD region.

In the Renewable and Lifestyle 1.5 °C scenarios, CO_2 emissions decrease to 1.1 Gt and 1.3 Gt in 2050, respectively. The contributing factors to these reductions are similar for both 1.5 °C scenarios, with efficiency and CO_2 intensity improvements (mainly electrification) having the most impact. Electrification is a crucial aspect because passenger cars are responsible for the lion's share of energy demand. The electricity share in total passenger travel increases to 65% in 2050 in the Renewable scenario and 49% in the Lifestyle scenario (Figure 7). Mode shift contributes strongly to emission reduction in the 1.5 °C Lifestyle scenario as well. This is mainly due to a shift away from flying (share of total travel distance from 9% in 2015 to 5% in 2050, compared to 12% in baseline in 2050) and bus (from 26% to 20%) to train (from 6% to 16% share of travel distance).

Perhaps counterintuitively, the impact of activity reduction on emissions is lower in the Lifestyle scenario than in the Renewable scenario—even though total passenger kilometres are less in the Lifestyle scenario. The reason for this is that the travel modes are more emission-intensive (due to less electrification) in the Lifestyle scenario than in the Renewable scenario (Figure 7).

In the non-OECD region, energy efficiency improvement of cars and buses (relative to Baseline) is a more important contributing factor for emission reductions than in the OECD region.

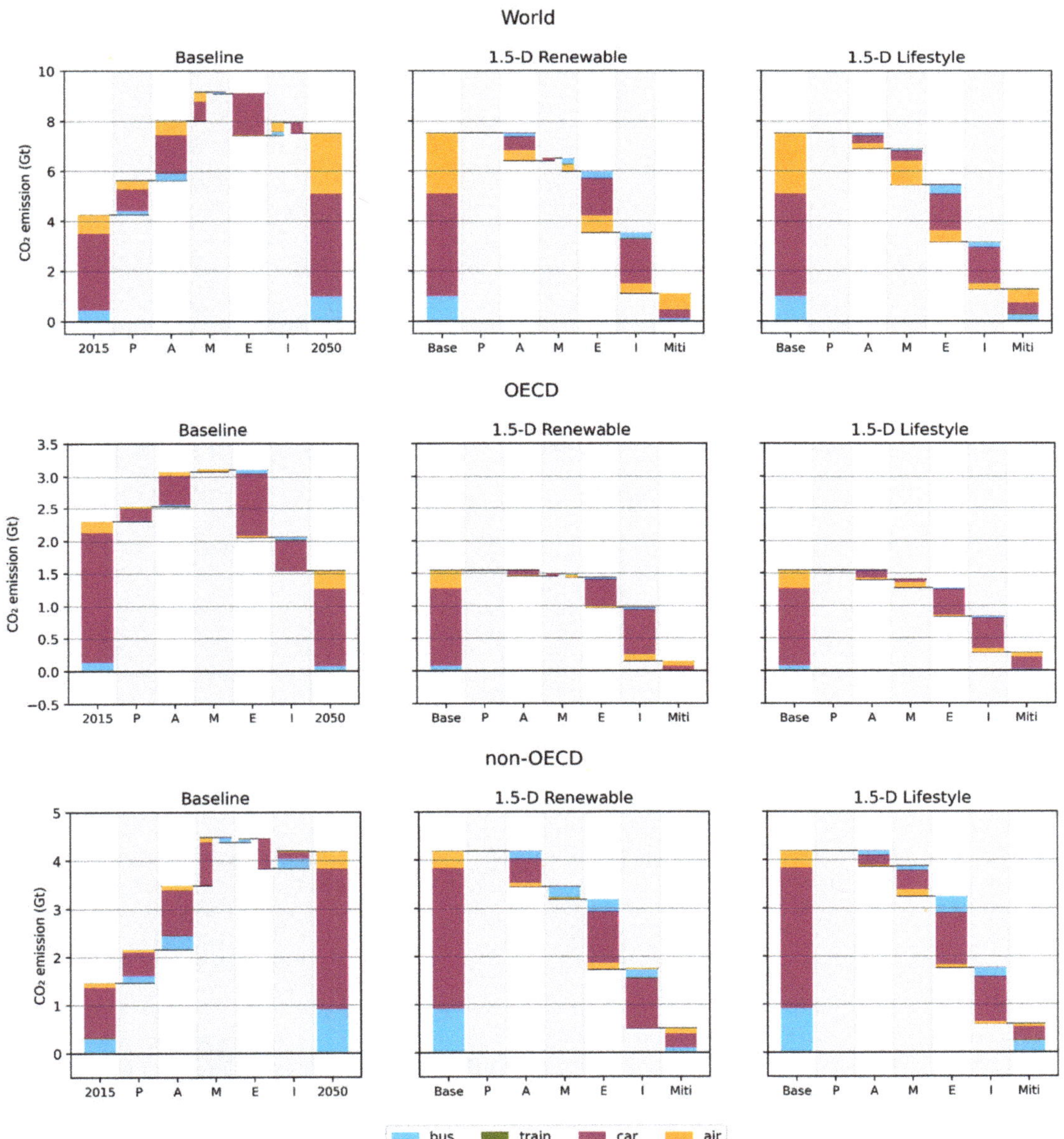

Figure 6. Changes in passenger travel emissions between 2015 and 2050 (Baseline) and between the Baseline and mitigation scenarios (in 2050) due to different drivers. P: population; A: activity; M: mode shift, E: efficiency, I: carbon intensity (includes the impact of electrification). Emissions from international bunkers are shown in global graph only.

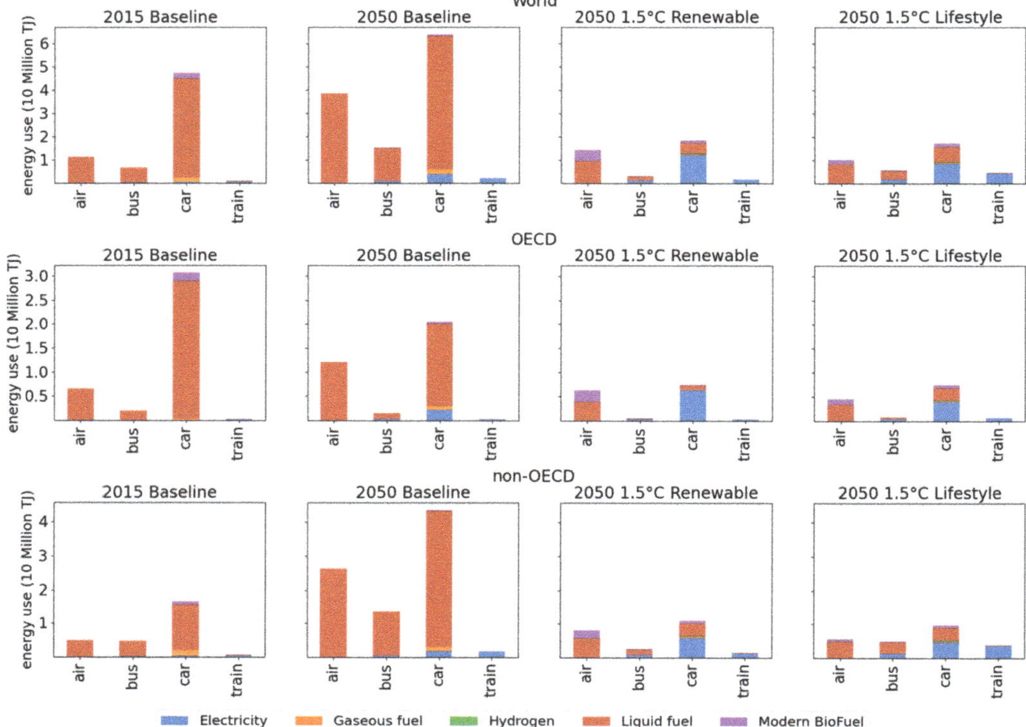

Figure 7. Energy mix by passenger travel mode.

Freight Transport

CO_2 emissions from freight transport are projected to increase strongly in non-OECD countries and decrease in OECD countries in the Baseline, leading globally to a small total increase by 2050. The increase in activity levels, mainly from trucks, is almost completely offset by efficiency improvements of these trucks (Figure 8). In OECD countries, CO_2 intensity improvement contributes substantially to emission reduction in the baseline, mainly due to the increase in the share of plug-in electric trucks.

In the 1.5 °C Renewable scenario, emissions are reduced to almost zero, whereas some remain in the 1.5 °C Lifestyle scenario. Switching to less carbon-intensive fuels for medium and heavy trucks is the largest contributing factor for emission reduction in both scenarios, but especially in the Renewable scenario. In 2050, 75% of truck fuel consists of hydrogen in the 1.5 °C Renewables scenario, with an additional 11% of modern biofuels and 6% electricity (Figure 9). The Lifestyle scenario has a stronger impact of efficiency in reducing emissions, but this cannot offset the stronger impact of switching to non-fossil energy in the Renewable scenario.

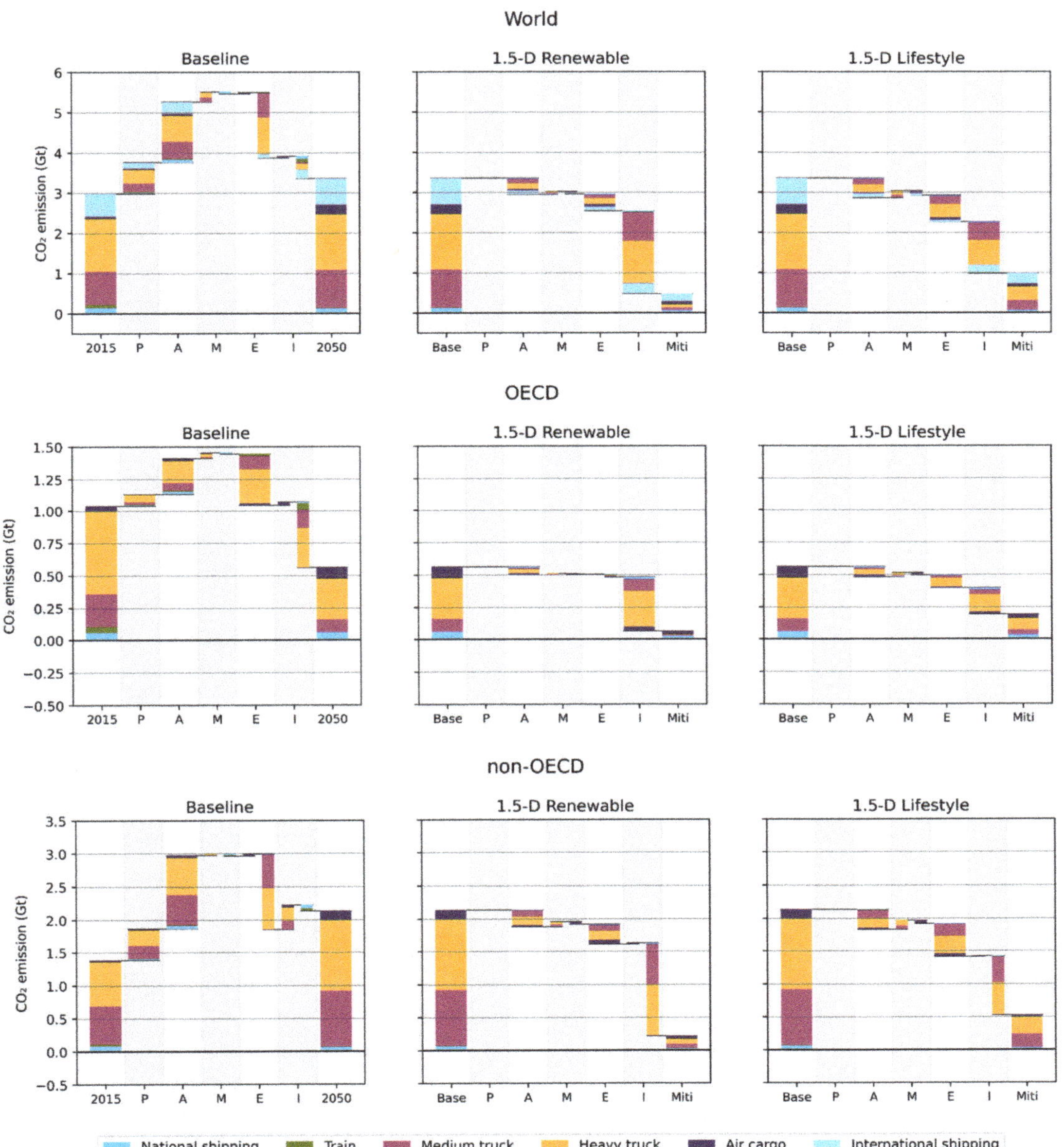

Figure 8. Changes in freight transport emissions between 2015 and 2050 (Baseline) and between the Baseline and mitigation scenarios (in 2050) due to different drivers. P: population; A: activity; M: mode, E: efficiency, I: carbon intensity (includes the impact of electrification). Emissions from international bunkers are only shown in global graph.

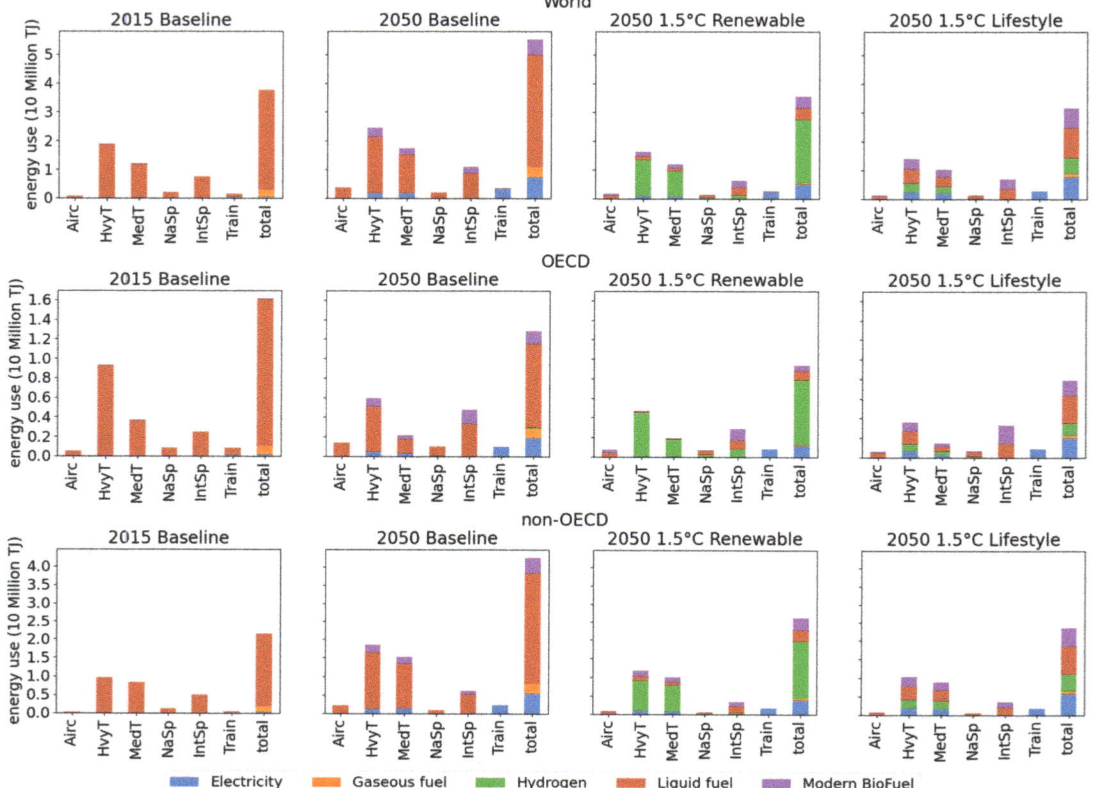

Figure 9. The energy mix of freight transport. Ship: national shipping, Train: freight train, MedT: medium truck, HvyT: heavy truck, Airc: air cargo.

3.2.4. Synthesis

Figure 10 synthesises the results of the decomposition analyses. It shows the reductions relative to the Baseline scenario caused by each factor. By far, the largest reductions take place in power generation, with renewables, CCS, and improving the carbon intensity (mainly by switching from coal to natural gas) all being major drivers. CCS is also an important contribution to emission reduction in industry.

Improving energy efficiency and electrification are the major contributors to reduce emissions in transport (note that electrification and carbon intensity have been analysed as one decomposition factor in transport—but its energy mix shows that electrification is the most important factor). Shifting to more climate-friendly modes of transport contributes significantly to reducing transport emissions in the Lifestyle scenario as well (note that emissions from international air are only visible in the global results, as these are not allocated to regions).

In all sectors, electrification plays a larger role in reducing emissions in the Renewables scenario, which is reflected by the positive effect of changes in activity on emissions from power generation in this scenario. Changes in efficiency also lead to an increase in emissions from power generation, as bioenergy has a relatively low efficiency.

While the OECD and non-OECD regions overall show a similar trend, there are some important differences as well. Efficiency improvement is more important in reducing emissions in industry and transport in the non-OECD region, while electrification has a more important role for reducing passenger transport in the OECD region. Renewables have a stronger impact in reducing emissions in power generation in the non-OECD region and CCS in the OECD region.

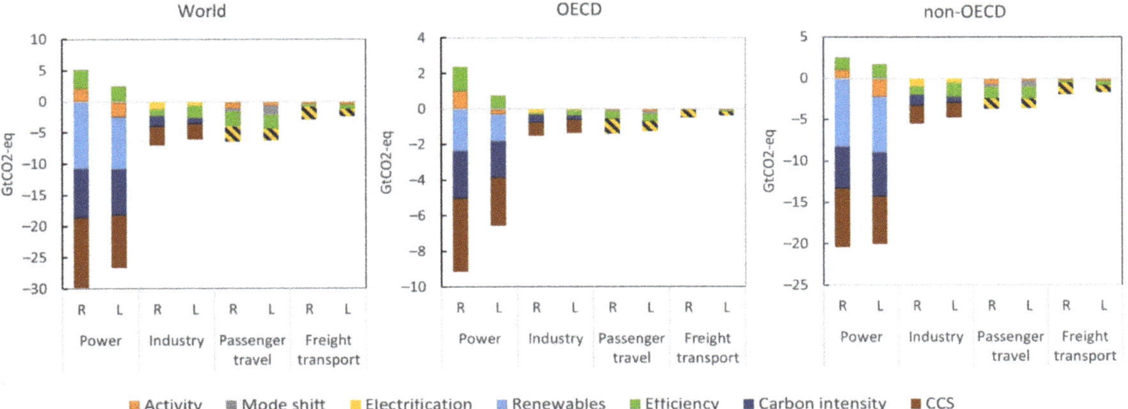

Figure 10. Changes in global emissions in mitigation scenarios by driving factor relative to the Baseline, 2050. R: Renewable, L: Lifestyle. Emissions from international bunkers are only shown in the global graph. Electrification and carbon intensity are one decomposition term in transport and therefore shown together.

4. Discussion

This study applied a decomposition analysis on the three highest CO_2 emitting sectors: power generation, industry, and transport. The IMAGE output, on which the decomposition was applied, allowed the extraction of relevant variables to decompose the changes in emissions in these sectors. However, other sectors such as land use, other energy sources, the residential sector, industrial processes, and non-CO_2 emissions also have a large potential to reduce greenhouse gas emissions. Therefore, we suggest that future work can focus on these sectors, especially the high CO_2 emitting sources such as land and other energy (including energy use for hydrogen and second heat production, biomass processing, and agriculture), to understand the factors contributing to mitigation and the possible mitigation strategies in these sectors as well.

In addition, we have applied decomposition analyses on an aggregate level for the whole industry and large world regions. For specific industry sectors (e.g., steel, cement, pulp and paper, food, and chemical industries) and regions, mitigation strategies may look very different [26]. Sectoral decomposition gives us a good overall view, but a more detailed decomposition analysis within the sector or for specific regions can be useful for more specific policy advice.

We only focused on one baseline and two 1.5 °C scenarios in this study. Other pathways may be analysed as well, e.g., 2 °C or current-policy scenarios. We have chosen to focus on 1.5 °C scenarios since we were mainly interested in how net-zero emission ambitions could be met. Future work could, for instance, focus on comparing decomposition results between 1.5 °C scenarios and current-policy scenarios, where the current-policy scenario reflects the drivers that are now being considered, and the 1.5 °C scenarios can show where the main progress can be made.

5. Conclusions

This study presents several decomposition techniques to focus on the most important contributions to emission reductions in recently developed deep mitigation scenarios and analyse the differences across scenarios, sectors and regions. Our main conclusions are as follows.

Decomposition is a useful tool for identifying the main contributions to changes in emissions in different sectors and regions, allowing easy comparison of scenarios. The analysis can easily show the largest contributions to emission reductions and compare trends across scenarios. Providing that sufficient data from the scenarios is available, this tool can also be used for other models and sectors.

CCS can play an important role in reducing emissions in power generation and industry. The IMAGE scenarios emphasise the role of renewables and lifestyle change in reducing emissions. However, CCS and especially BECCS are important drivers in reducing emissions in power generation and industry. BECCS provides the required negative emissions needed to offset remaining emissions in other sectors. This also means that if emissions from these latter sources can be reduced, the need for CCS—including BECCS—is reduced, as shown by the lower contribution of CCS in reducing emissions on the 1.5 °C Lifestyle scenario. CCS contributes more strongly to emission reductions in the 1.5 °C Renewable scenario, partly due to the relatively high electricity demand in this sector, driven by the high electrification rate.

Other technological measures are also important for reducing emissions, but their relative importance differs among the sectors. Reducing the overall carbon intensity of non-renewable fuels (switching from coal to gas and bioenergy) is an important contribution to reducing emissions in power generation and industry, while electrification is especially important in reducing transport emissions.

Efficiency improvement is an important contributing factor for reducing emissions in industry and transport. While efficiency is already improving strongly in the Baseline scenario, further improvements in the mitigation scenarios contribute substantially to emission reductions in industry and passenger travel. In power generation, changes in efficiency have an increasing impact on emissions, which is due to a switch to relatively less-efficient biomass. It is, of course, not a certainty that the efficiency improvements shown in the baseline occur without additional policies, which means that this remains an important aspect to focus on in mitigation scenarios—both for research and policymaking.

Changes in activity levels and mode shift contribute significantly to emission reductions in transport. While technology plays a strong role in reducing emissions, as the above conclusions show, lowering and changing passenger travel activity has a strong impact on reducing emissions from transport. This emphasises the importance of changing consumption patterns to reduce emissions.

While the factors contributing to emission reductions are similar in the OECD and non-OECD region, there are some crucial differences as well. These differences relate especially to the higher importance of efficiency improvement in the non-OECD region to reduce passenger travel and industry emissions. At the same time, electrification plays a more critical role in the OECD region. For power generation, renewables have a more substantial impact in reducing emissions in the non-OECD region.

Author Contributions: Conceptualization, H.-H.C., A.F.H. and D.P.v.V.; methodology, H.-H.C., A.F.H., O.Y.E. and D.P.v.V.; software, H.-H.C., K.-I.v.d.W. and M.v.d.B.; validation, H.-H.C., A.F.H. and D.P.v.V.; formal analysis, H.-H.C.; investigation, H.-H.C.; resources, H.-H.C., O.Y.E., A.F.H., D.P.v.V., V.D., H.S.d.B., K.-I.v.d.W. and D.P.v.V.; data curation, H.-H.C.; writing—original draft preparation, H.-H.C. and A.F.H.; writing—review and editing, H.-H.C., A.F.H., V.D., H.S.d.B., K.-I.v.d.W., O.Y.E. and D.P.v.V.; visualization, H.-H.C. and K.-I.v.d.W.; supervision, A.F.H.; project administration, A.F.H. All authors have read and agreed to the published version of the manuscript.

Funding: This research received no external funding. The APC was funded by IRENA.

Institutional Review Board Statement: Not applicable.

Informed Consent Statement: Not applicable.

Data Availability Statement: The dataset for the figures in this study can be find here: https://github.com/hsinghs/Decomposition-paper-energies.

Conflicts of Interest: The authors declare no conflict of interest.

References

1. Rogelj, J.; Luderer, G.; Pietzcker, R.C.; Kriegler, E.; Schaeffer, M.; Krey, V.; Riahi, K. Energy system transformations for limiting end-of-century warming to below 1.5 °C. *Nat. Clim. Change* **2015**, *5*, 519–527. [CrossRef]
2. Clarke, L.; Jiang, K.; Akimoto, K.; Babiker, M.; Blanford, G.; Fisher-Vanden, K.; Hourcade, J.-C.; Krey, V.; Kriegler, E.; Löschel, A.; et al. Assessing Transformation Pathways. In *Climate Change 2014: Mitigation of Climate Change*; Edenhofer, O., Pichs-Madruga, R., Sokona, Y., Farahani, E., Kadner, S., Seyboth, K., Adler, A., Baum, I., Brunner, S., Eickemeier, P., et al., Eds.; Contribution of Working Group III to the Fifth Assessment Report of the Intergovernmental Panel on Climate Change; Cambridge University Press: Cambridge, UK; New York, NY, USA, 2014.
3. Van Vuuren, D.P.; Stehfest, E.; Gernaat, D.E.H.J.; Van Den Berg, M.; Bijl, D.L.; De Boer, H.S.; Daioglou, V.; Doelman, J.C.; Edelenbosch, O.Y.; Harmsen, M.; et al. Alternative pathways to the 1.5 °C target reduce the need for negative emission technologies. *Nat. Clim. Change* **2018**, *8*, 391–397. [CrossRef]
4. Le Quéré, C.; Korsbakken, J.; Wilson, C.; Tosun, J.; Andrew, R.; Andres, R.J.; Canadell, J.G.; Jordan, A.; Peters, G.P.; Van Vuuren, D.P. Drivers of declining CO_2 emissions in 18 developed economies. *Nat. Clim. Change* **2019**, *9*, 213–217. [CrossRef]
5. Marangoni, G.; Tavoni, M.; Bosetti, V.; Borgonovo, E.; Capros, P.; Fricko, O.; Gernaat, D.E.H.J.; Guivarch, C.; Havlik, P.; Huppmann, D.; et al. Sensitivity of projected long-term CO_2 emissions across the Shared Socioeconomic Pathways. *Nat. Clim. Change* **2017**, *7*, 113–117. [CrossRef]
6. Riahi, K.; van Vuuren, D.P.; Kriegler, E.; Edmonds, J.; O'Neill, B.C.; Fujimori, S.; Bauer, N.; Calvin, K.; Dellink, R.; Fricko, O.; et al. The Shared Socioeconomic Pathways and their energy, land use, and greenhouse gas emissions implications: An overview. *Glob. Environ. Change* **2017**, *42*, 153–168. [CrossRef]
7. Van den Berg, N.J.; Hof, F.A.; van der Wijst, K.-I.; Akenji, L.; Daioglou, V.; Edelenbosch, O.Y.; van Sluisveld, M.A.E.; van Vuuren, D.P. Decomposition analysis of per capita emissions: A tool for assessing consumption changes and technology changes within scenarios. *Environ. Res. Commun.* **2021**, *3*, 15004. [CrossRef]
8. Sharmina, M.; Edelenbosch, O.Y.; Wilson, C.; Freeman, R.; Gernaat, D.E.; Gilbert, P.; Larkin, A.; Littleton, E.W.; Traut, M.; van Vuuren, D.P.; et al. Decarbonising the critical sectors of aviation, shipping, road freight and industry to limit warming to 1.5–2 °C. *Clim. Policy* **2021**, *21*, 455–474. [CrossRef]
9. Edelenbosch, O.Y.; van Vuuren, D.P.; Blok, K.; Calvin, K.; Fujimori, S. Mitigating energy demand sector emissions: The integrated modelling perspective. *Appl. Energy* **2020**, *261*, 114347. [CrossRef]
10. Llop, M. Changes in energy output in a regional economy: A structural decomposition analysis. *Energy* **2017**, *128*, 145–151. [CrossRef]
11. Karmellos, M.; Kosmadakis, V.; Dimas, P.; Tsakanikas, A.; Fylaktos, N.; Taliotis, C.; Zachariadis, T. A decomposition and decoupling analysis of carbon dioxide emissions from electricity generation: Evidence from the EU-27 and the UK. *Energy* **2021**, *231*, 120861. [CrossRef]
12. Marcucci, A.; Fragkos, P. Drivers of regional decarbonisation through 2100: A multi-model decomposition analysis. *Energy Econ.* **2015**, *51*, 111–124. [CrossRef]
13. Guan, D.; Hubacek, K.; Weber, C.L.; Peters, G.P.; Reiner, D.M. The drivers of Chinese CO_2 emissions from 1980 to 2030. *Glob. Environ. Change* **2008**, *18*, 626–634. [CrossRef]
14. Jiang, X.-T.; Su, M.; Li, R. Decomposition Analysis in Electricity Sector Output from Carbon Emissions in China. *Sustainability* **2018**, *10*, 3251. [CrossRef]
15. Van Vuuren, D.; Stehfest, E.; Gernaat, D.; de Boer, H.; Daioglou, V.; Doelman, J.; Edelenbosch, O.; Harmsen, M.; van Zeist, W.J.; van den Berg, M.; et al. The 2021 SSP Scenarios of the IMAGE 3.2 Model. Available online: https://eartharxiv.org/repository/view/2759/ (accessed on 15 November 2021).
16. Ang, B.W. Decomposition analysis for policymaking in energy: Which is the preferred method? *Energy Policy* **2004**, *32*, 1131–1139. [CrossRef]
17. Törnqvist, L.; Vartia, P.; Vartia, Y.O. How Should Relative Changes be Measured? *Am. Stat.* **1985**, *39*, 43–46.
18. Albrecht, J.; François, D.; Schoors, K. A Shapley decomposition of carbon emissions without residuals. *Energy Policy* **2002**, *30*, 727–736. [CrossRef]
19. Sun, J.W. Changes in energy consumption and energy intensity: A complete decomposition model. *Energy Econ.* **1998**, *20*, 85–100. [CrossRef]
20. Van Vuuren, D.P.; Stehfest, E.; Gernaat, D.E.; Doelman, J.C.; van den Berg, M.; Harmsen, M.; De Boer, H.S.; Bouwman, L.F.; Daioglou, V.; Edelenbosch, O.Y.; et al. Energy, land-use and greenhouse gas emissions trajectories under a green growth paradigm. *Glob. Environ. Change* **2017**, *42*, 237–250. [CrossRef]
21. Hof, A.F.; Esmeijer, K.; de Boer, H.; Daioglou, V.; Doelman, J.C.; den Elzen, M.G.; Gernaat, D.E.H.J.; van Vuuren, D.P. Regional energy diversity and sovereignty in different 2 °C and 1.5 °C pathways. *Energy* **2022**, *239*, 122197. [CrossRef]

22. Van Sluisveld, M.A.; Martínez, S.H.; Daioglou, V.; Vuuren, D.P. Exploring the implications of lifestyle change in 2 °C mitigation scenarios using the IMAGE integrated assessment model. *Technol. Forecast. Soc. Change* **2016**, *102*, 309–319. [CrossRef]
23. Stehfest, E.; Bouwman, L.; Vuuren, D.; Elzen, M.; Eickhout, B.; Kabat, P. Climate benefits of changing diet. *Clim. Change* **2009**, *95*, 83–102. [CrossRef]
24. IPCC. Climate Change 2007: Working Group I: The Physical Science Basis. Available online: https://archive.ipcc.ch/publications_and_data/ar4/wg1/en/ch2s2-10-2.html (accessed on 15 November 2021).
25. Olivier, J.; Peters, J. *Trends in Global CO$_2$ and Total Greenhouse Gas Emissions: 2020 Report*; PBL Netherlands Environmental Assessment Agency: The Hague, The Netherlands, 2020.
26. Van Sluisveld, M.A.; de Boer, H.; Daioglou, V.; Hof, A.F.; Vuuren, D.P. A race to zero—Assessing the position of heavy industry in a global net-zero CO$_2$ emissions context. *Energy Clim. Change* **2021**, *2*, 100051. [CrossRef]

Article

A Probabilistic and Value-Based Planning Approach to Assess the Competitiveness between Gas-Fired and Renewables in Hydro-Dominated Systems: A Brazilian Case Study

Felipe Nazaré [1], Luiz Barroso [2] and Bernardo Bezerra [3,*]

[1] Electrical Engineering Department, Pontifícia Universidade Católica do Rio de Janeiro (PUC-RJ), Rio de Janeiro 22541-041, Brazil; felipelucasnazare@gmail.com
[2] Instituto de Investigación Tecnológica, Escuela Técnica Superior de Ingeniería (ICAI), Universidad Pontificia Comillas, 28015 Madrid, Spain; luiz.barroso@comillas.edu
[3] Omega Geração S.A., R. Elvira Ferraz, 68, São Paulo 04552-040, Brazil
* Correspondence: bebezerra@gmail.com

Abstract: The main challenge with the penetration of variable renewable energy (VRE) in thermal-dominated systems has been the increase in the need for operating reserves, relying on dispatchable and flexible resources. In the case of hydro-dominated systems, the cost-effective flexibility provided by hydro-plants facilitates the penetration of VRE, but the compounded production variability of these resources challenges the integration of baseload gas-fired plants. The Brazilian power system illustrates this situation, in which the development of large associated gas fields economically depends on the operation of gas-fired plants. Given the current competitiveness of VRE, a natural question is the economic value and tradeoffs for expanding the system opting between baseload gas-fired generation and VRE in an already flexible hydropower system. This paper presents a methodology based on a multi-stage and stochastic capacity expansion model to estimate the optimal mix of baseload thermal power plants and VRE additions considering their contributions for security of supply, which includes peak, energy, and operating reserves, which are endogenously defined in a time-varying and sized in a dynamic way as well as adequacy constraints. The presented model calculates the optimal decision plan, allowing for the estimation of the economical tradeoffs between baseload gas and VRE supply considering their value for the required services to the system. This allows for a comparison between the integration costs of these technologies on the same basis, thus helping policymakers and system planners to better decide on the best way to integrate the gas resources in an electricity industry increasingly renewable. A case study based on a real industrial application is presented for the Brazilian power system.

Keywords: power system expansion; co-optimization of energy and reserve; associated natural gas; multi-stage stochastic programming; electricity-gas integration

Citation: Nazaré, F.; Barroso, L.; Bezerra, B. A Probabilistic and Value-Based Planning Approach to Assess the Competitiveness between Gas-Fired and Renewables in Hydro-Dominated Systems: A Brazilian Case Study. *Energies* **2021**, *14*, 7281. https://doi.org/10.3390/en14217281

Academic Editor: Dimitrios Katsaprakakis

Received: 16 September 2021
Accepted: 30 October 2021
Published: 3 November 2021

Publisher's Note: MDPI stays neutral with regard to jurisdictional claims in published maps and institutional affiliations.

1. Introduction

The integration between electricity and gas started in the 90s as a consequence of a widespread construction of new gas-fired power plants, both combined-cycle and open-cycle. As competitive, modular, and efficient, the technology was able to displace existing generation in many countries, mostly those already dominated by inefficient and more pollutant thermal generation. This had a perfect fit for investors with the creation of wholesale energy markets launched also in the 1990s: as baseload resources they could secure a volume in the energy spot markets and capture a revenue stream from spot-prices. The baseload dispatch of efficient gas plants also solved the commercial feasibility of the development of oil-gas fields, whose gas supply agreements demand take-or-pay clauses to secure a stable revenue stream to enable the financing of the new gas infrastructure. The combination of a large consumption market for power and non-power uses (industry, heating, etc.) engendered a massive integration of gas in the electricity industry in many developed countries and, most importantly, the baseload dispatch of thermal plants secured stable revenue streams for both electricity and gas investors.

The integration of electricity and gas integration was extended to multi-country electricity-gas markets, mostly by Europe. These were a natural evolution to the existing "official" cross-border interconnections, which were originally established by the countries' governments for sharing reserves and carrying out limited economic interchanges. Finally, the development of Liquefied Natural Gas (LNG) has introduced flexibility and removed many physical barriers for the integration of these two markets [1].

While the gas power technology supported the substitution of part of the inefficient thermal power generation in some countries, many others, with significant hydro power generation and access to gas resources (domestic, gas pipelines or LNG imports), have experienced more difficulty in integrating gas-fired plants in the electricity market. Since hydro generation may displace gas-fired plants (even the efficient ones) in the generation merit-order curves depending on meteorological conditions, the demand-side for gas products becomes very dependent on water inflow availability. Therefore, the dispatch of the gas-fired plants has become volatile (for compensating the water inflows), which ends up creating an undesirable (from the gas-sector point of view) variability in natural gas consumption [2].

Whilst the gas-market is still incipient, with a non-power gas use limited, gas supply agreements (GSA) for power generation contracts are typically of long-term with high "take-or-pay" clauses to ensure the financing of the gas production-transportation infrastructure [3]. From the power sector point of view, these clauses are undesirable; due to the uncertainty of dispatch, gas-based power generators aim to negotiate a higher flexibility with gas suppliers in order to become more competitive in the power market while maintaining the "guarantee" of the gas availability whenever the dispatch is needed. This "dilemma" has demanded the development of more flexible supply-demand options, such as LNG-supply with high take-or-pay clauses—to complement the more inflexible options for the gas supply agreements for power generation. This gas supply flexibility is better and easier handled when the demand side of gas industrial is also active, allowing for the explicit pricing of gas surplus by non-power consumers [4].

The increasing participation of variable renewables energy (VRE) resources in this power mix has intensified the issues of variability and uncertainty of the dispatch of all of the technologies, even in the thermal power systems. The increasing need for operating (spinning) reserves has highlighted the value of gas-fired plants as flexible assets. In hydro-dominated countries, the integration of renewables has also increased the value of hydropower as flexibility providers.

When it comes to power system planning, the competition for system expansion between renewables and gas-fired plants has increased. On the one hand, the increasing VRE participation implies the need for sustaining the energy balance through greater amounts of reliable and flexible power resources, which, from the gas-fired plants point of view, increases the variability of the dispatch, resulting in higher take-or-pay clauses on the gas supply agreements. This is also a characteristic of hydro-dominated systems. On the other hand, the competitiveness of "inflexible" gas-fired plants faces greater challenges, especially for those plants whereby the source of gas comes from associated gas fields, where a constant gas flow is required to ensure oil production, avoiding reinjection costs. Hence, defining the optimal tradeoff between variable resources with backup supply or inflexible power generation, also considering aspects of reliability and flexibility needs, became an interesting challenge.

This paper presents a methodology based on a multi-stage and stochastic capacity expansion planning model to determine the competitiveness of a given technology against an existing system, considering its reliability contribution, for peak, energy, and ancillary services. Our work applies this methodology to calculate the tradeoffs between base-loaded gas supply and VRE supply, considering their value for these adequacy and operating services in the system. This allows for a comparison between the integration costs of these technologies on the same basis, thus helping policymakers to better decide on the best way to integrate the gas resources in an electricity industry increasingly renewable. A case study based on a real industrial application is presented for the Brazilian power system.

1.1. The Brazilian Power System and Problem Description

Brazil is the largest country in Latin America with a power sector containing an installed capacity of 170,000 MW. In the 1990's, hydro plants were responsible for more than 90% of power generation in. These plants had some important characteristics for supply reliability, such as flexible output and cheap ancillary services—which is fundamental to any power system in handling unexpected imbalances in real-time operation. On the other hand, a hydro-dominated system needs firm energy backup to deal with dry years, and the Brazilian strategy in the early 2000's started to import natural gas from Bolivia with take-or-pay clauses to remunerate the gas infrastructure investments, while developing offshore natural gas fields.

This integration between natural gas and electricity faced some difficulties in the 2004–2006 period due to the thermal dispatch volatility [1,2,5] and, in 2007, the country started to import LNG to provide greater flexibility for the thermal power plants (TPP), enabling an energy backup during dry periods. This strategy was also adopted by other countries in Latin America, for geopolitical reasons and for a better integration between gas and power [1,4].

Well documented by the International Renewable Energy Agency (IRENA) [6], in the late 2000's Brazil started a renewable auction program for long-term energy contracts, which was responsible for a massive penetration of wind plants. One of the reasons for the auction success was the hydro-domination of the system [7], which reduced the integration cost and ensured very competitive prices for wind generation. In 2014, Brazil started long-term contract auctions for solar energy with similar success and a pathway to integrated renewables to the grid was initiated.

However, due to environmental and social constraints, Brazil's hydro expansion in the last 20 years has only been based on run-of-river hydro plants. The storage capacity of the system compared to the total energy consumption [8] has thus decreased. Along with increasing transmission bottlenecks in the country, this has adversely affected the system operation, increasing the need for other dispatchable alternatives to accommodate the increasing flexibility needs due to the ongoing and expanding integration VRE. Figure 1 below depicts the evolution of the supply mix between 1990 and 2021 considering consolidated values.

In recent years, Brazil has been considering different alternatives to (re)introduce resources for flexibility in the power sector [9]. From alternatives of flexible power plants to the development of green hydrogen and increasing the transmission system in view of diminishing the variability from generation and demand sides and using the power reserves from distant areas, the planning studies focused on alternatives for integrating VRE into the power sector.

On the other hand, the country has giant offshore oil reserves, in the so-called pre-salt fields, with associated natural gas to the oil production. These are located up to 300 km from the coast and 3000 m below sea level. These gas and oil fields were discovered in 2008 and, in 2019, they contributed to 62% of Brazil's total oil production and 57% of natural gas production. The pre-salt is geographically highlighted in Figure 2.

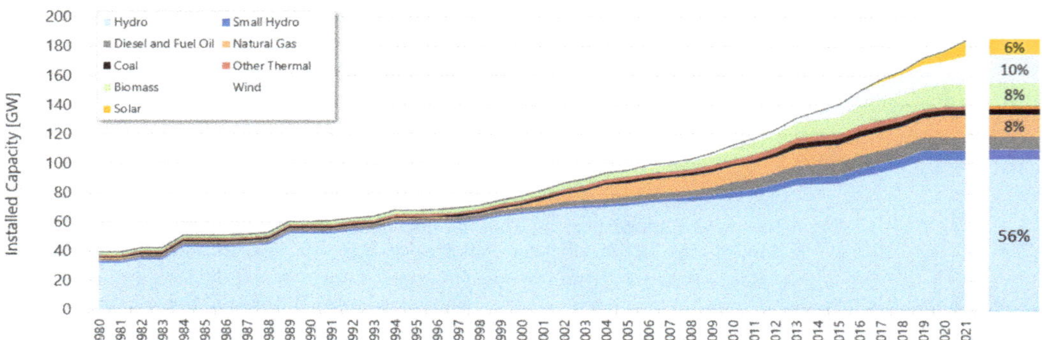

Figure 1. Installed capacity evolution in Brazilian power system (Source: Authors, utilizing data from EPE).

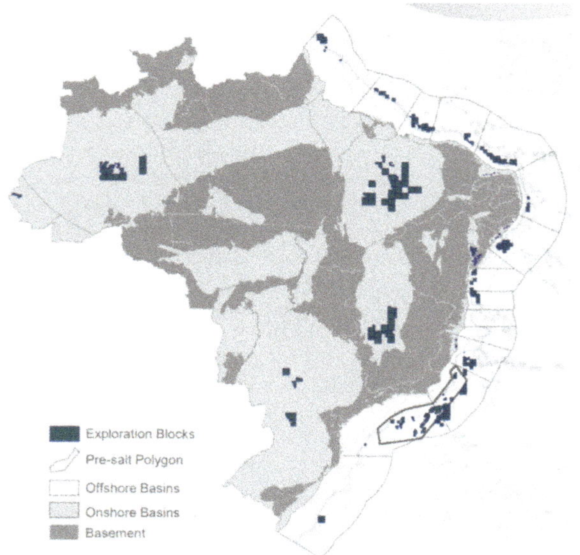

Figure 2. Brazilian oil fields map (Source: ANP).

The massive oil-gas production of these fields is expected to start sometime before 2030. Since natural gas is a byproduct of the oil production, it may be leveraged by oil production, creating an opportunity for Brazil to develop new competitive gas-fired plants to complement the VRE integration and meet its increasing load growth. Nevertheless, for the development of pre-salt oil fields, the utilization of the associated gas is required, that might be through consumption, reinjection or flaring. This problem has been addressed in past decades in [10] in Norway, that developed associated gas fields similar to those found in Brazilian coast. In terms of gas utilization, the power sector might become a relevant consumer in order to accommodate the gas production through power generation and baseload thermal dispatch, depending on its final cost. The needed flexibility in this case would be indirectly provided by increasing the hydroelectric reservoir levels, recovering the system's storage capacity and providing flexibility and ancillary services to accommodate VRE integration. The issue addressed in this technical article is to estimate maximum gas costs delivered to the power plants that still allows for its development regarding the construction of other alternatives. The uncertainties around the cost of pre-salt natural gas, which depend on the distance from the coast, the amount of CO_2 concentration on the fields and the opportunity cost of the oil production will define the competitiveness of the gas-fired plants for new investments.

In this sense, Brazilian policymakers, energy planners and oil-gas majors (that have concessions rights of those fields) are facing a debate regarding to what extent it is economic to introduce baseload gas-fired plants from these fields and enable their exploration and production. This is a very relevant policy decision for the country and for the owner of the concession rights of these fields, which are the major oil companies: Equinor, Shell, Galp and Exxon. This question is directly connected to a more general tradeoff faced by the power sector globally to address generation expansion: should new capacity be secured based mainly on VRE, apparently cheaper in pure \$/MWh terms but that will increase the usage of existing sources (hydro, in Brazil's case) to provide security services and/or of building new costly flexible thermal plants, or should investments in baseload gas-fired plants which enable existing plants to provide flexibility to accommodate VRE integration be the choice of direction, or at least part of it?

1.2. Objectives of This Work

This work, then, addresses this practical research question by means of a methodology based on a multi-stage and stochastic capacity expansion model to estimate the optimal mix of baseload thermal power plants and VRE additions to the system expansion portfolio, considering their reliability contribution for the supply of peak, energy and operating (spinning) reserves. We represent the operating reserves as time-varying and dynamic requirements, endogenously defined by our proposed optimization model. This means that reserve requirements are not static, defined as, for example, a percentage of peak load, but vary per hour of the day and each hour may have different reserve requirements, sized based on renewable forecast-errors and on the portfolio of existing and candidate power plants. Our model, then, calculates the tradeoffs between baseload gas and VRE supply considering their value for these services in the system. For the sake of completeness, our model also considers a set of adequacy constraints, which represents the need of system planners to have a firm capacity margin to supply peak demand [11–13]. A case study based on a real industrial application is presented for the Brazilian power system.

1.3. Literature Survey and Paper Contributions

The literature on electricity and gas integration is vast. Most papers cover integration issues on the operation side [14–19]. Most of these papers discuss the representation of the gas supply and network in the electricity operations models aiming to an integrated schedule. Other papers discuss similar issues on the planning side [20–23] and the literature survey is also strong on risk management and market design [3–5,24] issues between these two industries. The literature on capacity expansion planning models is even vaster. The development of generation expansion planning in optimization process first started with [25], which considered linear programming as a tool to solve the expansion problem. [26–28] provide a detailed analysis of the generation expansion planning tools history and how they have evolved. In the most general form, the capacity expansion models minimize investment and operation costs and the incorporation of reliability constraints and security criteria, as discussed in [29–33]. This work considers the contribution value of each candidate to the criteria, which is reviewed in [33].

The development of VRE has brought new challenges for optimization tools, with requirements for a more granular representation of the timescale and a greater variability on the supply-side representation. More recently, a great effort has been made on the definition of operating reserves to couple with VRE integration for operations planning. Since the sizing of reserves depends on the renewable production, the dynamic sizing of reserves to reduce procurement's cost has gained momentum [34–37]. The authors in [34] show that the operating reserves costs might decrease by about 20% in German system. The same conclusions are reached by the authors of [37]. They presented a dynamic sizing method that determines the required capacity on a daily basis, using the estimated probability of facing a system imbalance during the next day. A gradual implementation of dynamic reserves in Belgium since 2020 has been decided based on the results of this study. This methodology is at the core of our work.

From generation expansion planning purposes, uncertainties have been considered in different ways: the authors of [28] focus on detailing the expansion planning regarding largescale renewable participation. Ref. [38] developed a stochastic, multistage optimization tool in order to obtain the optimal transmission and energy storage expansion. [39] solved the optimal placement of storage equipment in systems with a high penetration of wind power systems. Moreover, the integration of dynamic sizing of the spinning reserve in the generation expansion planning model was developed in [40], treating it as an endogenous variable. This is an important development that has formed the basis of the research in our work as it allows for the capacity planning model to dictate the generation expansion options that also minimizes reserves costs. This model considers the variability of the hourly differences of the production from each renewable scenario. Ref. [41] raised the importance of representing greater granularity characteristics in power system modelling, showing an underestimation of costs in the Belgian power system of up to 58% in case of neglecting the low temporal constraints [42].

Our work fits into the planning and investment sides and covers the specific application of valuing baseload gas-fired power plants in a hydro-dominated system with increasing penetration of renewables. The contributions of our work are threefold: (i) we develop a methodology to determine the breakeven gas prices that turn the baseload gas plants and renewables equivalent under the hydro-dominated system standpoint, that is, to satisfy the same set of constraints (providing equivalent economic value); (ii) for this, we utilize a capacity expansion planning model that addresses energy, peak and the endogenous definition of time-varying and dynamic operating reserves in a probabilistic way. Our model is based on multistage, multiarea, stochastic MIP problem with hourly timescale resolution with co-optimization between energy and reserves and adequacy constraints; (iii) we apply the methodology to real-life case studies for the Brazilian system, where this problem is current under discussion. To the best of our knowledge, this is the first time this type of analysis is conducted in the literature for a hydro-dominated system. With our analysis, we also expect to contribute to the discussion on how to compare renewables and gas-fired plants on the same basis, i.e., to supply the same set of system services (or "attributes") and of the value aggregated for them.

This paper is structured as follows: Section 2 provides the overview of the proposed methodology. Section 3 presents a real case study addressing this problem in Brazilian power system. Section 4 discusses the observed results from the proposed methodology. Finally, Section 5 concludes the work's observations and analysis.

2. Methodology

The capacity expansion problem demands the selection of the generation and transmission options to meet future power load requirements at the least possible cost. Since different resources have different characteristics, the value of each project to the system depends on how the production profile and services provision match the system needs and on the correspondent cost. The Brazilian power system has many candidate options: a large resource potential for hydro is available, as it is for wind, solar and biomass. Gas-fired plants can be an alternative, with gas sourced from cross-border pipelines, LNG or from domestic the pre-salt fields. Each option has its own characteristic: distance to load centers, transmission investment needs, flexibility, intermittency, etc.

Our goal is to calculate the value of baseload gas-fired generation against other portfolio options when integrated into the Brazilian system. This paper uses an approach founded on the gas opportunity cost calculation by the point of view of the power sector. In other words, the attractiveness in terms of installed capacity of the pre-salt power plants depending on their natural gas cost and the maximum gas price that the power sector is willing to pay based on the other expansion alternatives. This is represented in the following procedure:

Step 1: Start from an existing generation-transmission system configuration over a given horizon;

Step 2: Define a load growth scenario and a set of generation expansion candidates, as well as their CAPEX & OPEX and technical characteristics;

Step 3: for a given gas price delivered at the power plant, utilize the solution strategy defined in the capacity expansion planning model explained in the next section to determine the sizing of the gas-fired plants in the system to supply the load growth.

Figure 3 depicts this procedure.

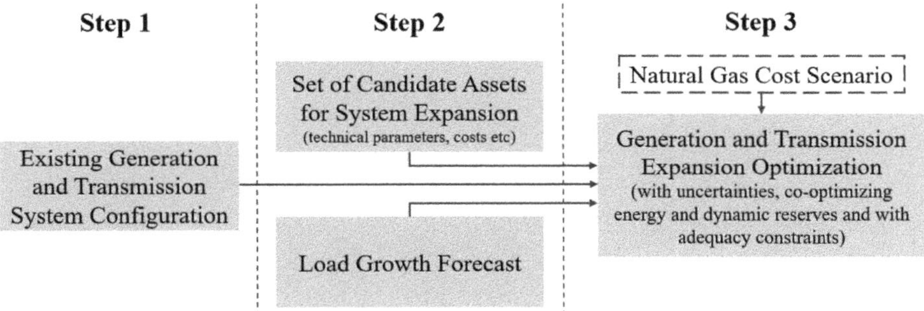

Figure 3. Main steps of the methodology (Source: Authors' elaboration).

Steps (1) to (3) can be repeated for a set of candidate gas prices for the pre-salt gas, depicting the attractiveness of the baseload power generation as a function of the natural gas price. The value that enables gas-fired plants to be selected by the optimal planning model defines how the power sector can breakeven. For gas prices higher than this value, the expansion based on greater amounts of VRE with other flexible options is more cost-effective. This exercise can be repeated for different modeling features of the capacity expansion model and assess how the supply of different services affect the competitiveness of the gas-fired plants. For example, we can assess the competitiveness with respect to the energy supply and to the supply of other reliability constraints and reserves. In this paper, this exercise was conducted twice, representing different constraints in the expansion planning model.

2.1. Solution Strategy

For addressing the answers of the questions raised questions in this work, a generation, transmission and reserve requirements co-optimization model was developed, which aims to minimize the system total cost, represented by the sum of investment and the expected operating costs, satisfying a set reliability constraints (supply of energy, peak and endogenously-refined operating reserves) under uncertainty in water availability and wind and solar generation. This is conducted through a mixed integer programming (MIP) strategy, with hourly resolution.

Since power systems expansion planning is usually carried out over time spans of several years, some simplifying strategies are required to make its solution feasible, especially when representing the intraday operation. To do so, the current work uses some time-clustering assumptions. The first step of this process is clustering some of the months into seasons, which should be defined based on rainy and dry periods and the demand profiles. Once the seasons are defined, the representative days within each of them must be estimated, here referred to as typical days.

This type of representation aims to reduce problem size, capturing the main characteristics within each common day in each season. The work developed in [43] uses a clustering concept to define the typical days to be used by the proposed generation expansion model. For the modelling presented in this work, two typical days were defined for each of the four seasons. The definition of the seasons was based on three-months clusters. For each season, the days were separated into two groups: weekdays and weekends. Figure 4 summarizes the discussed clustering strategy.

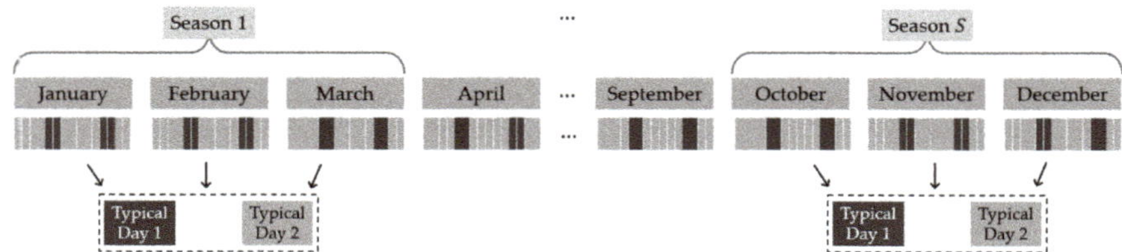

Figure 4. Example of seasons and typical days clustering strategy (Source: Authors' elaboration).

The optimization developed in this paper also contemplates the operating reserve constraints as a variable of the decision process, which will depend on the generation variability of renewable energy sources. The endogenous sizing of the spinning reserve calculation considers, for each hour of each typical day, a convex combination between the average and Conditional Value at Risk (CV@R) of the differences between the real and the expected variation between hours of the production of renewable assets. For the linear programming problem of the CV@R, we refer to [44]. Figure 5 illustrates the process of Dynamic Probabilistic Reserve (DPR) sizing scheme, similar to the ones proposed by [34,35,40].

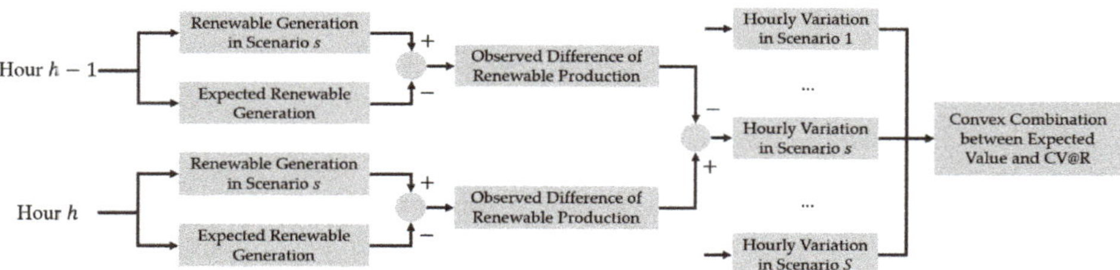

Figure 5. Dynamic Probabilistic Reserve sizing scheme (Source: Authors' elaboration).

Therefore, this approach ensures that the reserve requirements will be sized for each specific hour, considering every scenario used in optimization problem. Since the spinning reserve requirement will be different for each hour, it is also considered as dynamic, as well as probabilistic once it uses different renewable production scenarios in its sizing.

2.2. Problem Formulation

The optimization model used in this simulation is based on [40] and can be formulated as follows:

- *Objective Function:*

$$Min \sum_{k\in\mathcal{K}} I_k \cdot x_k + \sum_{s\in\mathcal{S}} p_s \sum_{t\in\mathcal{T}} \sum_{l\in\mathcal{L}} \beta_l \sum_h \left(\sum_{i\in\mathcal{I}} c_i \cdot g_{i,t,l,h,s} + \sum_{k\in\mathcal{K}_i} c_k \cdot g_{k,t,l,h,s} + c^d \cdot \sum_{b\in\mathcal{B}} \sigma_{b,t,l,h,s} \right) \quad (1)$$

- *Load Balance*:

$$s.t.: \sum_{i \in \mathcal{J}_b} g_{i,b,t,l,h,s} + \sum_{k \in \mathcal{K}_{i,b}} g_{k,b,t,l,h,s} + \sum_{j \in \mathcal{J}_b} \left(f^+_{b,j,t,l,h,s} - f^-_{b,j,t,l,h,s} \right)$$
$$+ \sum_{k \in \mathcal{K}_{j,b}} \left(f^+_{b,k,t,l,h,s} - f^-_{b,k,t,l,h,s} \right) + \sum_{n \in \mathcal{N}_b} \left(\theta^+_{b,n,t,l,h,s} - \theta^-_{b,n,t,l,h,s} \right) \tag{2}$$
$$+ \sum_{k \in \mathcal{K}_{n,b}} \left(\theta^+_{b,k,t,l,h,s} - \theta^-_{b,k,t,l,h,s} \right) + \sigma_{b,t,l,h,s} = d_{b,t,l,h}, \ \forall b,t,l,h,s$$

- *Operative Variables Limits*:

$$0 \leq \delta_{b,t,l,h,s} \leq d_{b,t,l,h}, \forall b,t,l,h,s \tag{3}$$

$$0 \leq g_{i,t,l,h,s} + r_{i,t,l,h,s} \leq \overline{g}_i, \ \forall i,b,t,l,h,s \tag{4}$$

$$\underline{g}_i \leq g_{i,t,l,h,s}, \ \forall i,b,t,l,h,s \tag{5}$$

$$0 \leq f^+_{b,j,t,l,h,s}, f^-_{b,j,t,l,h,s} \leq F_{b,j}, \ \forall j,b,t,l,h,s \tag{6}$$

$$0 \leq g_{k,t,l,h,s} + r_{k,t,l,h,s} \leq \overline{g}_k \cdot x_k, \ \forall b,t,l,h,s, \forall k \in \mathcal{K}_i \tag{7}$$

$$\underline{g}_k \cdot x_k \leq g_{k,t,l,h,s}, \ \forall b,t,l,h,s, \forall k \in \mathcal{K}_i \tag{8}$$

$$0 \leq f^+_{b,j,t,l,h,s}, f^-_{b,j,t,l,h,s} \leq F_{b,k} \cdot x_k, \ \forall b,t,l,h,s, \forall k \in \mathcal{K}_j \tag{9}$$

- *Ramp-up and Ramp-down Limits*:

$$g_{k,t+1,l,h,s} - g_{k,t,l,h,s} \leq \overline{\Delta g_i}, \ \forall i,b,t,l,h,s \tag{10}$$

$$g_{k,t,l,h,s} - g_{k,t+1,l,h,s} \leq \Delta g_i, \ \forall i,b,t,l,h,s \tag{11}$$

$$r_{i,t,l,h,s} \leq \Delta g_i, \forall i,b,t,l,h,s \tag{12}$$

- *Operating Reserve and Adequacy Constraints*:

$$\sum_{i \in \mathcal{I}_b} r_{i,t,l,h,s} + \sum_{k \in \mathcal{K}_{i,b}} r_{k,t,l,h,s} = R_{b,t,l,h}, \ \forall b,t,l,h,s \tag{13}$$

$$\underline{\Phi} \leq \sum_{i \in \mathcal{I}} \phi_i + \sum_{k \in (\mathcal{K}_i U \mathcal{K}_n)} \phi_k \cdot x_k \leq \overline{\Phi} \tag{14}$$

- *Hydro Power Plants Constraints*:

$$v_{i,t+1,l,h,s} = v_{i,t,l,h,s} + a_{i,t,l,h,s} - \left(u_{i,t,l,h,s} + u'_{i,t,l,h,s} \right) + \sum_{m \in \mathcal{M}} \left(u_{m,t,l,h,s} + u'_{m,t,l,h,s} \right)$$
$$- \eta_{i,t,l,h,s}, \forall i,t,l,h,s \tag{15}$$

$$v_{i,T+1,l,h,s} = v_{i,1,l,h,s}, \forall i,l,h,s \tag{16}$$

$$0 \leq u_{i,t,l,h,s} \leq \overline{u}_i, \forall i,t,l,h,s \tag{17}$$

$$\underline{v}_i \leq v_{i,t,l,h,s} \leq \overline{v}_i, \forall i,t,l,h,s \tag{18}$$

$$u_{i,t,l,h,s} + u'_{i,t,l,h,s} \geq \underline{q}_i, \forall i,t,l,h,s \tag{19}$$

$$g_{i,t,l,h,s} = \rho_i \cdot u_{i,t,l,h,s}, \forall i,t,l,h,s \tag{20}$$

- *Energy Storage Equipment Constraints:*

$$v^b_{n,t+1,l,h,s} = \epsilon \cdot v^b_{n,t,l,h,s} + \mu \cdot \theta^+_{n.t,l,h,s} - \theta^-_{n,t,l,h,s}, \ \forall n,t,l,h,s \tag{21}$$

$$0 \leq \theta^+_{n,t,l,h,s}, \theta^-_{n,t,l,h,s} \leq \Theta_n, \ \forall n,t,l,h,s \tag{22}$$

$$v^b_{n,T,l,h,s} = v^b_{n,1,l,h,s}, \ \forall n,t,l,h,s \tag{23}$$

$$0 \leq v^b_{n,t,l,h,s} \leq \overline{v^b}_n, \ \forall n,t,l,h,s \tag{24}$$

$$0 \leq v^b_{n,t,l,h,s} \leq \overline{v}^b_n \cdot x_k, \ \forall k,t,l,h,s \tag{25}$$

- *Dynamic Probabilistic Reserve Formulation:*

$$\hat{g}_{i,b,t,l,h} = \sum_{s \in S} p_s \cdot g_{i,b,t,l,h,s}, \ \forall i,b,t,l,h \tag{26}$$

$$\hat{g}_{k,b,t,l,h} = \sum_{s \in S} p_s \cdot g_{k,b,t,l,h,s}, \ \forall b,t,l,h, \forall k \in \mathcal{K}_{i,ren} \tag{27}$$

$$\delta_{b,s,t,l,h} = \sum_{i \in \mathcal{I}_{ren}} \left(g_{i,b,s,t,l,h} - \hat{g}_{i,b,t,l,h} \right) + \sum_{k \in \mathcal{K}_{i,ren}} \left(g_{k,b,s,t,l,h} - \hat{g}_{k,b,t,l,h} \right) \cdot x_k, \ \forall b,s,t,l,h \tag{28}$$

$$\Delta_{b,s,t,l,h} = \left| \delta_{s,t,l,h} - \delta_{s,t,l,h-1} \right|, \ \forall b,s,t,l,h \tag{29}$$

$$R_{b,t,l,h} \geq (1-\lambda) \cdot \mathbb{E}[\Delta_{b,s,t,l,h}] + \lambda \cdot CVaR_\alpha(\Delta_{b,s,t,l,h}) + 0.05 \cdot d_{b,t,l,h}, \ \forall b,t,l,h \tag{30}$$

- *Binary Variables:*

$$x_k \in \{0,1\}, \ \forall k \tag{31}$$

Equation (1) defines the objective function, which aims to minimize the total system cost, that contains two main parcels: the investment cost and the expected operational cost. The first parcel of the objective function formulated above comprises an investment cost decision. In this parcel, k represents a candidate from the set of available candidates, $\mathcal{K}$, composed by generation candidates, $\mathcal{K}_i$, transmission candidates, $\mathcal{K}_j$, and energy storage equipment candidates, $\mathcal{K}_n$. For each of the candidates, there are defined the annualized investment cost, I_k. Moreover, the decision of investing in each of available candidate is based on the binary decision variable, $x_k \in \{0,1\}$.

The second parcel represents the expected value of the operating costs. Hence, for each scenario, s, in the set of simulated scenarios, $\mathcal{S}$, it is defined a probability, p_s, that multiplies the total operational cost of this scenario. The seasons are represented by t and the set of seasons by $\mathcal{T}$. The typical day is defined as l and the set of typical days as $\mathcal{L}$. To represent the correct duration of each typical day, the variable β_l indicates the weight of the typical day in its season. The variable $g_{i,t,l,h,s}$ represents the generation of an existing power plant, i, in the set of all existing power plants, which is multiplied by its operative cost, c_i. Similarly, the cost of generation of each candidate is calculated by multiplying the generation, $g_{k,t,l,h,s}$, by its operative cost, c_k. Finally, the loss of load cost is represented by c^d, while $\sigma_{t,l,h,\,s}$ indicates the depth of loss of load at each moment.

For each bus b, that belongs to the group of buses, $\mathcal{B}$, the Equation (2) represents its load balance. It states that at every moment the load balance must be satisfied through generation of the assets connected in this bus, energy imports/exports, $f^+_{b,j,t,l,h,s}, f^-_{b,j,t,l,h,s}$, battery charge/discharge, $\theta^+_{b,n,t,l,h,s}, \theta^-_{b,n,t,l,h,s}$, and load-shedding. Equation (3) limits the load shedding at each bus to its own load.

Equation (4) indicates the maximum generation capacity of an existing power plant. This constraint limits the sum of generation and reserve, $r_{i,t,l,h,s}$, that was allocated to the generation asset to its maximum generation, $\overline{g}_i$. It guarantees the generation asset will be able to produce the necessary energy if requested. The minimum generation of a power plant is defined as g_i in Equation (5). The transmission capacity is limited by the Equation (6) in both directions. The right-hand side of this equation is the maximum flow through the circuits j that connect the bus b, thus represented by $F_{b,j}$. Equations (7)–(9) have the same concept as Equations (4)–(6), however for candidates. It is noteworthy to mention that by multiplying the maximum generation by the decision variable of constructing the candidate will allow for the limitation of the maximum generation to zero in case of not deciding to construct it.

The maximum ramp-up and ramp-down are represented by $\overline{\Delta g_i}$ and Δg_i, respectively, and limit the output variation of generation assets in Equations (10) and (11). In order to accommodate the allocated spinning reserve and guarantee that the generator will be able to provide the expected production if necessary; the spinning reserve is also restrict to the ramp-up and ramp-down maximum values in Equation (12).

Equation (13) represents the operating reserve balance per bus, allowing for the allocation of the reserve to the available generators, satisfying the reserve requirements, which are dynamically defined by the variable $R_{b,t,l,h}$ that will be further explained. Equation (14) represents adequacy constraint, which allows the possibility for the system planner to exogenously incorporate minimum volumes of firm capacity requirements on top of the peak loads. This has been of increasing interest in many systems all around the globe. The contribution of each power plant to the firm capacity is represented by ϕ and the minimum and maximum firm capacity requirements by $\underline{\Phi}$, $\overline{\Phi}$, respectively. The set of constraints comprised of Equations (13) and (14) reinforces the supply of operating reserves and system adequacy. Thus, hereafter, this set will be also referred to security constraints.

Equation (15) shows the water balance in each hydro reservoir. The variable $v_{i,t+1,l,h,s}$ is the reservoir level by the end of the hour h, while $v_{i,t,l,h,s}$ is the reservoir level on the beginning of the hour h. The inflow is $a_{i,t,l,h,s}$, the water discharged into turbines $u_{i,t,l,h,s}$, the spilled water $u'_{i,t,l,h,s}$ and the water losses $\eta_{i,t,l,h,s}$ (necessary to represent the irrigation and evaporation, for example). The upstream reservoirs comprise the set $\mathcal{M}$ (only those right before the analyzed reservoir), and the sum of their discharged water is added to the current reservoir.

Equation (16) associates the final volume to the initial volume, representing a steady state strategy for the reservoir in this model. Its intention is to represent a steady-state operation, in which the initial point of operation does not matter, however, the net energy balance should be zero in the study horizon. Equation (17) limits the water discharged into turbines and Equation (18) the minimum and maximum reservoir levels, which are limited respectively by $\underline{v_i}$, $\overline{v_i}$. Equation (19) guarantees the river flow by adding a constraint that imposes the minimum outflow from reservoir, $\underline{q_i}$. Finally, the generation of this power plant is related to the discharged water by Equation (20), where ρ_i is the energy production function of this power plant. In this case, the function is defined as a constant.

The operation of batteries is represented by Equations (21)–(25). Equation (21) states the energy balance in storage equipment (similar to hydro power plants), where $v^b_{i,t,l,h,s}$ is the volume of the battery, ϵ represents the energy loss from one period to the other, $\theta^+_{n,t,l,h,s}$, $\theta^-_{n,t,l,h,s}$ mean, respectively, the charging and discharging variables and μ the energy loss in charging process.

Equation (22) limits the range of charging and discharging variables to the maximum output capacity of the battery, Θ_n. Equation (23) equalizes the initial volume of the final volume of the storage equipment. Finally, Equations (24) and (25) limit the volume of the batteries (both existing and candidates) to the maximum volume, $\overline{v}^b_n$.

Finally, the DPR representation in optimization problem is defined through Equations (26)–(30). The Equations (26) and (27) are used to calculate the expected generation, $\hat{g}$, of the existing and candidate renewable assets, respectively. Equation (28), then, calculates the difference between the observed renewable generation for each scenario to the expected generation. It is noteworthy to mention that, in case the renewable candidate is not selected, its contribution to the increment of this variable in null. Hence, the Equation (29) is used to calculate the absolute difference of the variation of the total renewable production between hours, $\Delta_{b,s,t,l,h}$.

Since this optimization problem considers the convex combination of expected value and conditional value at risk of hourly difference variability to calculate the operations reserve requirement of the system, it is necessary to represent the CV@R as a linear programming problem in order to integrate it into the original problem. Therefore, we refer the linear formulation of the CV@R to [44]. Finally, the final spinning reserve requirement is defined in Equation (30). It is composed by the convex combination of the expected value and CV@R of the $\Delta_{b,s,t,l,h}$ added to the 5% of the load. In this equation, the λ represents the convex combination parameters, which provides the CV@R weight in convex combination function and α is the CV@R parameter corresponding to the percentile of the scenarios.

Based on this formulation, this model is able to capture the intermittency and correlation associated to VRE, since it represents the generation of power plants in hourly steps. Furthermore, due to the usual daily pattern of VRE, the spinning reserve requirements is also defined in hourly steps through this model.

2.3. Solution Approach

Our model is formulated as a mixed-integer linear program (MILP) and solved by commercially available optimization solvers.

3. Case Study: Assessing the Competitiveness of Base-Load Gas Generation from Pre-salt Gas Fields

Assumptions

For the purpose of this paper, the natural gas opportunity cost of pre-salt projects is obtained through the calculation of the generation and transmission expansion optimization in order to supply the load growth for the 2024 to 2030 period. The expansion results are presented for the year 2030. Additionally, for the sake of simplicity, we represent this opportunity cost as a flat number along the year. Table 1 depicts the system's capacity breakdown per technology for the 2024 configuration of the study.

Table 1. Existing installed capacity per technology in 2024 (Source: EPE).

Technology	Existing Installed Capacity (GW)
Hydro	119.0
Biomass	18.2
Wind	28.5
Solar	9.7
Diesel	0.0
Nuclear	3.4
Natural Gas	27.2
Fuel Oil	1.2
Coal	3.4

The yearly average load consumption and yearly load peak projected for 2030 are, respectively, 1149 TWh and 174.5 GW. One can notice that the installed capacity of 2024 (211 GW) is much greater than the average energy demand of 2030 (131 average GW). This is typical from hydro systems, designed to supply load under very adverse hydrological conditions (dry seasons), which do not occur often. Hence, it is natural to have an excess capacity with respect to the energy load (the energy supply is limited by hydrological conditions). Renewables compound this challenge with their own intermittency, thus requesting more spare capacity to make up for the reserves.

For the purpose of this analysis, this model considers stochastic analysis for the energy production of renewable power plants (hydro, solar and wind). The hydro availability was modelled as periodic autoregressive model (PAR) [45], using the available monthly historical records from 1931 to 2018 from the Brazilian system operator dataset. For renewable power plants, the scenarios were created using a Bayesian network approach [46], which were correlated to the hydrological inflows in order to capture the real characteristics of the resources' availability. These renewable scenarios are considered in hourly timescale, which grants to capture the effect of hourly constraints in system expansion planning.

Table 2 provides the capital expenditure (CAPEX) and operational expenditure (OPEX) costs for each candidate source used for this simulation, which is an input for the optimization model to economic evaluate the new capacity sources and build the least-cost energy portfolio, satisfying the adequacy requirements.

Table 2. CAPEX and OPEX assumptions (Source: PSR Energy Consulting and Analytics, using data from IRENA and IEA).

Technology	CAPEX (USD/kW)	OPEX (USD/kW·Year)
Wind	1385	28
Solar	1108	14
Biomass	1108	25
Open-cycle gas turbine	720	75
Close-cycle gas turbine (pre-salt)	831	47
Close-cycle gas turbine (LNG)	942	47

Besides the investment costs, it is necessary to calculate the operating variable cost for those indicative TPP projects. Table 3 presents the assumptions for the operating variable cost and flexibility of each TPP project, considering the gas price of 3 USD/MMBtu for the pre-salt gas fields as an example.

Table 3. Operating cost assumptions for candidates' gas-fired thermal powerplants (Source: PSR Energy Consulting and Analytics, using data from IRENA and IEA).

Candidate	Gas Price [1] (USD/MMBtu)	Heat Rate (MMBtu/MWh)	Operating Cost [2] (USD/MWh)	Flexibility
Open-cycle gas turbine	12.60	8.50	138.50	Flexible
Combined-cycle gas turbine (pre-salt)	3.00	6.00	25.05	Baseload
Combined-cycle gas turbine (LNG)	6.80 [3]	6.00	54.22	Flexible

[1] With taxes and charges of the gas sector. [2] Includes variable O&M cost and taxes and charges of the power sector. [3] Considers 2 USD/MMBtu from liquefaction, 1 USD/MMBtu form shipping, 1 USD/MMBtu for regasification and 1.15 · *Henry Hub*.

4. Results and Discussion

The value of the pre-salt gas fired plants was assessed for two different simulations: (i) competitiveness based only on the supply of energy needs in the capacity planning model and (ii) analysis also considering the need to supply reliability constraints (dynamic operating reserves and peak), in addition to the energy demand.

The CPU time for each simulation ranged from 656 s to 23,029 s, taking 7416 s on average, since this optimization model is based on MIP, which depends on heuristics and convergence methods to obtain its solution.

4.1. Value of Pre-Salt Natural Gas Power Plants: Energy-Only Cost Analysis

In the first analysis, the model does not consider the set of constraints defined in Equations (13) and (14) on peak and operating reserves in the model formulation. Consequently, the model calculates an optimal expansion plan based on a pure-economical tradeoff between building new capacity and using the existing system to meet the energy demand growth (with the purpose to reduce investments plus operative costs). Since transmission is also considered, there is another tradeoff (not explored in this paper) that is to build candidates close to the load center or to invest in new technology far away from it, which would require the construction of new transmission lines

Figure 6 presents the amount of installed capacity for pre-salt gas-fired projects for a range of pre-salt gas prices, including all taxes, sector charges and gas distribution cost.

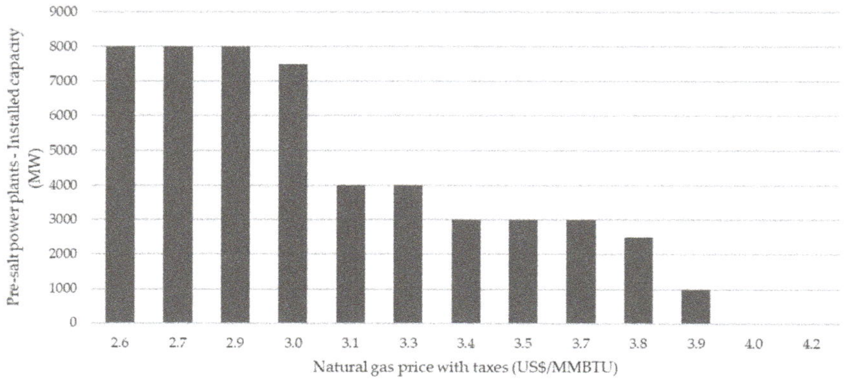

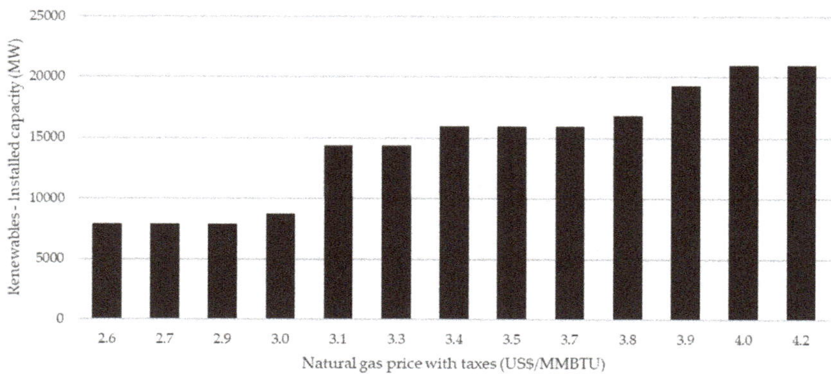

Figure 6. Additions of pre-salt natural gas power plants for different pre-salt natural gas prices.

As shown, the maximum price for base-loaded pre-salt natural gas that can charged to the power system and that would enable the economic construction of any volume is 3.9 USD/MMBTU. As expected, the reduction in natural gas price increases the attractiveness of these projects, as their volume is increased. While the decision for 3.9 USD/MMBTU gas price is to install 1 GW of those pre-salt power plants, the 3.3 USD/MMBTU and 2.9 USD/MMBTU gas prices increase this amount to 4 GW and 8 GW—reaching approximately 40 Mm^3/day of gas consumption in total.

Since pre-salt projects are only evaluated by their energy production, it is noticeable that their main competitors were renewable sources. The increment in baseload gas-fired plants implies a in reduction in other alternatives, as it is possible to observe in Figure 7, where the total VRE capacity is presented. Figure 7 shows total installed capacity for non-conventional renewable sources (which means solar, wind and biomass) considering different pre-salt natural gas prices.

Figure 7. Additions of VRE power plants for different pre-salt natural gas prices.

For pre-salt natural gas price of 2.9 USD/MMBTU, the total renewable capacity expansion is about 8 GW. If the natural gas price were 3.7 USD/MMBTU, the total renewable expansion is duplicated (16 GW). After 4.0 USD/MMBTU, the natural gas price does not affect the system expansion since the pre-salt power plants would not be competitive anymore.

Accordingly, as expensive pre-salt natural gas is, more competitive non-conventional renewable sources are. It should be noted, however, that in terms of installed capacity, the displacement of pre-salt power plants by renewable projects would not be equal due to the differences between their dispatch factors.

Another important result of this model is the total system cost, which incorporates the investment and operating costs. For propose of this paper, the annualized total system cost was presented a range of pre-salt natural gas prices in Figure 8. Noticeably, the system increases its total cost accordingly to the growing profile of the pre-salt natural gas price, until the cost in which these baseload powerplants are not economical interesting to the power sector.

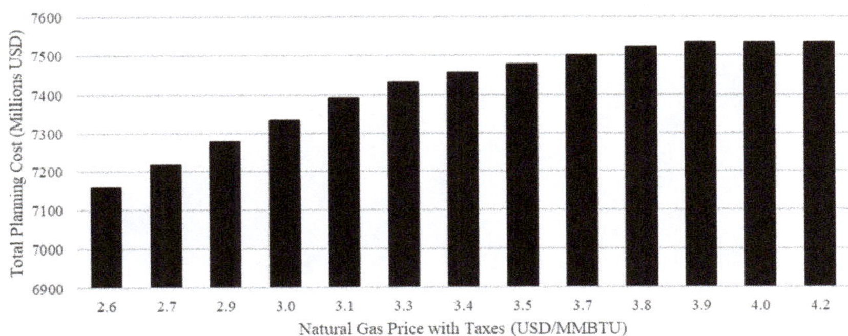

Figure 8. Total cost (sum of operating and investment) for different pre-salt natural gas prices—Energy analysis.

It is important to highlight that since part of the renewable projects represents wind farms, which are mainly from Northeast region, the optimization model has also considered transmission system expansion reinforcements in its results, so each expansion includes the transmission cost in its tradeoff.

4.2. Pre-Salt Natural Gas Breakeven Price—Considering Security and Adequacy Constraints

This section presents results of simulations considering Equations (13) and (14), which are dynamic operating reserves and firm capacity constraints. In this alternative scenario, the optimization model needs to cope with those constraints, so demand growth is not the only driver for system expansion. The operating reserve requirements are dynamically and endogenously defined by the optimization model itself for each model run, that determines a new supply expansion scenario for each gas price hypothesis. The representation of the adequacy constraint (Equation (14)) is, however, trickier. Each generation candidate has a different firm contribution for capacity, and it beyond the scope of this work to define their calculation, which is carried out in Brazil by the Ministry of Energy supported by the Energy Planning Company. All of the resources contribute in one way or the other for firm capacity. The typical values for the Brazilian power system are described below:

The firm capacity contribution of each candidate is defined in Table 4.

We understand these constraints are exogenously defined and may be considered as rather arbitrary and determinant to influence the results. This is true, however, our practical work shows that system planners have had great interest for this type of representation in the planning models. However, inn our runs, these constraints were not binding.

Figure 9 presents the total installed capacity of gas-fired TPPs using pre-salt natural gas for a range of gas prices.

Table 4. Firm Capacity Contribution of each candidate.

Technology	Capacity MW	De-rating Factor for Firm Capacity % Available Capacity
Wind	100	45%
Solar	100	29%
Biomass	100	55%
Open-cycle gas turbine	200	95%
Close-cycle gas turbine (pre-salt)	500	95%
Close-cycle gas turbine (LNG)	500	95%

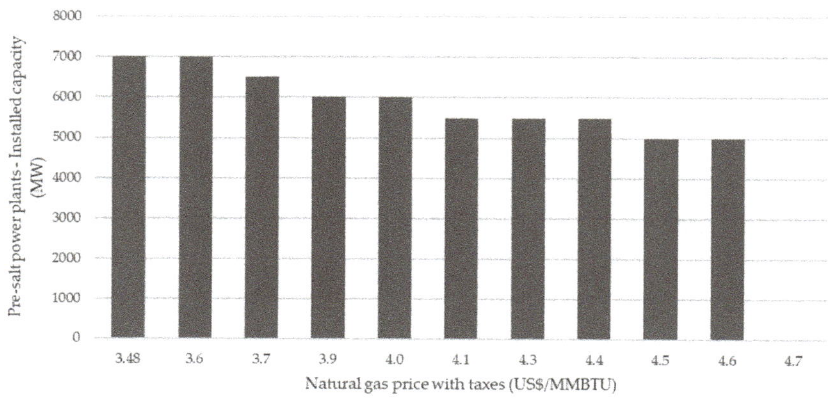

Figure 9. Additions of pre-salt power plants for a range of prices when reliability constraints are considered.

For this simulation, the breakeven price increases from 3.9 to 4.6 USD/MMBTU. In order to have at least 6 GW of new projects (an equivalent consumption of 30 Mm^3 of natural gas per day), natural gas price must be lower than 4.0 USD/MMBTU. For a natural gas price of 3.5 USD/MMBTU, the added capacity grows from 3 GW in only energy evaluation analysis to 7 GW in the security-constrained simulation.

With a gas price higher than 4.6 USD/MMBTU, pre-salt is replaced by 2.2 GW of open cycle natural gas and 1 GW of TPPs using LNG.

Noticeably, the attractiveness of these thermal plants grows substantially due to three main reasons:

1. TPPs contribute for operating reserve (due to its small flexible portion) and firm capacity requirements;
2. The growth of VRE increases the operating reserve requirements;
3. The optimal volume of capacity additions of baseload dispatch increases the hydro storage levels, thus enabling hydro plants to supply, in a cost-effective way, the operating reserves dynamically defined. An interesting discussion—but out of the scope of this paper—is how to share the benefits associated to the reserves provision between hydro ("executers") and the base load gas plants ("enablers").

Figure 10 illustrates the installed capacity of VRE for different pre-salt natural gas prices.

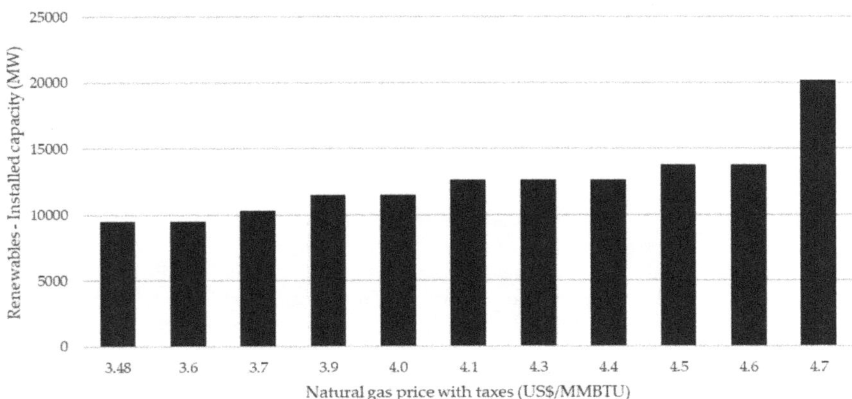

Figure 10. Addition of VRE power plants for a range of different gas prices when reliability constraints are considered.

Noticeably, the total amount of VRE reduces drastically when compared to the energy simulation, which is a consequence of their low contribution to firm capacity and increasing of spinning reserve requirements. Another consequence of both constraints is the need of other back-up sources, such as open cycle thermopower plants, to bring more flexibility to the system.

The total system cost is presented in Figure 11. Again, the growing cost of the natural gas for the power sector results in higher total cost. Besides that, the breakeven price for the attractiveness of the baseload thermopower in the Brazilian power system entailed a drastic increase in total investment cost.

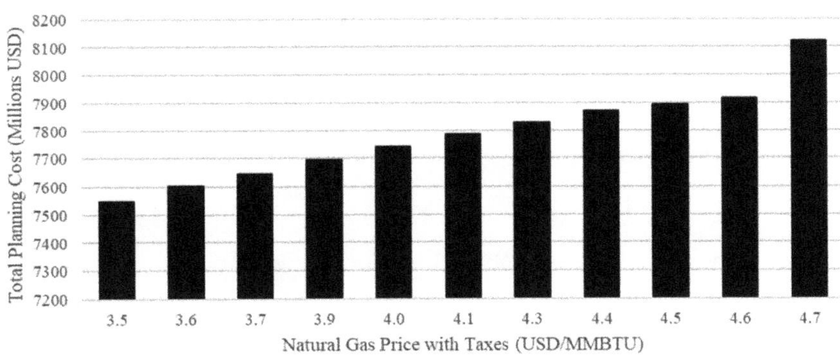

Figure 11. Total cost (sum of operating and investment) for different pre-salt natural gas prices—Reliability constraints analysis.

5. Conclusions

This paper presented a methodology based on a multi-stage and stochastic capacity expansion planning model to determine the competitiveness of a given technology against an existing system considering its reliability contribution, for peak, energy, and ancillary services. Our work applies this methodology to calculate the tradeoffs between base-loaded gas supply and VRE supply considering their value for these adequacy and operating services to the system. This allows for a comparison between the integration costs of these technologies on the same basis, thus helping policymakers to better decide on the best way to integrate the gas resources in an electricity industry which is increasingly renewable.

A case study based on a real industrial application is presented for the Brazilian power system.

Our results were expressed in terms of the maximum gas price the power system is willing to pay to integrate MW from base-loaded gas plants. When applying our methodology for the Brazilian system, we show that if adequacy constraints are represented and enforced in the probabilistic planning model a maximum gas price of 4.6 USD/MMBTU (CIF) still makes base-loaded gas plants competitive. If adequacy and operating constraints are not represented, the maximum gas price becomes 3.9 USD/MMBTU. Therefore, it is possible to say that the value of the adequacy and operating services—in addition to the value of energy production—provided by baseload gas plants (such as those from the pre-salt oil and gas fields) is approximately 0.7 USD/MMBTU.

The approach proposed by this work can be used to assess the value of other technologies, as we did with the gas-fired generation, and for other power systems, not exclusively the hydro-dominated ones. The main principle is simple: factor in the planning model constraints that represent planning and operation needs and use a planning framework to assess their value by different technologies.

Furthermore, this paper did not explore many other issues that are relevant to the discussion, such as who benefits from the reserve provision of the integration of the gas-fired plants and renewables, the reserve costs and allocation costs between market participants. These remain topics for future work.

Author Contributions: Conceptualization, L.B. and B.B.; methodology, B.B., F.N.; software, F.N.; validation, F.N., B.B.; formal analysis, L.B., B.B.; investigation, L.B., F.N., B.B.; resources, PSR Energy Consulting and Analytics; data curation, F.N.; writing—original draft preparation, F.N., B.B.; writing—review and editing, L.B.; supervision, B.B.; project administration, B.B.; funding acquisition, PSR Energy Consulting and Analytics. All authors have read and agreed to the published version of the manuscript.

Funding: This research received no external funding.

Institutional Review Board Statement: Not applicable.

Informed Consent Statement: Not applicable.

Conflicts of Interest: The authors declare no conflict of interest.

Nomenclature

a	Water inflow	$\left[\mathrm{m}^3\right]$
$b \in \mathcal{B}$	Bus index	
c	Operative cost	[\$/MWh]
c^d	Loss of load cost	[\$/MWh]
f^+, f^-	Energy imports/exports	[MWh]
F	Maximum energy transfer	[MWh]
g	Power generation	[MWh]
$\overline{g}, \underline{g}$	Maximum/minimum generation	[MWh]
$\hat{g}$	Expected generation	[MWh]
h	Hour index	
i	Existing power plant index	
I_k	Annualized investment cost	[\$]
$k \in \mathcal{K}$	Candidate index	
$\mathcal{K} = \left\{\mathcal{K}_i, \mathcal{K}_j, \mathcal{K}_n\right\}$	Set of candidates for generation, transmission and energy storage equipment	
$l \in \mathcal{L}$	Typical day index	
$\mathcal{M}$	Set of upstream reservoirs	
p_s	Scenario probability	[%]

$\underline{q}$	Minimum outflow from reservoir	$[\text{m}^3]$
r	Power reserve	$[\text{MWh}]$
R	Dynamically reserve requirements	$[\text{MWh}]$
$s \in \mathcal{S}$	Scenario index	
$t \in \mathcal{T}$	Seasons index	
u	Water discharged into the turbines	$[\text{m}^3]$
u'	Water spillage	$[\text{m}^3]$
$\underline{v}, \overline{v}$	Maximum/minimum reservoir levels	$[\text{hm}^3]$
v_t, v_{t+1}	Reservoir level by the beginning/ end of the period	$[\text{hm}^3]$
v^b	Volume of the storage equipment	$[\text{MWh}]$
$\overline{v}^b$	Maximum storage equipment volume	$[\text{MWh}]$
$x_k \in \{0,1\}$	Decision of investing in a candidate	
α	Percentile of the scenarios	
β_l	Weight of the typical day in its season	$[\%]$
Δ	Absolute difference of the variation of the total renewable production between hours stages	$[\text{MWh}]$
$\overline{\Delta g}, \underline{\Delta g}$	Maximum ramp-up/ramp-down	$[\text{MWh}]$
ϵ	Stage energy loss	$[\text{MWh}]$
η	Water losses	$[\text{m}^3]$
θ^+, θ^-	Battery charge/discharge	$[\text{MWh}]$
Θ	Maximum output capacity	$[\text{MWh}]$
$\lambda \in [0,1]$	Convex combination parameter	
μ	Energy loss in charging process	$[\text{MWh}]$
ρ	Energy production function	$[\text{MWh}/(\text{m}^3)]$
σ	Depth of loss of load	$[\text{MWh}]$
ϕ	Firm capacity	$[\text{MWh}]$
$\overline{\Phi}, \underline{\Phi}$	Maximum/minimum firm capacity requirements	$[\text{MW}]$

References

1. Barroso, L.A.; Rudnick, H.; Mocarquer, S.; Kelman, R.; Bezerra, B. LNG in South America: The markets, the prices and the security of supply. In Proceedings of the IEEE Power and Energy Society General Meeting: Conversion and Delivery of Electrical Energy in the 21st Century, PES, Pittsburgh, PA, USA, 20–24 July 2008.
2. Bezerra, B.; Barroso, L.A.; Kelman, R.; Flach, B.; Latorre, M.; Campodonico, N.; Pereira, M. Integrated Electricity–Gas Operations Planning in Long-term Hydroscheduling Based on Stochastic Models. In *Hanbook of Power Systems I, Energy Systems*; Springer: Berlin/Heidelberg, Germany, 2010; Volume 1, pp. 3–32.
3. Chabar, R.M.; Barroso, L.A.; Granville, S.; Pereira, M.V.; Iliadis, N. Optimization of Fuel Contract Management and Maintenance Scheduling for Thermal Plants in Hydro-Based Power Systems. In Proceedings of the Powergen Europe, Conference & Exhibition, Milan, Italy, 28–30 June 2005.
4. Street, A.; Barroso, L.A.; Chabar, R.; Mendes, A.T.S.; Pereira, M.V. Pricing Flexible Natural Gas Supply Contracts under Uncertainty in Hydrothermal Markets. *IEEE Trans. Power Syst.* **2008**, *23*, 1009–1017. [CrossRef]
5. Barroso, L.A.; Flach, B.; Kelman, R.; Bezerra, B.; Binato, S.; Bressane, J.M.; Pereira, M.V. Integrated gas-electricity adequacy planning in Brazil: Technical and economical aspects. In Proceedings of the IEEE Power Engineering Society General Meeting, San Francisco, CA, USA, 16 June 2005.
6. IRENA. *Renewable Energy Auctions: Status and Trends Beyond Price*; International Renewable Energy Agency: Abu Dhabi, United Arab Emirates, 2019.
7. Porrua, F.; Bezerra, B.; Barroso, L.A.; Lino, P.; Ralston, F.; Pereira, M. Wind power insertion through energy auctions in Brazil. In Proceedings of the IEEE PES General Meeting, Minneapolis, MN, USA, 25–29 July 2010.
8. Bezerra, B.; Barroso, L.A.; Brito, M.; Porrua, F.; Flach, B.; Pereira, M.V. Measuring the hydroelectric regularization capacity of the Brazilian hydrothermal system. In Proceedings of the IEEE PES General Meeting, Minneapolis, MN, USA, 25–29 July 2010; pp. 1–7.
9. De Marreco, J.M.; Carpio, L.G.T. Flexibility valuation in the Brazilian power system: A real options approach. *Energy Policy* **2006**, *34*, 3749–3756. [CrossRef]
10. Hinderaker, L.; Njaa, S. Utilization of associated petroleum gas (APG)—The Norwegian experience. In Proceedings of the Society of Petroleum Engineers—SPE Russian Oil and Gas Technical Conference and Exhibition, Moscow, Russia, 26 October 2010; Volume 2, pp. 837–848.

11. Crespo, L. STE Plants: Beyond Dispatchability Firmness of Supply and Integration with VRE. *Energy Procedia* **2015**, *69*, 1241–1248. [CrossRef]
12. Pérez-Arriaga, I. Security of electricity supply in Europe in a short, medium and long-term perspective. *Eur. Rev. Energy Mark.* **2007**, *2*, 1–28.
13. Faria, E.; Barroso, L.A.; Kelman, R.; Granville, S.; Pereira, M.V. Allocation of Firm-Energy Rights among Hydro agents using Cooperative Game Theory: An Aumann-Shapley approach. *IEEE Trans. Power Syst.* **2009**, *24*, 541–551. [CrossRef]
14. Unsihuay, C.; Lima, J.W.M.; De Souza, A.C.Z. Modeling the integrated natural gas and electricity optimal power flow. In Proceedings of the IEEE Power Society General Meeting, Tampa, FL, USA, 24–28 June 2007; pp. 1–7.
15. Unsihuay-Vila, C.; Marangon-Lima, J.W.; De Souza, A.C.Z.; Perez-Arriaga, I.J.; Balestrassi, P.P. A model to long-term, multiarea, multistage, and integrated expansion planning of electricity and natural gas systems. *IEEE Trans. Power Syst.* **2010**, *25*, 1154–1168. [CrossRef]
16. De Mello, O.D.; Ohishi, T. An integrated dispatch model of gas supply and thermoelectric systems. In Proceedings of the 15th Power Systems Computation Conference 2005, Liege, Belgium, 22–26 August 2005.
17. He, C.; Wu, L.; Liu, T.; Shahidehpour, M. Robust Co-Optimization Scheduling of Electricity and Natural Gas Systems via ADMM. *IEEE Trans. Sustain. Energy* **2017**, *8*, 658–670. [CrossRef]
18. Chaudry, M.; Jenkins, N.; Strbac, G. Multi-time period combined gas and electricity network optimisation. *Electr. Power Syst. Res.* **2008**, *78*, 1265–1279. [CrossRef]
19. An, S.; Li, Q.; Gedra, T.W. Natural Gas and Electricity Optimal Power Flow. In Proceedings of the IEEE Power Engineering Society Transmission and Distribution Conference, Dallas, TX, USA, 7–12 September 2003; Volume 1, pp. 138–143.
20. Salimi, M.; Ghasemi, H.; Adelpour, M.; Vaez-ZAdeh, S. Optimal planning of energy hubs in interconnected energy systems: A case study for natural gas and electricity. *IET Gener. Transm. Distrib.* **2014**, *9*, 695–707. [CrossRef]
21. Zhou, X.; Guo, C.; Wang, Y.; Li, W. Optimal expansion co-planning of reconfigurable electricity and natural gas distribution systems incorporating energy hubs. *Energies* **2017**, *10*, 124. [CrossRef]
22. Barati, F.; Seifi, H.; Sepasian, M.S.; Nateghi, A.; Shafie-Khah, M.; Catalao, J.P.S. Multi-Period Integrated Framework of Generation, Transmission, and Natural Gas Grid Expansion Planning for Large-Scale Systems. *IEEE Trans. Power Syst.* **2015**, *30*, 2527–2537. [CrossRef]
23. Qiu, J.; Yang, H.; Dong, Z.Y.; Zhao, J.H.; Meng, K.; Luo, F.J.; Wong, K.P. A Linear Programming Approach to Expansion Co-Planning in Gas and Electricity Markets. *IEEE Trans. Power Syst.* **2016**, *31*, 3594–3606. [CrossRef]
24. Chen, H.; Baldick, R. Optimizing short-term natural gas supply portfolio for electric utility companies. *IEEE Trans. Power Syst.* **2007**, *22*, 232–239. [CrossRef]
25. Massé, P.; Gibrat, R. Application of Linear Programming to Investments in the Electric Power Industry. *Manag. Sci.* **2016**, *3*, 149–166. [CrossRef]
26. Oree, V.; Sayed Hassen, S.Z.; Fleming, P.J. Generation expansion planning optimisation with renewable energy integration: A review. *Renew. Sustain. Energy Rev.* **2017**, *69*, 790–803. [CrossRef]
27. Sadeghi, H.; Rashidinejad, M.; Abdollahi, A. A comprehensive sequential review study through the generation expansion planning. *Renew. Sustain. Energy Rev.* **2017**, *67*, 1369–1394. [CrossRef]
28. Ringkjøb, H.K.; Haugan, P.M.; Solbrekke, I.M. A review of modelling tools for energy and electricity systems with large shares of variable renewables. *Renew. Sustain. Energy Rev.* **2018**, *96*, 440–459. [CrossRef]
29. Van Stiphout, A.; De Vos, K.; Deconinck, G. The Impact of Operating Reserves on Investment Planning of Renewable Power Systems. *IEEE Trans. Power Syst.* **2017**, *32*, 378–388. [CrossRef]
30. Aghaei, J.; Akbari, M.A.; Roosta, A.; Gitizadeh, M.; Niknam, T. Integrated renewable-conventional generation expansion planning using multiobjective framework. *IET Gener. Transm. Distrib.* **2012**, *6*, 773–784. [CrossRef]
31. Jirutitijaroen, P.; Singh, C. Reliability and cost tradeoff in multiarea power system generation expansion using dynamic programming and global decomposition. *IEEE Trans. Power Syst.* **2006**, *21*, 1432–1441. [CrossRef]
32. Gorenstin, B.G.; Campodonico, N.M.; Costa, J.P.; Pereira, M.V.F. Power System Expansion Planning Under Uncertainty. *IEEE Trans. Power Syst.* **1993**, *8*, 129–136. [CrossRef]
33. Løken, E. Use of multicriteria decision analysis methods for energy planning problems. *Renew. Sustain. Energy Rev.* **2007**, *11*, 1584–1595. [CrossRef]
34. Bucksteeg, M.; Niesen, L.; Weber, C. Impacts of dynamic probabilistic reserve sizing techniques on reserve requirements and system costs. *IEEE Trans. Sustain. Energy* **2016**, *7*, 1408–1420.
35. Zhou, Z.; Botterud, A. Dynamic scheduling of operating reserves in co-optimized electricity markets with wind power. *IEEE Trans. Power Syst.* **2014**, *29*, 160–171. [CrossRef]
36. Bruninx, K.; Van den Bergh, K.; Delarue, E.; D'haeseleer, W. Optimization and allocation of spinning reserves in a low-carbon framework. *IEEE Trans. Power Syst.* **2015**, *31*, 872–882. [CrossRef]
37. De Vos, K.; Stevens, N.; Devolder, O.; Papavasiliou, A.; Hebb, B.; Matthys-Donnadieu, J. Dynamic dimensioning approach for operating reserves: Proof of concept in Belgium. *Energy Policy* **2019**, *124*, 272–285. [CrossRef]
38. Qiu, T.; Xu, B.; Wang, Y.; Dvorkin, Y.; Kirschen, D.S. Stochastic Multistage Coplanning of Transmission Expansion and Energy Storage. *IEEE Trans. Power Syst.* **2017**, *32*, 643–651. [CrossRef]

39. Ghofrani, M.; Arabali, A.; Etezadi-Amoli, M.; Fadali, M.S. A framework for optimal placement of energy storage units within a power system with high wind penetration. *IEEE Trans. Sustain. Energy* **2013**, *4*, 434–442. [CrossRef]
40. Soares, A.; Perez, R.; Morais, W.; Binato, S. Addressing the Time-Varying Dynamic Probabilistic Reserve Sizing Method on Generation and Transmission Investment Planning Decisions. *arXiv* **2019**, arXiv:1910.00454.
41. Poncelet, K.; Delarue, E.; Duerinck, J.; Six, D.; D'haeseleer, W. The Importance of Integrating the Variability of Renewables in Long-Term Energy Planning Models. TME Working Paper—Energy and Environment. 2014, pp. 1–18. Available online: https://www.mech.kuleuven.be/en/tme/research/energy_environment/Pdf/wp-importance.pdf (accessed on 1 September 2021).
42. International Renewable Energy Agency. Planning for the Renewable Future: Long-Term Modelling and Tools to Expand Variable Renewable Power in Emerging Economies. 2017. Available online: https://www.irena.org/publications/2017/Jan/Planning-for-the-renewable-future-Long-term-modelling-and-tools-to-expand-variable-renewable-power (accessed on 1 September 2021).
43. Pineda, S.; Morales, J.M. Chronological time-period clustering for optimal capacity expansion planning with storage. *IEEE Trans. Power Syst.* **2018**, *33*, 7162–7170. [CrossRef]
44. Rockafellar, R.T.; Uryasev, S. Optimization of Conditional Value-at-Risk. *The Magazine of the Fine Arts*, 2 January 1999; 1–26.
45. Noakes, D.J.; McLeod, A.I.; Hipel, K.W. Forecasting monthly riverflow time series. *Int. J. Forecast.* **1985**, *1*, 179–190. [CrossRef]
46. Dias, J.A.; Machado, G.; Soares, A.; Garcia, J.D. Modeling Multiscale Variable Renewable Energy and Inflow Scenarios in Very Large Regions with Nonparametric Bayesian Networks. *arXiv* **2020**, arXiv:2003.04855.

Article

Reliability Metrics for Generation Planning and the Role of Regulation in the Energy Transition: Case Studies of Brazil and Mexico

Ana Werlang [1],[*], Gabriel Cunha [1], João Bastos [1], Juliana Serra [1], Bruno Barbosa [1] and Luiz Barroso [2]

[1] PSR Energy Consulting & Analytics, Rio de Janeiro 22250-040, Brazil; gabriel@psr-inc.com (G.C.); joao@psr-inc.com (J.B.); julianaxavier@psr-inc.com (J.S.); brunopeixoto@psr-inc.com (B.B.)

[2] Instituto de Investigación Tecnológica, Escuela Técnica Superior de Ingeniería (ICAI), Universidad Pontificia Comillas, 28015 Madrid, Spain; luiz.barroso@comillas.edu

[*] Correspondence: anabeatriz@psr-inc.com

Abstract: In recent years electricity sectors worldwide have undergone major transformations, referred to as the "energy transition". This has required energy planning to quickly adapt to provide useful inputs to the regulation activity so that a cost-effective electricity market emerges to facilitate the integration of renewables. This paper analyzes the role of system planning and regulations on two specific elements in the energy market design: the concept of firm capacity and the presence of distributed energy resources, both of which can be influenced by regulation. We assess the total cost of different regulatory mechanisms in the Brazilian and Mexican systems using optimization tools to determine optimal long-term expansion for a given regulatory framework. In particular, we quantitatively analyze the role of the current regulation in the total cost of these two electricity systems when compared to a reference "efficient" energy planning scenario that adopts standard cost-minimization principles and that is well suited to the most relevant features of the new energy transformation scenario. We show that two very common features of regulatory designs that can lead to distortions are: (i) renewables commonly having a lower "perceived cost" under the current regulations, either due to direct incentives such as tax breaks or due to indirect access to more attractive contracts or financing conditions; and (ii) requirements for reliability are often defined more conservatively than they should be, overstating the hardships imposed by renewable generation on the existing system and underestimating their potential to form portfolios.

Keywords: regulation; energy transition; Brazil; Mexico; renewables; reliability; generation system expansion; efficient energy planning

Citation: Werlang, A.; Cunha, G.; Bastos, J.; Serra, J.; Barbosa, B.; Barroso, L. Reliability Metrics for Generation Planning and the Role of Regulation in the Energy Transition: Case Studies of Brazil and Mexico. *Energies* **2021**, *14*, 7428. https://doi.org/10.3390/en14217428

Academic Editor: Dolf Gielen

Received: 2 October 2021
Accepted: 6 November 2021
Published: 8 November 2021

Publisher's Note: MDPI stays neutral with regard to jurisdictional claims in published maps and institutional affiliations.

1. Introduction

Optimization and simulation models are often used in energy policymaking to portray future system scenarios, used as a basis for the definition of long-term policy goals and the most economic investment pathways to them [1]. The idealized optimization and simulation models used by policymakers, system operators and scholars to represent and study the electricity sector tend to agree that the fundamental objective of electricity system planning is to pursue the minimization of the total cost of the system (or, equivalently, maximizing total social welfare)—despite differences in terms of representation and solution strategy. Generally speaking, the problem of optimizing system expansion involves choosing the best possible mix of technologies to meet system needs. Knowledge of how well each technology's physical attributes align with those needs is thus at least in principle sufficient to determine the desirability of investing in that particular technology.

There are many pathways through which new generators may come into the system: enabled by the spot market directly, free market bilateral contracting [2], long-term centralized auctions [3] and direct consumer-side investment. All methods have their strengths

Energies **2021**, *14*, 7428. https://doi.org/10.3390/en14217428   https://www.mdpi.com/journal/energies

and potential weaknesses and therefore, regardless of the main strategy that a country picks for organizing system expansion, there is the potential for efficient or inefficient outcomes. In an efficient market that follows this ideal cost minimization objective, the role of regulation is simply to facilitate the most socially desirable outcome coming to fruition: for example, by minimizing transaction costs, correcting market failures or externalities and making sure agents have the proper incentives. In practice, however, regulatory design is a challenging task, and it is possible that regulations may yield less than ideal outcomes by introducing distortions in the market and in the contracting routes.

Indeed, in the context of a rapidly changing electricity sector, it is possible that regulations may not evolve fast enough to adapt to the new reality—for example, by using outdated assumptions to address system needs and technological contributions, therefore leading to a non-optimal expansion. Sometimes, market distortions are intentionally made in order to promote specific technologies, such as distributed generation, renewable energy or "strategic" projects such as reservoir hydro or nuclear. Intentionally or not, most regulation and reforms lead to some impact on the market agents' perceived cost and, therefore, on the development of the system [4].

In recent years, the electricity sector has undergone major transformations, often referred to as the "energy transition". There is an increasing role for variable renewable energy resources and for distributed demand-side services, as well as a strong incentive for net-zero scenarios [5]. Therefore, policy makers are responsible for ensuring that long-term energy targets are achieved without compromising the system's reliability and safety, and that the long-term costs of the energy transition are appropriately assessed [6]. Nevertheless, it seems likely that in some countries regulations may already be operating more as an obstacle to optimal system expansion than as a facilitator as intended. Even if regulators are fully benevolent, given the rapidly changing context, it is easy for regulation to lead to inefficient expansion due to assumptions that are simply out of date. Additionally, the regulatory process is further complicated in practice by the existence of legacy costs that need to be recovered, legacy contracts and commitments that need to be honored and special interest groups that may attempt to influence policymakers.

This paper analyzes the role of regulations regarding firm capacity and distributed energy resources in guiding long-term system expansion by comparing the outcomes in terms of total system cost of current regulatory practices in Brazil and Mexico to a reference "efficient energy planning" scenario. In terms of firm capacity, the "efficient energy planning" criterion involves ensuring that the system's total generation supply is at least three standard deviations greater than demand at all times as a reliability criterion ("three sigma" rule). This criterion was implemented as an iterative process in the optimization tool for determining optimal generation system expansion, repeated until convergence was reached at the desired reliability level. In terms of distributed generation, the "efficient energy planning" representation involves simply eliminating cross-subsidies and other regulatory benefits for adopters of these systems, and thus ensuring that adopters are truly motivated by their individual preferences and not external motivations.

With this quantitative exercise, we show that renewables commonly have a lower perceived cost under the current regulations, either due to some direct incentive or due to indirect access to more attractive contracts or financing conditions—not only in the case of distributed generation but also for centralized generation applications. Furthermore, we find that requirements for reliability are often defined more conservatively than they should be, overstating the hardships imposed by renewable generation on the existing system and underestimating their potential to form portfolios.

2. Materials and Methods

2.1. Core Methodology

This section describes the core methodology used to model the system's long-term equilibrium for both case studies (Brazil and Mexico), which is the basis of this paper's methodology for quantitatively determining the impacts of regulatory practices. In summary, the authors use a combination of the three simulation and optimization modules: (i) distributed generation, (ii) reliability and (iii) expansion and operation. Together these modules envision a minimization of the system's total costs, while ensuring the predefined reliability requirements. Each module's methodology, as well as inputs, adopted for the efficient scenario and regulatory constraints, will be further detailed throughout the chapter. It is worth highlighting that the representation in each of these modules can be affected by regulations—as a country's policies change the methodology for assessing different technologies' contribution to system reliability, change incentives that end consumers may perceive for adopting distribution generation, and/or change the "perceived cost" different system expansion candidates by offering preferential tax treatment and/or financing.

The main goal of the present paper is to address how the regulations currently implemented in Brazil and Mexico would lead to deviations in the long-term equilibrium relative to an idealized "distortion-free" scenario. Finding the "true" distortion-free expansion result is evidently a challenging task, and even though the present paper provides a robust methodology for these benchmarks, methodological refinements could be implemented as potential future work. Nonetheless, it seems undeniable that regulatory practices in many countries incorporate significant deviations from an "ideal" representation, due to political influence, legacy contracts, methodological simplifications, lack of data and other reasons—see, for example [6,7].

Figure 1 illustrates the general scheme of the methodology adopted, highlighting all modules and connections between them.

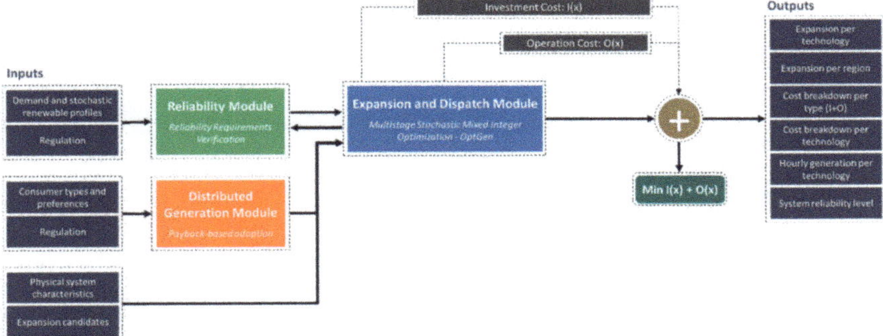

Figure 1. General scheme of the methodology adopted.

2.1.1. Distributed Generation Module

Assuming efficient price signals, the market equilibrium achieved from utility-maximizing agents would be equal to the result of a cost-minimization problem—which would allow these small-scale solutions to be simply incorporated into the optimization model as an additional expansion candidate. However, regulatory incentives to DER tend to result in consumers perceiving radically different price signals compared to what the dispatch model suggests. Therefore, distributed energy resources (DER) adoption by individual consumers needs to be considered separately from the optimal system expansion.

In this paper, the authors have implemented an iterative process that aims to simulate the interactions between DER adoption and market-driven system expansion [8]. The adoption decision was based on a payback-based adoption curve, which is a methodology vastly used in the existing literature [9,10]. The "payback" represents the number of years until the system "pays for itself", considering the upfront cost of the distributed generation system and the yearly benefit corresponding to the avoided cost of purchasing electricity from the grid at the electricity tariff. The smaller the payback, the more economically attractive the distributed generation investment, and therefore the higher the share of consumers that will ultimately choose to adopt this alternative.

2.1.2. Reliability Module

For the reliability module in the "efficient energy planning" scenario (that is, in the absence of regulatory distortions), the authors have selected the "three sigma" (3σ) criterion, which ensures that the system's total generation supply is at least three standard deviations greater than demand at all times. Assuming a normal distribution of the net supply, the 3σ criterion leads to a probability of 99.7% of being able to supply the demand without issue.

In order to ensure that the supply would meet the 3σ criterion, an iterative process was implemented. For the first run of the model, an initial expansion with no reliability restriction was used. Then, the variability of the net demand (demand minus renewable generation) was measured and compared to the firm capacity of the system—as defined by the regulation of each country and further detailed in Section 2.5. If the criterion was met, the optimization stopped, otherwise the contribution of the renewable technologies would be recalculated based on the results and adjusted for the next iteration. At each iteration, the expansion and dispatch model is called, and the same analysis and check of the 3σ criterion is carried out. If the criterion is met, the optimization stops, otherwise a new iteration begins.

2.1.3. Expansion and Dispatch Module

In liberalized competitive electricity markets, system expansion is driven by generators acting with the goal of maximizing their own profits. Using standard microeconomic competitive market assumptions, the system expansion induced by market equilibrium of these profit-seeking agents would be equal to the one chosen by a central planner seeking to maximize total welfare [11,12]. Based on this fundamental principle, it is possible to estimate the generation system expansion in a liberalized market environment through a specialized computational tool that determines the minimum cost expansion plan for an electrical system.

For the simulations, the authors used a long-term expansion planning model that determines the least-cost decisions for the construction, retirement and reinforcement of generation and transmission projects. This optimization model is integrated with a dispatch simulation tool that represents the details of the production of all plants in the system, taking into account operational flexibilities and constraints and ensuring that supply and demand remains balanced at all times (a requirement of the electricity network). In this manner, the model optimizes the trade-off between investment costs to build new projects and the expected value of operative costs obtained from the transmission-constrained dispatch model [13,14].

One important aspect of the model that should be highlighted is that the hydrological and renewable generation uncertainties are handled explicitly with a stochastic Monte Carlo representation followed by the stochastic optimization of the utilization of the system's resources. In practice, hourly renewable energy stochastic scenarios gathered from georeferenced databases along with historical hydro inflows are fed to a statistical model in order to obtain correlated probability distributions for various locations and renewable resources, which in turn are used to produce the representative stochastic series used by the optimization software.

The representation of system dispatch involves an hourly resolution of the supply-demand balance—a particularly important feature in scenarios with high renewable share, representing operational constraints within each day. The model represents chronological links between the seasons (representing the management of hydro reservoirs between wet and dry seasons) but not between years, where a "cyclic" representation ensures that volumes stored at hydro reservoirs in the beginning of the year must coincide with volumes at the end of the year for each scenario.

Overall, this simulation approach, with the chronological decision-making, the stochastic modeling of hydrology and renewable generation, the hourly temporal granularity, among others, is compatible with recommended reference methodologies for energy planning—and, more specifically, for energy planning in the context of the energy transition [6].

2.2. Expansion Optimization Paradigm

As a benchmark for the system expansion planning, the analysis considered a reference year whose demand is twice the current demand, where the system portfolio resulting from the expansion model would be operated to meet this required load in a "steady state" (or "static") manner in the very long term. Note that, considering a demand growth rate of 2% per year, for example, our demand benchmark would be reached in 35 years (~2055), whereas with a growth rate of 4% per year, this benchmark would be reached within 18 years (~2038). Therefore, for ease of reference, we have considered the reference horizon of the simulations to be representative of year 2040. It should be highlighted that the proposed system expansion methodology only looks at the target year, building the entire amount of new capacity needed at once to meet the target demand. This is not entirely realistic in practice, seeing that the ultimate expansion outcome may depend on the incremental decisions made in each of the intermediate years—this notion of path-dependency can be especially prominent in a context of sharp changes over time (such as cost decreases, regulatory changes and phase-out of policy incentives). Nonetheless, analyzing the "optimal" long-term system breakdown without these path-dependency constraints can yield interesting insights about the systems. It should also be noted that, in order to allow maximum flexibility in the choice of expansion technologies, it was assumed that most plants of the existing system can be decommissioned—they would be replaced by the construction of new ones, if this is economically desirable given the least-cost criterion.

Furthermore, in order to reduce the computational effort required by the expansion problem while maintaining a detailed hourly resolution representation, the concept of seasons and typical days was used in the modelling. The first step of this strategy consists of grouping the months of the year into sequential seasons—in this analysis, standard seasons with a length of three months each were used. All "weekdays" were grouped together as one representative day, and all Saturdays, Sundays and holidays as a second "weekend" type representative day—taking into account that, within each season, all days that belong to each of these two categories tend to be not so different from one another and thus can be represented as being drawn from the same probability distribution. Even though refinements could be added (in particular, a distinction between Saturdays and Sundays), the authors found that the impact of such refinements on the optimization results was extremely minor.

Regarding the role of regulation, first of all we assume that regulations are carried out under conditions of perfect competition, which implies that the optimal solution from the expansion model can be interpreted as resulting from market equilibrium between generators competing in the electricity market [11,12]. From this construction, the optimal system expansion from the central planner's perspective is the same as the competitive market equilibrium, and the role of an efficient regulatory design would be simply to minimize "frictions" in order to ensure that this optimal outcome would be reached. However, regulatory initiatives can also introduce frictions and distortions, which the authors represent using two alternative approaches (which together can account for the impacts of most regulatory implementations):

- Changing the perceived costs of specific technologies: making them appear cheaper or costlier than they actually are for the purpose of system expansion decisions due to subsidies or surcharges, respectively; or
- Introducing new constraints in the optimization problem.

In both cases, the optimization model is used to find a new equilibrium expansion strategy, and the cost of the modified optimization problem is expected to increase with the introduction of these policies.

2.3. Modeling of Candidates for System Expansion

Five key representative technologies were used as candidates for the system expansion:

(i)   Utility-scale solar power plant (assumed to have one-axis tracking);
(ii)   Utility-scale wind power plant;
(iii) Combined cycle gas-fired plant, highly efficient but with a preference for a more predictable dispatch profile (CCGT);
(iv)   "Peaker" type gas-fired plant, prioritizing operational flexibility over thermal conversion efficiency (OCGT); and
(v)   Battery storage technology.

The candidates' attractiveness for system expansion in the absence of special regulations was determined from a purely economic standpoint, and the optimization model determines whether their investment costs, fixed costs and operating variable costs are compensated by their corresponding benefit to the system (based on avoided costs of dispatching costlier plants and avoiding electricity shortages). The final parameters were based on international benchmarks, especially "Lazard's Levelized Cost of Energy Analysis—Version 12.0" [15] and "Lazard's Levelized Cost of Storage Analysis—Version 4.0" [16].

Even though solar and wind technologies have been observing a continuous decreasing price trend for several years [17], there is a significant degree of uncertainty with regards to how long this trend may continue and therefore a conservative assumption of not representing any additional cost decreases in the long-term expansion was used. For battery storage technology, on the other hand, a decreasing cost curve was considered—since the technology is currently not sufficiently cost-competitive for large-scale applications in the electricity sector and it is at a much earlier stage in its technology life cycle than wind or solar, suggesting it still has ways to go before achieving maturity. Synergies with other economic segments (such as consumer electronics and electric vehicles) are also likely to contribute to pressuring battery prices downwards. The battery candidates were modeled as batteries with 4 h storage capacity at a long-term cost circa 60% lower than the average current price from Lazard (same cost in both case studies). Additionally, for transmission, only the main corridors between regions were represented, using the distance between the regions and a cost benchmark in USD/km for high voltage networks in order to estimate the cost of expanding interconnection capacity as an additional candidate technology for the expansion model.

Ultimately, Tables 1 and 2 summarize the assumptions for each technology used in the analysis for the two countries.

Table 1. Assumptions for expansion technologies in Brazil.

Technology	CC Gas	Gas Peaker	Solar	Wind	Battery
Investment cost (USD/kW)	700	600	680	1000	680
Fixed OPEX (USD/kW·year)	25	15	6.8	24	10.2
Variable cost (USD/MWh)	2.2	3.3	-	-	-
Useful life (years)	20	20	25	25	10
Investment cost (USD/kW)	700	600	680	1000	680

In order to properly represent the maximum generation potential for renewable energy sources along with the corresponding profiles, a plurality of potential plants was created based on wind and solar scenarios from different locations with various resource quality levels. Typically, the quality of resource becomes a limiting factor as a greater amount of total capacity is developed: the best available areas tend to be developed first, though in practice, in our simulations the potential was very rarely completely exhausted. Figures 2 and 3 illustrate the different representative wind generation daily profiles (pictures on the left, representing the average profile over 24 h and highlighting regional variabilities), the zones with high wind potential (pictures in the middle, color-coded to represent the quality of the resource) and high solar potential (pictures on the right, similarly color-coded) in each country.

Table 2. Assumptions for expansion technologies in Mexico.

Technology	CC Gas	Gas Peaker	Solar	Wind	Battery
Investment cost (USD/kW)	700	600	600	1000	680
Fixed OPEX (USD/kW·year)	25	15	6	24	10.2
Variable cost (USD/MWh)	2.2	3.3	-	-	-
Useful life (years)	20	20	25	25	10
Investment cost (USD/kW)	700	600	600	1000	680

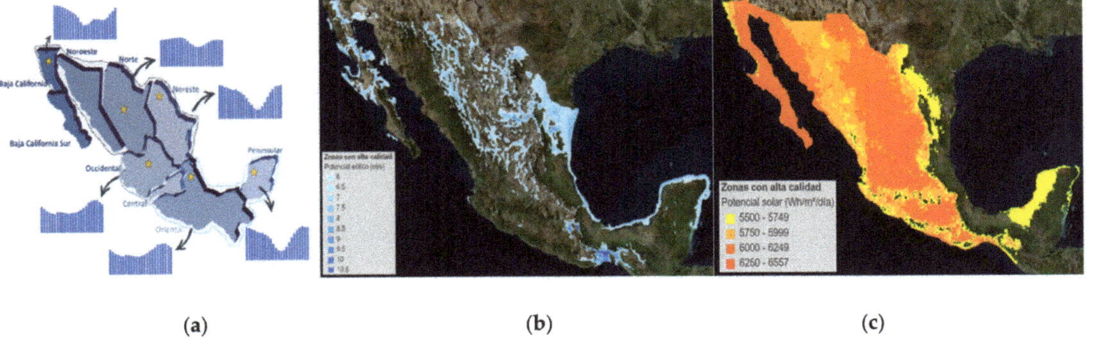

(a)   (b)   (c)

Figure 2. Wind profiles (**a**) and zones with high wind (**b**) and solar (**c**) potential in Mexico.

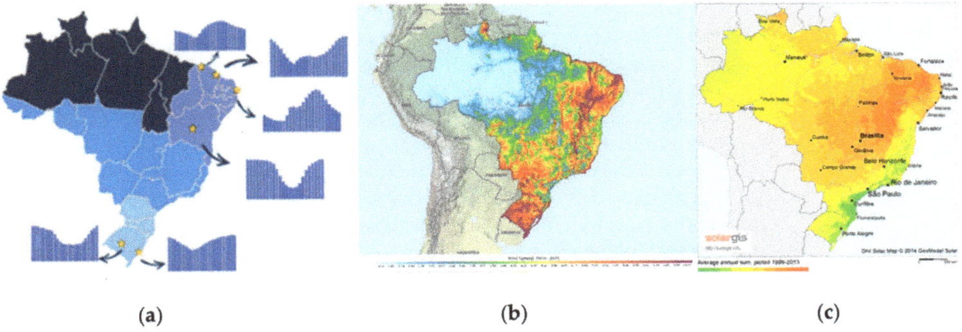

(a)   (b)   (c)

Figure 3. Wind profiles (**a**) and zones with high wind (**b**) and solar (**c**) potential in Brazil.

It is also crucial to properly represent the uncertainty and variability of renewable energy sources and in particular, the historical correlations among hydrology, wind, solar and other variables in the power system in order to properly incorporate portfolio effects into the optimization. These spatial dependencies were modeled through a Bayesian network, which automatically identifies the dependency relationships between the various time series of interest [18]. The result is a set of coherent probabilistic scenarios for the resource availability of inflows and renewables that can be used for the calculation of the stochastic operation policy with hourly representation.

Fuel prices are another key driver of electricity prices and are an extremely relevant input for system expansion decisions, since they directly impact the operational cost of thermal power plants and, consequently, their competitiveness with other technologies. The opportunity costs of hydro power plants are also highly affected by fuel prices, though indirectly. Generally speaking, if the domestic fuel market is efficient, fuel pricing should be driven mostly by the international fuel price—as this represents a "netback" price at which fuel can be imported or fuel surpluses can be exported. Therefore, a direct relation of fuel prices with international dynamics is assumed for all fuels in the efficient energy planning scenario. In order to ensure that long-term international fuel price forecasts are coherent (despite the inherent uncertainty given in the long-term horizon of the analysis), the projections of the U.S. Energy Information Administration (EIA) were adopted as a reference to the international price dynamics, taking 2040 as a base date [19]. This assumption leads to a long-term gas cost in the USA (Henry Hub, USA) of 4.3 USD/MMBtu in the long term (in the reference year 2040).

It is necessary to further incorporate additional costs for transportation, losses and similar services that must be considered in the final price of natural gas. In Brazil, the natural gas that sets the marginal price for gas-fired expansion is imported via LNG, which leads to an assumption of loss factor of 15% and additional costs of 3 USD/MMBtu— yielding a final gas price of 7.3 USD/MMBtu to be used in the simulations. In Mexico, in turn, Henry Hub natural gas is typically imported via pipeline, leading to negligible losses and additional costs (in line with what is reported by PEMEX) of approximately 1.65 USD/MMBtu—yielding a final gas price of 5.95 USD/MMBtu.

Regulatory Constraints

Under the efficient energy planning scenario, the attractiveness of each technology is evaluated based on its levelized cost. However, it is possible for regulation to introduce distortions that may result in a perceived cost for certain technologies that can be different from their true cost—usually through some type of (direct or indirect) incentive or subsidy. Generally speaking, a technology with a perceived cost that is higher than its true cost is disincentivized and becomes less likely to be built, while conversely a technology with a perceived cost that is lower than its true cost is incentivized and becomes more likely to be built—increasing the likelihood of suboptimal system expansion choices and cost overruns.

For this analysis, the notion of the pre-tax weighted average capital cost (WACC) was used to represent the financial attractiveness of a particular investment by evaluating only the project's (pre-tax) cashflow. The WACC consolidates information on financing and taxation, allowing cost-benefit analyses to be made on the project's cashflow directly without requiring further assumptions on company strategy, cost of debt and other parameters. For the purpose of the efficient energy planning scenario, all generation sources had the same WACC of 9% per year. For the current regulation scenario, however, typical market practices were used to estimate a perceived WACC that is allowed to vary for each technology [10].

In Brazil, investors and financiers have long required higher interest rates for infrastructure projects when compared to Mexico, which translated into a higher WACC. The role of these uncertainties in increasing the WACC, however, is offset in terms of perceived cost by the availability of cheap loans by Brazilian public banks for renewable projects. In Mexico, renewables tend to be favored by the current contracting mechanisms, since they have been the only ones allowed to offer all three products in the long-term auctions, which is one of the most relevant drivers to the system expansion and the other mechanisms do not introduce important distortions. This is reflected as a lower perceived cost for these technologies. Additionally, from the results of the auctions that have taken place in the country, it is possible to infer very low WACCs.

Market information on typically practiced debt-to-equity ratios, interest rate on debt and price practices in previous auctions were used to calibrate the generators' perceived WACC. In this analysis, the main differences identified were that renewable projects that were able to secure long-term contracts in auctions and had special conditions for loans tended to achieve higher leverage ratios (higher D/E) and lower return rates—both from the financier (debt) and from the sponsor (equity). Conversely, in a similar analysis, higher WACC rates were identified for thermal projects, mainly driven by the restrictions to the contracting alternatives available to them. The final pre-tax WACC in the efficient scenario and in the current regulation scenario is presented in Figure 4.

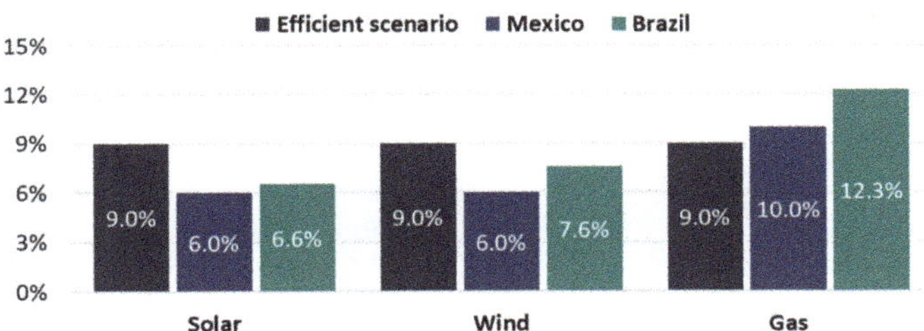

Figure 4. Pre-tax WACC for Brazil and Mexico in the current and efficient regulation scenarios.

2.4. Distributed Energy Resources Expansion

Distributed generation is playing an increasingly important role in modern electricity systems, and thus it merits evaluating to what extent regulation may be facilitating (or making things difficult) for this type of consumer-driven initiative to flourish. Regulation is necessary to ensure that consumers with distributed generation (DG) facilities are properly rewarded for the energy they provide to the grid, and in an efficient energy planning scenario the incentive passed through to consumers is equal to the benefit that these installations provide to the grid—which in turn depends on their generation profile and on their role in reducing costs of transmission, if applicable. It is worth highlighting that the methodology used in our assessments was limited to solar distributed generation, which has achieved a degree of maturity that allows for modeling adoption with a reasonable level of accuracy. It would be possible, however, to extend the methodology or several other types of distributed energy resource.

In order to estimate the DER expansion, a payback-based adoption curve was used, following practices commonly adopted in the existing literature [9,10]. Generally speaking, DER plays a role whenever it becomes desirable for individual consumers to invest in their own system rather than purchasing electricity from the grid, which is incorporated into the payback metric (representing the number of years necessary to recoup the initial investment due to their monthly savings on the electricity bill). Even though this decision can be different for each individual consumer, in aggregate the share for the market will adopt a larger amount of DER units if the payback is lower (implying that the system pays for itself in a relatively short time). The analysis was focused on small-scale solar systems (assuming a mix of residential, commercial and industrial systems), which tend to be the most prominent DER adopters, with significant market penetration even today. Figure 5 illustrates a range of possible adoption curves, as compiled by Sigrin [9]. The vertical axis shows the total market share ultimately achieved by distributed generation as a function of the payback on the horizontal axis.

The analysis considered the "RW Beck" curves as the key benchmark, which follows the simple exponential formula represented in Equation (1). Note that this methodology is in line with what has been commonly used by EPE (the planning entity in Brazil) in their forecasting studies—and, even though slightly different "payback sensitivity" parameters have been tested based on actual adoption data in different consumer classes in Brazil, in practice they deviate little from the 0.3 benchmark [20,21].

$$\text{Long-term Adoption Rate } [\%] = \exp\left(-0.3 \times \text{Payback [years]}\right) \tag{1}$$

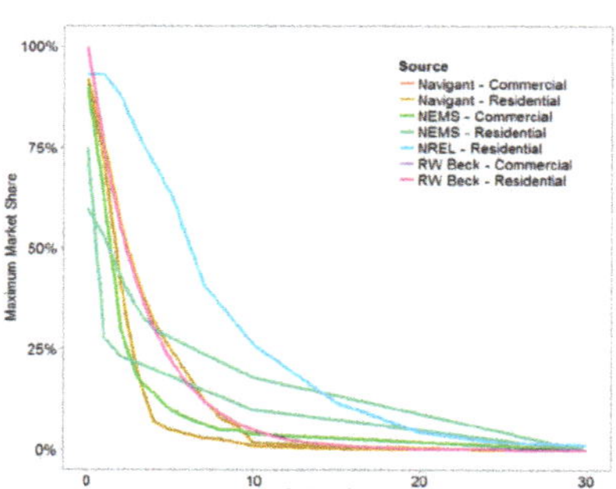

Figure 5. Possible adoption curves as compiled by Sigrin (2016) [9].

Even though, by assumption, all consumer classes were represented as having the same adoption curve, they perceive different payback levels, which created heterogeneity within each market.

Even though the solar generation technology is well-known for being relatively modular, meaning that economies of scale are less significant than with more traditional generation sources, residential-scale systems still tend to be around 25% costlier than commercial-scale systems, which in turn tend to be around 10% costlier than utility-scale systems (though with significant variations on those ratios). This is illustrated, for example, by comparisons of the cost of a utility-scale solar system (several thousand kW), a commercial-scale system (a few hundred kW) and a residential-scale system (a few kW) in different regions, as obtained from IRENA's renewable energy cost database [22] and illustrated in Figure 6.

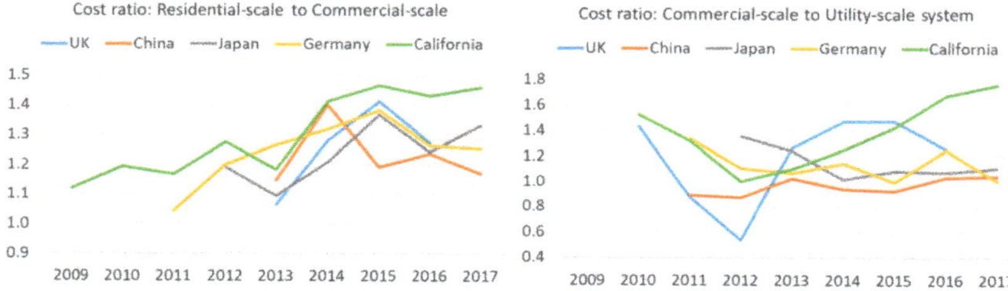

Figure 6. Comparisons of the cost of a utility-scale solar system, a commercial-scale system (a few hundred kW) and a residential-scale system in different regions (authors' analysis based on IRENA's renewable energy cost database) [22].

Additionally, rooftop solar implementations also have lower performance on average than utility-scale ones, as they tend to undergo cleaning and maintenance less frequently and to have suboptimal orientation towards the sun (as they usually use the roof's inclination to save on the cost of the structure). In the modelling, an additional 7% loss in the performance ratio was assumed for commercial-scale rooftop systems and a 15% loss for residential-scale systems (when compared to utility-scale) in addition to the higher costs described earlier to estimate the payback of those system sizes. This differentiation is applied on top of the regional differentiation based on the quality of the solar resource (which also affects payback).

Finally, the most important component for the analysis of the effect of regulation is the electricity tariff perceived by different consumer classes. The representation of "current regulations" in each of the reference countries was based on historical tariff data of each consumer class, and an "efficient regulations" scenario was constructed by applying multipliers that seek to represent whether consumers are able to offset a payment in USD/MWh that is higher than the true benefit of their DG installations for the system. The main goal of these multipliers is to capture the effect of regulations on the price incentives perceived by potential adopters even in the long term as system expansion and marginal prices interact: for example, consumers that can offset costs corresponding to transmission and distribution cost components of the electricity tariff effectively benefit from a regulatory distortion, and the lack of time-of-use tariff distinctions also tends to benefit DG adopters as the share of solar power in the system increases. The idea is that each individual consumer makes the choice that makes the most economic sense for themselves (given their preferences), which if their incentives are efficient would be exactly in line with what would be optimal for the system as a whole (after incorporating all possible externalities, such as reducing technical losses in the distribution network). If regulations over-incentivize DG adoptions, however, the share of consumers that will opt for owning a distributed generation system will increase, with consequences for the system expansion decisions.

In our model, the adoption rate resulting from the payback curve determines a fixed amount of distributed generation capacity to be part of the final expansion (complemented by the decisions of the system expansion module). It is worth highlighting that, as a general rule of thumb, distributed solar generation tends to offset centralized solar generation in the optimal system expansion, as it has similar characteristics (such as a generation profile peaking around the midday hours). However, there are a few key differences between utility-scale solar and rooftop solar from a system planning point of view, which have been incorporated into the model's parameterization. The first one is the location of the projects: distributed generation projects are usually located near load centers, while centralized projects tend to be placed in the locations with the best resource potential. The second is the lower capacity factor of the distributed generation projects, due to a less reliable maintenance of the solar panels and lack of solar tracking.

2.4.1. Regulatory Constraints: Brazil

Perhaps the most significant consequences of regulation on the DG market are felt when distributed generation allows consumers to offset not only the tariff corresponding to the costs of energy but also other costs such as transmission and distribution costs and system charges—which is the case for Brazil. In the Brazilian case study, another distortionary effect is that taxation of electricity in most states is also offset from consumers' electricity bill in proportion to distributed generation, thus strengthening the incentive by a significant amount (around 35% given steep electricity tax rates). Adding together these contributing factors, the end result in terms of DG adoption under current regulations for Brazil is shown Table 3, both as a fraction of the demand within each category and converted into the corresponding total capacity that would be built (in total, the model suggests 6 GW of distributed generation capacity).

Table 3. Adoption of distributed generation in the current regulation scenario by class and region in Brazil.

Class		Southeast	South	Northeast	North
Residential	% demand	8.79%	8.29%	10.86%	11.98%
	MW	1805	523	646	353
Low-voltage commercial/ industrial	% demand	16.18%	16.19%	19.73%	20.92%
	MW	1150	353	406	213
Medium-voltage commercial/industrial	% demand	2.19%	2.22%	2.64%	3.23%
	MW	308	96	108	65

Overall, low-voltage commercial and industrial consumers present the most significant cost-benefit ratio and therefore they are the ones with the highest distribution generation adoption levels—though the residential market, due to its substantial size, still accounts for most of the capacity additions according to this model. Medium-voltage consumers have a two-part tariff, and the capacity portion (proportional to peak demand) is in nearly all cases unaffected by the installation of distributed generation, as they typically occur at night.

2.4.2. Regulatory Constraints: Mexico

In Mexico, there are also regulatory incentives that reduce the payback of the distributed generation systems as perceived by end consumers (though not as profoundly as in Brazil). There is a program for residential consumers and small and medium-sized enterprises that provides an economic incentive equivalent to 10% of the total cost of the system, with the remaining 90% financed with FIDE resources, whereas for the agricultural sector the Shared Risk Program grants up to 50% of the value of the generation projects. There is also a support program for low-income families for the installation of ecotechnologies such as photovoltaic systems, among other initiatives [23]. Additionally, public policies have been implemented to guarantee open access, not unduly discriminatory against distributed generation. However, this policy seems to follow a reasonable economic rationale and offers no undue benefits—as the reinforcements of distribution network necessary for connecting distributed generation plants whose capacity exceeds the current limits of the maximum allocation capacity determined by the distributor will be borne by the applicant.

Distributed generation is also directly related to the tariff level in Mexico. Even though prior to the tariff reform in 2018, tariffs had been set below marginal price (thus disincentivizing rooftop solar), they have been readjusted in 2018, becoming more cost-reflective. It should be noted that, although a "Time-Of-Use" tariff (distinguishing between base, intermediary and peak hours) is available for large consumers from the industrial and commercial sectors, most low-voltage consumers typically only perceive an average monthly tariff that is applied equally to all hours. As a consequence, these consumers may potentially overvalue distributed generation delivered at midday in case there is an oversupply of solar power (which is also contemplated in the payback variable in our model). As low-voltage consumers do not perceive a time-dependent tariff, they are likely to be overcompensated for generation delivered at midday hours (the benefit to the system as a whole is low if there is a sufficiently large solar installed capacity, but the low-voltage consumer will be remunerated according to the average tariff). Adding together these contributing factors, the end result is shown in Table 4, both as a fraction of the demand within each category and converted into the corresponding total capacity that would be built. In total, the model suggests 1.6 GW of equilibrium installed capacity of rooftop solar.

Table 4. Adoption of distributed generation in the current regulation scenario by class and region in Mexico.

Class		Big North	Big Central	Big South
Residential	% demand	0.66%	0.70%	0.67%
	MW	47	60	28
Low-voltage commercial/industrial	% demand	6.18%	7.23%	5.04%
	MW	112	157	53
Medium-voltage commercial/industrial	% demand	3.97%	4.36%	3.65%
	MW	405	536	218

As in the case of Brazil, low-voltage commercial and industrial consumers in Mexico present the most significant cost-benefit ratio, thus being the ones with the highest distribution generation adoption levels—with the medium-voltage commercial market dominating the additions due to their size. Adoption levels are typically higher in the Central region, mainly motivated by a higher loss factor passed through to the tariff incentive. The residential sector, in turn, has the lowest incentive due to a combination of a relatively low tariff and higher investment costs.

2.5. Reliability Requirements

Contrary to most markets in classic microeconomics, where there is a possibility of short- to medium-term storage at various points of the supply chain, the electricity grid is very sensitive to fluctuations, and instability can provoke outages with substantial social impact. This characteristic implies that electricity systems must ensure that supply and demand are balanced at any given point in time, which in turn requires procuring some amount of excess capacity to protect against supply inadequacy. As renewables have been increasing their share in most countries at a very fast pace, this topic has been rapidly increasing in importance, as the variability of intermittent generation sources compounds with the uncertainty of variations in the non-controllable demand and equipment outages to potentially increase the system's need for robustness. In particular, several power systems have explicit regulations on "firm capacity" requirements (or similar metrics) to ensure that the system is operating with enough flexible dispatchable capacity to overcome even high-stress situations—typically implying high-demand hours in a high-demand season (potentially with additional contingencies). These regulations will be further detailed and modeled in the "current regulation" scenario for the Brazil and Mexico case studies.

In the "efficient energy planning" benchmark, the authors searched for a proxy for the ideal requirement level. Although this subject is broadly discussed worldwide, there is currently no absolute consensus among planning entities and system operators regarding the best methodology to calculate system needs in order to ensure reliability. Therefore, instead of using the explicit ad hoc constraints commonly applied by regulators and system operators, the authors sought to design a simple methodology from first principles that fairly represents the system needs even in a context of very high expected renewables penetration in the energy mix.

The main starting point is the principle of technology neutrality—that is, all technologies ought to be treated equally and their net effect on system reliability should be assessed exclusively based on (i) how much they contribute to increasing variance and uncertainty in the supply-demand balance (that needs to be accommodated by other units) and (ii) how much they contribute with flexibility that can be used to accommodate other agents' uncertainty. Note that the neutrality principle used to guide the efficient energy planning representation is intuitive: if certain types of variability (e.g., climatic events such as El Niño) are treated differently from others, the system may prioritize these types of uncertainty (investing "too much" in being protected against these events) while possibly neglecting other sources of uncertainty, which means the system would not be as robust to these types of events. Therefore, the proposed methodology for calculating the system's "efficient" reliability requirement focuses on the probability distribution of the net supply margin, defined as the difference between the available capacity and the net demand, and contemplating all possible sources of uncertainty equally. The net demand is defined as the demand discounted from the non-controllable renewable generation—in this study, the solar and wind generation, as per Equations (2) and (3). The index ω represents each scenario (or potential outcome) in the space of possibilities, seeing that these quantities are represented as random variables.

$$\text{NetDemand}\,(\omega) = \text{Demand}\,(\omega) - \text{SolarGeneration}\,(\omega) - \text{WindGeneration}\,(\omega) \qquad (2)$$

$$\text{NetSupply}\,(\omega) = \text{AvailableCapacity}\,(\omega) - \text{NetDemand}\,(\omega) \qquad (3)$$

One common metric used in the context of system reliability analysis is the loss of load probability (LOLP) [24], defined as the probability that the net supply is negative (that is, the probability that the system's capacity is insufficient to meet demand). Reliability requirements can be constructed based on this metric, by first defining a target LOLP level $\overline{P}$ and calculating what is the minimum required firm capacity k to ensure this reliability level is met. Note that the firm capacity representation is necessarily a simplification, seeing that k is not a random variable (does not depend on ω), even though in practice all technologies do involve some degree of uncertainty. It is straightforward, however, to check whether the reliability criterion is satisfied for all clusters (or groups of scenarios) Ω, as depicted in Equation (6): if the probability is found to be much lower than $\overline{P}$ for all Ω, this is a sign that the system is oversupplied.

$$\text{LOLP}: \ \mathbb{P}\left[\text{NetSupply}\left(\omega\right) < 0\right] \tag{4}$$

$$\text{Firm capacity (for given } \overline{P}) : \ \min k \text{ such that } \mathbb{P}\left[k\text{-NetDemand}\left(\omega\right) < 0\right] \leq \overline{P} \tag{5}$$

$$\text{Efficient firm capacity condition}: \ \mathbb{P}\left[\text{NetSupply}\left(\omega\right) \ < \ 0 \,|\, \omega \in \Omega\right] \ \leq \ \overline{P} \text{ for all } \Omega \tag{6}$$

In practice, our methodology did not incorporate uncertainties in the available capacity (e.g., generator failures) into the representations of the joint probability distribution: dispatchable resources (such as thermal plants and batteries) were assumed to have negligible uncertainty and were represented as "pure" firm capacity values after discounting their expected unavailability rates. In practice, generator failures could have an effect in creating "fatter tails" in the probability distribution, and a more refined representation could be explored in future work.

Another key simplification made is to assume that, after properly subdividing scenarios into clusters Ω (as will be described further), the LOLP within each cluster is chiefly described by the standard deviation σ of the probability distribution of net demand. This standard deviation, in turn, requires estimating the standard deviations and correlations between the individual components of net demand—namely, the demand side, solar output and wind output. Under this simplification, as depicted in Equation (7), it suffices for the expected value of net supply to be greater than three times its standard deviation ("3σ rule") in order to ensure that the probability that net supply is greater than zero is at most $\overline{P}$.

$$\text{Adapted firm capacity}: \ \min k \text{ such that } \mathbb{E}\left[k\text{-NetDemand}\left(\omega \ \in \Omega\right) \ < \ 0\right] \geq 3 \times \sigma\left[k\text{-NetDemand}\left(\omega \ \in \Omega\right) \ < \ 0\right] \text{ for all } \Omega \tag{7}$$

In order to obtain consistent descriptions of the probability distribution of each component, the "clusters" for each scenario are defined by:

(i)   The season, which once again has known patterns for both demand and renewables; and
(ii)   The hour (highlighting daily profile patterns of demand and renewables).

Note that this paper focuses on weekdays for the presentation of the analysis of the variability of demand, though it would be straightforward to define an additional weekday versus weekend separation of clusters. There are, therefore, a total of 96 clusters (24 h and 4 seasons) modeled individually—each of which is represented individually. However, there is some structure to the time series data beyond pure classification into clusters: for example, two days in the summer of 2020 are expected to be "more similar" to one another than two days in the summer of different years, and the amount in hour 2 and in hour 3 of the same day are expected to be correlated despite belonging to different clusters. To account for this effect, the variation between samples of the same cluster are defined by three components, each of which is modeled as an autoregressive time series (that is, the "day" effect has some memory from the previous day, the "year" effect has some memory from the previous year, but they are otherwise unrelated):

(i)   The year, which may have a higher or lower than expected electricity demand (typically due to economic shocks or particularly harsh or mild summers or winters) and may also be subject to renewable resource effects (with hydrological multi-year cycles being notably pronounced);
(ii)   The day, which represents the fact that resource availability within each day is correlated across hours;
(iii) The hour, which in practice represents a statistical residue (that is, the component of variation that cannot be explained by either yearly or daily correlations).

To make things intuitive, the aggregate supply margin will be broken down into separate timescale components in order to focus on each of those variables that describe the "clusters" of variability X: the annual effect, the daily effect (between days after eliminating the effect of the year and season) and the residual effect (between individual hours after eliminating the hourly profile effect and other previous effects). The final net supply margin for each cluster is therefore a random variable equal to the sum of the random variables for each timescale component and each technology component X (solar, wind, hydro and demand).

$$\text{NetSupplyComponent}_X(\omega) = X_y(\text{year}(\omega)) + X_s(\text{day}(\omega)) + X_h(\text{hour}(\omega)) \tag{8}$$

For each of the individual clusters, the total standard deviation σ_X is written as a sum of components X_j representing contributions from the demand side or specific generation technologies (hydro, wind and solar) j—the sum of all components X_j yields X. Note that there is a fundamental relationship between the second-order moments that allow for describing the standard deviation of the net supply as a whole by considering the variance $\mathbb{V}$, covariance $\mathbb{C}$, correlation ρ and standard deviation σ of each of the components that add up to it. Equation (9) shows a derivation of this property, where σ_j represents the standard deviation of each component X_j of net supply, σ_X represents the standard deviation of net supply as a whole (with $X = \sum_j X_j$), and $\rho_{X_j,X}$ represents the coefficient of correlation between the component X_j and the whole X. Note that all equalities in Equation (9) are exact: the only key assumption required, as described earlier, is that the second-order moment is sufficient to describe the system's reliability needs to a reasonable level of precision.

$$\begin{aligned}
\sigma_X^2 = \mathbb{V}X = \mathbb{V}\left[\sum_j X_j\right] = \sum_j \mathbb{C}\left(X_j, \sum_j X_j\right) = \sum_j \mathbb{C}\left(X_j, X\right) \\
= \sum_j \rho_{X_j,X} \cdot \sqrt{\mathbb{V}X_j \cdot \mathbb{V}X} = \sum_j \rho_{X_j,X} \cdot \sigma_X \cdot \sigma_{X_j} \\
\sigma_X = \sum_j \rho_{X_j,X} \cdot \sigma_j
\end{aligned} \tag{9}$$

Note that the aggregation was made first among technologies within each timescale and then among timescales—the index j in Equation (9) implicitly represents both types of aggregation. The sum, considering the covariances between the technologies and timescales, is the total variability that must be accounted for when designing the reliability constraint for efficiently guiding system expansion. A small additional caveat with regards to this representation is that the sequential nature of the optimization problem was imperfectly represented—each sample within each cluster is drawn from a probability distribution that may depend on the hours, days and years that came before it via autoregressive models, but when assessing reliability requirements in the long term, this correlation is not explicitly incorporated. This is in fact a reasonable approximation, seeing that at the expansion planning stage it is not possible to obtain special knowledge on short-term dynamics and it makes sense to consider a reliability criterion that weights all possible outcomes equally.

In summary, the impact of each timescale component and each technology was separated, and the covariance among these factors was calculated, reaching the total variability of the system's net supply. This variability is then used to create the efficient reliability requirement, which should be met by the firm capacity in the system. A commonly used and reasonably conservative requirement involves a 3σ criterion, implying that the system's expected excess supply (that is, total supply minus demand) is at least three standard deviations greater than zero in all clusters. If the probability distribution of the net supply was normally distributed, the 3σ criterion would yield a probability of being able to meet the demand without issue of 99.7%—reflecting a relatively conservative criterion. The probability distribution for the net demand could in principle be more fat-tailed, although in practice for very large numbers of generators and consumer units the distribution tends to approach the normal curve. It should be noted that, due to Chebyshev's inequality [25], even if the true probability distribution had the worst possible shape, the 3σ criterion still ensures that the LOLP cannot possibly be higher than 11.1%—and it would be possible to apply a higher sigma (σ) multiplier in order to obtain even more conservative rules to add a "buffer" against more fat-tailed distributions. In order to comply with this 3σ requirement, the planning model utilized in this work was used with the analysis of the net supply margin in a loop, until convergence was reached.

In contrast to this "efficient" methodology for reliability requirements, which ensures that all technologies are treated in the same manner and only as a function of their variability parameters, different countries use very different approaches for defining reliability requirements. Despite these methodologies' intent of promoting greater security of supply, a less than optimal methodology to define generators' firm capacity (for example) can lead to inefficiency in the system expansion choices. An over-conservative criterion, for example, can lead the market to overcommit new capacity to attend reliability requirements, overcharging the consumer segment. On the other hand, an overly optimistic criterion or the lack of a proper periodic revision of firm capacities can lead to a serious violation of the system's desired reliability levels. We discuss below the key aspects of the regulations currently applied in the two countries of the case study.

2.5.1. Regulatory Constraints: Brazil

In Brazil, the central focus of the institutional framework for planning of the electric system is the security of supply, which is guaranteed by two basic rules that are enforced on a 12-month basis: (i) every consumer must have 100% of its consumption covered by registered contracts [26]; and (ii) every contract must be backed up by a power plant capable of sustainably producing the contracted volumes, as measured by a "Physical Guarantee" value assigned to each power plant by the Ministry based on their physical characteristics [27]. In a simplified way, the physical guarantee calculation process of hydro and thermal generators can be summarized in two main steps. First, the maximum demand that the existing physical system can supply (according to a predefined security of supply criterion) is calculated. This number will ultimately correspond to the total sum of the physical guarantees of all plants in the system, ensuring that the system has a comfortable supply-demand balance (again according to the pre-established criterion) if and only if the system's total physical guarantee is enough to cover the entirety of demand. Then, in the second step, the total physical guarantee is allocated among individual generation facilities.

For renewables, however, this approach is different—their contributions are calculated only for the plant itself without taking into account synergies with the existing system. The physical guarantee of wind power plants, for example, is calculated based on the energy expected to be yearly produced in, at least, 90% of years (P90), discounting the expected unavailability and losses up to the plant's connection point [28]. The P90 value is assessed by a specialized company that certifies the wind measurements and associated calculations. For solar power, the methodology is similar, but the statistic used to determine the physical guarantee is simply the expected production value assessed by the certifying entity, rather than the more conservative P90 [29]. It should be noted that wind power physical guarantee is treated in a conservative fashion: a practice that could lead to overburdening consumers with the cost of too much unneeded extra capacity in the long term. On the other hand, reassessments of the robustness of the system as a whole are not carried out as often as they should be, and there is evidence that hydro plants have been generating less than their joint physical guarantees for several years [30]—possibly suggesting the opposite, that Brazil is being less conservative than it should be in its assessments of the country's supply-demand balance.

Moreover, the Brazilian system is historically "energy-constrained" (as opposed to "capacity-constrained") and the peak demand has easily been met with cheap instantaneous power provided by large hydro power plants. Therefore, all physical guarantee requirements mentioned are currently enforced only for energy production targets for the long term. However, as the penetration of renewables grows in the Brazilian electricity market, the requirements of peak demand supply are becoming more relevant due to the variable hourly pattern of these energy sources. In this context, developing security rules based on instantaneous power requirements may be a need in the near future, which is already on the regulator's agenda. The concept of "peak physical guarantee", focusing instead on ensuring that there would be enough capacity for generators to provide power during peak hours, was introduced in some early regulations and even contracts after the market reform in the early 2000s, though it has not been officially enforced. Using this regulation as a starting point, the modelling of the Brazilian current regulation scenario will include both a "Firm Energy" constraint (represented by the classic physical guarantees mechanism), which must be greater than average demand; and a "Firm Capacity" constraint (represented by the "peak physical guarantee"), which must be greater than peak demand. Figure 7 illustrates the concept of these constraints, as well as provides an estimation of the contribution of each technology to each criterion in terms of share of total installed capacity.

2.5.2. Regulatory Constraints: Mexico

In Mexico, a distinguishing feature is the existence of a separate market for "capacity product" involving yearly settlements based on generators' contribution in the 100 critical hours of each year, verified ex post [31]. The "yearly spot price" of capacity is calculated based on;

(i)   System-critical capacity margins calculated by the system operator;
(ii)   The cost of building a new peaker plant (calculated by the system operator using a reference thermal technology); and
(iii) The surplus revenue that this reference thermal technology would have received selling its energy in the spot market under ideal conditions (which is discounted from the final capacity payment).

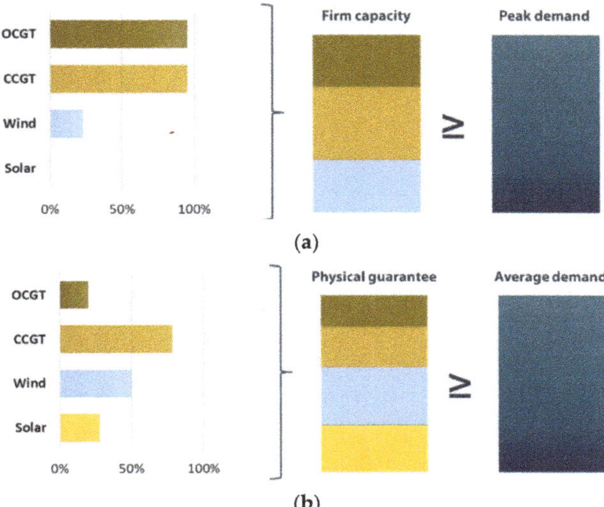

Figure 7. Reliability constraints in Brazil: firm capacity (**a**) and physical guarantee (**b**) criterion.

The yearly capacity payment is thus calculated to be complementary to energy spot market revenues, contributing to stabilizing generators' yearly cashflows.

The Mexican capacity market has the key features of a regulatory reliability mechanism, by focusing on the ability to supply demand in the most extreme circumstances (as represented by the 100 critical hours). Interestingly, it does not operate as a "hard" constraint, but rather as the imposition of a "soft" financial incentive: if the minimum capacity margin drops below the minimum (defined as 7.7% in the current regulation), for example, generators would be allowed to recover twice their fixed costs, thus incentivizing the construction of new capacity capable of supplying the system during the critical hours. The end result of this incentive, therefore, is in a way similar to what can be achieved with "hard" firm capacity constraints, as it aims to incentivize a certain reliability level to be met. Figure 8 summarizes the price formation used for the mechanism. A curious feature of the Mexican reliability mechanism is that, because spot market revenues are used in the calculation, this means that generators may end up not receiving capacity revenues at all in years when the market is exceptionally tight (though they will still receive them during high-supply years). In practice, the adopted methodology accounts for the average expected capacity revenue (considering all types of supply-demand balance), which is expected to be the most reliable signal for system expansion.

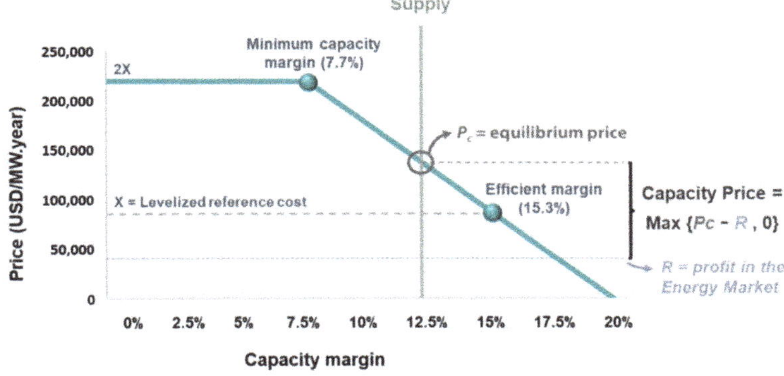

Figure 8. Reliability constraints in Mexico: capacity market price formation.

Renewables are remunerated according to their measured output in the critical hours, whereas hydro and thermal plants are remunerated according to their availability (maximum potential output) at these same moments. In this sense, the Mexican reliability mechanism is relatively progressive, ensuring that (despite their stochastic nature) renewables' contributions under critical conditions are valued by the mechanism. Despite this positive feature of all technologies being properly contemplated by the mechanism, the incentives put in place by the Mexican capacity market still tend to slightly favor conventional generators. For example, the fact that the price of capacity is dependent on the fixed cost and assumed energy market revenues of a peaker thermal plant in particular means that this technology tends to have less risk in its capacity market revenues. In addition, and perhaps most noticeably, hydro and thermal plants usually have their contributions during the critical hours equal to their available capacity even if they are not dispatched, whereas renewables have contribution equal to their actual generation—implying that they may be penalized if they need to be curtailed during critical demand hours (for example due to transmission bottlenecks or to accommodate ramping of thermal generators).

The Mexican capacity market was represented by altering expansion candidates' "perceived cost" for choosing optimal system expansion. This was conducted by subtracting the expected capacity market revenues from the annualized investment cost for each technology. It should be noted that determining expected capacity revenues is an iterative process, as the capacity prices and the critical hours themselves shift depending on the expansion mix (which in turn is decided by the perceived costs of the technologies). In order to estimate this interplay, the modelling accounted for how system expansion alters the identification of which hours are likely to be considered "critical". Following the current regulation, the firm capacity of renewable technologies was adjusted to reflect the expected generation in these critical hours, while for thermals and hydros it was assumed to be equal to their availability.

3. Results

3.1. Brazil Results

The final expansion mix, both in terms of capacity and share of the expansion, is summarized in Figure 9 for the two scenarios: efficient and current regulations.

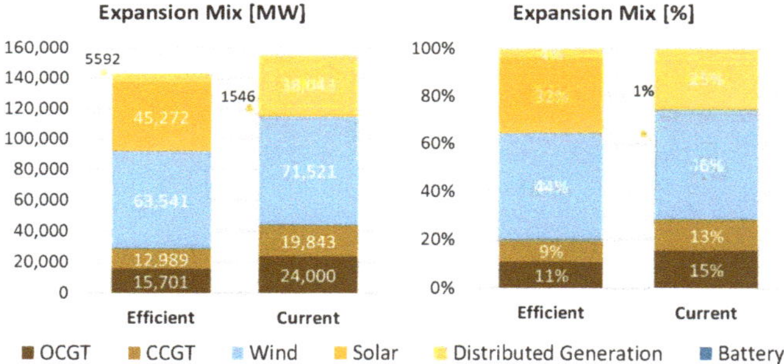

Figure 9. Comparison of total expansion between scenarios.

The first point that stands out is the remarkably larger expansion of distributed generation in the current regulation scenario compared to the efficient energy planning one, caused by the numerous incentives to this source. Consequently, the utility-scale solar expansion is drastically reduced—for the most part it is substituted by the rooftop solar alternative, which has a similar production profile, but does not contribute to the system reliability criterion as detailed earlier. In contrast, wind technology increases its share in total expansion. Also noticeable is the higher share of natural gas sources in the current regulation scenario, motivated by the higher share of intermittent sources and by the regulatory reliability requirements. Battery capacity does not participate in any of the scenarios simulated—this is because the large amount of existing hydropower in the Brazilian system is already sufficient to shift demand between hours of each day, providing a service that in other systems would need to be delivered by batteries.

Another interesting comparison pertains to the expansion among electrical regions [32], as depicted in Figure 10. In both regulation scenarios, solar capacity is mostly concentrated in the Southeast region, since it has a great potential and is located close to demand; whereas wind capacity is greatly focused in the Northeast, where the highest capacity factors lie. Regarding natural gas, in the optimal scenario, this expansion is mostly concentrated in the South, while in the current regulation one, it is spread out among the regions. In addition, as the distributed generation potential is proportional to the regional demand, it is mostly concentrated in the Southeast.

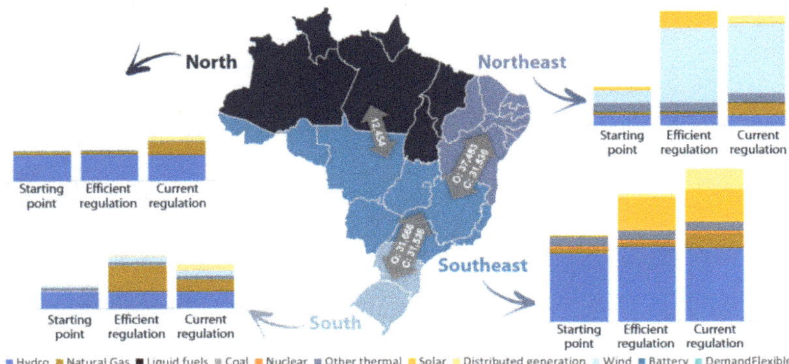

Figure 10. Comparison of the regional expansion between scenarios.

It is also worth analyzing the impact of the distinct expansions in the system operation, which is depicted in Figures 11 and 12. It should be noted that this figure represents only the average values for the main typical day of each season (typical weekday). Each scenario incorporates data on hydrological inflows (that vary by season) and solar and wind production (that varies hourly).

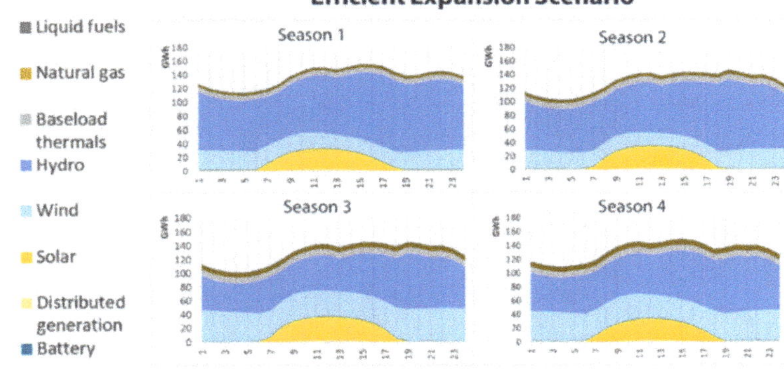

Figure 11. Generation profile per season in the efficient regulation scenario.

Notably, solar and wind generation, as relatively inflexible resources in the intraday period, are the "base" of the generation mix and usually require flexible sources to accommodate their daily pattern. This flexibility is chiefly provided by hydro generation, which fluctuates along the day to ensure energy balance, while thermal generation is very smooth during the day in this average profile. During season 3, for example, there is an almost 50 GW ramp in hydro generation in only a couple of hours caused by solar and wind power increment.

Subsequently, the reliability of the system is analyzed. For this analysis, the consultants compared the net demand (that is, demand discounted by renewable generation) and its variability with the total available capacity of the system, as presented in Figure 13. Remarkably, in both expansion scenarios the reliability margin is very close to the $\pm 3\sigma$ criterion, indicating good equilibrium of supply-demand and a well-adjusted system. Nonetheless, the volatility of the net demand in the current regulation scenario is substantially higher when compared to the efficient case, caused by a greater wind and distributed generation expansion—which in turn implies a higher need for flexible thermal capacity.

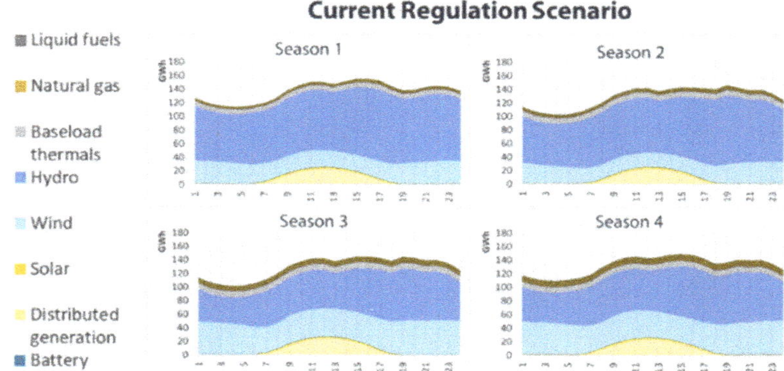

Figure 12. Generation profile per season in the current regulation scenario.

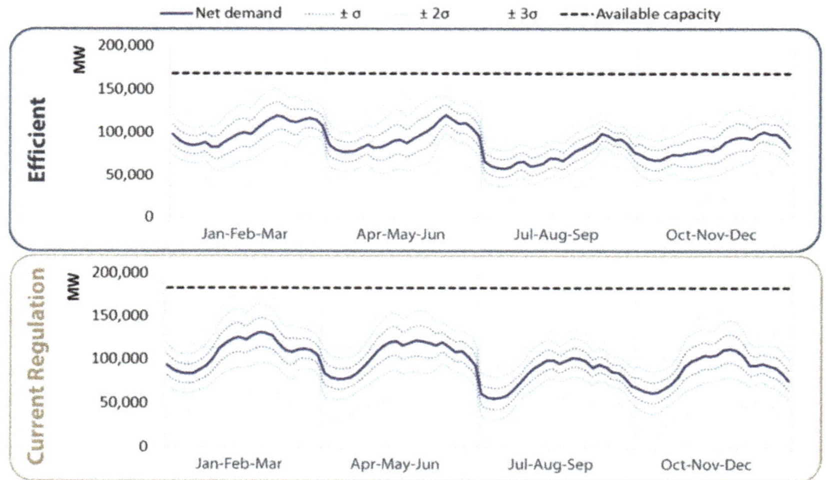

Figure 13. Comparison of the system reliability level between scenarios.

Figures 14 and 15 highlight the contrast between the two scenarios in terms of total cost needed to remunerate the existing generation system (only operational costs) as well as the system expansion represented. Note that the total social surplus perceived by consumers (from having their electricity demand met) may be shared among market agents in several ways—implicitly, whenever a given sector (such as transmission, generation or trading) receives positive profits, they are allowed to capture a greater share of this social surplus. However, as the regulations that govern this cost allocation across agents can be very complex and the number of assumptions required to make a long-term assessment is extremely high, the assessment of this idealized total cost view is limited to focusing on the aggregate outcome. The total cost is proportional to the area of the curve, with the width representing installed capacity of each technology and the height representing the cost per unit of installed capacity. Note that, generally speaking, because the expansion mix was selected by an optimization model, it is to be expected that costlier units also have proportionally higher contributions to system reliability and/or flexibility, justifying this higher cost. In addition, in order to allow the direct comparison between the efficient energy planning and current regulation scenarios, note that all costs represented reflect true costs (rather than perceived costs).

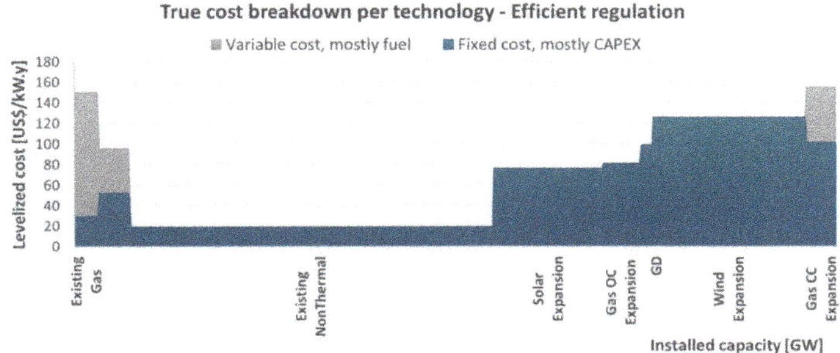

Figure 14. Cost breakdown in the efficient regulation scenario. Total cost (variable + fixed) equal to USD 20.8 billion.

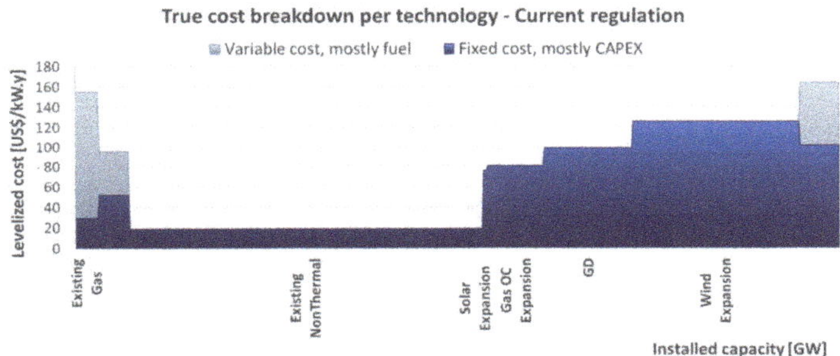

Figure 15. Cost breakdown in the current regulation scenario. Total cost (variable + fixed) equal to USD 24.2 billion.

One main impact that stands out is the greater deployment of distributed generation and a drastic reduction in large-scale solar expansion in the current regulation scenario, when compared to the efficient energy planning one. Another remarkable point is the greater expansion of natural gas power plants, both open cycle and combined cycle. These differences resulted in greater investment costs and also in more pronounced operational costs. Overall, the regulatory distortions led to a 16% higher total cost than the one obtained in the efficient energy planning scenario.

3.2. Mexico Results

The final expansion mix, both in terms of capacity and share of the total expansion, is summarized in Figure 16 for the two scenarios: efficient and current regulation.

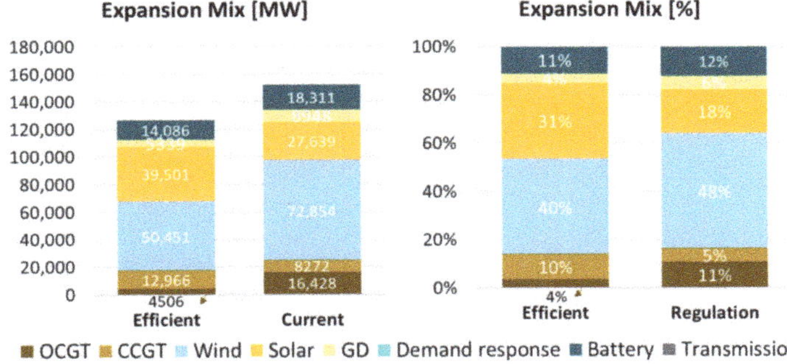

Figure 16. Comparison of total expansion between scenarios.

Notably, the current regulation scenario promotes a significantly higher insertion of wind power and natural gas, both in terms of absolute capacity and in terms of share of the expansion mix. In contrast, due to the low recognized contribution of solar for regulatory firm capacity requirements in the long-term, this technology tends to have a reduced representation in the current regulation scenario. Distributed generation is also greater in the current regulation case, largely due to the existing incentives. Battery capacity and transmission capacity are also slightly higher in the current regulation scenario.

Another interesting comparison is among regional expansions, depicted in Figure 17. It seems that both the reduction in solar capacity additions and the increase in wind capacity additions affected all three regions in a relatively uniform manner, maintaining relative proportions with the North and South being more prominent in wind power and Central concentrating most of the solar capacity. The increase in the natural gas expansion in the current regulation scenario, on the other hand, is mostly concentrated in the North, even though there is also a perceptible increase in the South. As distributed generation is proportional to the regional demand, it is mostly concentrated in the Central region.

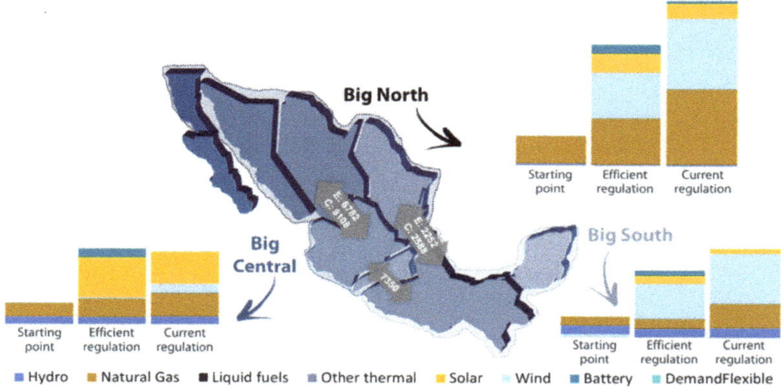

Figure 17. Comparison of the regional expansion between scenarios.

The significant high-quality potential of wind in the North and South of the country leads to a great expansion of this technology in both regions. In the Central region, the less attractive wind resources, high-quality solar potential and proximity to the main load centers contribute to a greater solar expansion. Regarding natural gas, the expansion is mostly concentrated in the North, due to the easy access to cheap North American natural gas. There is also a significant expansion of gas in the Central region, as this is the region with the greatest demand and natural gas can help form a portfolio with the large amounts of solar power built. Also noticeable is the expansion of the interconnection lines between the North and the South and Central zones.

Another interesting output from the simulation is the daily profile of generation per technology, illustrated in Figures 18 and 19. Most of the generation during the day is provided by wind and solar power plants, while at night natural gas becomes more relevant. It is also interesting that wind generation is higher in the night period—when demand is also more pronounced. Batteries play a role mostly by moving solar generation from the daytime (when supply is abundant) to the night period, reducing the need for thermal generation at night.

Subsequently, the reliability of the system is analyzed. For this analysis, the net demand—demand with renewable generation discounted—and its variability is compared with the available capacity of the system, as presented in Figure 20. The system expansion obtained in the efficient regulations scenario remains very close to 3σ in all four seasons, meaning that the system is well-balanced for delivering the desired level of reliability. In the current regulation scenario, the system expansion obtained highly surpasses the 3σ criterion in all four seasons, reaching a level very close to 6σ—indicating a substantial oversupply driven by the incentives put in place by the current regulations.

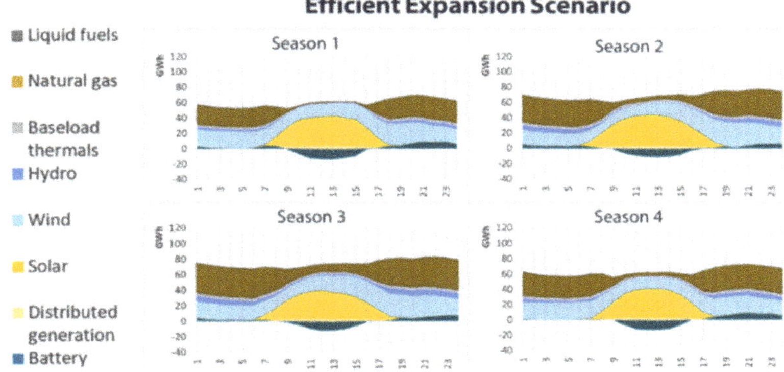

Figure 18. Generation profile per season in the efficient regulation scenario.

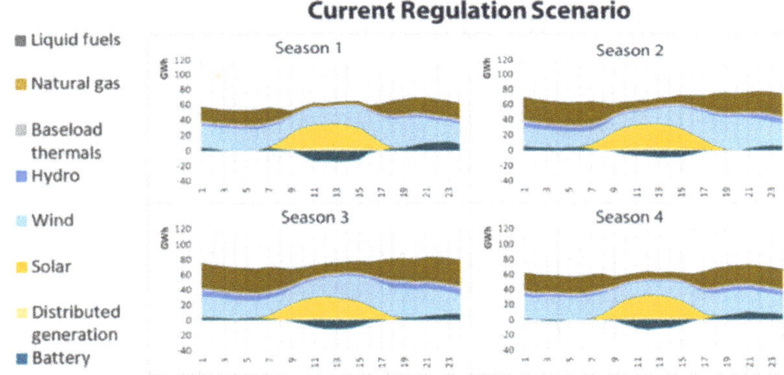

Figure 19. Generation profile per season in the current regulation scenario.

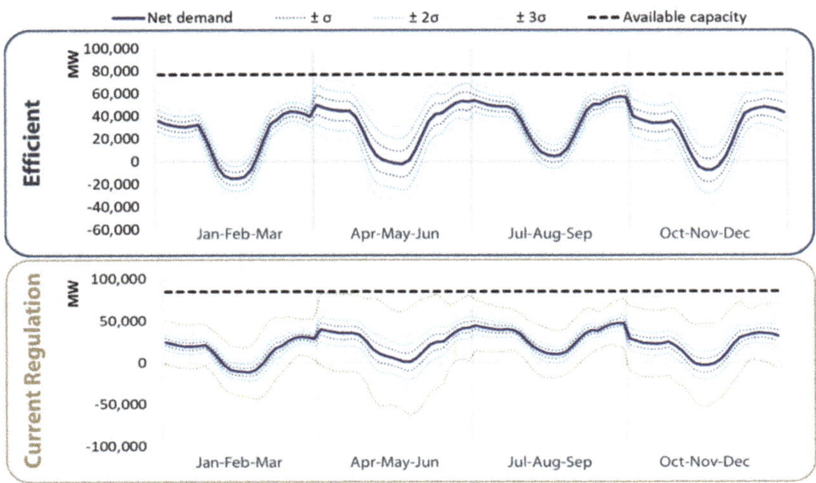

Figure 20. Comparison of the system reliability level between scenarios.

Figures 21 and 22 presented next highlight the contrast between the two scenarios in terms of total cost needed to remunerate the existing generation system for the Mexican case. The differences in expansion found, such as the higher overall amount added in the current regulation scenario, as well as its higher wind, natural gas (especially open cycle) and DG participation and lower solar generation, are translated into significantly higher total investment costs, though accompanied by a small reduction in the total operational costs largely due to the greater role of renewables. Overall, the regulatory distortions led to an 11% higher total cost when compared to the efficient energy planning case.

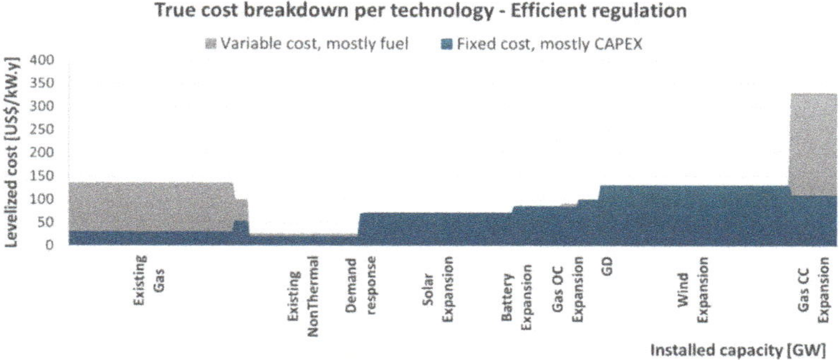

Figure 21. Cost breakdown in the efficient regulation scenario. Total cost (variable + fixed) equal to USD 22.8 billion.

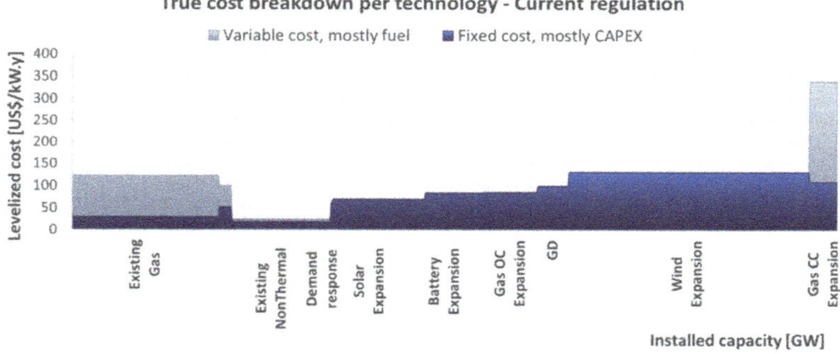

Figure 22. Cost breakdown in the current regulation scenario. Total cost (variable + fixed) equal to USD 25.3 billion.

4. Discussion

For each of the two case studies proposed (Brazil and Mexico), the authors contrasted the system expansion in a "current regulation" scenario with the expansion in an "efficient energy planning" scenario, analyzing the results of the optimization model. In the "current regulation" case, the authors have modeled the main regulations that directly impact the system expansion. In the "efficient energy planning" case, on the other hand, the authors have used a proxy for "ideal" policies with regards to reliability requirements and distributed generation, which involves ensuring *cost-reflectiveness* of all technologies. The analysis highlights how the countries' regulation impacts the systems' expansion, in addition to how the physical features of the different markets play a role in the results. Our chief conclusion is that imperfections in the regulatory incentives for system expansion led to a deadweight loss of significant magnitude, in the order of USD 3 billion: the Brazilian case study indicated a 16% surcost in the current regulation scenario compared to the efficient regulations one, and the Mexican case study indicated an 11% surcost. These differences are attributed to a combination of the three effects assessed in the present paper: a distorted representation of expansion candidates' perceived relative cost, a distorted incentive for adopters of distributed energy resources and a distorted representation of different technologies' contributions to system reliability and/or the system's reliability needs.

In Brazil, the current regulation led to a remarkably larger distributed generation expansion, caused by the numerous incentives to this source, and to substituting the utility-scale solar expansion from the efficient energy planning scenario. In contrast, wind technology increased its share in total expansion. Additionally, there was a higher share of natural gas sources in the current regulation scenario, motivated by the higher share of intermittent sources and regulatory reliability requirements. Battery capacity did not participate in any of the scenarios simulated, motivated by the large amount of existing hydropower in the system, which is already sufficient to shift demand between hours of a single day, providing a service that in other systems would be delivered by batteries. Another interesting comparison is among regional expansions: the natural gas expansion in the efficient scenario was mostly concentrated in the South, while in the current regulations one, it was spread out among the regions. These differences resulted in greater investment costs and also in more pronounced operational costs. Overall, the distortions attributed to the current Brazilian regulations led to a 16% higher total cost than the optimal scenario.

In Mexico, the current regulations scenario promoted a significantly higher insertion of wind power and natural gas, both in terms of absolute capacity and in terms of share of the expansion mix. In contrast, due to the low recognized contribution of solar for regulatory firm capacity requirements, this technology had a reduced representation in the current regulation scenario. Battery capacity was also slightly higher in the current regulation scenario. Also remarkable were the greater capacity additions of open-cycle natural gas, while combined cycle capacity was slightly reduced. These differences translated into significantly higher total investment costs, though accompanied by a small reduction in the total operational costs, largely due to the greater role of renewables. Overall, the regulatory distortions led to an 11% higher total cost when compared to the optimal regulation case.

Even though each system is different, both from a physical and a regulatory standpoint, the present work proposes a framework that makes it possible to make meaningful assessments of the role of regulation in a context of rapidly transforming electricity sectors, especially in terms of system expansion and planning. In particular, two very common features of regulatory designs in electricity markets (observed both in Mexico and in Brazil) are: (i) renewables commonly having a lower perceived cost under the current regulations, either due to some direct incentive or due to indirect access to more attractive contracts or financing conditions; and (ii) requirements for reliability are often defined more conservatively than they ought to be, overstating the difficulties imposed by renewable generation to the existing system and underestimating their ability to contribute. Together, these features tend to lead to an overcapacity situation, as the first driver seeks to purchase as many renewables as possible while the second one promotes contracting additional conventional generators to guarantee system firmness.

This impact is very noticeable in the results presented for Mexico, where the current regulation scenario led to an extremely conservative outcome in terms of reliability (as measured by the number of standard deviations between system net demand and firm capacity). In the case of Brazil, the current regulation also led to a higher total expansion, though interestingly not a more reliable system—as the higher capacity levels were counterbalanced by the higher variance of the expansion mix driven by current regulation. In both cases, although there is an incentive for the development of renewables, due to (i), there is also an incentive from the planning regulation for the expansion of thermal technologies from (ii)—as the contribution of renewables to system reliability is assumed to be very low and the regulation that defines the attributes of batteries or other storage technologies (for reliability and as a non-polluting technology) is still poorly developed. This indicates that planning regulations usually lead to a non-optimal expansion scenario, especially due to over-conservative requirements and the need for updates and revisions. Additionally, it raises an important obstacle to a net-zero scenario, which needs substantial changes in the regulation to be feasible in the analyzed systems.

Author Contributions: Conceptualization, G.C. and L.B.; methodology, G.C., A.W. and J.B.; software, J.S. and B.B.; validation, G.C. and A.W.; writing—original draft preparation, A.W. and G.C.; writing—review and editing, L.B. and A.W.; supervision, G.C.; project administration, A.W. and J.S.; funding acquisition, PSR Energy Consulting and Analytics. All authors have read and agreed to the published version of the manuscript.

Funding: This research received no external funding.

Institutional Review Board Statement: Not applicable.

Informed Consent Statement: Not applicable.

Data Availability Statement: Not applicable.

Conflicts of Interest: The authors declare no conflict of interest.

References

1. Mai, T.; Logan, J.; Blair, N.; Sullivan, P.; Bazilian, M. RE-ASSUME: A Decision Maker's Guide to Evaluating Energy Scenarios, Modeling, and Assumptions. International Energy Agency—Renewable Energy Technology Deployment. 2013. Available online: http://iea-retd.org/wp-content/uploads/2013/07/RE-ASSUME_IEA-RETD_2013.pdf (accessed on 9 August 2021).
2. Lima, D.A.; Paula, D.N.T. Free contract environment for big electricity consumer in Brazil considering correlated scenarios of energy, power demand and spot prices. *Electr. Power Syst. Res.* **2021**, *190*, 106828. [CrossRef]
3. Tolmasquim, M.T.; de Barros Correia, T.; Porto, N.A.; Kruger, W. Electricity market design and renewable energy auctions: The case of Brazil. *Energy Policy* **2021**, *158*, 112558. [CrossRef]
4. Daglish, T.; de Bragança, G.G.F.; Owen, S.; Romano, T. Pricing effects of the electricity market reform in Brazil. *Energy Econ.* **2021**, *97*, 105197. [CrossRef]
5. Soares, Í.N.; Gava, R.; de Oliveira, J.A.P. Political strategies in energy transitions: Exploring power dynamics, repertories of interest groups and wind energy pathways in Brazil. *Energy Res. Soc. Sci.* **2021**, *76*, 102076. [CrossRef]
6. IRENA. *Planning for the Renewable Future*; International Renewable Energy Agency: Abu Dhabi, United Arab Emirates, 2017.
7. Gonzalez-Romero, I.C.; Wogrin, S.; Gomez, T. Transmission and storage expansion planning under imperfect market competition: Social planner versus merchant investor. *Energy Econ.* **2021**, *103*, 105591. [CrossRef]
8. Cunha, G.; Moreira, H. An Econometric Assessment of the Technological Innovation and Diffusion Model for Distributed Generation in Brazil. In Proceedings of the IAEE International Conference, Paris, France, 4–7 July 2020.
9. Sigrin, B.; Gleason, M.; Preus, R.; Baring-Gould, I.; Margolis, R. *The Distributed Generation Market Demand Model (dGen): Documentation*; Technical Report NREL/TP-6A20-65231; National Renewable Energy Laboratory (NREL): Golden, CO, USA, 2016.
10. Sigrin, B.; Drury, E. Diffusion into New Markets: Economic Returns Required by Households to Adopt Rooftop Photovoltaics. In Proceedings of the AAAI Energy Market Prediction Symposium, Arlington, VA, USA, 15–17 November 2014.
11. Schweppe, F.C.; Caramanis, M.C.; Tabors, R.D.; Bohn, R.E. *Spot Pricing of Electricity*; Springer: New York, NY, USA, 1988.
12. Ehrenmann, A.; Smeers, Y. Risk adjusted discounted cash flows in capacity expansion models. *Math. Program. Ser. B.* **2013**, *140*, 267–293. [CrossRef]
13. Barroso, L.A.; Granville, S.; Oliveira, G.C.; Thomé, L.M.; Campodónico, N.; Latorre, M.L.; Pereira, M.V.F. Stochastic Optimization of Transmission Constrained and Large Scale Hydrothermal Systems in a Competitive Framework. In Proceedings of the IEEE General Meeting 2003, Toronto, ON, Canada, 13–17 July 2003; Volume 2.
14. Gorenstin, B.G.; Campodonico, N.M.; Costa, J.P.; Pereira, M.V.F. Power system expansion planning under uncertainty. *IEEE Trans. Power Syst.* **1993**, *8*, 129–136. [CrossRef]
15. Lazard. *Levelized Cost of Energy Analysis—Version 13.0*; Lazard: New York, NY, USA, 2019.
16. Lazard. *Levelized Cost of Storage Analysis—Version 5.0*; Lazard: New York, NY, USA, 2019.
17. IRENA. *Global Landscape of Renewable Energy Finance 2020*; International Renewable Energy Agency: Abu Dhabi, United Arab Emirates, 2020.
18. Dias, J.; Machado, G.; Soares, A.; Garcia, J. Modeling Multiscale Variable Renewable Energy and Inflow Scenarios in Very Large Regions with Nonparametric Bayesian Networks. *arXiv* **2020**, arXiv:2003.04855.
19. U.S. Energy Information Administration. *Annual Energy Outlook*; U.S. Energy Information Administration: Washington, DC, USA, 2019.
20. Empresa de Pesquisa Energética—EPE. Plano Decenal de Expansão de Energia 2029. 2019. Available online: http://www.epe.gov.br/sites-pt/publicacoes-dados-abertos/publicacoes/PublicacoesArquivos/publicacao-422/PDE%202029.pdf (accessed on 9 August 2021).
21. Empresa de Pesquisa Energética—EPE. Modelo de Mercado da Micro e Minigeração Distribuída (4MD): Metodologia—Versão PDE 2029. 2019. Available online: http://www.epe.gov.br/sites-pt/publicacoes-dados-abertos/publicacoes/PublicacoesArquivos/publicacao-423/topico-488/NT_Metodologia_4MD_PDE_2029.pdf (accessed on 9 August 2021).
22. IRENA. *Renewable Power Generation Costs in 2017*; International Renewable Energy Agency: Abu Dhabi, United Arab Emirates, 2018.
23. Mexican Secretary of Energy. Política Pública para Promover la Generación Distribuida en México. 2018. Available online: https://www.gob.mx/sener/documentos/politica-publica-para-promover-la-generacion-distribuida-en-mexico (accessed on 9 August 2021).
24. Billinton, R.; Allan, R. *Reliability Evaluation of Engineering Systems: Concepts and Techniques*, 2nd ed.; Plenum Press: New York, NY, USA, 1992.
25. Grimmett, G.; Stirzaker, D. *Probability and Random Processes*, 3rd ed.; Oxford University Press: Oxford, UK, 2001.
26. Brazil. Law N° 10.848. 2004. Available online: http://www.planalto.gov.br/ccivil_03/_Ato2004-2006/2004/Lei/L10.848.htm (accessed on 9 August 2021).

27. National Electricity Agency—ANEEL. Normative Resolution N° 514. 2012. Available online: http://www2.aneel.gov.br/cedoc/ren2012514.pdf) (accessed on 9 August 2021).
28. Ministry of Mines and Energy. Ordinance N° 66. 2014. Available online: http://www.mme.gov.br/documents/20182/d0100d7d-c743-171d-0410-c7964f179f0c (accessed on 9 August 2021).
29. Ministry of Mines and Energy. Ordinance N° 101. 2016. Available online: https://www.legisweb.com.br/legislacao/?id=317807 (accessed on 9 August 2021).
30. PSR. *Energy Report—Segurança de Suprimento: O Estrutural e o Conjuntural*; PSR: Rio de Janeiro, Brazil, 2013.
31. Mexican Secretary of Energy. Manual de Balance Potencia. 2016. Available online: https://www.cenace.gob.mx/Docs/MarcoRegulatorio/Manuales/Manual%20de%20Balance%20Potencia%20DOF%202016-09-22.pdf (accessed on 9 August 2021).
32. ONS. O Sistema Interligado Nacional. 2021. Available online: http://www.ons.org.br/paginas/sobre-o-sin/o-que-e-o-sin (accessed on 10 August 2021).

Article

Points of Consideration on Climate Adaptation of Solar Power Plants in Thailand: How Climate Change Affects Site Selection, Construction and Operation

Kampanart Silva *, Pidpong Janta and Nuwong Chollacoop

Renewable Energy and Energy Efficiency Research Team, National Energy Technology Center, National Science and Technology Development Agency, Pathum Thani 12120, Thailand; pidpong.jan@entec.or.th (P.J.); nuwong.cho@entec.or.th (N.C.)
* Correspondence: kampanart.sil@entec.or.th; Tel.: +66-87-812-0502

Abstract: Solar energy is planned to undergo large-scale deployment along with Thailand's transformation to a carbon neutral society in 2050. In the course of energy transformation planning, the issue of energy infrastructure adaptation to climate change has often been left out. This study aims to identify climate-related risks and countermeasures taken in solar power plants in Thailand using thematic analysis with self-administered observations and structured interviews in order to propose points of consideration during long-term energy planning to ensure climate adaptation capacity. The analysis pointed out that floods and storms were perceived as major climate events affecting solar power plants in Thailand, followed by lightning and fires. Several countermeasures were taken, including hard countermeasures that require extensive investment. Following policy recommendations were derived from the climate-proofing investment scenario study. Policy support in terms of enabling regulations or financial incentives is needed for implementation of climate-proofing countermeasures. Public and private sectors need to secure sufficient budget for fast recovery after severe climate incidents. Measures must be taken to facilitate selection of climate-resilient sites by improving conditions of power purchase agreement or assisting winning bidders in enhancing climate adaptability of their sites. These issues should be considered during Thailand's long-term energy planning.

Keywords: climate adaptation; adaptive capacity; solar power plants; thematic analysis; long-term energy scenarios (LTES); site selection; power purchase agreement

Citation: Silva, K.; Janta, P.; Chollacoop, N. Points of Consideration on Climate Adaptation of Solar Power Plants in Thailand: How Climate Change Affects Site Selection, Construction and Operation. *Energies* **2022**, *15*, 171. https://doi.org/10.3390/en15010171

Academic Editor: Dolf Gielen

Received: 15 November 2021
Accepted: 20 December 2021
Published: 28 December 2021

Publisher's Note: MDPI stays neutral with regard to jurisdictional claims in published maps and institutional affiliations.

1. Introduction

The 26th UN Climate Change Conference of the Parties (COP26) [1] drew the world a bit closer to net zero greenhouse gas (GHG) emission, or so-called climate neutrality [2]. From the Kyoto Protocol [3] in 1998 to the Paris Agreement [4] in 2015, voluntary emission reduction targets changed to Nationally Determined Contributions (NDCs) that should be updated every five years. COP26 was the first opportunity for countries to update their NDC, where top runners came up with ambitious emission reduction target in 2030 in order to achieve net zero emission by or before 2050 [5,6], and consequently maintain the global warming to 1.5 °C [7]. With a new financial support scheme to facilitate more proactive emission reduction in developing countries [8], a number of developing countries decided to declare a more stringent target than the NDC updated in 2020 in COP26. There was no exception for Thailand. Thailand stated in the updated NDC that it will reduce the GHG emission by 25% by 2030 with adequate access to technological, financial, and capacity building support [9]. In COP 26, apart from updating to the more ambitious emission reduction target of 40% by 2030, Thailand revealed its plan to achieve carbon neutrality by 2050 and consequently net zero emission by or before 2065 [10]. These updated national targets reflect the achievement of an important milestone of the global roadmap toward climate neutrality. At the same time, they oblige all countries to put forward enabling policies that can support GHG emission reduction. The energy sector is the sector that most countries set a strict emission reduction target on since it accounts for one-fourth of the total global emissions [11]. International and regional entities, e.g., International Renewable Energy Agency (IRENA) [12], Asia-Pacific Economic Cooperation (APEC) [13], Association of Southeast Asian Nations (ASEAN) [14], have been promoting increase in share of renewable energy in electricity and heat production portfolio of each country to replace carbon-intensive energy sources. In order to get in line with this trend, the new National Energy Plan of Thailand to be launched in 2022 includes an ambitious goal to increase the share of renewable energy in electricity and heat production to at least 50% by 2050 [15,16]. The share of renewable energy can be even higher to accommodate the updated targets in COP26.

Long-term energy scenarios (LTES) is an important tool to facilitate the achievement of the aforementioned renewable energy goals. It enables usage of scenario planning for strategic decision-making and helps account for potential transformational changes in energy sector [17]. Thanks to the updated NDCs mentioned above, renewable energy will play the most significant role ever in global energy transformation [18]. This leads to several additional considerations during the long-term energy planning. On the downside, intermittent nature of some renewable energy sources introduces a new challenge in ensuring operational flexibility in the planning of energy transition [19]. Energy storage can be one of the solutions to it, though balance between cost and flexibility should be carefully considered [20]. On the other hand, apart from mitigating the climate change impact, transition from non-renewable energy to renewable energy can contribute to energy security and job creation [21]. Particularly in developing countries, it can also accelerate energy access [22] and contribute to economic growth [23]. All these aspects have to be appropriately taken into account during long-term energy planning, and they have been done so in many economies, including Thailand [24].

One of the issues that have often been left out during energy transformation planning is the adaptation of energy infrastructure to climate change. Goal 13: Climate Actions of the UN SDGs requires the nations to *strengthen resilience and adaptive capacity* of critical infrastructure, including energy infrastructure, *to climate-related hazards and natural disasters* [25]. To date, 125 out of 154 developing countries are formulating and implementing national climate adaptation plans in which key economic sectors and services, including electricity generation, is incorporated in the highest priority areas of the plans [26]. This is because increasing intensity and frequency of climate hazards due to climate change can significantly affect the ability of the energy system to meet the electricity demand [27]. This results in additional investment not only to ensure the fulfillment of electricity demand, but also to enhance the energy infrastructure resilience [28]. Efforts have also been made to capture and monitor resilience of the energy infrastructure and its surroundings toward climate hazards during the transition toward renewable energy-intensive system [29] to ensure a smooth transformation. Therefore, a clear picture of effects from climatic events on energy infrastructure and the ways that the system can adapt to this change to make itself more resilient should be correctly illustrated in order to incorporate climate adaptation strategy into energy transformation planning.

Climate adaption has rarely been discussed under the framework of energy planning in Thailand. The latest Power Development Plan revised in late 2020 focuses on designing energy portfolio in order to ensure energy security, maintain appropriate power generation cost, and limit carbon dioxide emission [30]. On the other hand, energy infrastructure is only included as a part of the human settlements and security sector in Thailand's National Adaptation Plan (NAP). The plan promotes adoption of adaptive design in energy facilities to become climate resilient architecture, and development of emergency power production systems to ensure quality living during natural disasters [31]. However, this has not yet been adequately considered in national energy policymaking. Therefore, this study aims to trigger a dialogue between energy and climate policymakers on climate adaptation of energy infrastructure during the transition to a renewable energy-based society. The discussion on the ways to equip climate adaptive capacity to the renewable energy infrastructure to serve the ambitious GHG emission reduction target announced at COP26 could serve as a good starting point.

In a preparatory meeting for COP26 [32], several scenarios were proposed for energy sector in order to meet the target, including:

1. 50% renewable energy by 2050 and 69% electric vehicles by 2035;
2. 50% renewable energy by 2050 and 100% electric vehicles by 2035, and;
3. 75% renewable energy by 2050 and 69% electric vehicles by 2035.

All scenarios largely rely on solar energy, making its share as large as 37–56% of the total electricity and heat generation in 2050. This is due to the fact that Thailand, which is located in tropical climate zone, has higher solar radiation intensity than countries in temperate or dry climate zones [33]. In addition, the average wind speed in Thailand is relatively low which limits the suitable area for wind power plants [34], even though the global average levelized cost of electricity production by wind is equally low as solar photovoltaic [12,35]. The costs of bioenergy and waste-to-energy power plants are also too high to encourage a large-scale country-wide penetration. Hence, it seems reasonable to start the dialogue with the discussion on climate adaptation capacity of solar photovoltaic systems.

Scholars and practitioners have been using quantitative risk assessments as the basis for the consideration of climate adaptability of solar power plants and other energy infrastructures. Projections of precipitation [36–38], natural hazards' returning periods [39,40], resulting economic losses [41,42], are among the quantitative results that have been used to define the risks to solar power plants. Climate adaptation countermeasures have been derived combining these climate risks to the exposures of power plants and the vulnerabilities of those power plants toward climate risks [43–45]. Nevertheless, solar power plants in Thailand still face disruptions from climate-related events, which leads to a problem statement of whether there are other appropriate approaches to capture the climate-related risks of solar power plants in Thailand and design adaptive strategies for them to survive climate change. Since most past studies adopted a quantitative approach, this study will employ a qualitative approach. Therefore, the objective of this study is to use a qualitative approach to capture climate-related risks that are associated to solar power plants in Thailand, particularly solar farms, and the countermeasures that have been taken to adapt to the new climatic normal, in order to propose points of consideration during long-term energy planning to ensure climate adaptation capacity of the systems. The study adopted thematic analysis to identify climate-related risks and possible countermeasures, and uses triangulation of observations, interviews, and literatures to confirm the validity of the results. It then compared scenarios with and without climate-proofing investment in order to derive recommendations on the ways to contemplate climate adaptation of solar power plants during the long-term energy planning.

2. Methodology

This study adopted qualitative methods, namely structured in-depth interviews and self-administered observation. While a number of past studies [46–50] applied quantitative methodology, for example, cost-benefit analysis (CBA), to assess economic feasibility, resulting in net present value (NPV), benefit-cost ratio (B/C ratio) or return period, few studies attempted using qualitative methodology to examine perspectives and knowledge grounded in human experiences [51]. This kind of methodology offers a perspective contrary to quantitative methodology to realize insights on why people interact or engage in a particular action [52]. Human insights emerge and can be grasped when qualitative methodology applies, and aspect and engagement experience are derived from targeted persons. Thematic analysis is considered easy to grasp. It highlights key issues, and creates reflective insights from large dataset through flexible, yet well-structured approach with ability to adapt to different studies, proving richness in description [51]. Therefore, thematic analysis can be seen complementing CBA, giving perspectives from primary data gathered from person in charge. As two qualitative methods were employed to collect data, it is important to conduct the analysis in rigorous and methodical manners to obtain robust, constructive, and useful findings [53].

2.1. Data Collection

To capture climate-related risks and countermeasures, data collection involves two phases as shown in Figure 1. Before proceeding to the first phase, targeted solar power plants for the self-administered observation were selected out of the Thailand's Energy Regulatory Commission (ERC) database which cover 564 solar power plants (as of August 2021). The first phase aimed to gain the bird-eye view of solar power plants' climate adaptive capacity across all four regions in Thailand where 35 sites were selected.

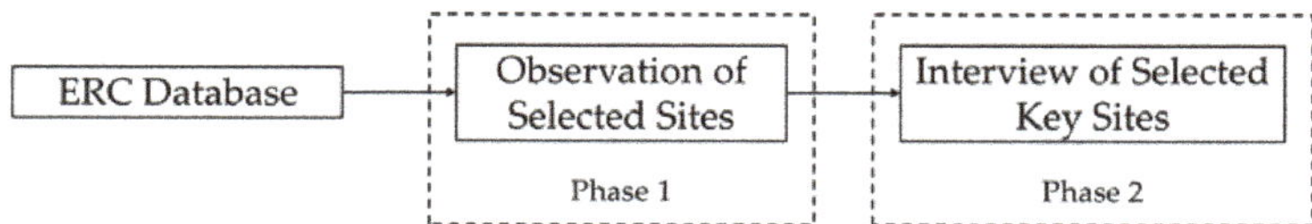

Figure 1. Data collection flow.

The criteria to select solar power plant sites involved verbal or written consent from the enterprise and COVID-19 regulatory approval to enter the site. Summary of the attributes of selected sites are shown in Table 1. This first phase was carried out through free discussion and field observations without interview structure and observation form, but only notes of synopsis of relevant climate issues. Table 2 shows observed items that could influence climate adaptability of the targeted solar power plants which were categorized into power plant-related, climate-related, and others. The notes of each item are used as inputs for the construction of the interview structure in the second phase.

Table 1. Selected solar power plants for self-administered observation.

No.	Region	No. of Plant	Range of Contracted Capacity (MW)
1	Northern	6	0.02–90 MW
2	Northeastern	13	1.02–8 MW
3	Central	8	0.05–2 MW
4	South	8	2–8 MW

The findings of free discussion and field observation could vary, ranging from climate issues/risks, vulnerabilities, responses, to future operation plans. The second phase involved structured in-depth interviews and self-administered observation in which the structure referred to the findings from the first phase's synopsis. The structured interviews shown in Table 3 were designed based on not only the synopsis from the first phase but also literatures on climate risks in solar power plants for the second phase. The interviews were planned and aimed only for key sites.

Table 4 illustrates key solar power plant sites and their characteristics. Each key site is established and located in different regions in Thailand with variation in temperature, season, precipitation, climate-related risks, and frequency of climatic events. Climate adaptation countermeasures applied to each key site could be diverse. Essentially, selected commercial sites are taken responsibility by experienced operators who are accountable for several sites under respective enterprises. The results from in-depth interviews are expected to be fruitful and diverse in detail. As a result, key targeted solar power plant sites comprised of a demonstration site (state-owned) in northern region and three commercial sites in northeastern, central, and south regions.

Table 2. Observed items from solar power plant self-administered observations.

No.	Category	Item
1	Power plant-related	Monitoring system Control room Inverter (brand/capacity) Junction box Photovoltaic panel (brand/capacity) Damaged or aged component (e.g., PV, mounting structure) Grid-connected system
2	Climate-related	Climate alert/alarm Consequences of climate events (if any) Pre-disaster adaptive countermeasure Post-disaster adaptive countermeasure
3	Others	Surroundings (farm/field/accessible road) Canal (if any) Water pond (if any) Lightning conductor (if any) Area under panel (waterlogged/weeds/dried soil/cement) Terrain (slope/flat)

Table 3. Areas, sub-areas and supportive references for structured interviews.

No.	Area	Sub-Area	Item	Reference
1	General information	Power plant	1. Location 2. Capacity	[54–57]
2	Establishment	Condition	1. Solar radiation 2. Location selection 3. Cost-effectiveness 4. Geographical condition 5. Weather condition	[54,56–59]
3	Risk	Natural Risk	1. Flood 2. Lightning strike 3. Tropical storm 4. Heavy rainfall 5. Forest fire 6. Extreme heat 7. Cloud cover	[54,60,61]
		Human-made risk	1. Chemical use 2. Celebration firework 3. Animal hunt 4. Lawn mowing accidents	[54,62]
4	Damage	Natural incident consequence Human-made incident consequence	1. Property 2. Economic loss 3. Opportunity loss 4. Others	[55,56,63]
5	Strategies and actions towards climate events	Implemented strategies and actions Unimplemented strategies and actions Unknown strategies and actions		[61,64,65]

Table 3. *Cont.*

No.	Area	Sub-Area	Item	Reference
6	Climate adaptation countermeasures			[63,64,66–75]

Table 4. Key solar power plants for structured interviews.

Key Solar Power Plant	Characteristics
Key site 1 (KS1)	Demonstration plant (state-owned), northern region
Key site 2 (KS2)	Commercial plant, northeast region
Key site 3 (KS3)	Commercial plant, central region
Key site 4 (KS4)	Commercial plant, south region

2.2. Data Analysis

As stated above, this study employed thematic analysis [76] to analyze and convert raw data into useful and applicable results. Thematic analysis is beyond simply summarizing obtained data, but interprets into decent themes [77]. This study adopted the thematic analysis proposed by Braun and Clarke [76] comprises of six procedures: reviewing raw data, creating initial codes, categorizing codes into themes, polishing theme, defining themes, and reporting results. To conduct such analysis, consistency of procedures should be systematically fostered from phase to phase [78]. To ensure reliability of data obtained from interviews and observation, trustworthiness criteria created by Lincoln and Guba [79] involving creditability, transferability, dependability, confirmability, and audit trails, was adopted and applied to each phase of the analysis. The criteria are known as essential qualitative instruments to enhance the worthy of research finding, being able to grab researchers and readers attention [80].

2.3. Confirming Validity of Study

By nature, validity in analysis process and findings of a qualitative study is always questioned. Data triangulation was employed ensure data accuracy, reflecting actual context with adequate creditability [81]. The method is generally used to help qualitative researchers establish validity. The method involves integrating sources of data/information to create strong foundation. In this study, structured interview form was designed based on the synopsis of climate-related issues from previous free discussion and with support of literature studies in order to ensure that in-depth interviews are conducted meticulously and cover essential points of discussion. As can be seen in Table 3, each interview item derived from the first phase and firmly supported by literature. They were categorized into five areas and nine sub-areas. Additionally, climate adaption countermeasures were included to actively interact with interviewees and efficiently carry out the interviews. Data triangulation in Figure 2 demonstrates a mixture of self-administered observation, thematic analysis-derived insights of in-depth interviews, and literature, that was used to produce research findings in this study.

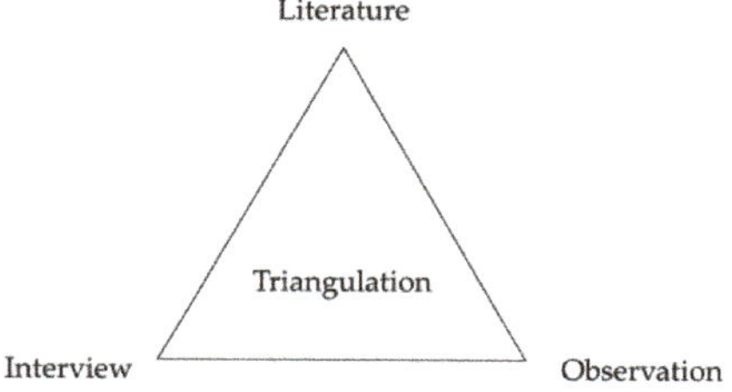

Figure 2. Data triangulation employed in the study.

3. Results

To realize climate-related risks in solar power plants and how the risks are managed, self-administered observation on 35 solar power plants were carried out which aimed to investigate operation strategies on climate-related risks. The 35 solar power plants belong to seven well-known solar power enterprises that make up the majority of solar power capacity in Thailand, as well as three small enterprises and one university-owned power plant. Synopsis of the discussion revealed that site selection is an important issue in solar power plant investment. Plant operators indicated that there had been several essential factors to determine site locations, for example, solar radiation, weather and geographical conditions, vulnerabilities to natural climate events, and availability of power distribution grid. Among the factors, solar radiation is the most important criterion to be taken into account [82]. Amount of solar energy should meet minimum standard [83]. In terms of economic aspect, solar power plants are encouraged to be located near power grid and substations to minimize investment cost [49], and to make it convenient for maintenance [84]. On the other hand, plant operators claimed that environmental factors such as weather and geographical conditions, and vulnerabilities of natural climate events are among critical factors which aligned well with literature [34,35,46,82,83]. Additionally, studies indicated that solar power plants should not be located in forest, agricultural area, water pathways, dams, and flood areas [54,85]. This is not only to minimize of environmental impact, but also to avoid natural climate events [86], especially flood and storm which are the major climate events in Thailand [87].

The self-administered observation and open informal conversation in the first phase were insufficient to reveal the true rationale of climate adaptability of solar power plants. The study was furthered to examine site selection criteria that contribute to prevention and mitigation of climate events. In-depth interviews with four key sites and further field observations were conducted, and thematic analysis were employed and systematically applied to all transcribed data obtained from the interviews and observations. Data consisted of a 93-page-transcript from 3-h and 12-min-long interviews. The results from thematic analysis are presented in Table 5. Two key theme clusters: *Climate-related risks* and *Location selection* were derived from the process.

3.1. Climate-Related Risks

As shown in Table 6, flood and storm, forest fire, and lightning were climate-related risks indicated from four key solar power plant sites. All interviewees mentioned risks associated to human activities which are sometimes linked to climate risks; hence, human-made incidents were also included in the table.

Table 5. Result of thematic analysis.

Themes	Sub-Themes	Nodes
Baseline	Efficiency	Capacity factor Efficiency Peak time
	Maintenance	Maintenance Panel cleaning
	Weather	Solar radiation Weather condition
Countermeasure	Corporate social responsibility	Corporate social responsibility Local employment
	Firebreak	Firebreak Wet forest
	Flood-related actions	Canal cleaning Dike construction

Table 5. *Cont.*

Themes	Sub-Themes	Nodes
Causes	Flood and storm	Dam break Flooding Tree falls Tropical storm
	Fire	Forest fire
	Human-made	Car accident Damaged panel Human-made incidents Lawn mowing accident Public resistance
	Lightning	Lightning strike
Damage	Flood	Damage of flood Damaged cables Flood level Flooded area Water drainage Water level Waterlogged
	Human-made	Damage of human-made incidents Damaged panel Trip record
	Lightning	Damage of lightning strike
Result	Result	Expected results of unimplemented actions Results of implemented actions Success in implementation
Location	Condition	Geographical condition Land title deeds Ownership Site area Site location Solar radiation
	Contract	Feeder Power purchase agreement
	Justification of location	Justification of location
	Preparation	Preparation of land

Table 6. Word counts of climate and human-made risks in the scripts of in-depth interviews.

No.	Risk	Word Counts				Total Counts
		KS1	KS2	KS3	KS4	
1	Flood	1	19	6	9	35
2	Storm	9	3	2	0	14
3	Forest fire	3	0	2	1	6
4	Lightning	0	5	5	1	11
5	Human-made	5	9	3	4	21

3.1.1. Flood and Storm

The analysis indicates that key sites experienced climatic events such as floods, generally due to extreme rainfall and storm. Since intensity of extreme climate event can potentially affect plant operation [88], KS2 calmly stated "The 2011 Thailand Flood was the biggest ever flood we have ever experienced." This is one of the technological or human-made hazards rooted in increasing intensity of natural disasters, so-called natural hazards triggering technological disaster (NATECH) [89]. It is evident that extreme precipitation drove the flood which damaged the power plant [90]. The same informant further revealed the causes of the flood by saying "The cause of flood was exactly due to the low elevation of the plant." and "The reservoir (Ban Morasuap reservoir) was cracked and holed. It could not bear with enormous water body from heavy rain. I remembered it was a raging storm." According to literature, flat terrain is favorable for solar power plants [84]. For sites with slope or low elevation, there is a possibility of flood [86]. Awareness on necessity for land leveling and grading should be raised as it is suggested that development of solar power plant should be taking place outside flood risk area [86]. Not every enterprise realizes its importance, but KS3 described "Our executives and board members prioritized land leveling and grading; therefore, we raised the land level by 50 cm." In the solar power plant located on high slope, extreme rainfall and tropical storm still can harm plant operation even after leveling the land, since heavy rain not only cause flood, but also waterlogged and landslides. To prevent the consequences, literature suggested agrivoltaic as a solution [55]. KS1 told "Water erosion made no damage to the panel nor mounting structure, but left holes on ground." KS1 added "Wind and storm as well. They didn't cause damages to our solar farm, but to the (electricity) authority's transmission lines. This affected the generation output." These informant's insights can be considered robust and comprehensive as they are supported by academic literature [91]. Heavy rain and tropical storms can lead to extensive flooding [43], bringing about severe damages to energy infrastructure and subsequently plant operation disruption [44]. KS2 stated "The panels and inverters were submerged and caused disruption of a two-day long and it took us two days to drain." and added "Due to the event, we stopped our system for a month before full recovery." Additionally, KS4 responded to the assumptions "There were windstorms occurred and trees fell upon transmission lines. This happened to another site under our responsibility." and "That tree falls caused the trip of the transmission line." Although some power plants were well-prepared, extreme rain fall incurs waterlogged [44]. KS3 stated "We were not able to drain out water since the rain was too heavy." In some cases, energy infrastructure and system components were not designed to bear with instantaneous and simultaneous floods [44]. Power generation systems could be disrupted. KS2 indicated "30 boxes (array junction box) were severely damaged, and it costed us approximately 3000 US dollars." and added "Over 15,480 PV panels were submerged. Unfortunately, only 4000 panels can be covered by insurance." Consequences from flooding was not only to the energy infrastructure but also transportation facilities such as road to power plants, which suspended plant operation and maintenance activities. KS4 added "During the season (rain), water overflowed the road (affecting the accessibility of the plant)."

To minimize consequences from floods, solar power plants implemented several climate adaptation countermeasures. Many means, concepts and guidelines were studied and developed covering a number of countermeasures, for example, project site selection [65], disaster risk management application, water level measurement [62], improvement of power system specifications, implementation of emergency response, construction of dikes, embankments, dams and reservoirs [61], land leveling [85], installation of concrete structure with bar for module mounting [85], installation of water-proofing equipment, upgrade of transmission sub-stations [62]. From the interviews, informants pledged "We do have budget for land grading." "It (budget) was included in the capital investment. (KS1)", KS1 added "The wet forest consisting of vetiver grass will be planted on slope ground. It will help prevent landslide." and added "Soil compaction was done roughly. When it rained, the ground decayed and mounting structure was eroded." KS2 stated "In the past, there were embankments around the site." then "After the flood event, we decided to increase the level (of the embankment).". KS2 added "We also increase the mounting structure level up to 3 m and 30 array junction boxes was raised up to 1.2–1.5 m over the ground which took considerable period of time and caused additional investment." In the third and fourth key sites, climate adaptation countermeasures were adopted as well. KS3 stated "We built up dikes around plant border and drilled a well for water drainage." and KS4 "We elevated panel mounting structures over the water level." then "With the height of structure, we decided to attach the wires to the upper part of the structure which resulted in approximately 20% additional expenses, but the cost was acceptable." Given the attempts to avoid and/or minimize flood risks and consequences, additional expenses were required and should be made available to put the strategies in place [62]. Since these strategies are costly, they should be as practicable as possible without sacrificing the economic feasibility [92].

3.1.2. Fire

Due to climate change, extreme weather events caused severity frequency of the event is expected to increase [43], causing damage to solar infrastructure [93]. The closer solar power plant is located to dried forest, the higher the risk that the plant could be affected by forest fire. Informant reported that power plant sites were located nearby dried forest and in the area where locals light fires in forests for several purposes, for example, to make a living and animal hunts (accelerating germination of mushrooms, bamboo shoots and forcing animals from hideouts). KS1 reported "Fires can be easily observed in the area." However, KS1 declared no effects from forest fire since the fires were small and far from the site. Additionally, KS2 informed "Locals light fire from time to time." Based on finding from key sites, there was no fire-related hazards reported to create any damage and losses. Nevertheless, key sites were well-prepared and applied forest fire countermeasures to tackle with the hazard. By creating firebreak and surrounding the solar power plant with wet forests, KS1 believed that the strategies are practicable against forest fire. "The wet forest was costly. The number is about 4600 to 6100 US dollars for a 430-m long." However, this kind of risk is considered difficult to quantify, but can be avoided by consideration of plant's location at the first place [61].

3.1.3. Lightning

Depending on its location, literature indicated that energy infrastructure can be exposed to lightning, sometimes called atmospheric over-voltages, which is one of harmful natural events [45]. Lightning strikes can critically damage significant electric component of energy systems, leading to short-time power cut and disruption in plant operation [90] as well as failure in monitoring system. Informants reported unrecognized lightning strikes causing damages on the solar panels. For years that key sites under KS2 responsibility have experienced panel damage caused by this kind of natural events. KS2 reported "It (lighting) stroke in another branch of ours but didn't make serious damage." KS4 added "Since it (the site) is an open field, there were a few strikes yearly." then losses from the incidents were reported "To date, the damaged panels (from lightning strikes) were counted up to 100." by KS1. On the other hand, the informant's experience with lightning argued against literature that indicated that all devices in the power plant could be damaged by lightning strikes. KS2 claimed "Other equipment were damaged (by lightning strikes), but not PV panels." However, another key site faced lightning strikes on panels, inverters, and the monitoring system. KS3 pledged "The cause was mainly from surge issues, resulting in damage in monitoring system in damage. Direct hit from lightning may affect inverters and PV panels." KS3 added that lightning strikes had significant impact on inverter. "Another site under our responsibility were also damaged due to the issue (lightning). 30 of 600 inverters were severely damaged." This is strong evidence that support the importance of deployment of lightning protection system (LPS) which can prevent insulation breakdown [88]. However, inadequate protection against lightning may delay return of investment on power generation system [88]. Key sites indicated strategic ways to reduce the risk. KS2 stated "grounding and earthing were implemented in all structures to protect lightning." KS4 supported "We attached lighting conductors to the top of the mounting structure and the utility pole."

3.1.4. Human-Made Incidents

From the discussion with informants over several topics, it was found that human-made actions are one of the crucial issues that should never be overlooked. Accidents derived from plant operation activities can damage critical parts of energy infrastructure [94]. In key sites, several human-derived issues were reported, and consequences were found. Starting from KS1, it was found that soil issue such as inadequate soil compaction is a reason that landslide and waterlogged occurred, as KS1 confirmed "Soil compaction was done roughly. When it rained, the ground decayed, and mounting structure was eroded." Additionally, photovoltaic panel with underconditioned quality control causes hotspots on the panel surface. These defects are technological failures that reduces output power [95]. This is not just a technical issue, but an inappropriate utilization of tools for periodic maintenance which can result in opportunity loss of power generation. Three key sites indicated that several panels were cracked due to lawn mowing activities. KS1, KS2, and KS3 reported "Lawn mowing was an issue" "At least 6–7 PV panels were damaged by the pebbles" "We were using razor blades. After that, we changed to trimmer lines." respectively. Changing to trimmer lines seems to be a good countermeasure to prevent pebbles from bouncing to panels. KS2 stated "For such activity, we needed to use the (trimmer) lines rather than blade." In addition, "We used machines in bigger area and put the cover on." Later, KS1 reported that cracked of flawed panels are not only caused by lawn mowing activities, but also occasion celebration events and animal hunts. Since solar power plants are often set up at location where local communities surround, fireworks, skyrocket, bullets from local activities such as New Year's Eve celebration (shooting at sky), and animal hunts spotted cracked panels. These incidents harm plant operating. KS3 confirmed "We didn't receive damages from such activities. However, there were skyrocket cases in other sites under our responsibility, especially in northeast." To assure safety, key sites approached communities and requested for awareness of safety and re-direction of fireworks, skyrocket, and bullets. The requests for cooperation were carried out and ended up successfully throughout community participation.

3.2. Location Selection

Thematic analysis revealed location selection as another important issue. There are three themes under this cluster, including site selection conditions, land preparation, and justification of location were highlighted, as illustrated in Table 5.

3.2.1. Site Selection Conditions

Site selection is indicated by many studies [54,82–86] as a crucial procedure for the development of a solar power plant. Several necessary criteria were identified and encouraged to be taken into account, for example, solar radiation, proximity to transmission line [96], weather pattern, land slope [92], land use, and flood-sensitive area [82]. Based on the analysis of in-depth interviews, it was indicated that geographical conditions were primary criteria in site selection to minimize high capital investment. It was reported that preferred conditions were flat terrain, proximity to transport infrastructure (road), flood-avoided areas, large areas with land deeds. KS1 reported that flat terrain prevents shading effect from adjacent panels. "Our site was located on flat terrain for the reason to prevent the shading (KS1)." KS2 pledged it is important to find sufficiently large area to accommodate the contracted capacity. For 1 MW, the site should be a minimum of 4 acres. It was indicated "This site comprised of 3 branches, and the (land) size is 80 acres." and "this branch was 8 MW, 30 acres." KS3 claimed that solar power plant site should prioritize access to transport infrastructure such as road which facilitates operation. The informant stated that "Our solar sites was next to major roads, since executives gave the priority to it." In addition, key site under KS4 responsibility aim for flood-avoided areas, since the available feeder were in flood-sensitive area. The informant claimed that "We took over most solar (power plant) sites from other enterprises." "We only picked one where flood cannot reach." On the other hand, "There were also other areas, where we can develop solar power plant, but it was difficult to reach out, since it was quite far away from roads and was essentially flooded area."

3.2.2. Land Preparation

Since land slope can affect solar radiation and initial project costs [82], land preparation should be taken into account when deploying solar power plants. IRENA indicated absence of consensus regarding accepted percentage of slope [97]. KS1 indicated suitable area for solar requires land preparation. "We got the land prepared and graded. Trees in the site needed to be cut and removed (KS1)." and "we needed to flatten the surface of land as much as possible to make it suitable for installation of PV panels." This evidence aligned with literature's suggestion that steep slope of land is not preferable due to shadow projection [84]. Therefore, there is a need to facilitate land leveling, resulting in an increase in project cost. It was suggested that solar power plant sites should be installed on barren land, refraining agricultural or fertile areas [82]. However, with the land size requirements, land preparation must be conducted. "Sugar palm trees were cut and removed, and farm fields were backfilled with soil", KS2 spoke. In some cases, enterprises studied and experienced flood events in solar power plant sites. Land leveling was carried out even if the flood risk is small. KS3 informed that the strategy aimed to encounter unexpected events such as a flash flood. Ever since the launch of the site, it found no risks towards incidents from water. KS3 stated "Plan for land leveling was prepared since the initial stage." and "We needed to elevate it by 50 cm because it helped reduce the (flood) risks." KS4 was backfilled, but not leveled due to unaffordable investment and unworthiness within the time frame of contract license. KS4 informed "The site was farmland. We did grade but did not elevate due to high cost."

3.3. Justification of Location

All informants reported different reasons on justification of their site location. KS1 claimed that total available area required for development of a solar power plant was large. The selected location was less steep and require smaller land preparation budget compared to other areas. KS1 informed "This area is the most suitable land for grading. Other areas are high and with steep slope. Indeed, land preparation would be more costly." In case of KS2, the justification of site location mainly involved availability of the transmission line and the sub-station. The selected area is able to install adequate panels and other equipment according to the power purchase agreement and contract license, as well as the voltage level is acceptable when compared to other large available lands. KS2 claimed "Transmission lines were available and able to support 24 MW output to sub-station." KS3 reported in the same direction to KS2 where power purchase agreement and contract license played a critical role in site location. With limited choices of areas, KS3 reported difficulty in location search. KS3 described "Power purchase agreement would indicate location where we can invest on solar power plants." and "due to that (agreement) we started to seek for a suitable land, and some were recommended by agencies." Lastly, KS4's solar power plant prioritized flood sensitivity and solar radiation during the location selection. Still, available area is limited by the conditions in the contract. KS4 explained "Since the agreement determined locations in the city, our site here was the most suitable to set up a power plant. Flood sensitivity and solar radiation were indicators for site selection."

4. Discussion

4.1. Key Insights from Thematic Analysis

4.1.1. Climate Risks and Countermeasures

The thematic analysis above revealed climate risks associated with solar power plants, particularly flood and storm risk, fire risk, and lightning risk, and the countermeasures that have been taken to prevent or alleviate the risks, as shown in Figure 3. To prioritize the climate-related risks of solar power plants based on the analysis, word frequencies of each risk are summarized in Table 6. It could be observed that "flood" is the only natural hazard that appears in all interviews and has the highest counts among the keywords, followed by storms, lightning, and fires, respectively. This aligns well with past quantitative studies that indicated that floods and storms are the major natural hazards affecting countries around the world [42], including Thailand [41], and they are the main reasons to power outages [98]. Thailand is listed among countries where power generation has the highest exposure to natural hazards causing high annual damage and generation losses [42]. It seems natural that the power plants are concerned about flooding since the 2011 Thailand Floods made Thailand the only middle-income country that is listed in the top 10 countries most affected by natural disaster from 2000 to 2019 [99]. Precipitation projections also indicated that the intensity of daily rainfall events will be higher in the future, in Bangkok Metropolitan [36], in Thailand [37], and in Asia [38], which leads to more frequent floods with potentially longer duration [100]. To make things worse, climate change will shorten the return period of a flood similar to the 2011 Thailand Floods [40] of which the return period is expected to be 10–20 years [39]. These quantitative flood and storm risks back up the fact that all interviewees recognize these risks and make considerable efforts to address them.

Based on the thematic analysis results, consequences to the power plants vary with size, location and existing countermeasures, starting from submersion of panels, inverters, and/or other apparatus [44], to soil erosion at the basement of the structure of the solar panels. Storms can also topple the trees or rip the branches near the transmission line [101], causing the outage of the electricity supply system which stops the power plant's sales of electricity. Some owners build dikes or fill the land to avoid being flooded, some decide to leave with floods by increasing the height of the solar panel structure by 1–2 m. Soil compaction and agrivoltaic are the two available options to address the issue of soil erosion at the basement of the structure by helping with the soil retention [102]. Lightning which usually comes with storm is also another concern of the power stations. It may lead to insulation breakdown, grounding potential rise, and panel and/or inverter destruction [45]. Lightning conductors are thus installed at many power plants to minimize the damage from lightning. Though there have not been any fire events in all four interviewed key sites, three of them possess or have considered installation of firebreaks since there are lit fires during dry season within the proximity of the power plants. Even though the interview was centered around climate hazards, it can be seen from Table 6 that most interviewees touched upon human-made incidents, e.g., fireworks, skyrockets, and bullets. Most sites successfully mitigate consequences from these events by building good relationships with surrounding communities through corporate social responsibility (CSR) activities or employment of locals.

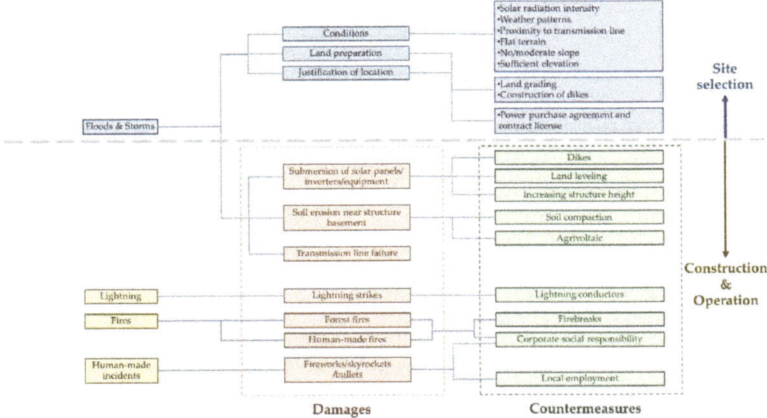

Figure 3. Key insights from thematic analysis on climate risks, countermeasures, and points of consideration during location selection.

4.1.2. Relationship between Climate Risks and Site Selection

The other important part of Figure 3 covers the relationship between climate risks and site selection. The results from the thematic analysis indicated that climate adaptability of solar power plants also depend on the site selection which is the phase before the construction of the plants. Apart from solar radiation intensity [59], average temperature [103], weather patterns, and proximity to transmission line [104], which are the main points of consideration for most power plants, there are several more aspects based on past studies that limit the possible sites for solar power plants, especially large-scale solar farms. Considering economic factors such as land cost and construction cost is unavoidable [105], land with slopes which is generally unused and cost saving is preferable at first sight [106]. However, land grading and construction can significantly increase capital investment. Hence, construction of large-scale solar power plant often required vast flat land with low slope, since high slopes make it difficult for logistics, constriction, and the right angle of panels [107]. Moreover, land surface temperature was indicated as important since it prevents energy loss derived from heat under PV panels [59]. It is sometimes argued that climatic conditions are crucial and can affect the power plant performance [83]. These conditions for site selection perfectly resonate with the insights from the thematic analysis. While others focused on climatic, social and economic factors for site selection, a few studies suggested strategic application of climate-proofing countermeasures to address natural risks, e.g., embankments, dikes, and reservoirs. However, these countermeasures significantly increase the capital investment. Some power plant owners may end up selecting climate-vulnerable sites and taking the risk of being flooded during severe flood events without implementing any countermeasures. Another important element that the enterprise use to justify the site selection is the conditions of power purchasing in Thailand. Solar power plants can only be installed where the feeders are available. Auction is used to choose the companies that propose the lowest selling prices and give them the right to build the power plants. This scheme prevents enterprises from investing on site selection before being awarded. Furthermore, the enterprises should plan carefully to start operating before the scheduled commercial operation date (SCOD) in order not to be penalized. This significantly shortens the period of consideration for site selection process.

4.2. Scenario Development for Consideration of Climate Adaptation

Figure 3 shows four different scenarios generally considered during the discussion on feasibility of climate-proofing countermeasures [108]. They are basically the combination of scenarios with/without climate change and with/without climate proofing countermeasure(s). Climate change is expected to increase the frequency and the intensity of climate events, e.g., high precipitation, storms, drought [42]. In case of solar power plants, this can potentially disrupt the electricity production or cause damages to equipment. Most climate-proofing countermeasures, whether hard or soft, result in additional expenses. Most assessment has been focusing on how to ensure economic feasibility of the countermeasures [108–110], and how to facilitate the climate-proofing investment [111]. However, as can be seen from the results from thematic analysis, this study focuses more on the details of the actual effects from climate change on solar power plants and the actual or planned countermeasures in order to derive policy recommendation to accelerate climate-proofing investment.

Scenarios without climate change normally serves as a baseline case for economic analysis. However, the Intergovernmental Panel on Climate Change (IPCC) claimed with high confidence that climate change is real [7]; hence, there is no need to consider the case where climate change does not happen. As discussed above, climate change will increase the precipitation intensity [36–38], increase the frequency and the severity [100], and shorten the returning period of floods and storms [40]. This study will focus on the right-hand side of Figure 4 which compare the cases that climate change happens with and without climate-proofing investment. For the scenario without climate-proofing countermeasures, the capital investment will not be different from current situation. Enterprises would still focus on solar radiation intensity, average temperature, weather patterns, and proximity to transmission line during site selection. They would spend extensive money on site preparation and climate-proofing countermeasures during power plant construction and operation, only if one of the climate events severely affected one of their power plants in the past, or other facilities in the proximity. Climate change will result in power plants being more frequently affected by climate incidents, especially floods and storms, and consequently larger total expenditure throughout the lifecycle in most power plants [42]. When the solar energy share gradually increases to meet the NDC, the number of power plants being out of order during and after disasters will escalate. This poses a great impact on energy security of the country [112]. Consequently, government will need to spare additional budget for disaster relief and for recovery of solar power plants. On the other hand, if all climate-proofing countermeasures are implemented, the capital investment will be very high, and the cost to operate and maintain the power plants will also be significantly higher than the former scenario. However, as most cost-benefit analyses of investment on climate proofing show that the benefit-cost ratio of most countermeasures is larger than 1 [42,113], the aforementioned investment should be much smaller than their total benefits in reducing expenses on climate-derived damages of affected stakeholders. The main issue is that the paying parties and the benefitting parties may not be the same.

Both extreme scenarios will never happen in the future. In reality, government policies supporting climate adaptation of energy infrastructure and those unintentionally discouraging climate-proofing investment, as well as the responses of private sector to those policies will create a spectrum of possible futures between the two scenarios. From the triangulation of observations, interviews, and literatures, it was found that the statements of interviewees align well with the observations by the interviewers and the findings from past literatures and other quantitative studies. Effects from climate incidents, especially floods and storms, on solar power plants are increasing due to climate change. In addition, selection of suitable location will be even more difficult when the number of solar power plants rises, which means climate adaptation strategies will play a more significant role in the future. Therefore, policies in the domain of solar energy should be designed in order to move closer to the scenario with climate-proofing measures.

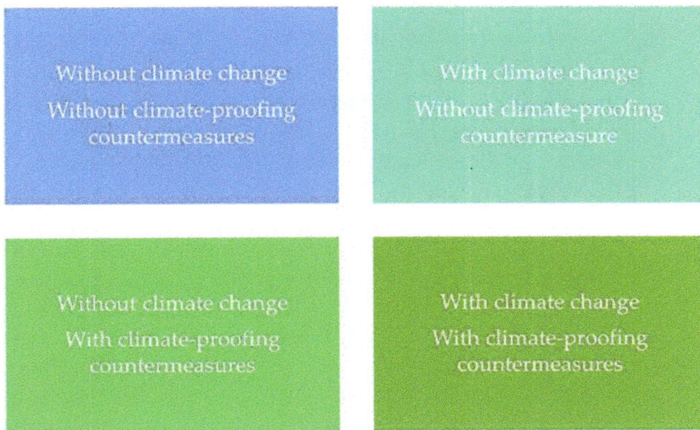

Figure 4. General scenarios for discussion on climate adaptation.

4.3. Policy Recommendation for Facilitation of Solar Energy Large-Scale Deployment in Thailand

4.3.1. Support for Implementation of Climate-Proofing Countermeasures

It is evident from the previous section that climate risks identified by past quantitative risk assessments exist, and climate proofing countermeasures are needed to increase the climate adaptive capacity of the solar power plants. It is commonly known that policy support from national and local governments are inevitable to the promotion of these climate-proofing countermeasures [114]. Policy support can be in terms of legislation or regulations that ease the implementation of the countermeasures [115], or financial incentives [116]. For example, the government can set a rule that the elevation of the solar power plant sites should be at least at the same level as the adjacent roads. Financial incentives can be in the forms of subsidy, tax exemption or no interest loan for a large-scale investment on climate adaptation, e.g., land grading, firebreak construction. Financial disincentives, such as penalties for solar power plants that cannot recover within the specified period after a climate event. If these kinds of policy support exist, KS4 would have undergone countermeasures to prevent waterlogged during rainy season. Since mobilizing finance is one of the goals of COP26 [1], international financial support for climate-proofing investment projects can also be expected to backup financial incentivization of the Thai government [117,118].

4.3.2. Preparation for Fast Recovery after a Nationwide Disaster

Aforementioned policy support would be sufficient for furnishing solar power plants with climate-proofing countermeasures in order to adapt to climate change during the first few years of the long-term energy plan. However, the frequency and the severity of climate events will eventually increase with time, and the number of solar power plants will significantly grow according to the new National Energy Plan [16]. If the government is not able to secure extensive budget to support climate adaptive capacity building of the solar power plants, future sites that meet climate resilience requirements would become progressively limited. More and more power plants will fail to withstand climate incidents, and a number of power plants could be out of order simultaneously during nationwide disasters. National energy security would be disrupted during such disasters. Enterprises, with adequate support from national and local governments, would need to secure more budget for emergency preparedness and response in order to maintain the national electricity generation capacity during a severe climate event. On the other hand, more climate-proofing investment has to be made to increase the level of resilience of the vulnerable sites. In this regard, energy storages can increase the supply reliability by storing the generated electricity for emergency use [91], though they were not touched upon by any of the interviewees, potentially due to their current economic unfeasibility

4.3.3. Reconsideration of Conditions for Power Purchasing

Thailand adopts auction as the main scheme to award the contract to the bidders that are among those who proposed the lowest Feed-in-Tariff (FiT) rate for solar power wholesales to the electricity authorities [119]. This is beneficial to the consumers since the scheme can theoretically minimize the electricity unit price [120]. However, as stated above, the scheme unexpectedly hinders site selection process. Enterprises will not start seeking for the land if they are not certain that they will be awarded a solar power plant contract. On the other hand, the government wants to follow its plan of increasing the share of solar energy accordingly and needs the power plants to enter their commercial operation as soon as possible. Moreover, solar power plants require large flat terrain with adequate elevation, which significantly limits the land availability. This coincidence results in a very limited timeframe for site selection and makes many existing sites vulnerable to climate incidents. Changing from auction to fix rate power purchase is also not a good idea since it will put more burden on the shoulder of the public. There are three possible ways to ensure climate adaptability of the sites without significantly altering the bidding process.

1. Give priority to bidders that own or have a prospect to own a climate-resilient site. These bidders can be considered first or get a better electricity selling price.
2. If the winning bidders provide sufficient evidence to prove the difficulty in finding a climate-resilient site, the SCOD can be delayed by a certain period of time.
3. The Energy Regulatory Commission can coordinate with Department of Land to produce a map that indicates the areas that are suitable for solar power plants. The team can also assist the winning bidders with land mobilization.

Further study should be conducted to optimize the solutions for this issue. At this stage, it is important to recognize the influence of conditions for power purchasing on the climate adaptability of the solar power plants and the necessity to seek for good solutions in order to achieve climate-resilient solar-based electricity generation

5. Conclusions

This study captured climate-related risks associated to solar power plants in Thailand and climate-proofing countermeasures that were taken or planned, using thematic analysis. The findings of current study are derived from qualitative methodology capturing human insights and experience towards climatic incidents which could be difficult to obtain from quantitative methodology. The thematic analysis pointed out that floods and storms were perceived as major climate events affecting solar power plants in Thailand, followed by lightning and fires. Floods and storms could cause submersion of solar panels, inverters or equipment, soil erosion near structure basement of solar panels, and transmission line failure. Several countermeasures were taken, including dike construction, land leveling, increasing structure height, soil compaction, and agrivoltaic. Installation of firebreaks and corporate social responsibility activities helped alleviate fire risk, and lightning conductors were installed in many power plants to avoid lightning strikes. Flood and storm risks could also significantly affect site selection. The necessity of the land to be flat with no slope and with sufficient elevation in order to withstand future floods makes it difficult to find suitable sites within the provided timeframe. The sites might need to undergo land leveling or dike construction. It was also found that the auction scheme and the conditions under power purchase agreement could hinder the optimization of site selection.

Results from thematic analysis were used to develop two scenarios to consider climate adaptation of solar power plants: scenario without climate-proofing investment and scenario with climate proofing investment. It was recognized that possible future is somewhere in between the two scenarios, and that it is necessary get as close as possible to the latter scenario. Following recommendations to facilitate solar energy large-scale deployment were then derived from the scenarios.

1. National and local government should continue to provide policy support to solar power plant for climate-proofing investment in terms of enabling legislation or regulations, or financial incentives.
2. To accommodate climate events and the number of solar power plants that increase with time, enterprises, with adequate support from national and local governments, need to secure sufficient budget for fast recovery after severe climate incidents, especially nationwide disasters.
3. Since the current power purchasing conditions significantly affect the site selection process and consequently the climate adaptability of the solar power plants, measures must be taken to facilitate selection of climate-resilient sites by improving conditions of power purchase agreement or assisting winning bidders in enhancing climate adaptability of their sites.

The study indicated points of consideration from climate adaptation viewpoint that should be included in the long-term energy planning, apart from climate mitigation, i.e., greenhouse gas emission reduction. Significant increase in electricity generation capacity of solar energy in order to meet Thailand's updated NDC would result in more climate-vulnerable sites of solar power plants. If climate-proofing countermeasures are not appropriately and sufficiently implemented, the overall climate risks of all stakeholders would increase bringing about decrease in the power plants' climate adaptability. This confirms the necessity of considering both climate mitigation and climate adaptation during long-term energy planning. In addition, as renewable energy auctions which basically give priority to low electricity selling price can be an important reason of climate inadaptability of the solar power plants, policymakers should place more importance on the consideration of climate adaptive capacity of the power plants during the auctions and encourage the bidders to find climate-resilient sites or implement countermeasures to enhance the resilience of their solar power plant sites.

Though the study observed 35 solar power plants and interviewed only four of them, the power plants were carefully selected to cover power plants with different installed capacities scattered in all regions of Thailand. This ensures that the insights obtained from the study would be useful for most existing and future solar power plants in Thailand, and the recommendation on careful consideration of climate adaptability during site selection would be particularly helpful for countries employing renewable energy auctions.

Author Contributions: Conceptualization, K.S., P.J. and N.C.; methodology, P.J. and K.S.; validation, P.J.; formal analysis, P.J. and K.S.; investigation, K.S. and P.J.; resources, K.S. and P.J.; data curation, P.J.; writing—original draft preparation, K.S. and P.J.; writing—review and editing, K.S. and P.J.; visualization, K.S. and P.J.; supervision, N.C. and K.S.; project administration, K.S.; funding acquisition, N.C. All authors have read and agreed to the published version of the manuscript.

Funding: This research received no external funding.

Institutional Review Board Statement: Not applicable.

Informed Consent Statement: Informed verbal consent was obtained from all subjects involved in the study.

Acknowledgments: The authors would like to acknowledge solar power plant owners and operators who kindly cooperated with our self-administered observation and in-depth interviews, and thank our teammates, particularly Khemrath Vithean and Kampanat Thapmanee, who helped with figures and references.

Conflicts of Interest: The authors declare no conflict of interest.

References

1. UNFCCC. *COP26 Explained*; UNFCCC: Bonn, Germany, 2021.
2. UNFCCC. *Climate Neutral Now Guidelines for Participation*; UNFCCC: Bonn, Germany, 2021.
3. UN. *Kyoto Protocol to the United Nations Framework Convention on Climate Change*; UN: New York, NY, USA, 1998.
4. UN. *Adoption of Paris Agreement*; UN: New York, NY, USA, 2015.
5. UK Government. *United Kingdom of Great Britain and Northern Ireland's Nationally Determined Contribution*; UK Government: London, UK, 2020.
6. EU. *Update of the NDC of the European Union and its Member States*; EU: Berlin, Germany, 2020.
7. Allen, M.; Dube, O.; Solecki, W.; Aragón-Durand, F.; Cramer, W.; Humphreys, S.; Kainuma, M.; Kala, J.; Mahowald, N.; Mulugetta, Y. *Global Warming of 1.5 °C. an IPCC Special Report on the Impacts of Global Warming of 1.5 °C above Pre-Industrial Levels and Related Global Greenhouse Gas Emission Pathways, in the Context of Strengthening the Global Response to the Threat of Climate Change, Sustainable Development, and Efforts to Eradicate Poverty*; IPCC: Geneva, Switzerland, 2018.
8. UNDP. *UN Secretary-General Issues New Global Roadmap to Secure Clean Energy Access for All by 2030 and Net Zero Emissions by 2050*; UNDP: New York, NY, USA, 2021.
9. UNFCCC. *Thailand's Updated Nationally Determinded Contribution*; UNFCCC: New York, NY, USA, 2020.
10. MFA. The Prime Minister Participated in the World Leaders Summit during the 26th United Nations Framework Convention on Climate Change Conference of the Parties (UNFCCC COP26) in Glasgow, United Kingdom. Available online: https://www.mfa.go.th/en/content/cop26-glasgow?cate=5d5bcb4e15e39c306000683c (accessed on 7 December 2021).
11. Edenhofer, O.; Pichs-Madruga, R.; Sokona, Y.; Farahani, E.; Kadner, S.; Seyboth, K.; Adler, A.; Baum, I.; Brunner, S.; Eickemeier, P. *Climate Change 2014: Mitigation of Climate Change. Working Group III Contribution to the Fifth Assessment Report of the Intergovernmental Panel on Climate Change*; IPCC: Geneva, Switzerland, 2014.
12. IRENA. *Renewable Power Generation Costs 2020*; IRENA: Abu Dhabi, United Arab Emirates, 2021.
13. APERC. *APEC Energy Overview 2021*; APERC: Tokyo, Japan, 2021.
14. APAEC. *ASEAN Plan of Action for Energy Cooperation (APAEC) Phase II: 2021–2025*; APAEC: Jakarta, Indonesia, 2020.
15. Thailand Government. *Mid-Century, Long-Term Low Greenhouse Gas Emission Development Strategy*; Thailand Government: Bangkok, Thailand, 2021.
16. EPPO. National Energy Plan. Available online: http://www.eppo.go.th/epposite/index.php/th/petroleum/oil/link-doeb/item/17093-nep (accessed on 7 December 2021).
17. IRENA, J. *Benchmarking Scenario Comparisons: Key Indicators for the Clean Energy Transition*; IRENA: Abu Dhabi, United Arab Emirates, 2021.
18. Gielen, D.; Boshell, F.; Saygin, D.; Bazilian, M.D.; Wagner, N.; Gorini, R. The role of renewable energy in the global energy transformation. *Energy Strategy Rev.* **2019**, *24*, 38–50. [CrossRef]

19. Oree, V.; Hassen, S.Z.S.; Fleming, P.J. Generation expansion planning optimisation with renewable energy integration: A review. *Renew. Sust. Energy Rev.* **2017**, *69*, 790–803. [CrossRef]

20. Panos, E.; Kober, T.; Wokaun, A. Long term evaluation of electric storage technologies vs alternative flexibility options for the Swiss energy system. *Appl. Energy* **2019**, *252*, 113470. [CrossRef]

21. Hong, J.H.; Kim, J.; Son, W.; Shin, H.; Kim, N.; Lee, W.K.; Kim, J. Long-term energy strategy scenarios for South Korea: Transition to a sustainable energy system. *Energy Policy* **2019**, *127*, 425–437. [CrossRef]

22. Ouedraogo, N.S. Africa energy future: Alternative scenarios and their implications for sustainable development strategies. *Energy Policy* **2017**, *106*, 457–471. [CrossRef]

23. Mohsin, M.; Kamran, H.W.; Nawaz, M.A.; Hussain, M.S.; Dahri, A.S. Assessing the impact of transition from nonrenewable to renewable energy consumption on economic growth-environmental nexus from developing Asian economies. *J. Environ. Manag.* **2021**, *284*, 111999. [CrossRef]

24. Chaichaloempreecha, A.; Chunark, P.; Limmeechokchai, B. Assessment of Thailand's Energy Policy on CO2 Emissions: Implication of National Energy Plans to Achieve NDC Target. *Int. Energy J.* **2019**, *19*, 47–60.

25. UN. *Transforming Our World: The 2030 Agenda for Sustainable Development*; UN: New York, NY, USA, 2015.

26. UN. *The Sustainable Development Goals Report 2021*; UN: New York, NY, USA, 2021.

27. Fu, G.; Wilkinson, S.; Dawson, R.J.; Fowler, H.J.; Kilsby, C.; Panteli, M.; Mancarella, P. Integrated approach to assess the resilience of future electricity infrastructure networks to climate hazards. *IEEE Syst. J.* **2017**, *12*, 3169–3180. [CrossRef]

28. Miara, A.; Cohen, S.M.; Macknick, J.; Vorosmarty, C.J.; Corsi, F.; Sun, Y.N.; Tidwell, V.C.; Newmark, R.; Fekete, B.M. Climate-Water Adaptation for Future US Electricity Infrastructure. *Environ. Sci. Technol.* **2019**, *53*, 14029–14040. [CrossRef]

29. Binder, C.R.; Muhlemeier, S.; Wyss, R. An Indicator-Based Approach for Analyzing the Resilience of Transitions for Energy Regions. Part I: Theoretical and Conceptual Considerations. *Energies* **2017**, *10*, 36. [CrossRef]

30. EPPO. *Thailand Power Development Plan 2018–2037 (PDP2018)*; EPPO: Luxembourg, 2019.

31. MONRE. Thailand's National Adaptation Plan. Available online: https://datacenter.deqp.go.th/media/881465/th-national-adaptaion-plan-concept-and-progress.pdf (accessed on 7 December 2021).

32. MOE. Modelling results for five scenarios for the evaluation of expected CO2 emission. In Proceedings of the Meeting on Thailand's Greenhouse Gas Emission Reduction Framework for Energy Sector, Bangkok, Thailand, 4 June 2021.

33. Chaianong, A.; Pharino, C. Outlook and challenges for promoting solar photovoltaic rooftops in Thailand. *Renew. Sust. Energy Rev.* **2015**, *48*, 356–372. [CrossRef]

34. Chingulpitak, S.; Wongwises, S. Critical review of the current status of wind energy in Thailand. *Renew. Sust. Energy Rev.* **2014**, *31*, 312–318. [CrossRef]

35. Lorenczik, S.; Kim, S.; Wanner, B.; Bermudez Menendez, J.M.; Remme, U.; Hasegawa, T.; Keppler, J.H.; Mir, L.; Sousa, G.; Berthelemy, M. *Projected Costs of Generating Electricity*, 2020 ed.; Organisation for Economic Co-Operation and Development: Paris, France, 2020.

36. Cooper, R.T. Projection of future precipitation extremes across the Bangkok Metropolitan Region. *Heliyon* **2019**, *5*, e01678. [CrossRef]

37. Amnuaylojaroen, T. Projection of the precipitation extremes in thailand under climate change scenario RCP8. 5. *Front. Environ. Sci.* **2021**, 1–344. [CrossRef]

38. Westra, S.; Fowler, H.; Evans, J.; Alexander, L.; Berg, P.; Johnson, F.; Kendon, E.; Lenderink, G.; Roberts, N. Future changes to the intensity and frequency of short-duration extreme rainfall. *Rev. Geophys.* **2014**, *52*, 522–555. [CrossRef]

39. Gale, E.L.; Saunders, M.A. The 2011 Thailand flood: Climate causes and return periods. *Weather* **2013**, *68*, 233–237. [CrossRef]

40. Paltan, H.; Allen, M.; Haustein, K.; Fuldauer, L.; Dadson, S. Global implications of 1.5 C and 2 C warmer worlds on extreme river flows. *Environ. Res. Lett.* **2018**, *13*, 094003. [CrossRef]

41. UNDRR. *Disaster Risk Reduction in Thailand: Status Report 2020*; UNDRR: Geneva, Switzerland, 2020.

42. Hallegatte, S.; Rentschler, J.; Rozenberg, J. *Lifelines: The Resilient Infrastructure Opportunity*; World Bank Publications: Washington, DC, USA, 2019.

43. Cronin, J.; Anandarajah, G.; Dessens, O. Climate change impacts on the energy system: A review of trends and gaps. *Clim. Change* **2018**, *151*, 79–93. [CrossRef]

44. Shiradkar, N. Reliability and safety issues observed in flood affected PV power plants and strategies to mitigate the damage in future. In Proceedings of the 2019 IEEE 46th Photovoltaic Specialists Conference (PVSC), Chicago, IL, USA, 16–21 June 2019; pp. 3097–3102.

45. Damianaki, K.; Christodoulou, C.A.; Kokalis, C.-C.A.; Kyritsis, A.; Ellinas, E.D.; Vita, V.; Gonos, I.F. Lightning protection of photovoltaic systems: Computation of the developed potentials. *Appl. Sci.* **2021**, *11*, 337. [CrossRef]

46. Phasha, T.; Aderemi, B.; Chowdhury, S.; Olwal, T.; Abu-Mahfouz, A. Design strategies of an off-grid solar PV plant for office buildings: A case study of Johannesburg. In Proceedings of the 2018 IEEE PES/IAS PowerAfrica, Cape Town, South Africa, 26–29 June 2018; pp. 722–727.

47. Cavanini, L.; Ciabattoni, L.; Ferracuti, F.; Ippoliti, G.; Longhi, S. Microgrid sizing via profit maximization: A population based optimization approach. In Proceedings of the 2016 IEEE 14th International Conference on Industrial Informatics (INDIN), Poitiers, France, 18–21 July 2016; pp. 663–668.

48. Ciabattoni, L.; Ferracuti, F.; Ippoliti, G.; Longhi, S. Artificial bee colonies based optimal sizing of microgrid components: A profit maximization approach. In Proceedings of the 2016 IEEE Congress on Evolutionary Computation (CEC), Vancouver, BC, Canada, 24–29 July 2016; pp. 2036–2042.

49. Savic, N.; Katic, V.; Dumnic, B.; Milicevic, D.; Corba, Z.; Katic, N. Cost-benefit analysis of the application of a distributed energy sources in the university campus microgrid proposal. In Proceedings of the IEEE EUROCON 2019-18th International Conference on Smart Technologies, Novi Sad, Serbia, 1–4 July 2019; pp. 1–6.

50. Benis, K.; Turan, I.; Reinhart, C.; Ferrão, P. Putting rooftops to use–A Cost-Benefit Analysis of food production vs. energy generation under Mediterranean climates. *Cities* **2018**, *78*, 166–179. [CrossRef]

51. Sandelowski, M. Using qualitative research. *Qual. Health Res.* **2004**, *14*, 1366–1386. [CrossRef] [PubMed]

52. Rosenthal, M. Qualitative research methods: Why, when, and how to conduct interviews and focus groups in pharmacy research. *Curr. Pharm. Teach. Learn.* **2016**, *8*, 509–516. [CrossRef]

53. Attride-Stirling, J. Thematic networks: An analytic tool for qualitative research. *Qual. Res.* **2001**, *1*, 385–405. [CrossRef]

54. Georgiou, A.; Skarlatos, D. Optimal site selection for sitting a solar park using multi-criteria decision analysis and geographical information systems. *Geosci. Instrum. Methods Data Syst.* **2016**, *5*, 321–332. [CrossRef]

55. Faizi, M.A.; Sandipkumar, V.M.; Verma, A.; Jain, V. Design and optimization of solar photovoltaic power plant in case of agrivoltaics. In *Recent Trends in Materials and Devices*; Springer: Cham, Switzerland, 2020; pp. 59–69.

56. Thanarak, P.; Chiramakara, T. GHG emission and cost performance of life cycle energy on agricultural land used for photovoltaic power plant. *Int. J. Energy Econ. Policy* **2019**, *9*, 156–165.

57. Merrouni, A.A.; Elalaoui, F.E.; Mezrhab, A.; Mezrhab, A.; Ghennioui, A. Large scale PV sites selection by combining GIS and Analytical Hierarchy Process. Case study: Eastern Morocco. *Renew. Energy* **2018**, *119*, 863–873. [CrossRef]

58. Ong, S.; Campbell, C.; Denholm, P.; Margolis, R.; Heath, G. *Land-Use Requirements for Solar Power Plants in the United States*; National Renewable Energy Lab. (NREL): Golden, CO, USA, 2013.

59. Türk, S.; Koç, A.; Şahin, G. Multi-criteria of PV solar site selection problem using GIS-intuitionistic fuzzy based approach in Erzurum province/Turkey. *Sci. Rep.* **2021**, *11*, 5034. [CrossRef]

60. Solaun, K.; Cerdá, E. Climate change impacts on renewable energy generation. A review of quantitative projections. *Renew. Sustain. Energy Rev.* **2019**, *116*, 109415. [CrossRef]

61. Makky, M.; Kalash, H. Potential Risks of Climate Change on Thermal Power Plants. 2013. Available online: https://www. researchgate.net/publication/236174007 (accessed on 7 December 2021).

62. Crook, J.A.; Jones, L.A.; Forster, P.M.; Crook, R. Climate change impacts on future photovoltaic and concentrated solar power energy output. *Energy Environ. Sci.* **2011**, *4*, 3101–3109. [CrossRef]

63. ADB. *Guidelines For Climate Proofing Investment In The Energy Sector*; ADB: Manila, Philippines, 2013.

64. IRENA. *Overcoming Barriers to Authorizing Renewable Power Plants and Infrastructure*; IRENA: Abu Dhabi, United Arab Emirates, 2013.

65. Wu, Y.; Deng, Z.; Tao, Y.; Wang, L.; Liu, F.; Zhou, J. Site selection decision framework for photovoltaic hydrogen production project using BWM-CRITIC-MABAC: A case study in Zhangjiakou. *J. Clean. Prod.* **2021**, *324*, 129233. [CrossRef]

66. Roberts, K.; Dowell, A.; Nie, J.-B. Attempting rigour and replicability in thematic analysis of qualitative research data; a case study of codebook development. *BMC Med. Res. Methodol.* **2019**, *19*, 66. [CrossRef] [PubMed]

67. Maswabi, M.G.; Chun, J.; Chung, S.-Y. Barriers to energy transition: A case of Botswana. *Energy Policy* **2021**, *158*, 112514. [CrossRef]

68. Moorthy, K.; Patwa, N.; Gupta, Y. Breaking barriers in deployment of renewable energy. *Heliyon* **2019**, *5*, e01166.

69. De Desarrollo, B.A. *Climate Proofing: A Risk-Based Approach to Adaptation*; ADB: Manila, Philippines, 2005.

70. UN. *Climate Proofing Toolkit: For Basic Urban Infrastructure with a Focus on Water and Sanitation*; UN: New York, NY, USA, 2021.

71. Williamson, L.E.; Connor, H.; Moezzi, M. *Climate-Proofing Energy Systems*; Helio International: Paris France, 2009.

72. Sirasoontorn, P.; Koomsup, P. *Energy Transition in Thailand: Challenges and Opportunities*; Friedrich-Ebert-Stiftung Thailand Office: Bangkok, Thailand, 2017.

73. WHO. *Floods: Climate Change and Adaptation Strategies for Human Health: Report on a WHO Meeting: London, United Kingdom, 30 June–2 July 2002*; World Health Organization, Regional Office for Europe: København, Denmark, 2002.

74. UN. *Guidelines for Climate Change Proofing in UNDP Projects and Programmes in Armenia*; UN: New York, NY, USA, 2009.

75. UN. *Theme Report on Energy Access towards the Achievement of SDG 7 and Net-Zero Emissions*; UN: New York, NY, USA, 2021.

76. Braun, V.; Clarke, V. Using thematic analysis in psychology. *Qual. Res. Psychol.* **2006**, *3*, 77–101. [CrossRef]

77. Maguire, M.; Delahunt, B. Doing a thematic analysis: A practical, step-by-step guide for learning and teaching scholars. *Irel. J. High. Educ.* **2017**, *9*, 3.

78. Lo, Y.-C.; Janta, P. Resident's perspective on developing community-based tourism—A qualitative study of Muen Ngoen Kong Community, Chiang Mai, Thailand. *Front. Psychol.* **2020**, *11*, 1493. [CrossRef]

79. Lincoln, Y.S.; Guba, E.G. *Naturalistic Inquiry*; Sage: Thousand Oaks, CA, USA, 1985.

80. Nowell, L.S.; Norris, J.M.; White, D.E.; Moules, N.J. Thematic analysis: Striving to meet the trustworthiness criteria. *Int. J. Qual. Methods* **2017**, *16*, 1609406917733847. [CrossRef]

81. Guion, L.A.; Diehl, D.C.; McDonald, D. Triangulation: Establishing the validity of qualitative studies. *Edis* **2011**, *2011*, 3. [CrossRef]

82. Mensour, O.N.; El Ghazzani, B.; Hlimi, B.; Ihlal, A. A geographical information system-based multi-criteria method for the evaluation of solar farms locations: A case study in Souss-Massa area, southern Morocco. *Energy* **2019**, *182*, 900–919. [CrossRef]

83. Vafaeipour, M.; Zolfani, S.H.; Varzandeh, M.H.M.; Derakhti, A.; Eshkalag, M.K. Assessment of regions priority for implementation of solar projects in Iran: New application of a hybrid multi-criteria decision making approach. *Energy Convers. Manag.* **2014**, *86*, 653–663. [CrossRef]

84. Rios, R.; Duarte, S. Selection of ideal sites for the development of large-scale solar photovoltaic projects through Analytical Hierarchical Process–Geographic information systems (AHP-GIS) in Peru. *Renew. Sustain. Energy Rev.* **2021**, *149*, 111310. [CrossRef]

85. Aragonés-Beltrán, P.; Chaparro-González, F.; Pastor-Ferrando, J.; Rodríguez-Pozo, F. An ANP-based approach for the selection of photovoltaic solar power plant investment projects. *Renew. Sustain. Energy Rev.* **2010**, *14*, 249–264. [CrossRef]

86. Coe, R. Flood Risk Assessment On: Proposed Solar Farm, Mill Farm, Grantham, 2015. Rossi Long Consulting, UK. Available online: http://planning.southkesteven.gov.uk/SKDC/S15-2137/806061.pdf (accessed on 7 December 2021).

87. ENTEC. *Economic Analysis of Climate-Proofing Investment in Road and Rail Transport Sectors—UNDP Thailand/NDC Support Project: Delivering Sustainability through Climate Finance Actions in Thailand*; ENTEC: Pathum Thani, Thailand, 2021.

88. Perera, A.; Nik, V.M.; Chen, D.; Scartezzini, J.-L.; Hong, T. Quantifying the impacts of climate change and extreme climate events on energy systems. *Nat. Energy* **2020**, *5*, 150–159. [CrossRef]

89. UNDRR. *Asia-Pacific Regional Framework for NATECH (Natural Hazards Triggering Technological Disasters) Risk Management*; UNDRR: Geneva, Switzerland, 2020.

90. Dowling, P. The impact of climate change on the European energy system. *Energ Policy* **2013**, *60*, 406–417. [CrossRef]

91. Shaqsi, A.Z.A.; Sopian, K.; Al-Hinai, A. Review of energy storage services, applications, limitations, and benefits. *Energy Rep.* **2020**, *6*, 288–306. [CrossRef]

92. Heo, J.; Moon, H.; Chang, S.; Han, S.; Lee, D.-E. Case Study of Solar Photovoltaic Power-Plant Site Selection for Infrastructure Planning Using a BIM-GIS-Based Approach. *Appl. Sci.* **2021**, *11*, 8785. [CrossRef]

93. Pašičko, R.; Branković, Č.; Šimić, Z. Assessment of climate change impacts on energy generation from renewable sources in Croatia. *Renew. Energy* **2012**, *46*, 224–231. [CrossRef]

94. Touili, N. Hazards, Infrastructure Networks and Unspecific Resilience. *Sustainability* **2021**, *13*, 4972. [CrossRef]

95. Schuss, C.; Leppänen, K.; Remes, K.; Saarela, J.; Fabritius, T.; Eichberger, B.; Rahkonen, T. Detecting defects in photovoltaic cells and panels and evaluating the impact on output performances. *IEEE Trans. Instrum. Meas.* **2016**, *65*, 1108–1119. [CrossRef]

96. Zhang, X.; Huang, G.; Zhou, X.; Liu, L.; Fan, Y. A multicriteria small modular reactor site selection model under long-term variations of climatic conditions–A case study for the province of Saskatchewan, Canada. *J. Clean. Prod.* **2021**, *290*, 125651. [CrossRef]

97. IRENA. *Global Atlas for Renewable Energy: Overview of Solar and Wind Maps*; IRENA: Abu Dhabi, United Arab Emirates, 2014.

98. Panteli, M.; Mancarella, P. Influence of extreme weather and climate change on the resilience of power systems: Impacts and possible mitigation strategies. *Electr. Power Syst. Res.* **2015**, *127*, 259–270. [CrossRef]

99. Eckstein, D.; Künzel, V.; Schäfer, L. *Global Climate Risk Index 2021. Who Suffers Most from Extreme Weather Events?* Germanwatch: Bonn, Germany, 2021.

100. Ward, P.; Kummu, M.; Lall, U. Flood frequencies and durations and their response to El Niño Southern Oscillation: Global analysis. *J. Hydrol.* **2016**, *539*, 358–378. [CrossRef]

101. Maruyama Rentschler, J.E.; Obolensky, M.A.B.; Kornejew, M.G.M. *Candle in the Wind? Energy System Resilience to Natural Shocks*; Energy System Resilience to Natural Shocks (17 June 2019); World Bank Policy Research Working Paper; World Bank: Washington, DC, USA, 2019.

102. ADB. *Bangladesh: Spectra Solar Power Project*; ADB: Manila, Philippines, 2019.

103. Sánchez-Lozano, J.M.; Teruel-Solano, J.; Soto-Elvira, P.L.; García-Cascales, M.S. Geographical Information Systems (GIS) and Multi-Criteria Decision Making (MCDM) methods for the evaluation of solar farms locations: Case study in south-eastern Spain. *Renew. Sustain. Energy Rev.* **2013**, *24*, 544–556. [CrossRef]

104. Effat, H.A. Selection of potential sites for solar energy farms in Ismailia Governorate, Egypt using SRTM and multicriteria analysis. *Int. J. Adv. Remote Sens. GIS* **2013**, *2*, 205–220.

105. Kengpol, A.; Rontlaong, P.; Tuominen, M. Design of a decision support system for site selection using fuzzy AHP: A case study of solar power plant in north eastern parts of Thailand. In *Proceedings of the 2012 Proceedings of PICMET'12: Technology Management for Emerging Technologies, Vancouver, BC, Canada, 29 July–2 August 2012*; pp. 734–743.

106. Jung, J.; Han, S.; Kim, B. Digital numerical map-oriented estimation of solar energy potential for site selection of photovoltaic solar panels on national highway slopes. *Appl. Energy* **2019**, *242*, 57–68. [CrossRef]

107. Agyekum, E.B.; Amjad, F.; Shah, L.; Velkin, V.I. Optimizing photovoltaic power plant site selection using analytical hierarchy process and density-based clustering–Policy implications for transmission network expansion, Ghana. *Sustain. Energy Technol. Assess.* **2021**, *47*, 101521. [CrossRef]

108. Margulis, S.; Hughes, G.; Schneider, R.; Pandey, K.; Narain, U.; Kemeny, T. *Economics of Adaptation to Climate Change: Synthesis Report*; World Bank: Washington, DC, USA, 2010.

109. ADB. *Economic Analysis of Climate-Proofing Investment Projects*; ADB: Manila, Philippines, 2015.

110. UNDP. *Climate Change Benefit Analysis CCBA Guidelines*; UNDP: New York, NY, USA, 2016.

111. UNFCCC. Investment and Financial Flows to Address Climate Change. In Proceedings of the United Nations Framework Convention on Climate Change. 2007. Available online: http://unfccc.int/files/cooperation_and_support/financial_mechanism/application/pdf/background_paper.pdf (accessed on 7 December 2021).
112. Gupta, A. Energy Security and Resilience in South Asia. Available online: https://www.nbr.org/publication/energy-security-and-resilience-in-south-asia/ (accessed on 7 December 2021).
113. NIBS. *Natural Hazard Mitigation Saves: 2019 Report*; NIBS: Washington, DC, USA, 2019.
114. OECD. *Integrating Climate Change Adaptation Into Development Co-Operation: Policy Guidance*; Organisation for Economic Co-Operation and Development: Paris, France, 2009.
115. Australian Government. *The Role of Regulation in Facilitating or Constraining Adaptation to Climate Change for Australian Infrastructure*; Australian Government: Canberra, ACT, USA, 2011.
116. Mercure, J.-F.; Salas, P.; Vercoulen, P.; Semieniuk, G.; Lam, A.; Pollitt, H.; Holden, P.; Vakilifard, N.; Chewpreecha, U.; Edwards, N. Reframing incentives for climate policy action. *Nat. Energy* **2021**, *6*, 1133–1143. [CrossRef]
117. EU. *Commission Notice—Technical Guidance on the Climate Proofing of Infrastructure in the Period 2021–2027*; EU: Brussels, Belgium, 2021.
118. BOJ. *Outline of Climate Response Financing Operations*; BOJ: Tokyo, Japan, 2021.
119. Sagulpongmalee, K.; Therdyothin, A.; Nathakaranakule, A. Analysis of feed-in tariff models for photovoltaic systems in Thailand: An evidence-based approach. *J. Renew. Sustain. Energy* **2019**, *11*, 045903. [CrossRef]
120. IRENA. *Renewable Energy Auctions: Analysing 2016*; IRENA: Abu Dhabi, United Arab Emirates, 2017.

MDPI

St. Alban-Anlage 66

4052 Basel

Switzerland

Tel. +41 61 683 77 34

Fax +41 61 302 89 18

www.mdpi.com

Energies Editorial Office

E-mail: energies@mdpi.com

www.mdpi.com/journal/energies